Healthy, Digital and Sustainable Buildings and Cities

Healthy, Digital and Sustainable Buildings and Cities

Guest Editors

Lina Šeduikytė
Jakub Kolarik

Basel • Beijing • Wuhan • Barcelona • Belgrade • Novi Sad • Cluj • Manchester

Guest Editors

Lina Šeduikytė	Jakub Kolarik
Faculty of Civil Engineering and Architecture	Department of Civil and Mechanical Engineering
Kaunas University of Technology	Technical University of Denmark
Kaunas	Kgs. Lyngby
Lithuania	Denmark

Editorial Office
MDPI AG
Grosspeteranlage 5
4052 Basel, Switzerland

This is a reprint of the Special Issue, published open access by the journal *Buildings* (ISSN 2075-5309), freely accessible at: https://www.mdpi.com/journal/buildings/special_issues/0521A36F10.

For citation purposes, cite each article independently as indicated on the article page online and as indicated below:

Lastname, A.A.; Lastname, B.B. Article Title. *Journal Name* **Year**, *Volume Number*, Page Range.

ISBN 978-3-7258-3291-0 (Hbk)
ISBN 978-3-7258-3292-7 (PDF)
https://doi.org/10.3390/books978-3-7258-3292-7

Cover image courtesy of Jakub Kolarik

© 2025 by the authors. Articles in this book are Open Access and distributed under the Creative Commons Attribution (CC BY) license. The book as a whole is distributed by MDPI under the terms and conditions of the Creative Commons Attribution-NonCommercial-NoDerivs (CC BY-NC-ND) license (https://creativecommons.org/licenses/by-nc-nd/4.0/).

Contents

About the Editors vii

Indre Grazuleviciute-Vileniske, Aurelija Daugelaite and Gediminas Viliunas
Classification of Biophilic Buildings as Sustainable Environments
Reprinted from: *Buildings* **2022**, *12*, 1542, https://doi.org/10.3390/buildings12101542 1

Naglaa Sami Abdelaziz Mahmoud and Chuloh Jung
Analyzing the Bake-Out Effect in Winter for the Enhancement of Indoor Air Quality at New Apartments in UAE
Reprinted from: *Buildings* **2023**, *13*, 846, https://doi.org/10.3390/buildings13040846 16

Gvidas Plienaitis, Mindaugas Daukšys, Evi Demetriou, Byron Ioannou, Paris A. Fokaides and Lina Seduikyte
Evaluation of the Smart Readiness Indicator for Educational Buildings
Reprinted from: *Buildings* **2023**, *13*, 888, https://doi.org/10.3390/buildings13040888 36

Maria Hurnik, Joanna Ferdyn-Grygierek, Jan Kaczmarczyk and Piotr Koper
Thermal Diagnosis of Ventilation and Cooling Systems in a Sports Hall—A Case Study
Reprinted from: *Buildings* **2023**, *13*, 1185, https://doi.org/10.3390/buildings13051185 49

Aivaras Simonaitis, Mindaugas Daukšys and Jūratė Mockienė
A Comparison of the Project Management Methodologies PRINCE2 and PMBOK in Managing Repetitive Construction Projects
Reprinted from: *Buildings* **2023**, *13*, 1796, https://doi.org/10.3390/buildings13071796 70

Attila Kostyák, Szabolcs Szekeres and Imre Csáky
Investigation of Sensible Cooling Performance in the Case of an Air Handling Unit System with Indirect Evaporative Cooling: Indirect Evaporative Cooling Effects for the Additional Cooling System of Buildings
Reprinted from: *Buildings* **2023**, *13*, 1800, https://doi.org/10.3390/buildings13071800 101

Tomas Makaveckas, Raimondas Bliūdžius, Sigita Alavočienė, Valdas Paukštys and Ingrida Brazionienė
Investigation of Microclimate Parameter Assurance in Schools with Natural Ventilation Systems
Reprinted from: *Buildings* **2023**, *13*, 1807, https://doi.org/10.3390/buildings13071807 118

Aleksejs Prozuments, Anatolijs Borodinecs, Sergejs Zaharovs, Karolis Banionis, Edmundas Monstvilas and Rosita Norvaišienė
Evaluating Reduction in Thermal Energy Consumption across Renovated Buildings in Latvia and Lithuania
Reprinted from: *Buildings* **2023**, *13*, 1916, https://doi.org/10.3390/buildings13081916 138

Jakub Kolarik, Nadja Lynge Lyng, Rossana Bossi, Rongling Li, Thomas Witterseh, Kevin Michael Smith and Pawel Wargocki
Application of Cluster Analysis to Examine the Performance of Low-Cost Volatile Organic Compound Sensors
Reprinted from: *Buildings* **2023**, *13*, 2070, https://doi.org/10.3390/buildings13082070 155

Chuloh Jung, Naglaa Sami Abdelaziz Mahmoud, Nahla Al Qassimi and Gamal Elsamanoudy
Preliminary Study on the Emission Dynamics of TVOC and Formaldehyde in Homes with Eco-Friendly Materials: Beyond Green Building
Reprinted from: *Buildings* **2023**, *13*, 2847, https://doi.org/10.3390/buildings13112847 174

María Teresa Gómez-Villarino, María del Mar Barbero-Barrera, Ignacio Cañas, Alba Ramos-Sanz, Fátima Baptista and Fernando R. Mazarrón
Construction Solutions, Cost and Thermal Behavior of Efficiently Designed Above-Ground Wine-Aging Facilities
Reprinted from: *Buildings* 2024, *14*, 655, https://doi.org/10.3390/buildings14030655 193

Farah Shoukry, Sherif Goubran and Khaled Tarabieh
Enhanced Indoor Air Quality Dashboard Framework and Index for Higher Educational Institutions
Reprinted from: *Buildings* 2024, *14*, 1640, https://doi.org/10.3390/buildings14061640 209

Luis Cortés-Meseguer and Jorge García-Valldecabres
Proposal of a Sensorization Methodology for Obtaining a Digital Model: A Case Study on the Dome of the Church of the Pious Schools of Valencia
Reprinted from: *Buildings* 2024, *14*, 2057, https://doi.org/10.3390/buildings14072057 234

Dima Abu-Aridah and Rebecca L. Henn
Construction 4.0 in Refugee Camps: Facilitating Socio-Spatial Adaptation Patterns in Jordan's Zaatari Camp
Reprinted from: *Buildings* 2024, *14*, 2927, https://doi.org/10.3390/buildings14092927 249

Paulius Vestfal and Lina Seduikyte
Systematic Review of Factors Influencing Students' Performance in Educational Buildings: Focus on LCA, IoT, and BIM
Reprinted from: *Buildings* 2024, *14*, 2007, https://doi.org/10.3390/buildings14072007 275

Lina Seduikyte, Indrė Gražulevičiūtė-Vileniškė, Ingrida Povilaitienė, Paris A. Fokaides and Domantas Lingė
Trends and Interdisciplinarity Integration in the Development of the Research in the Fields of Sustainable, Healthy and Digital Buildings and Cities
Reprinted from: *Buildings* 2023, *13*, 1764, https://doi.org/10.3390/buildings13071764 296

About the Editors

Lina Šeduikytė

Lina Šeduikytė is a professor and chief researcher at the Faculty of Civil Engineering and Architecture at Kaunas University of Technology (KTU). Her research field is related to sustainable development, life cycle assessment (LCA), indoor air quality, thermal comfort, human well-being, and energy efficiency. Lina Šeduikytė has participated/participates in implementing international (H2020, Erasmus+, and COST) and national research projects. She is also the author and co-author of more than 60 research papers listed in the WOS and SCOPUS databases. Lina Šeduikytė is a member of the scientific committee of several international conferences.

Jakub Kolarik

Jakub Kolarik is an Associate Professor at the Technical University of Denmark (DTU), affiliated with the Department of Civil and Mechanical Engineering. His research focuses on indoor environments, air quality, and sustainable building systems. He explores human responses to these environments, investigating their impact on productivity, activity, and well-being, in addition to the interplay between the built environment and its occupants. Outside of his research endeavors, Jakub is actively involved in teaching architectural engineering students.

Article

Classification of Biophilic Buildings as Sustainable Environments

Indre Grazuleviciute-Vileniske *, Aurelija Daugelaite and Gediminas Viliunas

Faculty of Civil Engineering and Architecture, Kaunas University of Technology, LT-51326 Kaunas, Lithuania
* Correspondence: indre.grazuleviciute@ktu.lt

Abstract: Biophilic design approach aims at creating favorable conditions for humans in various types of anthropogenic environments, while at the same time restoring broken human–nature connection. The biophilic design guidelines and principles are general and flexible and allow wide array of architectural expressions. In order to better understand the architectural expression possibilities provided by biophilic design approach, the existing classifications of biophilic architecture and biophilic design examples were analyzed with the aim to develop the classification that would reflect the links between a building's architectural expression and biophilic qualities. Three categories of biophilic architecture were distinguished in the developed classification: mimetic, applied, and organic. The distinguished categories were illustrated with the characteristic building examples and the evaluation of biophilic qualities and human-nature collaboration potential of these example buildings was carried out using comprehensive system of criteria. The analysis has demonstrated that all three distinguished categories—mimetic, applied, organic—allow for the creation of biophilic environments and hold the potential for human–nature collaboration, although organic biophilic design would be currently considered as the least developed, although most promising category.

Keywords: biophilia; biophilic architecture; biophilic building; classifications of biophilic design; human–nature collaboration

1. Introduction

The term biophilia was coined as early as in 1964 by E. Fromm; however, it was developed and popularized in certain circles by biologist and naturalist and writer E. O. Wilson, who had developed and published in 1984, and later refined in his further works, what he has referred to as the biophilia hypothesis. According to this hypothesis, humans, as well as other species on Earth, had developed throughout their evolution and history surrounded by biodiversity and thus the interconnections with natural environment have persisted until this day [1–3]. The biophilia hypothesis states that human beings have an innate biological need to affiliate with nature; consequently, the biological diversity, the diversity of relations to nature, and diversity of landscape types are important for healthy human physical and psychological development [4]. Despite the benefits of connections with nature proven by environmental psychologists, medical researchers etc. [1,4], the human–nature connections and the biophilic qualities of our everyday environments continue to decline. Some researchers even identify our contemporary living environments as anti-biophilic [5]. According to A. Samalavičius [2], with the entrenchment of technologies and technological processes in human civilization, the human environment has strongly changed as well. Human habitats became closed and relatively sterile, even movement between locations happens in the closed environment of automobile. The mega-cities became inhospitable to nature and humans became distanced from their natural contexts, which has been their habitat for millennia. In order to restore broken human–nature connections and provide all the potential benefits of biophilic environments—the improvement of individual physical, psychological, and cognitive health and well-being as well as some social benefits like enhanced workers' productivity, improved public health, phytoremediation of industrial ruins [1,6]—the disciplines of biophilic design, biophilic urbanism, and diverse systems

of criteria and patterns (for example, Kellert et al. [7]; Browning et al. [1], Salingaros [5]) that facilitate biophilic projects implementation have emerged. E. L. M. Wolfs [6] distinguished several characteristics of biophilic approach to design that differentiate it from other environmentally-oriented design concepts: positive focus on enhancing instead of minimizing and the potential for mutually beneficial human–nature collaboration. E. L. M. Wolfs [6] biophilic design focuses on the actualization and enhancement of nature's ability to improve the quality of human experience and well-being instead of focusing on minimizing negative human impacts, and strives to develop the link between artificial and natural processes based on symbiotic interdependence.

However, some challenges could be identified related with the increasingly growing trend of biophilic design. One of these challenges is identified by A. Samalavičius [2] as the paradigm of thinking entrenched by the architectural ideology of the last century. The other challenge is design superficiality, mentioned by E. L. M. Wolfs [6], when the biophilic commitment of the creators is limited to "videos of cats, the rounded edges of a mobile phone or the digital representation of natural material". The relevance of biophilic design and distinguished challenges encourage directing the attention of designers and researchers towards the peculiarities of biophilic architectural form, understanding better how architectural form can engender biophilic qualities or/and how biophilic features can be integrated into architectural form.

The aim of the research is to analyze the existing classifications of biophilic architecture and biophilic building design examples and to develop a classification that would reflect the links between building's architectural expression and biophilic qualities. The categories of biophilic architecture distinguished in the developed classification are illustrated with characteristic building examples and the evaluation of biophilic qualities of buildings is carried out using a comprehensive set of criteria. The methodology of the research includes an analysis of the literature and architectural design examples, a comparison and systematization, and an assessment of architectural designs according to predefined criteria. The relevance and novelty of the research are determined by the contemporary challenges of entrenched modernist architecture ideology and design superficiality and the proposed and elaborated classification of biophilic architecture, as well as the evaluation of buildings not only from the point of view of biophilic qualities, but also from the perspective of human–nature collaboration as possible responses to these challenges. The problems and difficulties in the context of this research were the need to grasp the diversity of expression of biophilic design into a limited number of categories as well as finding the categories that would reveal the synergistic relation between the expression of the building and its biophilic qualities.

2. Review of Present Classifications of Biophilic Design

In order to understand better the extent of biophilic design, it is important to delve into the ways that biophilic design applications could be classified and categorized. The review of existing classifications of biophilic design included a search of the literature on the subject of biophilic design in the scientific literature databases Web of Science, Scopus, and Google Scholar. The publications on the biophilic design of buildings [6–10], biophilic design principles [1,5,7–10], and biophilic urbanism [8] were reviewed and examined for existing classifications or distinguished specific categories of biophilic architecture. The analysis of the literature has revealed that first of all, with the growing understanding of the benefits provided by biophilic environments, efforts have been made to distinguish between biophilic and non-biophilic or "business as usual" designs or even anti-biophilic environments [5]. For this purpose, different systems of criteria [5] and patterns [1] were formulated. For example, the company Terrapin Bright Green has elaborated 14 patterns of biophilic design subdivided into three categories: nature in the space; natural analogues; nature of the space [1]. The analysis of the literature clearly reveals that biophilic design applications can be categorized according to scale (for example, biophilic building, biophilic block, biophilic street, biophilic neighborhood, biophilic community, and biophilic region [8]) or object (for example, biophilic interior design [9]). This research primarily focuses on the architectural expression of biophilic buildings. After reviewing the literature, several existing classifications applicable to architectural expression of buildings were distinguished in the biophilic design discourse: those inspired by nature and traditional design trends, historic and contemporary biophilic architecture, natural and artificial biophilic environments, and explicit and implicit biophilic design (Figure 1).

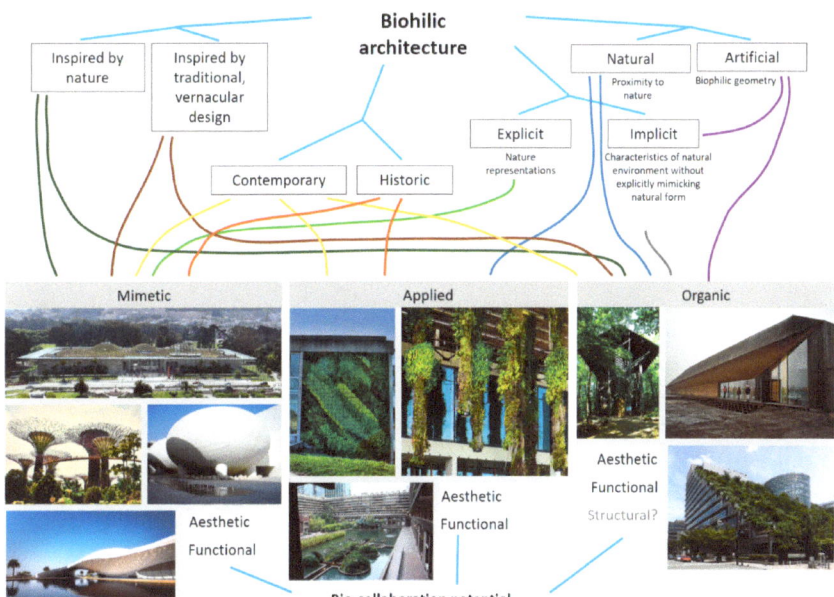

Figure 1. Existing classifications of biophilic architecture distinguished after a literature analysis and their links with proposed biophilic architecture categories: mimetic, applied, and organic. Each of three categories is illustrated by characteristic examples. Mimetic biophilic design: California Academy of Sciences (USA), Supertree Grove (Singapore), Weill Cornell Medical College (Qatar), BEEAH Headquarters (UAE). Applied biophilic design: green wall at Simon Fraser University (Canada), Perez Museum (Miami, Florida, USA), Khoo Teck Puat Hospital (Singapore), Barbican Estate (London, UK). Organic biophilic design: Thorncrown Chapel (USA), Wadden Sea Centre (Denmark), Acros building (Japan). All the images used in the illustration are from Wikimedia commons.

2.1. Inspired by Nature and Traditional Design

Two trends or dimensions towards which architects could orient the expression of biophilic buildings—inspired by nature, biological systems and natural shapes and inspired by traditional, vernacular, ethnic architectural forms, construction cultures and material applications—can be identified in the literature [7,10]. Both of these trends are generally characteristic for green design approaches since their inception and are identified in the book, Green Architecture by J. Wines [11]. For example, Snail House, designed by the Jersey Devil company in Forked River, New Jersey (USA) in 1972, is an example of architectural design inspired by nature [11]. S. Keller et al. identify Sydney Opera House designed by Jorn Utzon as inspired by natural shapes and forms [7]. An example of a building inspired by traditional design is the clay and straw Studio in the West Country, designed by David Lea (England) [11]. These broad trends allow conceptualizing biophilic shapes not only for natural and countryside landscapes, but also for urban and peri-urban areas.

2.2. Historic and Contemporary Design

Researchers analyzing biophilic design have noticed that both professional and vernacular architecture of the past eras was biophilic in its qualities even if the term itself was not known [1,5,6]. According to E. L. M. Wolfs [6], architects and designers have been inspired by nature since antiquity. W Browning et al. [1] notice that animal themes could be found even in prehistoric-built structures and nature was used as a source of symbolic and decorative ornamentation and directly brought into exterior and interior spaces in the form of plants, animals, natural materials, etc. S. Keller et al., note that many organic features are often encountered in Gothic architecture and mention the Gothic Revival Harkness Tower at Yale University, designed by James Gamble Rogers as example of biophilic architecture [7]. According to A. Salingaros [5], this human–nature relationship remains important from traditional cultures until today, but has increasingly been abandoned with the rise of industrialization.

2.3. Natural and Artificial

According to N. Salingaros [5], the positive effects of biophilic environments are induced either by proximity and visual contact with nature (plants, animals, scenic views, natural materials, or even other people) or by artificial environments that "follow geometrical rules for the structure of organisms". For example, proximity with nature is visible in the designs using vegetated surfaces, as in the case of botanists Patrick Blanc's green walls [8]. An example of an artificial biophilic environment is the interior design by Adolfsson & Partners in the King office complex in Stockholm providing artificial forest experience for its employees [12]. In other words—it is possible to design artificial shapes and spaces that stimulate the same responses as natural environments and biological forms. In the first case the positive effect of biophilic design can be defined as the "healing influence of nature" and the second case is identified as "biophilic geometry" by N. Salingaros [5]. Similar approach is visible in nature in the space and nature of the space categories in the 14 patterns of biophilic design [1]. The better results would be obtained if both the healing influence of nature and biophilic geometry were be integrated into the project.

2.4. Explicit and Implicit Representation of Nature

N. Salingaros [5] distinguishes explicit and implicit representations of nature. Explicit representations of nature include direct visual representations of natural forms in design. The foliated sculpture by Kent Bloomer in the Ronald Reagan National Washington Airport terminal is an example [7]. Implicit representations of nature would be "organized complexity—purposeful complication that is also accompanied by a high degree of organization". This abstract characteristic of natural world can be achieved in artificial environments in numerous ways, creating hierarchies similar to natural ones and providing an information-rich expedience as an intriguing balance between boring and overwhelming [1]. An example of a multilevel organized light and space complexity is Genzyme Center interior space designed by Behnisch Architekten in Cambridge, Massachusetts [13]. Similar categorizing could be inferred from the 14 patterns of biophilic design, where "biomorphic forms and patterns" and "complexity and order" are distinguished as "natural analogues" [1].

3. Proposed Classification of Biophilic Architecture

The above-presented analysis of existing classifications applicable to biophilic architecture has revealed the lack of universal classification that would be suitable both to innovative and traditional buildings or to buildings based on the biophilic geometry and the healing influence of nature, or which integrate both of those aspects. Moreover, it would be desirable that the classification would reflect the interconnections between the architectural form of building and its biophilic properties. In order to develop such a classification, the analysis of existing biophilic design classifications presented in Section 2 was complemented with an additional review of the literature on the classification of architectural objects and features in the above-mentioned scientific literature databases as well as available Internet search engines and encyclopedias, the general overview of biophilic design principles [1,5,7,9], and existing examples of biophilic buildings. The search and general overview of biophilic architecture examples was carried out using available Internet search engines and keyword combinations such as "biophilic architecture", "biophilic building", "biophilic design", etc. The criteria for selecting examples for analysis were that the buildings were referenced in prominent architectural online editions and corresponded to the criteria of biophilic design. Bearing in mind that buildings having biophilic qualities are not always explicitly identified as biophilic, an additional overview was carried out in architectural websites and databases, such as ArchDaily, Divisare, Dezeen, etc. The iconographic material (photographs, drawings, visualizations) and the descriptions of projects were overviewed in light of biophilic design criteria, presented by A. Salingaros [5], W. Browning et al. [1], and S. Kellert et al. [7]. The analysis revealed the diversity of scales, functions, and expressions of biophilic buildings. The examples demonstrate that biophilic qualities can be achieved using the internal and external layers of vegetation and natural materials to mimic forms of natural landscape or biological organisms by creating a complex organization of volumes and spaces characteristic to natural environments. The results of this general overview allowed parallels to be made between the interconnection between ornament and architecture and biophilic qualities and architecture. Bearing in mind this semblance, the literature on ornament in architecture was reviewed and the classification of architectural objects by A. Tikkanen [14], based on the character and integration of ornament, was viewed as having potential for adaptation to classifying biophilic buildings. A. Tikkanen [14] has distinguished three types of architecture: mimetic or imitative (symbolic ornaments imitating natural features, structural elements of preceding wooden structures etc.), applied (ornament, often without precise symbolic meaning, applied as a surface element for pure decorative purposes), and organic (ornamental effect of the inherent qualities of building materials). Such principle of classification—mimetic, applied, and organic—was modified and adapted to biophilic buildings (Figure 1).

3.1. Mimetic Biophilic Buildings

In the case of biophilic architecture, mimetic biophilic buildings are those that achieve biophilic qualities by using forms which "have certain definite meanings or symbolic significance" [14]. This can be either through botanical motifs, imitation or interpretation of traditional architectural forms, or interpretation of landscape features in the design of the building. For example, the New California Academy of Science Museum designed by Renzo Piano integrates the interpretation of the surrounding hills in its roof structure [6]. Some mimetic biophilic design features can also be identified in the structures designed by Arata Isozaki and Zaha Hadid Architects (Figure 1).

3.2. Applied Biophilic Buildings

In the case of biophilic architecture, applied biophilic buildings are such designs, where biophilic qualities are added as a layer that appears extrinsic to the structure itself. These can be buildings with vertical internal and external greenery, interior parks, or roof gardens that, in addition to these biophilic features, maintain a modernist, high-tech, or sleek architectural outlook. The example of such approach is Khoo Teck Puat hospital in Singapore, which integrates modernist design and lush greenery and viable ecosystems and numerous other biophilic qualities [6], similar to Perez Art Museum, Miami, by Herzog & de Meuron (Figure 1). The applied approach is typical and successful in situations when biophilic refurbishment of the existing structure is necessary, as in the cases of Barbican estate in London.

3.3. Organic Biophilic Buildings

In the case of biophilic architecture, organic biophilic buildings are such designs where the synergistic relation between the biophilic qualities and the structure is achieved. In this case the biophilic qualities are inherent in a building's shape and arrangement of spaces, materials, and functions. The creative works of Barry Wark [15], like the Glasgow School of Art Extension, can be mentioned as the attempts to create a biophilic architectural form which would be capable of generating human health and well-being benefits without directly emulating elements of natural world and at the same time providing a habitat for other species, thus fostering a human–nature connection.

According to E. L. M. Wolfs [6], biophilic architecture holds unprecedented potential for bio-collaboration, where the integration of natural elements goes beyond aesthetics or symbolism. Organic biophilic design could potentially create environments of distinctive architectural expression that positively affect human health and well-being and provide a habitat for a variety of natural systems, which in turn can "provide wide-ranging services that are integral to solving today's major ecological concerns" [6]. According to E. L. M. Wolfs [6] bio-collaboration in design could occur on aesthetic, functional, and structural levels. While functional and aesthetic bio-collaboration is widespread in biophilic architecture, the structural bio-collaboration, where "the design is primarily made by a living organism" [6] is still in the experimental stage and is referred to as bio-integrated design [16].

4. Evaluation of Selected Building Examples
4.1. Case Study Buildings Selection

In order to analyze in greater detail, the means of expression and design strategies of different categories of biophilic buildings, a sample of case study buildings were analyzed. It was decided that the examples to analyze would be located in the territory of Lithuania, as biophilic design is oftentimes associated with climate zones allowing lush exterior greenery. The Lithuanian climate, with clearly expressed seasons, requires different approaches for creating biophilic qualities, thus we decided to concentrate on the variety of design means and approaches available in such a climatic context. Moreover, biophilic design ideas are just taking the first steps in Lithuania and distinguishing these examples and analyzing them serves an important factor for the entrenchment of biophilic design culture in the Baltic Sea region. In order to select the examples for analysis, prominent Lithuanian architecture journals (both printed and online) aimed at both the professional community and the general public were reviewed, and the buildings were selected based on their correspondence to biophilic design criteria as presented by A. Salingaros [5], W. Browning et al. [1], S. Kellert et al. [7]. Additionally, the possibility to attribute the objects clearly to one of the three distinguished design categories: mimetic, applied, and organic was considered. Each of three categories was represented by one case study example.

The following examples were selected for analysis:

1. Example A—Kindergarten "Peledziukas" [17] (Figure 2); Type of biophilic design: mimetic; Design: "DO Architects" (G. T. Gylyte, D. Baltrunas, K. Ciplyte, V. Babij, S. Daugeliene, A. Baldisiute, A. Neniskis, M. Vysniauskas); Location: Pagiriai, Vilnius; Year of completion: 2021. This object was selected due to its architectural expression (volume and materials) both modern and recalling traditional architectural design in the urban context. Moreover, this is a reconstructed building located in the urban context, where renovation and re-use as well as biophilic quality of the environment are of high importance.
2. Example B—Vilnius University Kairenai Botanical Garden's Green Building-Plant [18] (Figure 3); Type of biophilic design: applied; Design: Paleko "ARCH studija" (R. Palekas, B. Puzonas, D. Zakaite, A. Palekiene, V. Linge); Location: Kairenų st. 43, Vilnius; Year of completion: 2016. This object was selected due to its direct correspondence to the trend of applied biophilic design as this is reconstructed building, the biophilic character of which is created with vegetated columns—an unusual and experimental solution in Lithuanian climate conditions.
3. Example C—Recreation and Water Center in Zarasai (Figure 4) [19]; Type of biophilic design: organic; Design: Archartele ir partneriai (H. Staude and A. Minkauskas); Location: The island of the great Zarasas, Zarasai; Year of completion: 2015. This object was selected due to the synergetic effect between building's shape, materials, and environment creating a biophilic experience.

Figure 2. Kindergarten "Peledziukas" located in Vilnius, made with materials and roof configuration recalling traditional wood architecture was selected as an example of mimetic biophilic design. Photographs by A. Daugelaite.

Figure 3. Vilnius University Kairenai Botanical Garden's Green Building-Plant with external layer of vegetation was selected as an example of applied biophilic design. Photographs by A. Daugelaite.

Figure 4. Recreation and Water Center in Zarasai with landscape-inspired volumes and nature-like spatial characteristics was selected as an example of organic biophilic design. Photographs by A. Daugelaite.

4.2. Assessment of Case Study Buildings

The overall biophilic design aim, to restore broken human–nature connections, encourages an analysis of biophilic buildings not only from the point of their aesthetic expression and human well-being benefits, but also from the points of view of human–nature integration and human–nature collaboration. As it was mentioned in the previous section, aesthetic and functional bio-collaboration distinguished by E. L. M. Wolfs [6] is widespread in biophilic architecture. In this research we apply the more general term human–nature collaboration, used in the fields of sustainability aesthetics [20] and regenerative sustainability [21–23]. The term human–nature collaboration encompasses bio-collaboration in its turn but is not limited to it and includes such factors as designs' engagement with environmental forces.

In order to evaluate the mimetic, applied, and organic biophilic designs from the points of view of biophilic qualities, aesthetic expression, and human–nature collaboration a series of questions was formulated. It was first based on the biophilic design criteria [1,5,7] and other sources (Table 1) and then applied for on-site evaluation of three selected design examples, representing the above-distinguished trends.

Table 1. Questions used for the assessment of selected buildings as a means for design evaluation from sustainability aesthetics, biophilic design, and human–nature collaboration points of view [1,5,7,20–26]. Buildings correspondence to the criterion is evaluated in the scale from 0 to 2: None = 0 (gray color in the table); Some = 1 (yellow color in table); Clearly expressed = 2 (green color in the table).

	Criteria of Architectural Expression	Architectural Means/Explanation/Hint	Examples A	B	C
Features of environment	Does the object adapt to local terrain and landscape conditions?	Prioritize real nature over simulated nature; Adaptation to local terrain forms; Preservation of vegetation; Response to landscape character	2	2	2
	Does the object express the engagement with environmental forces (water, air, sunlight...) in meaningful and visible way?	Sun, shade, reflections; Integration of waterbodies; Rainwater management; Integration of vegetation; Possibilities to feel airflow, etc.	2	2	2
	Does the object integrate ecosystems and habitats in a meaningful and visible way?	Flora: ecological systems, visual continuity, trees, shrubs, vegetated ground covers, habitats, rare plant species, nectar rich vegetation, flowering wild local herbs etc. Fauna: birds, insects, land animals and reptiles, fish, endangered species, etc.; Bird box, bat box, biotope for specified insects	1	1	1
	Does the object provide opportunities for seeing, hearing or touching of water?	Naturally occurring: river, stream, ocean, pond, wetland; Visual access to rainfall and flows; Seasonal flows Simulated or constructed: water wall, constructed waterfall, aquarium, fountain, constructed stream; Reflections of water (real or simulated) on another surface; Imagery with water in the composition	0	2	2
Materials	Does the object integrate natural (and local) materials?	Real materials are preferred over synthetic; Materials and elements from nature that, through minimal processing, reflect the local ecology or geology to create a distinct sense of place, sometimes stimulating to the touch	2	2	2
Visual interest	Are there visual connections between the object and its environment present?	A view to elements of nature, living systems and natural processes; Prospect—an unimpeded view over a distance; Quality views from the outside and inside	2	2	2
	Does the object contribute to scenic quality or landscape character?	Architectural object interacts with landscape (identical, similar, contrasting) and forms qualitative wholeness	2	2	2
	Does the object provide views to elements of nature, living systems, and other living things at all?	Naturally occurring: natural flow of a body of water; Vegetation, including food bearing plants, animals, insects, fossils, terrain, soil, earth Simulated or constructed: mechanical flow of a body of water; Koi pond, aquarium; Green wall; Artwork depicting nature scenes; Video depicting nature scenes; Highly transformed, designed landscapes	2	2	2
	Does the object correspond to other unique physical features?	Unique site elements are integrated into the design	2	2	2
	Is the object harmoniously integrated in landscape/cityscape and looks visually balanced?	Part/whole relationships that may include balance, coherence, concinnity, consonance, orchestration, proportion, symmetry, symphony, unity	2	2	2

Table 1. Cont.

	Criteria of Architectural Expression	Architectural Means/Explanation/Hint	Examples A	B	C
Shapes and forms	Does the object's design integrate/interpret natural forms and motifs?	Symbolic references to contoured, patterned, textured or numerical arrangements that persist in nature; Presence of natural (botanical, animal) motifs in the design	1	1	2
	Does the object's design mimic nature's forms (e.g., biomorphic shapes) in a functional way?	Functional biomimicry	0	0	2
	Is the object's design based on geomorphic shapes?	Relation to the form or surface features of the earth or landscape	0	0	2
	Does the object include spatial hierarchy similar to those encountered in nature?	Complexity that simultaneously stimulates senses of intrigue and order, and reduces stress	0	1	2
Light and space	Does the object integrate/provide natural light?	Architectural object provides users with natural lighting options	2	2	2
	Are light quality variations, such as diffused, filtered light, light and shadow, reflections present in the object?	Varying intensities of light and shadow that change over time to create conditions that occur in nature. Naturally occurring: daylight from multiple angles, direct sunlight, diurnal and seasonal light, firelight, moonlight and star light, bioluminescence. Simulated or constructed: multiple low glare electric light sources, illuminance, light distribution, ambient diffuse lighting on walls and ceiling, day light preserving window treatments, task and personal lighting; accent lighting. Personal user dimming controls; Circadian color reference	1	2	1
	Is the spatial diversity, variability and interest integrated in the object?	Curving edges; Dramatic shade and shadows; Winding paths; Partially revealed spaces; Translucent materials; Obscuring of the boundaries and a portion of the focal subject	2	1	2
Processes/Patterns *	Does the object create sensitive and cognitive variability and/or richness?	Information-richness, balance between boring and overwhelming	2	2	2
	Does the object express the process of co-creation with nature?	Construction using mycelium, technologies with algae for energy production and air quality improvement, "bio-concrete" made of moss and beef mushrooms in rainwater and allowing plants to be grown on the facades, salt slabs made of salt, sunflower and algae, bioplastics made of algae, etc.	1	2	1
	Does the object express the structural patterns related with fractality, centrality, part-whole integration?	Self-similarity across different scales. Integration or interpretation of naturally occurring fractals: branches of trees, animal circulatory systems, snowflakes, lightning and electricity, plants and leaves, geographic terrain and river systems, clouds, crystals; Nested fractal designs. Action of a central element in its periphery. Part-whole integration—relation of object's parts to the whole object itself; Application of the Fibonacci series, the Golden Mean	1	1	1
	Does the object express in a meaningful and visible way the behavior patterns characteristic to natural systems and organisms?	Change over time; Decaying—changing properties (rusting metal, wood changing color over time), natural patina of materials (leather, stone, copper, bronze, wood); Growing plants, moss	2	2	2
	Does the object express the stochastic and ephemeral connections with nature?	Integration, emphasis of naturally occurring phenomena: cloud movement, breezes, plant life rustling, water babbling, insect and animal movement, birds chirping, fragrant flowers, trees and herbs. Simulated or constructed: billowy fabric or screen, materials that move or glisten with light or breezes, reflections of water on a surface, shadows or dappled light that change with movement or time, nature sounds broadcasted at unpredictable intervals, mechanically released plant oils	1	2	1
	Does the object provide thermal and airflow variability?	Naturally occurring: solar heat gain, shadow and shade, radiant surface materials, space/place orientation, vegetation with seasonal densification. Simulated or constructed: HVAC delivery strategy, systems controls, window glazing and window treatment, window operability and cross ventilation	2	2	2

Table 1. Cont.

	Criteria of Architectural Expression	Architectural Means/Explanation/Hint	Examples A	B	C
Human–environment relations	Does the object maintain/contribute to the spirit of place?	The design maintains/contributes to tangible (buildings, sites, landscapes, routes, objects) and the intangible elements (memories, narratives, written documents, rituals, festivals, traditional knowledge, values, textures, colors, odors, etc.) of the spirit of place. The object connects to the essence of the place in ecological, cultural, historic, geographic dimensions	2	2	2
	Does the object involve restoration of the damaged environment in meaningful and visible way?	Improved ecological situation: surfaces are permeable to water, variety of vegetation, rainwater management (bioswales, raingardens, etc.), a section of the courtyard is left for natural succession (that is, to naturally grow and regenerate), composting biodegradable waste; Design prioritizes biodiversity over acreage, area or quantity	0	2	0
	Does the object employ/demonstrate self-healing qualities of nature?	Little maintenance is required, the site is self-operating like in natural places, like meadow or forest	0	0	0
	Does the object stimulate exploration and cognition?	The object creates the conditions that differentiate between surprise (i.e., fear) and pleasure, creates a sense of mystery, risk/peril, arouse interest of exploring Mystery created by the promise of more information achieved through partially obscured views or other sensory devices that entice the individual to travel deeper into the environment. e.g., Peek-a-boo windows that partially reveal, curving edges, winding paths. Risk/Peril is created as an identifiable threat coupled with a reliable safeguard: double-height atrium with balcony or catwalk, architectural cantilevers, infinity edges, façade with floor-to-ceiling transparency, experiences or objects that are perceived to be defying or testing gravity, transparent railing or floor plane, passing under, over or through water, proximity to an active honeybee apiary or predatory animals, life-sized photography of spiders or snakes	2	1	2
	Does the object stimulate sense of security in users and viewers perception?	Creating physical and mental safety, refuge—a place for withdrawal, from environmental conditions or the main flow of activity, in which the individual is protected from behind and overhead	2	2	2
	Does the object stimulate sense of attraction and emotional, spiritual connection with it and its place in users and viewers perception?	People are taking photographs, collect litter, spend their free time in and around the object	2	2	2
	Does the object stimulate experience of nature through senses?	Design stimulates auditory, haptic, olfactory, or gustatory stimuli referring to nature, living systems or natural processes. Naturally occurring: fragrant herbs and flowers, songbirds, flowing water, weather (rain, wind, hail), natural ventilation (operable windows, breezeways), textured materials (stone, wood), crackling fire/fireplace, sun patches, warm/cool surfaces Simulated or constructed: digital simulations of nature sounds, mechanically released natural plant oils, highly textured fabrics/textiles that mimic natural material textures, audible and/or physically accessible water feature, music with fractal qualities, horticulture/gardening, including edible plants, domesticated animals/pets, honeybee apiary	1	1	1
	Does the object stimulate connection with natural systems?	Naturally occurring: climate and weather patterns (rain, hail, snow, wind, clouds, fog, thunder, lightning), hydrology (precipitation, surface water flows and resources, flooding, drought, seasonal flows), geology (visible fault lines and fossils, erosion, shifting dunes), animal behaviors (predation, feeding, foraging, mating, habitation, migration), pollination, growth, aging and decomposition (insects, flowering, plants), diurnal patterns (light color and intensity, shadow casting, plant receptivity, animal behavior, tidal changes), night sky (stars, constellations, the Milky Way) and cycles (moon stages, eclipses, planetary alignments, astronomical events), seasonal patterns (freeze-thaw, light intensity and color, plant cycles, animal migration, ambient scents) Simulated or constructed: simulated daylighting systems that transition with diurnal cycles, constructed wildlife habitats (e.g., birdhouse, honeybee apiary, hedges, flowering vegetation), exposure of water infrastructure	1	2	2
		Total:	42	49	52

* Pattern—a form or model proposed for imitation.

31 questions subdivided into 7 categories—features of environment, materials, visual interest, shapes and forms, light and space, processes/patterns, and human–environment relations—were answered evaluating the answer in the scale from 0 to 2, evaluation 0 meaning that qualities are not present and 2 meaning qualities are clearly expressed. The highest

possible evaluation of the building using this approach is 62. Quantitative assessment of case study buildings has revealed that all of them can be considered as biophilic buildings, having features of sustainability aesthetics and human–nature collaboration as the evaluation score in all three cases has exceeded 30. Object A—Kindergarten "Peledziukas" was evaluated with the score 42, the lowest of all three case study objects with weakest evaluation in the categories of human–environment relations and shapes and forms. Object B—Vilnius University Kairenai Botanical Garden's Green Building-Plant was evaluated with the score 49 with weakest evaluations in shapes and forms category. The lower evaluation of shapes and forms of both buildings is determined by the fact that both objects are reconstructed Soviet era buildings. The facts of reconstruction and adaptive re-use give positive consideration from sustainability point of view. Object C—Recreation and Water Center in Zarasai received 52 scores from 62 and demonstrates the highest presence of biophilic qualities from all the evaluated case study objects. The weakest evaluation of this object is in the category of human–environment relations as well as in the first case study building. It is possible to conclude that the potential possibilities provided by restorative and regenerative approaches to design were not employed in these projects.

4.3. Descriptive Analysis and Discussion of Case Study Buildings

Descriptive qualitative analysis of case study objects provides the example of analyzing and discussing the buildings and their surroundings from biophilic design, sustainability aesthetics, and human–nature collaboration points of view offering a different angle for looking at projects and their implementation. The descriptive analysis of each object was elaborated based on the questions presented in the Table 1, demonstrating the suitability of this approach for both quantitative and qualitative analysis of buildings.

Kindergarten "Peledziukas". The object's terrain is flat, and the object is placed there without extreme changes in the terrain. The object is strongly engaged with the sun—it provides many opportunities of feeling the sun in different angles and places and provides shaded areas under the trees or tracery walls. The vertical timber panels cast changing shadows. The object provides a lot of open spaces, such as the inner garden, a rooftop terrace, playgrounds, etc. with a possibility to feel the air. However, there are no water features.

The project deserves the highest evaluation of the efforts to preserve the trees (the initial idea of the project was changed in order to save the old spruce tree, which even has a tale of origin). However, the area is poor with other parts of ecosystem, such as fauna habitats or wild herbaceous flowering plants. There is not any water element. The object is constructed of timber, which dominates in the interior and exterior and furniture design.

The object provides very strong visual connections among its spaces (for example, children can see the work at the canteen or cleaner's room) and to the outside with views to the pine grove from the roof terrace, which obviously add value to the project. The object definitely contributes to scenic quality of the area.

The object looks visually balanced and well placed. Strict lines and forms dominate in the building, which is rarely found in nature. Biomorphic forms are not directly visible in the design; aside from the color and shape of the roofing which recalls traditional wooden architecture as well as the stylized owl's ears that can be associated with the owl-themed name of the institution (Pelėdžiukas translates to owlet in English). The site's surface is flat, thus it is not applicable to the evaluation of geomorphic forms. Spatial hierarchy is expressed in the building's volume, but the object lacks fractality.

The object provides sunlight from different angles and in different daytimes, however lighting variations (interplay of light and shadow, diffused light, etc.) are rare. The object is rich with diverse and partially revealed spaces and their dynamics and creates cognitive variability. A co-creation with nature is expressed through the naturally aging wood cladding.

The object has strong relation between the whole and its parts. The centrality of the object is created through the central garden which forms the core of the whole project. Ephemeral connections with nature may be felt by seeing naturally occurring phenomena through the windows (like cloud movement, birds, etc.). There is a lack of other senses, like smells of plants, blooming flowers, water features or animals, insects, etc. life. The curtains in the corridor may rustle with light breeze. The windows are openable and the rooms can be ventilated, and air movement can be felt.

The object contributes to the spirit of place by enriching it with innovative architecture and improving the urban landscape of the area. Although the existing trees are involved beautifully in the design, other features of improving the local ecology, such as permeable surfaces, variety of vegetation, biodiversity, etc. are missing.

The object definitely stimulates exploration and cognition by involving "mystery" elements in partially revealed spaces, roof terrace, and walls with floor-to-ceiling transparency. The sense of safety and attraction is strong. The experience of nature and connection to the living systems could be stimulated through the senses even stronger. It is possible to feel warm/cool surface in sun-shaded areas, natural ventilation in the building, or feel the breeze while being in the courtyard, as well as see weather conditions through the large windows or touch the natural wood on the facade. However, auditory, olfactory, or gustatory stimuli are not reflected and the project could be enriched with flora and fauna.

Vilnius University Kairenai Botanical Garden's Green Building-Plant. The object's terrain is flat. The object is placed there without extreme changes in the terrain. The building engages with environmental forces by the vegetated façade that provides light and shadow interplay, the sound of wind through the plants, a little fountain is integrated near the entrance of the building, and large pond is located on the site. The object integration with local ecosystems and habitats is not visible despite the fact that it is located in the botanical garden. The project contains few habitats for the fauna in the backyard, there are shrubs growing on the premises to the building, however, in terms of habitat it is probably insufficient. Flowering plants for bees or butterflies are found further from the building premises. Concrete paving is hardly permeable surface, however, it only takes up a small area. The object provides views to the pond and fountain and it is possible to hear and touch the water on the building's site.

The object's façade is constructed of planted columns which encourages interest and desire to touch. The columns are constructed of local turf. The building is a reconstructed Soviet era apartment building. The views to the living systems and natural processes are obvious. The building provides an unimpeded view over a distance, views from the building are exceptional. The building definitely supplements the landscape. A unique site characteristic is the botanical motif which is transferred to the building. The building looks visually balanced itself and on the site.

Strict lines and forms dominate in the building, which is rarely found in nature. However, the planted columns soften the impression. The site's surface is flat, thus it is not applicable for the evaluation of geomorphic forms. Spatial hierarchy is not expressed and the main façade elements are of one size. The exception is the front garden which is a labyrinth that provides the full scale of fractals.

The spatial diversity, variability, and interest is high. The interplay of light and shadow is variable, however, these features could be expressed even stronger by more expressive loops of the paths and partially revealed spaces in the interior, etc. The process of co-creation with nature is strongly expressed. The object forms a strong relation of its parts to the whole object itself and event to its site which includes reference to fractal systems (planted surfaces), and provides the possibility to feel the airflow and hear nature sounds through open windows, feel the natural smells while being on the site, see water reflections, cloud movement, fountain water babbling, etc.

The object connects to the essence of the place by adding value to its character. Landscape restoration is not included and the surrounding lawn is poor in biodiversity terms. The project definitely improved the existing ecological situation. However, there is little information of how the rainwater is treated and if, for example, a section near the pond is left for natural succession (that is, to naturally grow and regenerate). These means could help to improve the richness of biodiversity. The site requires constant maintenance and self-healing qualities are not visible.

The object raises interest; however the stimulation of exploration and cognition may be expressed stronger by risk/peril and mystery means. A sense of security is strong. A conscious attempt to include auditory, haptic, olfactory, or gustatory stimuli into the design is not visible, however, some of these emerge from the special site itself. A connection to natural systems is stimulated through feeling (on the site) and observing (through windows) of naturally occurring processes like climate and weather, seasonal and diurnal patterns, and the feeling of the presence of vegetation and water. However, life of fauna is little expressed due to lack of wildlife habitats (e.g., birdhouse, honeybee apiary; hedges, flowering vegetation).

Recreation and Water Center in Zarasai. The object nicely integrates man-made structure and natural landscape. The architectural structure connects land and water. The question may arise about whether it is a building, a bridge, or a path. Structural variety offers possibilities to touch the water or to feel the wind breeze, shade, and sun. Although ecological systems such as habitats, rare plant species, nectar rich local vegetation, and others are not integrated in the project on purpose, it offers visual continuity of a man–nature made landscape, and opportunities to find fauna life in the trees or the lake.

Timber cladding reflects the local materials. Visual connections between the object and its environment are strong from both the inside and outside. The object definitely contributes to the scenic quality. The object provides views to elements of nature and living systems including the lake. The object provides the paths over water and roof terraces. The object takes advantage of the unique lake shores and existing tree line. The object looks harmoniously integrated in the landscape.

The object's design is based on an organic, naturally flowing form that looks like it is grown out of its site. Biomorphic shapes are repeated through the object's design—rooms or roof terraces are evolving out of the paths, etc. The object creates geomorphic forms and the image of the hills rising up or down. This feature expressed the spatial hierarchy as well. The object provides light from different angles and offers some dramatic shadows, however, light variations are not very rich. Curving edges and winding paths partially revealed spaces to offer spatial variability.

The object is information-rich and involves the process of co-creation with nature through decaying natural wood and strong connections with landscape. Fractality is not expressed. The object creates a central focal point of interest in the landscape and part-whole integration is nicely expressed. The object has a wonderful location; however, it expresses the stochastic and ephemeral connections with nature only partially. It could integrate more strongly the life of birds, insects, wild plants, and others, however, its impermeable asphalt surfaces, shortly cut lawn, and lack of surrounding biodiversity show the lack of landscape restoration means in a meaningful and visible way. It could be done additionally without changing the properties of the object itself. Thermal and airflow variability is rich and is provided by the possibilities of the variety of spaces. The object contributes to the spirit of place by enriching the landscape and providing strong attraction, as well as a meeting and recreational point in a small town. The site requires maintenance, although it would require less if the meadows would be left to bloom.

The object stimulates exploration and cognition by creating a sense of mystery, risk/peril and arouses the interest of exploring. Winding paths, terraces, and paths over water invite a visitor for a stroll. The sense of safety is strong as well as attachment to the area. As the object may be visited at night, it offers experiences of stargazing, watching the moon, etc. However, auditory, olfactory, or gustatory stimuli are not reflected and the project could be enriched with flora and fauna, especially those natural to its wild location.

It is possible to conclude that all three analyzed projects are strong at integrating environmental features, however struggling with the inclusion of fauna life, such as insects, and flora, like blooming local flowers. Biodiversity on the sites is not rich enough and surfaces are rarely permeable. Therefore, it leads to difficulties for implementing the design criteria of engagement to living things and other sensorial stimuli like smells. This confirms the results of quantitative evaluation, demonstrating the lack of human–nature collaboration and restorative and regenerative approaches in case study objects. Although projects are strong in creating good pieces of architecture, they could have more features to provide senses of mystery, risk/peril, and naturalness, as well as spaces requiring low maintenance and offering a variety of natural processes that enrich people's lives. However, all three case study objects confirm that it is possible to create biophilic buildings and biophilic interior and exterior experiences in the Lithuanian climate not only in natural, but in urban environments as well. Moreover, biophilic qualities were successfully created even in the cases of Soviet era buildings reconstruction and adaptive re-use.

5. Conclusions

The significance of the biophilia hypothesis and biophilic design in providing favorable conditions for human well-being and healthy development in anthropogenic environments, restoring human–nature connections, and potentially bringing the development of built environments and human habitats to human–nature collaboration level significant for regenerative sustainability encourages the analysis of possibilities of architectural expression of biophilic buildings.

Existing classifications of biophilic design distinguish such trends as inspired by nature and traditional biophilic design, historic and contemporary biophilic design, natural and artificial biophilic design solutions, and explicit and implicit representation of nature in biophilic design. Analysis of the literature has revealed the lack of a universal biophilic design trends classification that would be suitable both to innovative and traditional buildings or buildings based on the biophilic geometry and on the healing influence of nature or integrating both of those aspects.

The classification reflecting the interconnections between the architectural form of a building and its biophilic properties was developed in the course of this research. Biophilic buildings are categorized into mimetic, applied, and organic: mimetic biophilic design achieves biophilic qualities by using symbolic, mimetic forms related to nature or traditional architecture; in the case of applied biophilic designs biophilic qualities are added as a layer, which appears extrinsic to the structure itself; in case of organic biophilic design a synergistic relation between the biophilic qualities and the structure is achieved. The analysis of bio-collaboration and the human–nature collaboration potential of these trends has revealed that all three trends hold the potential in these fields with particular attention to organic design and its structural bio-collaboration possibilities. Evaluation of three selected building examples located in Lithuania corresponding to mimetic, applied, and organic trends according to a comprehensive set of biophilic design criteria confirmed the highest potential for the organic trend to create biophilic environments and the suitability of the applied trend for successful biophilic reconstruction of existing buildings. However, it is possible to conclude that application of each of these trends allows for the creation of biophilic buildings and biophilic experiences in different climatic conditions including in the temperate climate zone.

Author Contributions: Conceptualization, I.G.-V., A.D. and G.V.; methodology, I.G.-V., A.D. and G.V.; resources, I.G.-V. and A.D.; writing—original draft preparation, I.G.-V., A.D. and G.V.; writing—review and editing, I.G.-V. and A.D.; visualization, I.G.-V. and A.D.; supervision, I.G.-V. All authors have read and agreed to the published version of the manuscript.

Funding: This research received no external funding.

Institutional Review Board Statement: Not applicable.

Informed Consent Statement: Not applicable.

Conflicts of Interest: The authors declare no conflict of interest.

References

1. Browning, W.; Ryan, C.; Clancy, J. 14 Patterns of Biophilic Design: Improving Health & Well-Being in the Built Environment. 2014. Available online: https://www.terrapinbrightgreen.com/wp-content/uploads/2014/09/14-Patterns-of-Biophilic-Design-Terrapin-2014p.pdf (accessed on 17 December 2021).
2. Samalavičius, A. Biophilic architecture: Possibilities and grinders. *Logos* **2020**, *105*, 109–118.
3. Wilson, E.O. Biophilia and conservation ethics. In *The Biophilia Hypothesis*; Kellert, S., Wilson, E.O., Eds.; Shearwater Books: Washington, DC, USA, 1993.
4. Ode, Å.; Tveit, M.S.; Fry, G. Capturing landscape visual character using indicators: Touching base with landscape aesthetic theory. *Landsc. Res.* **2008**, *33*, 89–117. [CrossRef]
5. Salingaros, N. The Biophilic Healing Index predicts effects of the built environment on our wellbeing. *J. Biourbanism* **2019**, *8*, 13–34.
6. Wolfs, E.L.M. Biophilic design and bio-collaboration: Applications and implications in the field of industrial design. *Arch. Des. Res.* **2015**, *28*, 71–89.
7. Kellert, S.; Heerwagen, J.H.; Mador, M.L. *Biophilic Design. The Theory, Science, and Practice of Bringing Buildings to Life*; Wiley: Hoboken, NJ, USA, 2013.
8. Beatley, T. *Biophilic Cities: Integrating Nature into Urban Design and Planning*; Island Press: Washington, DC, USA, 2011.
9. McGee, B. Biophilic Interior Design. 2016. Available online: https://bethmcgee.wixsite.com/biophilicdesign (accessed on 17 December 2021).
10. DeGroff, H.; McCall, W. *Biophilic Design. An Alternative Perspective for Sustainable Design in Senior Living*; Perkins Eastman: New York, NY, USA, 2016.
11. Wines, J. *Green Architecture*; Taschen: Köln, Germany, 2000.
12. Vormittag, J.E. Back to the Future. Biophilic Design in the King Office Complex in Stockholm/SE. 2019. Available online: https://pld-m.com/en/article/lighting-design/back-to-the-future (accessed on 20 September 2021).
13. Gutiérrez, R.U.; De la Plaza Hidalgo, L. *Elements of Sustainable Architecture*; Routledge: London, UK, 2019.
14. Tikkanen, A. Ornament. Architecture. Available online: https://www.britannica.com/technology/ornament (accessed on 8 October 2021).
15. Wark, B. Glasgow School of Art Extension. Available online: https://www.barrywark.com/gsaextension (accessed on 28 December 2021).
16. Bio-Integrated Design. Available online: https://www.ucl.ac.uk/bartlett/architecture/programmes/postgraduate/bio-integrated-design-bio-id-marchmsc (accessed on 28 December 2021).
17. Kvepiantis Medžiu: "Pelėdžiukas" Rodo Valstybinio Vaikų Darželio Pavyzdį/Scented with Wood: "Peledziukas" Shows an Example of a State Kindergarten. Available online: https://pilotas.lt/2021/09/15/architektura/kvepiantis-medziu-peledziukas-rodo-valstybinio-vaiku-darzelio-pavyzdi/ (accessed on 28 December 2021).
18. VU Kairėnų Botanikos Sodo Žaliasis Pastatas Augalas/VU Kairenai Botanical Garden Green Building Plant. Available online: https://archiforma.lt/?p=2084 (accessed on 28 December 2021).
19. Poilsio ir Vandens Centras Zarasuose/Recreation and Water Center in Zarasai. Available online: https://archiforma.lt/?p=1967 (accessed on 28 December 2021).
20. Kagan, S. Aesthetics of sustainability: A transdisciplinary sensibility for transformative practices. *Transdiscipl. J. Eng. Sci.* **2011**, *2*, 65–73. [CrossRef] [PubMed]
21. Berardi, U. Clarifying the new interpretations of the concept of sustainable building. *Sustain. Cities Soc.* **2013**, *8*, 72–78. [CrossRef]
22. Du Plessis, C. Towards a regenerative paradigm for the built environment. *Build. Res. Inf.* **2012**, *40*, 7–22. [CrossRef]
23. Istiadji, A.D.; Hardiman, G.; Satwiko, P. What is the Sustainable Method Enough for our Built Environment? In Proceedings of the IOP Conference Series: Earth and Environmental Science, Semarang, Indonesia, 29 August 2018.
24. Zafarmand, S.J.; Sugiyama, K.; Watanabe, M. Aesthetic and sustainability: The aesthetic attributes promoting product sustainability. *J. Sustain. Prod. Des.* **2003**, *3*, 173–186. [CrossRef]
25. Vecco, M. Genius loci as a meta-concept. *J. Cult. Herit.* **2020**, *41*, 225–231. [CrossRef]
26. Daugelaite, A.; Dogan, H.A.; Grazuleviciute-Vileniske, I. Characterizing sustainability aesthetics of buildings and environments: Methodological frame and pilot application to the hybrid environments. *Landsc. Archit. Art* **2021**, *19*, 61–72. [CrossRef]

Article

Analyzing the Bake-Out Effect in Winter for the Enhancement of Indoor Air Quality at New Apartments in UAE

Naglaa Sami Abdelaziz Mahmoud [1,2] and Chuloh Jung [2,3,*]

1. Department of Interior Design, College of Architecture, Art and Design, Ajman University, Ajman P.O. Box 346, United Arab Emirates; n.abdelaziz@ajman.ac.ae
2. Healthy and Sustainable Buildings Research Center, Ajman University, Ajman P.O. Box 346, United Arab Emirates
3. Department of Architecture, College of Architecture, Art and Design, Ajman University, Ajman P.O. Box 346, United Arab Emirates
* Correspondence: c.jung@ajman.ac.ae

Abstract: Indoor air pollution has become a pressing issue in the United Arab Emirates (UAE) due to poor ventilation, inadequate airtightness, and using chemicals in building materials. Accordingly, the UAE is currently experiencing more cases of sick building syndrome (SBS) than any other country. This study aims to assess the effectiveness of the bake-out strategy in reducing indoor air pollutants in a new apartment building in the UAE. The study evaluated a reduction in toluene (C_7H_8), ethylbenzene (C_8H_{10}), xylene (C_8H_{10}), styrene (C_8H_8), and formaldehyde (HCHO) at room temperature and relative humidity. The airtight unit without winter bake-out had higher indoor concentrations of hazardous chemicals than the ventilated units, and the emission of dangerous substances increased with temperature. Moreover, harmful chemicals were only effectively reduced with ventilation times of at least seven days after the heating period. The release rate of contaminants after the bake-out was lower than before. The indoor concentration of hazardous chemicals was lower when bake-out and mechanical ventilation were combined, resulting in a reduction of 92.8% of HCHO. Furthermore, units with a certain amount of ventilation maintained a low indoor pollutant concentration, regardless of whether a bake-out was performed.

Keywords: bake-out; indoor air pollutants; hot desert climate; volatile organic compounds; formaldehyde; United Arab Emirates

Citation: Abdelaziz Mahmoud, N.S.; Jung, C. Analyzing the Bake-Out Effect in Winter for the Enhancement of Indoor Air Quality at New Apartments in UAE. *Buildings* **2023**, *13*, 846. https://doi.org/10.3390/buildings13040846

Academic Editors: Lina Šeduikytė, Jakub Kolarik and Christopher Yu-Hang Chao

Received: 31 January 2023
Revised: 7 March 2023
Accepted: 20 March 2023
Published: 23 March 2023

Copyright: © 2023 by the authors. Licensee MDPI, Basel, Switzerland. This article is an open access article distributed under the terms and conditions of the Creative Commons Attribution (CC BY) license (https://creativecommons.org/licenses/by/4.0/).

1. Introduction

Recently, the issue of indoor air pollution has become a significant concern in apartments and structures in the United Arab Emirates (UAE), where individuals spend 90% of their time indoors due to the sweltering summers and the absence of a clear demarcation between seasons [1–3]. This is largely due to the lack of ventilation, airtightness, and the utilization of new chemicals in building materials [4–6]. Reports have indicated that sick building syndrome (SBS) is becoming more prevalent in the UAE compared to other countries [7–9]. SBS is caused by the presence of volatile organic compounds (VOCs) and formaldehyde (HCHO), which are produced by finishing or construction materials [10–12]. Consequently, several methods have been proposed to enhance indoor air quality (IAQ), including the use of eco-friendly building materials, ventilation, the utilization of functional auxiliary materials, and management procedures such as bake-out after applying materials [13–15]. Recently, bake-out has gained attention as a strategy for decreasing indoor air pollutants [16–18].

It is worth noting that the concentration levels of VOCs and HCHO in the interiors of public housing in the UAE have surpassed the standard levels [19,20]. Additionally, air conditioning is a standard residential feature that is utilized nearly 24 h a day, year-round. Therefore, it is necessary to consider the specific circumstances in the UAE when implementing remediation techniques, especially since the outdoor concentrations of pollutants were lower than indoor concentrations due to the limited industrial activity in the surrounding area and the relatively low traffic volume during winter in the UAE [21]. While several techniques are available to improve indoor air quality, it is essential to compare the effectiveness of bake-out with other remediation approaches or no remediation at all [22,23]. However, a standardized method for bake-out has yet to be established, particularly for use in apartments [24,25]. The effectiveness of bake-out in reducing indoor air pollutants varies depending on factors such as heating time, temperature, and ventilation status [26,27]. The implementation method of bake-out may differ based on the season [28,29]. Therefore, this study proposes an effective implementation plan for bake-out by evaluating the reduction in indoor air pollutants using the winter bake-out method in a new apartment building in the UAE.

Bake-out is a technique that can reduce or eliminate pollutants and volatile compounds present in building materials by temporarily elevating the room temperature for a specific duration [30]. The buildings' bake-out concept is relatively straightforward. The term refers to the process of intentionally increasing the temperature within a building for a short period to hasten the drying of paints, mastics, or other finishes [31].

The bake-out process has shown promising results in reducing formaldehyde levels in raw wood composite panels. Longer bake-out durations may be required to achieve the desired reductions in formaldehyde concentrations. Given the significant variation in furniture and construction materials across different structures, it is crucial to investigate the bake-out treatment of numerous buildings under various conditions, such as temperature, time, and ventilation rate, before drawing conclusive results [32].

In one study, Park found that raising the room temperature from 30 °C to 50 °C, maintaining it for 24 h, and keeping the ventilation at 1.59 times per hour resulted in a 29% decrease in the total volatile organic compound (TVOC) concentration in both residential and office buildings [33]. Typically, various parameters, such as heating time, heating temperature, and ventilation rate during or after the bake-out, are measured to evaluate the effectiveness of the bake-out [34,35].

Shin et al. conducted a study on the effect of bake-out on a 14-story apartment building with a Reinforced Concrete (R/C) structure 4 months after completion, using a heating temperature of approximately 32 °C and a heating time of 24 h. The study revealed a 23–33% reduction in HCHO and 50% in TVOC concentrations [36]. Notably, indoor relative humidity control during bake-out can significantly affect its effectiveness [37,38]. Despite numerous studies on bake-out, a unified understanding of its efficacy in reducing chemical substances has not been reached [39]. Lv et al. demonstrated that formaldehyde emissions in construction materials mainly originate from water-based treatment agents and solvent-thinned adhesives and have low emission and delayed decomposition rates [40]. External heat sources have been proposed to supplement the existing HVAC system, including electric room heaters for truck-sized systems requiring portable generators [41]. This is due to the impact of different experimental conditions, the lack of a standardized evaluation method regarding heating conditions and evaluation methods, and the influence of finishing material specifications on reducing indoor chemical pollution by bake-out [42,43].

However, a study reported conflicting results from previous research on bake-out due to temperature, humidity, and ventilation variations that affect contaminant concentrations after the treatment [44]. Differences in ventilation requirements and rates during bake-out also contribute to these discrepancies [45,46]. Zuo et al. used simulation to identify optimal implementation conditions for bake-out, including increased ventilation rates during the treatment to minimize adsorption and desorption phenomena [47,48].

To reduce VOCs after construction and before occupancy, Leadership in Energy and Environmental Design (LEED) recommends using flush-out and limiting concentrations to 900 µg/m^3 for Styrene, 700 µg/m^3 for Xylene, 300 µg/m^3 for Toluene, 2000 µg/m^3 for Ethylbenzene, and 32 µg/m^3 for Formaldehyde during the first 14 days of occupancy in retail projects. These guidelines should be followed under typical occupancy ventilation conditions.

The research hypotheses are as follows:
- Bake-out reduces indoor air pollutants in apartments located in the UAE.
- Units with airtight construction have higher indoor concentrations of hazardous chemicals without bake-out.
- The emission of hazardous chemicals increases with temperature.

2. Materials and Methods

2.1. Methods for Measurement and Analysis of VOCs and HCHO

Toluene (C_7H_8), Ethylbenzene (C_8H_{10}), Xylene (C_8H_{10}), and Styrene (C_8H_8) were selected as representative VOCs, along with HCHO to evaluate the effectiveness of bake-out in reducing harmful indoor air pollutants [49,50]. In addition, the room temperature and relative humidity were also measured. The measuring equipment used in this experiment is outlined in Table 1.

Indoor air samples were collected after 30 min of ventilation, followed by 5 h of sealing [51,52]. Furthermore, indoor air samples were collected under natural and mechanical ventilation conditions, with front and rear balcony windows open by 10 cm [53]. VOCs were collected by attaching an adsorption tube filled with TenaxTA 200 mg in a 0.6 × 9 cm stainless tube to a low-flow sampling pump (SIBATA) at 150 mL/min for 30 min [54,55]. In the GC/MS system, an automatic thermal desorption device (ATD-400, Perkin Elmer, Waltham, MA, USA) was directly connected to a GC column and was used to analyze the VOCs in standard and emission test samples. Table 1 provides details of the analysis conditions for the GC/MS.

Table 1. The measuring instruments and the related analysis conditions.

	Measuring Items	Measuring Instruments	Analysis Conditions
VOCs	Toluene (C_7H_8)	- Gas Chromatograph (Agilent 8890 GC, Agilent, Santa Clara, CA, USA) - Mass Spectrometer (Agilent 5973N MSD, Agilent, Santa Clara, CA, USA) - Thermal Desorber (Agilent 7667A Mini Thermal Desorber, Agilent, Santa Clara, CA, USA)	- HP-1 Capillary Column (60 m × 0.32 μg/m × 0.25 nm) - Column Low: 1 mL/min - MS Ion Source Temperature: 260 °C - Column Temperature Rate: 60 °C (5 min) >> 5 °C/5 min to 260 °C - Split 20:1
	Ethylbenzene (C_8H_{10})		
	Xylene (C_8H_{10})		
	Styrene (C_8H_8)		
Formaldehyde (HCHO)		- High-Performance Liquid Chromatography (Shimadzu 10AVP Series HPLC System, Shimadzu, Kyoto, Japan)	- UV detector: 360 nm - Mobile ACN:H_2O = 60:40 - Extract Acetonitrile: 5 mL
Room Temperature Relative Humidity		- Digital Thermo-Hygrometer (TR-72U, Tecpel, New Taipei City, Taiwan)	

The collection of HCHO samples was carried out using an LpDNPH S10 cartridge (Supelco Inc., Bellefonte, PA, USA), which contains 350 mg of 2,4-DNPH-coated silica (1 mg DNPH) in a 1 cm (id) × 4 cm (total length) polypropylene tube [56]. A low-flow sampling pump (SIBATA) equipped with a flow control device was used, and the total sampling volume was 21 L. To remove interference caused by ozone, an ozone scrubber filled with potassium iodide (KI) in a polypropylene tube was connected to the front of the LpDNPH S10 cartridge to collect samples and then stored in a cold and dark place until extraction [57]. The carbonyl derivative of DNPH formed by the reaction with DNPH was extracted using 5 mL of HPLC-grade acetonitrile. The carbonyl compound from the extracted DNPH derivative was then analyzed using HPLC. The DNPH derivative has light absorption in the ultraviolet (UV) region, and the maximum wavelength was set at 360 nm [58].

2.2. Experiments Location

AQAAR constructed the Ajman Corniche Residence on Ajman Corniche Road (Figure 1) in the heart of Ajman Emirate, one of the UAE's Emirates [59,60]. The Residence comprises seven interlinked towers with one, two, and three-bedroom units on the same floor, same direction, and same environmental conditions [61].

Figure 1. Ajman Corniche Residence apartment.

Four apartment units in Tower A1102, Tower B1203, Tower C1501, and Tower D1302 at Ajman Corniche Residence did not undergo bake-out. Tower A1102, Tower B1203, and Tower C1501 maintained airtightness with natural ventilation or operated supply/exhaust ventilation units to keep airtightness [62,63]. Tower D1302, on the other hand, was an experimental unit that only used the exhaust fan of the kitchen range hood instead of the ventilation unit while maintaining airtightness by default, similar to Tower C1501. These experimental units were compared with those that underwent bake-out.

Meanwhile, Tower A1004, Tower B1103, Tower D1404, and Tower E1504 were the 4 units that underwent bake-out. Tower A1004 underwent bake-out as the primary method. Tower A1004 and Tower B1103 underwent bake-out while maintaining ventilation instead of continuous sealing after bake-out to review the appropriate ventilation period. Tower D1404 and Tower E1504 kept airtightness during the bake-out process, but Tower D1404 operated the supply/exhaust ventilation unit throughout the test period. Additionally, Tower E1504 used the exhaust fan of the kitchen range hood instead of the ventilation unit while undergoing bake-out and maintaining different ventilation types [64,65]. The sealing and ventilation conditions of the experimental units are shown in Figure 2. Figure 3 shows the measurement point on the plans of the 4 units consecutively: (a) Tower A1004, (b) Tower B1103, (c) Tower D1404, and (d) Tower E1504. These positions are similar in the 4 apartments at the same height level for the bake-out vacuum ovens (Thermo Scientific™ 3608, Waltham, WA, USA).

Figure 2. Sealed unit (Tower A1102) and ventilated unit (Tower B1203).

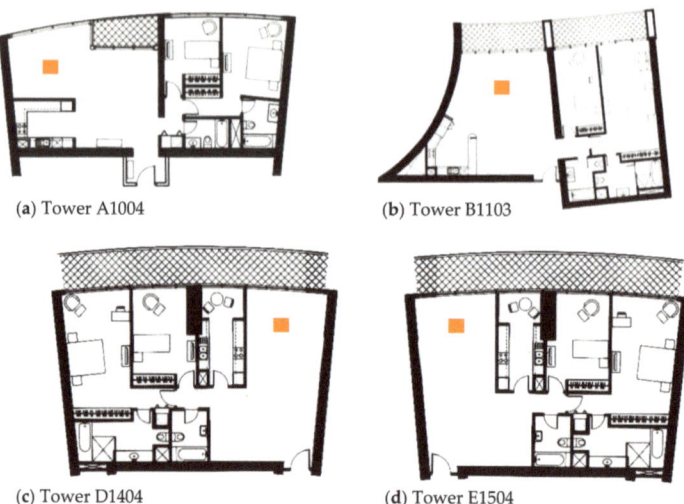

Figure 3. The measurement point on the 4 units' plans of the same floor level is (**a**) Tower A1004, (**b**) Tower B1103, (**c**) Tower D1404, and (**d**) Tower E1504.

2.3. The Experiment Design and Process

- Duration: the experiment was conducted between 2 December and 22 December 2020, in empty units prior to occupancy.
- Experimental units: there were 8 units, and Table 2 details each unit's composition and experimental content.
- Area and volume of units: the experiment units had an area of 1558 ft^2 (144.74 m^2) and a volume of 405.27 m^3.
- Ventilation rates: Tower C1501 and Tower D1404 were ventilated at approximately 6.23 ACH (air changes per hour) using the ventilation system. Tower D1302 and Tower E1504 were circulated at about 2.28 ACH through a kitchen exhaust fan.
- Cross ventilation: in the natural ventilation experiment, cross ventilation was achieved by opening the front and rear external windows in Tower B1203.
- Exhaust ventilation: units that employed exhaust ventilation (Tower C1501 and Tower D1404) supplied air from 5 points in each room (bedroom and living room) through a central air supply unit.
- Exhaust vent arrangement: Exhaust vents were placed in one location around the entrance and two areas in the living room to prevent exhaust from mixing with air supplied from the front. The exhaust was then expelled through the exhaust unit installed in the kitchen to the rear balcony.
- Local exhaust: in Tower D1302 and Tower E1504, the local exhaust was achieved using an exhaust fan in the kitchen range hood.

2.4. Bake-Out Experiment

The bake-out process involved heating the room to 32–34 °C for 4 days, as shown in Table 2 for Tower A1102, followed by 7 days of ventilation, considering the winter conditions in the UAE, and maintaining an airtight environment. Figure 4 displays the anticipated indoor concentration levels achieved through this method. C2b represents the average indoor concentration before the bake-out process, and C1b indicates the concentration level after 4 days have passed since reaching the target temperature. On the other hand, C2a represents the initial concentration in the room, and C1a represents the concentration level after the bake-out process [66,67]. The indoor concentration level was measured in the same position in all units, including the one where bake-out was not conducted, to compare the concentration variation with the essential bake-out generation [68–70].

Table 2. Compositions and experiment details.

#	Units	Before Bake-Out 2 December 2020	During Bake-Out 4 December 2020– 6 December 2020– 8 December 2020	Ventilation/Measurement 10 December 2020–15 December 2020	After Bake-Out 15 December 2020–22 December 2020	Bake-Out/ No Bake-Out
1	Tower A1102	Exterior Doors/Windows Kept Sealed				No Bake-out
2	Tower B1203	Exterior Doors/Windows Kept Open (Natural Ventilation)				
3	Tower C1501	Airtight with Air Supply/Exhaust Ventilation				
4	Tower D1302	Airtight with Kitchen Hood Exhaust Fan				
5	Tower A1004	Doors/Windows Sealed		Natural Ventilation	Airtight	Bake-out
6	Tower B1103	Doors/Windows Sealed		Natural Ventilation	Natural Ventilation	
7	Tower D1404	Exterior Doors/Windows are Kept Sealed, and Air Supply/Exhaust Ventilation				
8	Tower E1504	Exterior Doors/Windows are Kept Sealed, and Kitchen Hood Exhaust Fan				

All measurement points were based on the primary household of the bake-out in Tower A1102. Before baking (4 December), the concentration level was measured 4 days after reaching the target temperature and maintaining the temperature above 32 °C (8 December). Additionally, measurements were taken after 2 and 7 days of ventilation (10 and 15 December) and when the room was airtight after 7 days (22 December) [71,72]. During the experiment, "during bake-out" refers to the heating period, while "after bake-out" refers to the time after heating ceased and ventilation started, as shown in Figure 4. In total, there were five measurement points, including the initial measurement before the bake-out process, and measurements after four days of heating, two and seven days of ventilation, and seven days of airtight condition.

To summarize, the bake-out process involved heating the room for four days, followed by seven days of ventilation. Measurement points were taken after two and seven days of ventilation and seven days after the room was airtight. The initial measurement was taken before the bake-out process.

Five measurement points were taken in all units, including when the bake-out was not conducted. The initial measurement was taken before the bake-out process. Measures were taken after four days of heating, two and seven days of ventilation, and seven days of airtight condition. Therefore, the total number of VOC and formaldehyde samples collected and analyzed is five.

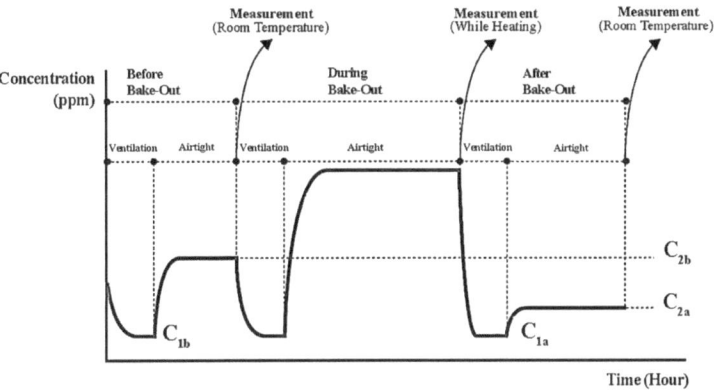

Figure 4. Rising pattern of the indoor concentration of the general bake-out.

3. Results

3.1. Temperature and Relative Humidity (RH) Distribution

Figures 5 and 6 display the indoor temperature and relative humidity (RH) in housing units that did not undergo bake-out, including Tower A1102, Tower B1203, Tower C1501, and Tower D1302. The indoor temperature ranged from 26.3 °C to 33.4 °C, lower than the outdoor temperature. Furthermore, the RH was maintained at approximately 60.5–77.8%, higher than most outside humidity as time passed. In contrast, in the housing units where bake-out was performed (Tower A1004, Tower B1103, Tower D1404, and Tower E1504), the temperature was generally between 26.7 °C and 35.6 °C, confirming the distribution was at the set temperature of 34 °C higher than the outside temperature until the heating period (4–8 December 2020), as depicted in Figure 7. The difference in the indoor temperature distribution between the two-unit groups was found when the temperature was heated for approximately four and seven days after turning off the heating. In the units where bake-out was performed, the RH decreased during the heating period, and there was no significant change as time passed after the heating was finished. The RH distribution was approximately 50.7–77.8%, as presented in Figure 8. Meanwhile, the indoor temperature was relatively higher, and the RH was lower in units that underwent bake-out than those that did not.

In Tower D1404, which had some ventilation and underwent bake-out, and Tower E1504, the temperature ranges were in the field of 31.2–34.5 °C and 32.8–35.2 °C, respectively. Both units reached a sufficient target temperature. Additionally, RH is one of the factors that affect the effectiveness of bake-outs. When the indoor temperature rises during the heating period, moisture in the concrete is released into the room, increasing the generation of hydrophilic substances such as HCHO. This can positively impact the abatement effect but is expected to impair hydrophobic substances such as C_7H_8 and C_6H_6.

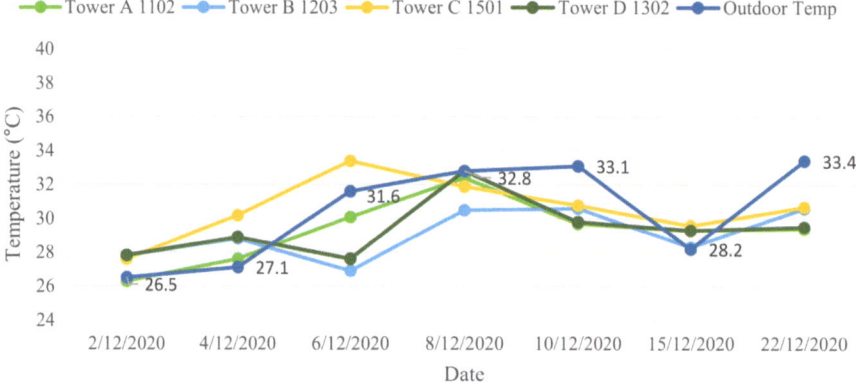

Figure 5. Temperature distribution of the units without bake-out.

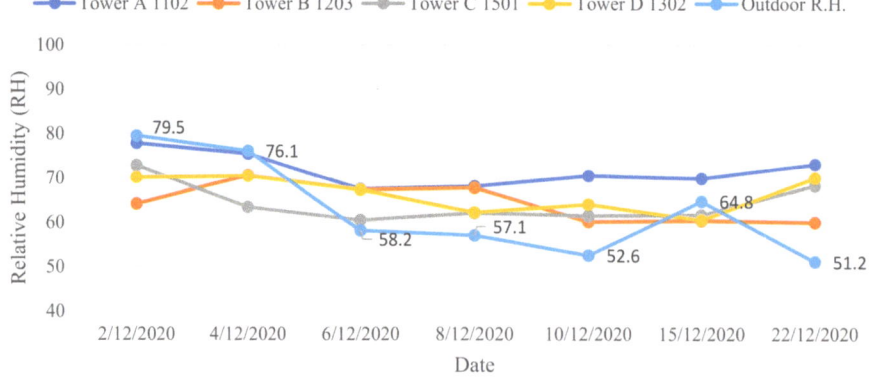

Figure 6. Relative humidity distribution of the units without bake-out.

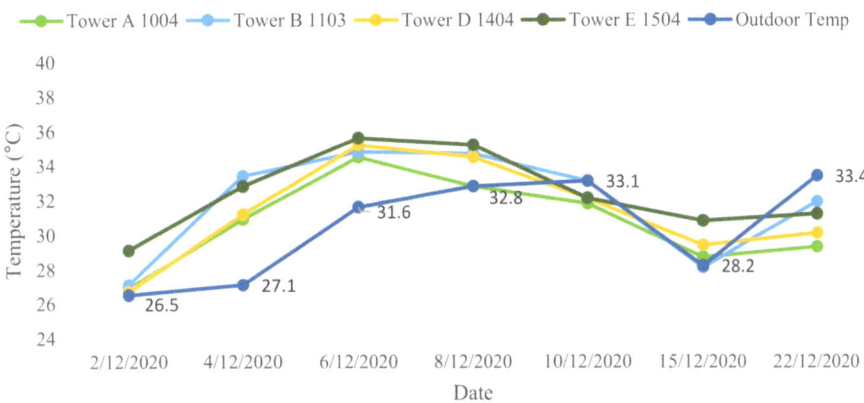

Figure 7. Temperature distribution of the bake-out units.

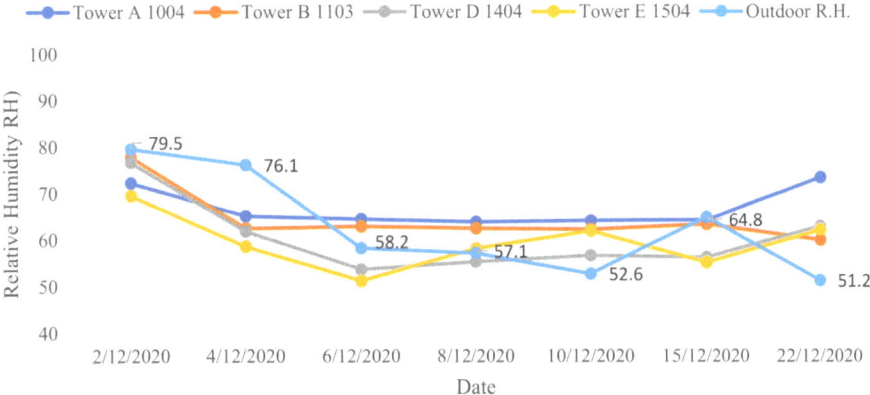

Figure 8. Relative humidity distribution of the bake-out units.

3.2. Characteristics of Changes in Indoor Air Pollutants by Housing Unit

3.2.1. Tower A1102 (Exterior Doors/Windows Kept Sealed)

During the experimental period, the changes in indoor air pollutant concentrations in Tower A1102 were measured by sealing all doors and windows in contact with the outside [51,52]. Figure 9 displays the changes in pollutant concentrations from the test start date (2 December 2020) to 20 days after (22 December 2020). The emission of HCHO increased with indoor and outdoor temperature, resulting in a high indoor concentration that exceeded the recommended standard of 210 µg/m³ during the experimental period. This confirmed the need to remove and manage HCHO during the winter [51,52]. Additionally, C_7H_8, ethylbenzene (C_8H_{10}), xylene (C_8H_{10}), and C_8H_8 also had high indoor concentrations. Specifically, C_7H_8 had a consistently high concentration that exceeded the recommended standard of 1000 µg/m³ for 10 days after the start of the test [51,52]. Although the difference between outdoor and indoor temperatures was slight due to winter conditions in the UAE and the openings were sealed during the experimental period, slight concentration reduction and reduction of concentration due to infiltration and adsorption, respectively, can still be predicted.

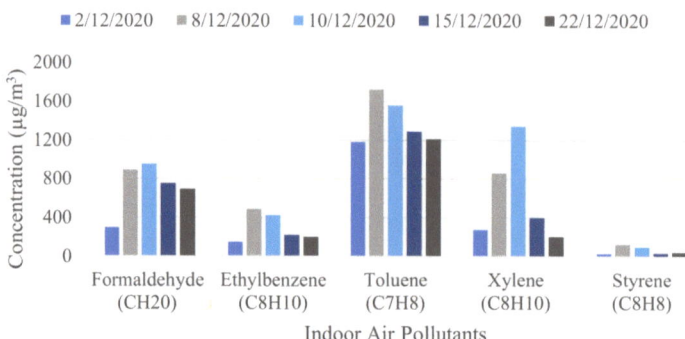

Figure 9. Changes in indoor air pollutants concentration over time (Tower A1102).

3.2.2. Tower (B1203) (Natural Ventilation: Exterior Doors/Windows Kept Open)

The indoor air pollutant concentrations in Tower B1203 were evaluated based on natural ventilation by opening the doors and windows facing the outside by 10 cm during the experimental period to assess the reduction or increase. Figure 10 displays the change in the concentration of each pollutant at the start of the test (2 December 2020) and when 20 days had elapsed (22 December 2020). The air was collected in the experimental unit while maintaining natural ventilation throughout the experiment. Figure 5 illustrates that the temperature distribution was generally lower than the outside temperature, enabling a certain amount of ventilation to ensure that the indoor concentration was much lower than that of the sealed Tower A1102. The measured values up to 22 December showed repeated increases and decreases for each substance. On 22 December, the indoor concentration of HCHO was 58.05 µg/m^3, and C$_7$H$_8$ was 298.55 µg/m^3. These values were relatively low compared to the sealed Tower A1102, where the indoor concentration of HCHO was 702.01 µg/m^3, and C$_7$H$_8$ was 1242.35 µg/m^3. The two units had different initial values for each substance.

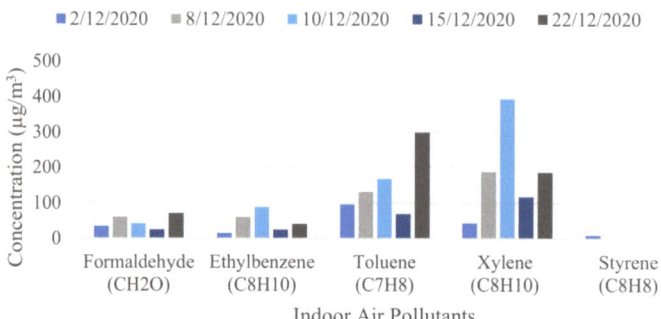

Figure 10. Changes in indoor air pollutants concentration over time (Tower B1203).

3.2.3. Tower C1501 (Airtight with Air Supply/Exhaust Ventilation)

Tower C1501 was evaluated by sealing all the doors and windows in contact with the outside, providing mechanical ventilation, and maintaining ventilation of 6.23 ACH. Figure 11 depicts the change in the concentration of each contaminant at the test start date (2 December 2020) and when 20 days had elapsed (22 December 2020). The measured values up to 22 December, 20 days after the start of the experiment, showed repeated increases and decreases for each substance but revealed a decreasing trend over time. For instance, after 16 days, HCHO slightly increased from the initial value of 130.2 to 141.2 µg/m^3, whereas C$_7$H$_8$ decreased from the initial value of 839.6 to 243.4 µg/m^3.

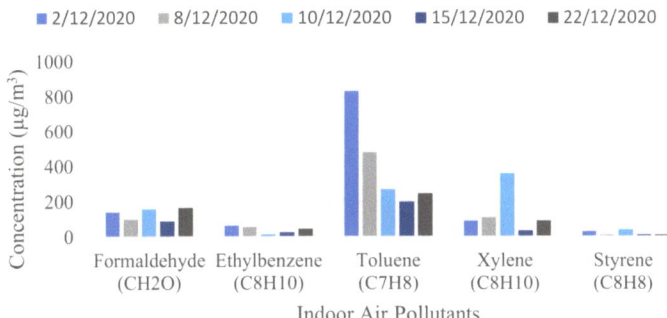

Figure 11. Changes in indoor pollutant concentration over time (Tower C1501).

3.2.4. Tower D1302 (Airtight with Kitchen Hood Exhaust Fan)

The study conducted in Tower D1302 measured the change in hazardous chemicals while maintaining ventilation of approximately 2.28 ACH by operating only the range hood exhaust fan in the kitchen, based on sealing the doors and windows facing the outside. Figure 12 illustrates the change in each contaminant's concentration from the experiment's start date (2 December 2020) to when 20 days had elapsed (22 December 2020). The initial concentration of 1346.29 µg/m^3 was reduced to 175.14 µg/m^3 after 20 days, showing a reduction of about 86.9%. The reduction in pollutants due to bake-out was also confirmed, with ethylbenzene (C_8H_{10}), xylene (C_8H_{10}), and C_8H_8 showing a 29.7% reduction of 55.0%, and 19.9%, respectively.

The indoor concentration of pollutants tended to increase slightly when airtight conditions were maintained for 7 days after 7 days of ventilation (22 December 2020). However, after 3 days of heating, 7 days of ventilation, and 7 days of airtight conditions, it showed a consistent re-increasing phenomenon with the indoor concentration prediction pattern of bake-out presented in Figure 4. This trend was also observed in the case of HCHO, with the release rate from the pollutant increasing during the heating period and the measured values of high contaminants repeatedly increasing and decreasing for up to five days. Overall, there was a decreasing trend over time, with HCHO showing 157.6 from an initial value of 177.9 µg/m^3 and C_7H_8 decreasing from an initial value of 1228.9 µg/m^3 to 429.1 µg/m^3 after 16 days.

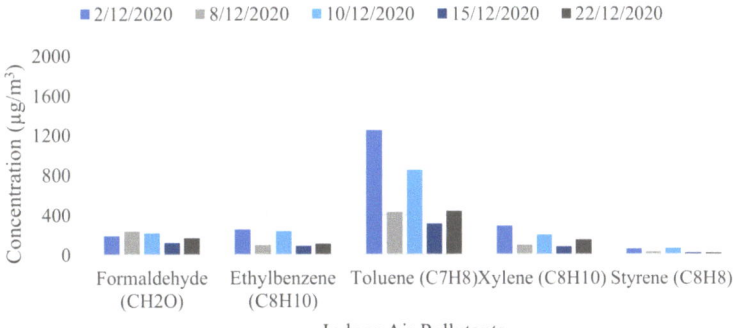

Figure 12. Changes in indoor pollutant concentration over time (Tower D1302).

3.2.5. Tower A1004 (Airtight after Bake-Out)

Tower A1004 serves as the control unit for the bake-out experiment, where the room is sealed for 7 days after the bake-out to measure the indoor pollutants' changes. The primary method of bake-out involves heating the room to a target temperature of 32–34 °C for 3 days, followed by natural ventilation for 7 days. Figure 13 depicts each substance's concentration change from the experiment's start date (2 December 2020) to when 20 days had elapsed (22 December 2020). All substances showed an increasing trend until the 20th day when the temperature was raised, indicating active emission of pollutants from building materials due to the higher target temperature, as shown in Figure 4. The concentration of C_7H_8 decreased after 2 days of ventilation following the bake-out, which was higher than the initial concentration before the bake-out, implying that 2 to 3 days of winter ventilation is insufficient in discharging emitted pollutants. However, the concentration was still high after seven days of ventilation. Additionally, a stable concentration reduction of about 75.6% from the initial concentration of 228.09 to 29.06 µg/m³ was observed after 7 days. Subsequently, a slight increase was observed after maintaining the seal for seven days.

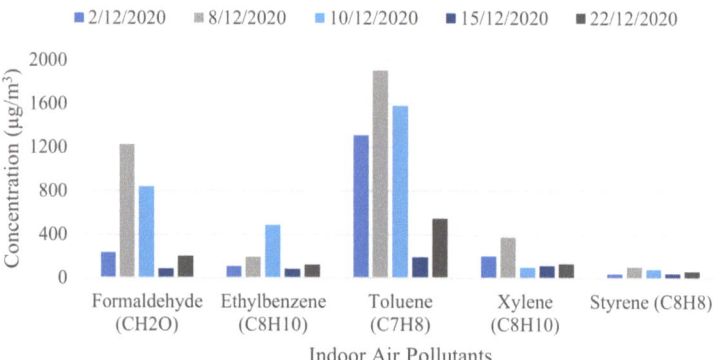

Figure 13. Changes in indoor pollutant concentration over time (Tower A1004).

3.2.6. Tower B1103 (Natural Ventilation after Bake-Out)

Tower B1103 measures the level of pollutant reduction by maintaining natural ventilation continuously after the bake-out. Figure 14 illustrates the changes in the concentration of pollutants at the experiment's start date (2 December 2020) and when 20 days had elapsed (22 December 2020). It was observed that the temperature rises during the bake-out resulted in an increase in the pollutant emission rate.

The concentration of pollutants reduced significantly during natural ventilation for seven days after the bake-out. HCHO was decreased by 92.8%, ethylbenzene (C_8H_{10}) by 70.0%, C_7H_8 by 94.1%, and xylene (C_8H_{10}) by 26.6%. Additionally, C_8H_8 was not detected on the 29th day, confirming that bake-out effectively reduces pollutant concentrations. Moreover, continuous ventilation was carried out throughout the experiment, and no significant concentration increase was observed during this period. Therefore, ongoing ventilation management was found to be necessary even after completing the bake-out process.

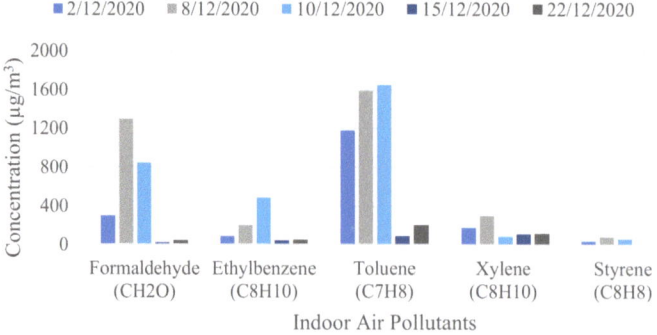

Figure 14. Changes in indoor pollutant concentration over time (Tower B1103).

3.2.7. Tower D1404 (Bake-Out and Air Supply/Exhaust Ventilation Simultaneously)

Tower D1404 performed the bake-out and utilized supply/exhaust ventilation during the experimental period, allowing for the identification of the effects of both bake-out and mechanical ventilation. Unlike other units, this housing unit did not use natural ventilation by opening windows facing the outside world after the bake-out, relying solely on mechanical ventilation. Figure 15 illustrates the change in the concentration of each substance at the beginning of the test and after 20 days.

Low concentrations of pollutants were observed even on the 23rd day of measurement, indicating that the number of pollutants emitted during the bake-out or heating period was minimal in this unit. However, it was confirmed in Figure 4 that the target temperature during bake-out was reached while maintaining mechanical ventilation of 6.23 ACH, which suggests that emissions from finishing materials may have increased. The indoor pollutant concentration remained low due to the continuous discharge of pollutants through the ventilation system. Compared to Tower A1004 and Tower B1103, which were baked out without ventilation, the indoor concentration in Tower D1404 remained relatively low during the experiment, indicating a lower possibility of re-adsorption.

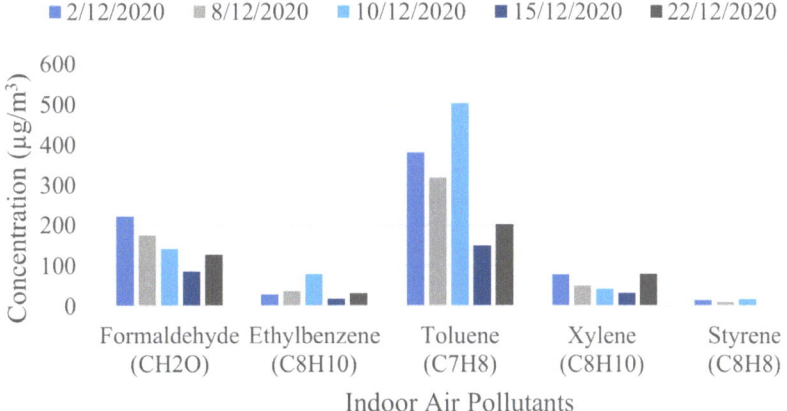

Figure 15. Changes in indoor pollutant concentration over time (Tower D1404).

3.2.8. Tower E1504 (Bake-Out and Kitchen Hood Exhaust Fan Simultaneously)

Tower E1504 conducted a bake-out while operating an indoor hood exhaust fan in the kitchen range. This unit aimed to identify the combined effect of the bake-out and the indoor exhaust fan on indoor pollutants. Instead of opening the outside window, this unit used the indoor exhaust fan for ventilation after the bake-out. Figure 16 illustrates the concentration changes of each contaminant at the beginning of the experiment and after 20 days, and all pollutants demonstrated a stable decrease.

At the start of the experiment, the concentration of C_7H_8 was very high but decreased after the exhaust fan was operated during the heating period of the bake-out. As shown in Figure 4, the experimental unit reached the target temperature due to the characteristics of the UAE winter while maintaining a ventilation rate of 2.28 ACH through the operation of the range hood exhaust fan. It was observed that the pollutants released during the heating period were effectively discharged to the outside through ventilation, similar to the results of Tower D1404.

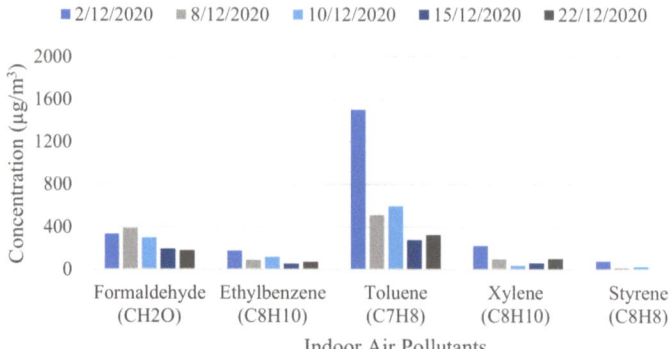

Figure 16. Changes in indoor pollutant concentration over time (Tower E1504).

4. Discussion

The experiments' results show significant changes in indoor air quality in all the units which applied or did not use the bake-out procedure in the new apartment in the hot desert climate in the UAE.

4.1. Units without Bake-Out

Figures 17 and 18 display the results of an experimental study conducted in Tower A1102, where the building was sealed. The levels of HCHO and C_7H_8 were measured and compared to those of Tower B1203, which had natural ventilation. The study observed that in Tower A1102, HCHO increased from 311 µg/m^3 to 670 µg/m^3, and C_7H_8 increased from 1189 µg/m^3 to 1224 µg/m^3. Conversely, Tower B1203 with natural ventilation had lower concentrations, with HCHO increasing from 35 µg/m^3 to 71 µg/m^3 and C_7H_8 increasing from 96 µg/m^3 to 299 µg/m^3 over time.

The study also investigated the effects of different ventilation conditions on the concentrations of HCHO and C_7H_8. When Tower C1501 had a sealed ventilation unit, HCHO slightly increased from 140 µg/m^3 to 165 µg/m^3, while C_7H_8 decreased from 830 µg/m^3 to 245 µg/m^3. Similarly, when Tower D1302 had a closed hood exhaust fan, HCHO slightly increased from 140 µg/m^3 to 165 µg/m^3, while C_7H_8 slightly decreased from 190 µg/m^3 to 170 µg/m^3.

These results support previous research indicating that airtight conditions increase HCHO concentrations. However, the study also found that exhaust ventilation positively impacts the concentration of C_7H_8. The findings suggest that using exhaust ventilation systems can effectively reduce C_7H_8 concentrations but may not have the same effect on HCHO concentrations.

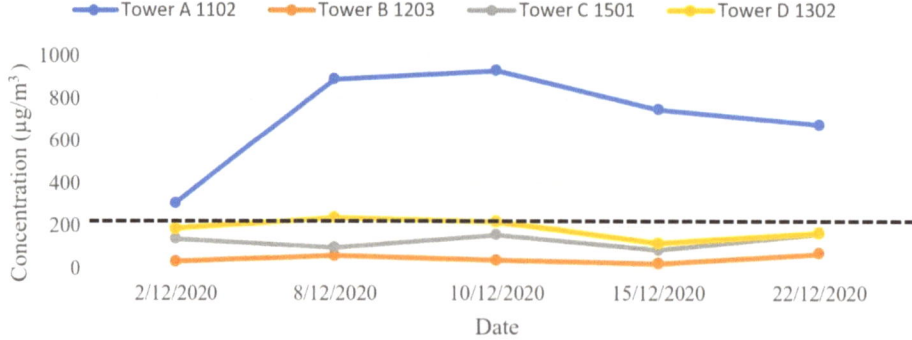

Figure 17. Comparison of HCHO concentration by units without bake-out (the dashed line shows the international standard level).

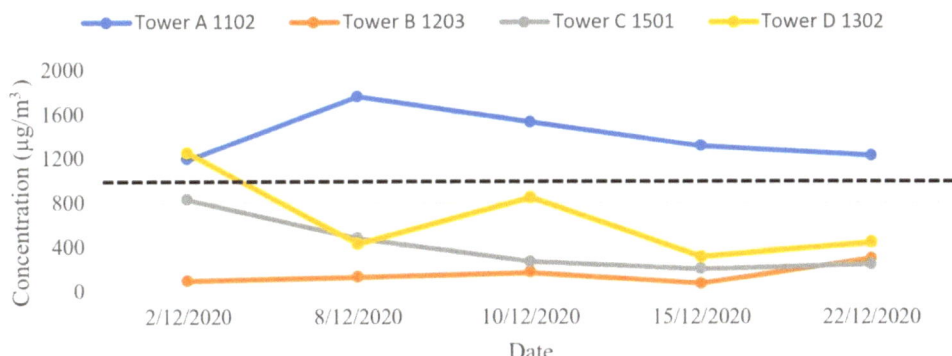

Figure 18. Comparison of C_7H_8 concentration by units without bake-out (the dashed line shows the international standard level).

4.2. Units with Bake-Out

Figures 19 and 20 present the results of bake-out treatments conducted in different units. In Tower A1004, which was sealed after the bake-out, HCHO decreased from 236 μg/m³ to 198 μg/m³, with an interstitial measurement of 1224 μg/m³, and C_7H_8 decreased from 1309 μg/m³ to 546 μg/m³, with an interstitial height of 1905 μg/m³. Meanwhile, in Tower B1103, which had natural ventilation continuing after the bake-out, HCHO decreased from 301 μg/m³ to 42 μg/m³, with an interstitial measurement of 1298 μg/m³, and C_7H_8 decreased from 1178 μg/m³ to 201 μg/m³, with a measurement of 1650 μg/m³. A comparison of pollutant concentration changes is shown for Tower D1404, which used bake-out and exhaust ventilation, and Tower E1504, which used both exhaust ventilation and the kitchen range hood's exhaust fan. In Tower D1404, HCHO decreased from 222 μg/m³ and 334 μg/m³ to 127 μg/m³ and 180 μg/m³, and C_7H_8 decreased from 381 μg/m³ and 1502 μg/m³ to 201 μg/m³ and 325 μg/m³.

The effectiveness of bake-out treatments depends on various factors, such as the temperature dependence of pollutant emission rates from building materials, adsorption and desorption phenomena, and ventilation rates. The emission rate of pollutants from building materials increases with temperature, and ventilation rates play a critical role in reducing pollutant concentrations. Additionally, the initial indoor pollutant concentrations varied across units, with Tower B1203 relying on natural ventilation, while Tower C1501 and Tower D1302 had mechanical ventilation of 2.28 ACH from a hood fan and 6.23 ACH from exhaust ventilation, respectively.

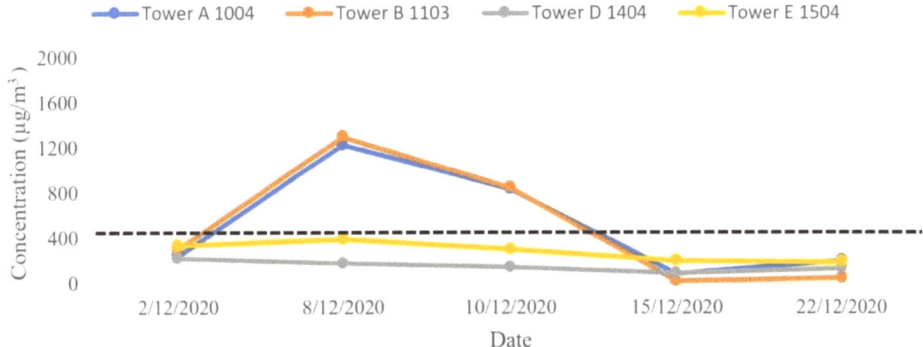

Figure 19. Comparison of HCHO concentration by units with bake-out (the dashed line shows the international standard level).

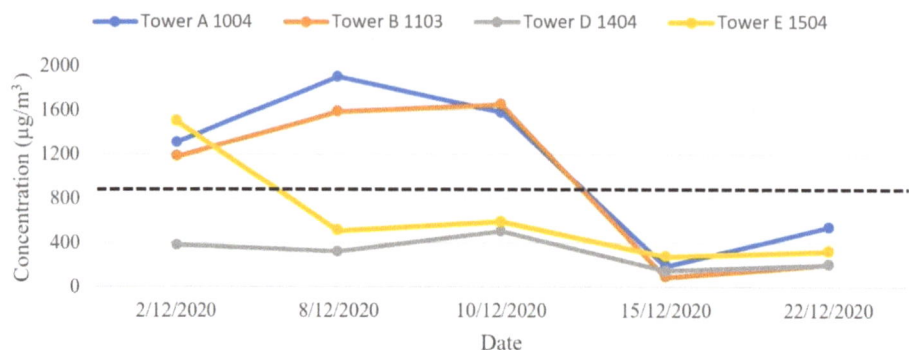

Figure 20. Comparison of C_7H_8 concentration by units with bake-out (the dashed line shows the international standard level).

During the experimental period, Tower A1102, which is an airtight unit, had indoor pollutant concentrations that exceeded recommended standards [51,52]. This highlights the need for specific management strategies to maintain indoor air quality during winter in the UAE, as the small temperature difference between indoor and outdoor environments makes ventilation difficult and causes pollutants to dissipate slowly [2,19].

In contrast, Tower D1404 and Tower E1504, which utilized mechanical ventilation systems (including ventilation units and range hood exhaust fans) and bake-out simultaneously, exhibited a continuous decrease in indoor pollutant concentrations, resulting in relatively low levels. On the other hand, Tower A1004 and Tower B1103, which lacked ventilation during the bake-out, exhibited extremely high concentrations during the heating period.

Furthermore, Tower D1404 and Tower E1504 reached the target temperature in the UAE winter, even with a certain level of ventilation during the bake-out heating period. Additionally, these towers had significant pollutant emissions from finishing materials. However, using ventilation systems ensured that pollutant concentrations remained low, as indicated by the analysis. Therefore, paying close attention during the bake-out process makes it possible to prevent the re-emission of contaminants effectively.

Figures 19 and 20 provide evidence that Tower A1004 and Tower B1103, which underwent bake-out, increased indoor air temperature and subsequent emissions from building materials that demonstrated temperature dependence. In Tower A1004, indoor pollutant concentrations slightly increased after 7 days without bake-out and sealing, indicating that reduced ventilation volume resulting from sealing led to an increase in the release rate of pollutants from the source. Furthermore, the indoor pollutant concentration pattern caused by the bake-out showed that the release rate of indoor contaminants decreased after the bake-out, as shown in Figure 4.

During the bake-out process, no indoor hazardous chemicals were discharged on 8 December 2020, when the air conditioner was turned off, or on 10 December 2020, when ventilation was performed for 2 days in Tower A1004 and Tower B1103. It was only on 15 December 2020, after 7 days of ventilation, that pollutant concentrations were sufficiently reduced.

This study provides important insights into the effectiveness of various methods for reducing the concentration of hazardous chemicals in indoor air during the winter in the UAE. However, it is essential to acknowledge the study's limitations to ensure that its findings are interpreted within their appropriate context.

Firstly, the study only measured indoor pollutant concentrations during limited ventilation periods, with the most extended period being seven days. Further sub-division of the intervals is necessary to determine the appropriate ventilation period to reduce indoor pollutants. Therefore, while the study provides valuable insights into the effectiveness of bake-out and mechanical ventilation in reducing indoor pollutants' concentration, more extensive research is required to establish the optimal ventilation periods.

Secondly, the study's findings indicate that the reduction effect of hazardous chemicals through bake-out is limited under certain conditions, and pollutant concentrations may increase when ventilation is stopped. Although the study's results suggest that sufficient ventilation time after bake-out is critical to reducing hazardous chemicals in indoor air, the effects of re-diffusion after bake-out were not fully understood. Thus, further research is needed to better understand the re-diffusion phenomenon and determine its implications for indoor air quality.

Finally, the study also highlights the need for future research to elucidate the mechanisms affecting building pollutant concentration, such as the surface layer of building materials, temperature, and ventilation volume. Obtaining more reliable results from field experiments requires a better understanding of these factors and their effects on indoor air quality.

In conclusion, while this study's findings are valuable, they should be interpreted in light of its limitations. Addressing these limitations through further research can lead to more comprehensive and reliable findings, informing future efforts to improve indoor air quality.

5. Conclusions

The study offers valuable insights into the effectiveness of different methods for reducing the concentration of hazardous chemicals in indoor air during winter in the UAE. The results demonstrate that implementing bake-out can effectively lower the concentration of harmful substances in units, but the effectiveness is limited when maintaining a temperature of 32–34 °C for 72 h and providing ventilation for only 7 days by opening windows and doors. The study emphasizes the need for sufficient ventilation time after bake-out completion to reduce hazardous chemicals in indoor air effectively. However, the study only measured indoor concentration during ventilation periods of two and seven days, and further subdivision of the intervals is necessary to determine the appropriate ventilation period.

Furthermore, the study found that the concentration of hazardous chemical substances increased when the seal was maintained for seven days compared to the case where ventilation was continued after bake-out completion. Further research is needed to understand the re-diffusion phenomenon after bake-out by observing the changing trend over seven days or more. The study also suggests that future research should investigate the mechanisms affecting the surface layer of building materials, temperature, and ventilation volume to obtain more reliable results based on the findings of these field experiments.

- The study documents a significant reduction of more than 70% for HCHO, highlighting the need for intensive management of emitted pollutants under high indoor temperatures to reduce harmful chemicals in indoor air in the UAE during winter.
- In Tower A1102, without bake-out during winter, the indoor concentrations of hazardous chemicals were HCHO = 931 µg/m^3 and C_7H_8 = 1761 µg/m^3.
- Tower B1203, Tower C1501, and Tower D1302 with ventilation had significantly lower concentrations of HCHO and C_7H_8 than Tower A1102.
- Performing bake-outs in Tower A1004 and Tower B1103 during winter in the UAE resulted in reduced hazardous chemicals, but the reduction effect is limited.
- To effectively reduce harmful chemical substances during winter in the UAE, sufficient ventilation time of at least seven days after bake-out is necessary.
- Concentrations of hazardous chemical substances increase when ventilation is not continued after bake-out, and future research is necessary to understand the re-diffusion phenomenon.
- During winter in the UAE, indoor pollutant concentration can be maintained low through ventilation, even during heating.

- Units with natural or mechanical ventilation maintained low indoor pollutant concentrations regardless of whether they performed bake-out.

After conducting the study, it can be concluded that managing the pollutants emitted in indoor air during high indoor temperatures is crucial to reduce the concentration of harmful chemicals in indoor air during winter in the UAE. In addition, the study indicates that a combination of bake-out and mechanical ventilation can effectively reduce indoor pollutant concentration during winter in the UAE. Moreover, it was observed that units with specific ventilation types, including natural or mechanical ventilation, maintained low indoor pollutant concentration levels, regardless of whether they underwent bake-out.

Furthermore, the study highlights the importance of securing sufficient ventilation time after bake-out. This is because hazardous chemical substances can only be effectively reduced when sufficient ventilation time of at least seven days is provided after bake-out during winter in the UAE. It was also observed that the concentration of hazardous chemical substances increases when ventilation is not continued after the bake-out. Therefore, future research is necessary to understand the re-diffusion phenomenon.

To obtain more reliable results based on the field experiments, the study suggests conducting future research to elucidate the mechanisms affecting the surface layer of building materials, temperature, and ventilation volume. This would help to develop more effective methods for reducing indoor pollutant concentration in the UAE during winter.

Author Contributions: C.J. and N.S.A.M. identified and secured the example buildings used in the study. The data acquisition system was designed, and C.J. and N.S.A.M. installed the sensors. N.S.A.M. was responsible for data collection. C.J. and N.S.A.M. performed data analysis. The manuscript was compiled by C.J. and reviewed by N.S.A.M. All authors have read and agreed to the published version of the manuscript.

Funding: This research received no external funding.

Data Availability Statement: New data were created and analyzed in this study, which can be shared upon request at the authors' consideration.

Acknowledgments: The authors would like to express their gratitude to Ajman University for APC support and to the Healthy and Sustainable Buildings Research Center at Ajman University for providing an excellent research environment.

Conflicts of Interest: The authors declare no conflict of interest.

References

1. Ababutain, I.M. Aeromycoflora of some eastern provinces of Saudi Arabia. *Indoor Built Environ.* **2013**, *22*, 388–394. [CrossRef]
2. Jung, C.; Alqassimi, N.; El Samanoudy, G. The comparative analysis of the indoor air pollutants in occupied apartments at residential area and industrial area in Dubai, United Arab Emirates. *Front. Built Environ.* **2022**, *8*, 998858. [CrossRef]
3. Farrag, N.; Abou El-Ela, M.A.; Ezzeldin, S. Sick building syndrome and office space design in Cairo, Egypt. *Indoor Built Environ.* **2022**, *31*, 568–577. [CrossRef]
4. Jung, C.; Awad, J. Improving the IAQ for learning efficiency with indoor plants in university classrooms in Ajman, United Arab Emirates. *Buildings* **2021**, *11*, 289. [CrossRef]
5. Mannan, M.; Al-Ghamdi, S.G. Indoor Air Quality in Buildings: A Comprehensive Review on the Factors Influencing Air Pollution in Residential and Commercial Structure. *Int. J. Environ. Res. Public Health* **2021**, *18*, 3276. [CrossRef]
6. Amoatey, P.; Omidvarborna, H.; Baawain, M.S.; Al-Mamun, A. Indoor air pollution and exposure assessment of the gulf cooperation council countries: A critical review. *Environ. Int.* **2018**, *121*, 491–506. [CrossRef]
7. Amoatey, P.; Omidvarborna, H.; Baawain, M.S.; Al-Mamun, A.; Bari, A.; Kindzierski, W.B. Association between human health and indoor air pollution in the Gulf Cooperation Council (GCC) countries: A review. *Rev. Environ. Health* **2020**, *35*, 157–171. [CrossRef]
8. Hosseini, M.R.; Fouladi-Fard, R.; Aali, R. COVID-19 pandemic and sick building syndrome. *Indoor Built Environ.* **2020**, *29*, 1181–1183. [CrossRef]
9. Jung, C.; Awad, J. The improvement of indoor air quality in residential buildings in Dubai, UAE. *Buildings* **2021**, *11*, 250. [CrossRef]
10. Nakaoka, H.; Todaka, E.; Seto, H.; Saito, I.; Hanazato, M.; Watanabe, M.; Mori, C. Correlating the symptoms of sick-building syndrome to indoor VOCs concentration levels and odour. *Indoor Built Environ.* **2014**, *23*, 804–813. [CrossRef]
11. Awad, J.; Jung, C. Evaluating the indoor air quality after renovation at the greens in Dubai, United Arab Emirates. *Buildings* **2021**, *11*, 353. [CrossRef]

12. Jung, C.; Awad, J.; Mahmoud, N.S.A.; Salameh, M. An analysis of indoor environment evaluation for the Springs development in Dubai, UAE. *Open House Int.* **2021**, *46*, 651–667. [CrossRef]
13. Gallego, E.; Roca, F.J.; Perales, J.F.; Guardino, X. Experimental evaluation of VOC removal efficiency of a coconut shell activated carbon filter for indoor air quality enhancement. *Build. Environ.* **2013**, *67*, 14–25. [CrossRef]
14. Lu, Y.; Liu, J.; Yoshino, H.; Lu, B.; Jiang, A.; Li, F. Use of biotechnology coupled with bake-out exhaust to remove indoor VOCs. *Indoor Built Environ.* **2012**, *21*, 741–748. [CrossRef]
15. Kim, J.T.; Yu, C.W. Hazardous materials in buildings. *Indoor Built Environ.* **2014**, *23*, 44–61. [CrossRef]
16. Lee, J.H.; Kim, J.; Kim, S.; Kim, J.T. Thermal extractor analysis of VOCs emitted from building materials and evaluation of the reduction performance of exfoliated graphite nanoplatelets. *Indoor Built Environ.* **2013**, *22*, 68–76. [CrossRef]
17. Yu, C.; Crump, D. Indoor Environmental Quality—Standards for Protection of Occupants' Safety, Health and Environment. *Indoor Built Environ.* **2010**, *19*, 499–502. [CrossRef]
18. Xu, B.; Chen, X.; Xiong, J. Air quality inside motor vehicles' cabins: A review. *Indoor Built Environ.* **2018**, *27*, 452–465. [CrossRef]
19. Bani Mfarrej, M.F.; Qafisheh, N.A.; Bahloul, M.M. Investigation of indoor air quality inside houses from UAE. *Air Soil Water Res.* **2020**, *13*, 1178622120928912. [CrossRef]
20. Wei, W.; Ramalho, O.; Mandin, C. Indoor air quality requirements in green building certifications. *Build. Environ.* **2015**, *92*, 10–19. [CrossRef]
21. Boldi, R.A. A comparison of the indoor and outdoor concentrations of fine particulate matter in various locations within Dubai, UAE. Smart, Sustainable and Healthy Cities. In *Proceeding of the 1st International Conference of the CIB Middle East & North Africa Research Network (CIB-MENA 2014), Abu Dhabi, United Arab Emirates, 14–16 December 2014*; Abu Dhabi University: Abu Dhabi, United Arab Emirates, 2014; p. 511.
22. Lee, J.H.; Jeong, S.G.; Kim, S. Performance evaluation of infrared bake-out for reducing VOCs and formaldehyde emission in MDF panels. *BioResources* **2016**, *11*, 1214–1223. [CrossRef]
23. Lee, Y.K.; Kim, H.J. Effect of temperature and bake-out on formaldehyde emission from UF bonded wood composites. *J. Korean Wood Sci. Technol.* **2012**, *40*, 91–100. [CrossRef]
24. Schütze, A.; Baur, T.; Leidinger, M.; Reimringer, W.; Jung, R.; Conrad, T.; Sauerwald, T. Highly sensitive and selective VOC sensor systems based on semiconductor gas sensors: How to? *Environments* **2017**, *4*, 20. [CrossRef]
25. Kolarik, B.; Andersen, H.V.; Frederiksen, M.; Gunnarsen, L. Laboratory investigation of PCB bake-out from tertiary contaminated concrete for remediation of buildings. *Chemosphere* **2017**, *179*, 101–111. [CrossRef]
26. Rickards, W.B.; Young, P.A.; Keniry, J.T.; Shaw, P. Thermal bake-out of reduction cell cathodes—Advantages and problem areas. In *Essential Readings in Light Metals*; Springer: Cham, Switzerland, 2016; pp. 694–698.
27. Shen, X.; Chen, Z. Numerical study of the effect of bake-out on the formaldehyde migration in a floor heating system. In *Proceedings of the 8th International Symposium on Heating, Ventilation and Air Conditioning*; Lecture Notes in Electrical Engineering; Li, A., Zhu, Y., Li, Y., Eds.; Springer: Berlin/Heidelberg, Germany, 2014; pp. 411–420.
28. Kim, E.H.; Kim, S.; Lee, J.H.; Kim, J.; Han, Y.; Kim, Y.M.; Kim, G.B.; Jung, K.; Cheong, H.K.; Ahn, K. Indoor air pollution aggravates symptoms of atopic dermatitis in children. *PLoS ONE* **2015**, *10*, e0119501. [CrossRef] [PubMed]
29. Padmavathi, P.; Sireesha, A. Indoor air quality in schools-an architectural perspective. *Int. J. Eng. Bus. Manag.* **2015**, *5*, 31–36.
30. Kang, D.H.; Choi, D.H.; Lee, S.M.; Yeo, M.S.; Kim, K.W. Effect of bake-out on reducing VOC emissions and concentrations in a residential housing unit with a radiant floor heating system. *Build. Environ.* **2010**, *45*, 1816–1825. [CrossRef]
31. Lu, Y.; Liu, J.; Lu, B.; Jiang, A.; Wan, C. Study on the removal of indoor VOCs using biotechnology. *J. Hazard. Mater.* **2010**, *182*, 204–209. [CrossRef]
32. Jung, C.; Mahmoud, N.S.A.; Alqassimi, N. Identifying the relationship between VOCs emission and temperature/humidity changes in new apartments in the hot desert climate. *Front. Built Environ.* **2022**, *8*, 1018395. [CrossRef]
33. Girman, J.R. Volatile organic compounds and building bake-out. *Occup. Med.* **1989**, *4*, 695–712.
34. Thevenet, F.; Debono, O.; Rizk, M.; Caron, F.; Verriele, M.; Locoge, N. VOC uptakes on gypsum boards: Sorption performances and impact on indoor air quality. *Build. Environ.* **2018**, *137*, 138–146. [CrossRef]
35. Yu, C.W.F.; Kim, J.T. Building environmental assessment schemes for rating of IAQ in sustainable buildings. *Indoor Built Environ.* **2011**, *20*, 5–15. [CrossRef]
36. Shin, H.; Park, W.; Kim, B.; Ji, K.; Kim, K.T. Indoor air quality and human health risk assessment for un-regulated small-sized sensitive population facilities. *J. Environ. Health Sci. Eng.* **2018**, *44*, 397–407.
37. Torpy, F.R.; Irga, P.J.; Burchett, M.D. Reducing indoor air pollutants through biotechnology. In *Biotechnologies and Biomimetics for Civil Engineering*; Springer: Cham, Switzerland, 2015; pp. 181–210.
38. Yun, J.S.; Lee, M.H.; Eom, S.W.; Kim, M.Y.; Kim, J.H.; Kim, S.D. Emission characteristics of volatile organic compounds from building flooring materials. *J. Korean Soc. Environ. Eng.* **2010**, *32*, 973–978.
39. Park, S.; Seo, J. Bake-out strategy considering energy consumption for improvement of indoor air quality in floor heating environments. *Int. J. Environ. Res. Public Health* **2018**, *15*, 2720. [CrossRef]
40. Lv, Y.; Liu, J.; Wei, S.; Wang, H. Experimental and simulation study on bake-out with dilution ventilation technology for building materials. *J. Air Waste Manag. Assoc.* **2016**, *66*, 1098–1108. [CrossRef]
41. Seo, J.H.; Park, S.H.; Lee, S.W. Application of sorptive building materials reducing indoor air pollution for improving indoor air quality. In *Applied Mechanics and Materials*; Trans Tech Publications Ltd.: Bach, Switzerland, 2015; Volume 749, pp. 358–361.

42. Park, S.; Seo, J. Optimum installation of sorptive building materials using contribution ratio of pollution source for improvement of indoor air quality. *Int. J. Environ. Res. Public Health* **2016**, *13*, 396. [CrossRef]
43. Shrubsole, C.; Dimitroulopoulou, S.; Foxall, K.; Gadeberg, B.; Doutsi, A. IAQ guidelines for selected volatile organic compounds (VOCs) in the UK. *Build. Environ.* **2019**, *165*, 106382. [CrossRef]
44. Arar, M.; Jung, C. Improving the Indoor Air Quality in Nursery Buildings in United Arab Emirates. *Int. J. Environ. Res. Public Health* **2021**, *18*, 12091. [CrossRef]
45. Jeon, J.; Park, J.H.; Wi, S.; Yun, B.Y.; Kim, T.; Kim, S. Field study on the improvement of indoor air quality with toluene adsorption finishing materials in an urban residential apartment. *Environ. Pollut.* **2020**, *261*, 114137. [CrossRef]
46. Park, S.I.; Kim, J.H.; Park, J.S. Effects of flush-out in the reduction of formaldehyde in newly built residential buildings. *Korean J. Air Cond. Refrig. Eng.* **2018**, *30*, 116–122. [CrossRef]
47. Zuo, Z.; Wang, J.; Lin, C.H.; Pui, D.Y.H. VOC outgassing from baked and unbaked ventilation filters. *Aerosol Air Qual. Res.* **2010**, *10*, 265–271. [CrossRef]
48. Al Horr, Y.; Arif, M.; Katafygiotou, M.; Mazroei, A.; Kaushik, A.; Elsarrag, E. Impact of indoor environmental quality on occupant well-being and comfort: A review of the literature. *Int. J. Sustain. Built Environ.* **2016**, *5*, 1–11. [CrossRef]
49. Kamal, M.S.; Razzak, S.A.; Hossain, M.M. Catalytic oxidation of volatile organic compounds (VOCs)—A review. *Atmos. Environ.* **2016**, *140*, 117–134. [CrossRef]
50. Xiong, J.; Zhang, P.; Huang, S.; Zhang, Y. Comprehensive influence of environmental factors on the emission rate of formaldehyde and VOCs in building materials: Correlation development and exposure assessment. *Environ. Res.* **2016**, *151*, 734–741. [CrossRef]
51. Lu, X.; Yang, T.; O'Neill, Z.; Zhou, X.; Pang, Z. Energy and ventilation performance analysis for CO_2-based demand-controlled ventilation in multiple-zone VAV systems with fan-powered terminal units (ashrae RP-1819). *Sci. Technol. Built Environ.* **2021**, *27*, 139–157. [CrossRef]
52. McNulty, M.; Moua-Vargas, P.; Abramson, B. Further simplifying ASHRAE standard 62.1 for Application to Existing Buildings: Comparing Informative appendix D and section 6.2. 5.2 with Real-World Data. *ASHRAE Trans.* **2019**, *125*, 579–587.
53. Taheri, S.; Razban, A. Learning-based CO_2 concentration prediction: Application to indoor air quality control using demand-controlled ventilation. *Build. Environ.* **2021**, *205*, 108164. [CrossRef]
54. Badji, C.; Beigbeder, J.; Garay, H.; Bergeret, A.; Bénézet, J.C.; Desauziers, V. Under glass weathering of hemp fibers reinforced polypropylene biocomposites: Impact of volatile organic compounds emissions on indoor air quality. *Polym. Degrad. Stab.* **2018**, *149*, 85–95. [CrossRef]
55. Chen, Y.; Yang, J.; Yang, R.; Xiao, X.; Xia, J.C. Contribution of urban functional zones to the spatial distribution of urban thermal environment. *Build. Environ.* **2022**, *216*, 109000. [CrossRef]
56. Markowicz, P.; Larsson, L. Influence of relative humidity on VOC concentrations in indoor air. *Environ. Sci. Pollut. Res.* **2015**, *22*, 5772–5779. [CrossRef] [PubMed]
57. Kain, G.; Stratev, D.; Tudor, E.; Lienbacher, B.; Weigl, M.; Barbu, M.C.; Petutschnigg, A. Qualitative investigation on VOC-emissions from spruce (*Picea abies*) and larch (*Larix decidua*) loose bark and bark panels. *Eur. J. Wood Wood Prod.* **2020**, *78*, 403–412. [CrossRef]
58. Torpy, F.; Clements, N.; Pollinger, M.; Dengel, A.; Mulvihill, I.; He, C.; Irga, P. Testing the single-pass VOC removal efficiency of an active green wall using methyl ethyl ketone (MEK). *Air Qual. Atmos. Health* **2018**, *11*, 163–170. [CrossRef]
59. AM. His Highness Sheikh Rashid bin Humaid Al Nuaimi Launches the AED750 Million Corniche Residences Project. Available online: https://www.am.gov.ae/media-center/press-news/his-highness-sheikh-rashid-bin-humaid-al-nuaimi-launches-the-aed-750-million-corniche-residences-project (accessed on 28 July 2021).
60. Aqaar. Ajman Corniche Residences. Available online: https://www.aqaar.com/acr (accessed on 20 August 2021).
61. DXBoffplan. Ajman Corniche Residents at Ajman Corniche. Available online: https://dxboffplan.com/properties/ajman-corniche-residences/ (accessed on 1 September 2021).
62. Shah, K.W.; Li, W. A review on catalytic nanomaterials for volatile organic compounds VOC removal and their applications for healthy buildings. *Nanomaterials* **2019**, *9*, 910. [CrossRef]
63. Sun, Z.; Wang, J.; Chen, Y.; Lu, H. Influence factors on injury severity of traffic accidents and differences in urban functional zones: The empirical analysis of Beijing. *Int. J. Environ. Res. Public Health* **2018**, *15*, 2722. [CrossRef]
64. Jiang, C.; Li, D.; Zhang, P.; Li, J.; Wang, J.; Yu, J. Formaldehyde and volatile organic compound (VOC) emissions from particleboard: Identification of odorous compounds and effects of heat treatment. *Build. Environ.* **2017**, *117*, 118–126. [CrossRef]
65. Choi, Y.J.; Lee, J.K.; Cha, Y.R. Analysis on characteristics and related factors of indoor air quality in newly built Wooden houses. *J. Korean Hous. Assoc.* **2015**, *26*, 23–32. [CrossRef]
66. Park, S.; Seo, J.; Kim, J.T. A study on the application of sorptive building materials to reduce the concentration and volume of contaminants inhaled by occupants in office areas. *Energ. Build.* **2015**, *98*, 10–18. [CrossRef]
67. Chen, Z.; Shi, J.; Shen, X.; Ma, Q.; Xu, B. Study on formaldehyde emissions from porous building material under non-isothermal conditions. *Appl. Therm. Eng.* **2016**, *101*, 165–172. [CrossRef]
68. Kozicki, M.; Guzik, K. Comparison of VOC emissions produced by different types of adhesives based on test chambers. *Materials* **2021**, *14*, 1924. [CrossRef] [PubMed]
69. Shen, X.; Shi, J.; Chen, Z. Experimental study on the formaldehyde emission under non-isothermal conditions. *Procedia Eng.* **2015**, *121*, 590–595. [CrossRef]

70. Khoshnava, S.M.; Rostami, R.; Mohamad Zin, R.; Štreimikienė, D.; Mardani, A.; Ismail, M. The role of green building materials in reducing environmental and human health impacts. *Int. J. Environ. Res. Public Health* **2020**, *17*, 2589. [CrossRef]
71. Tong, Z.; Liu, H. Modeling in-vehicle VOCs distribution from cabin interior surfaces under solar radiation. *Sustainability* **2020**, *12*, 5526. [CrossRef]
72. Zhang, C.; Pomianowski, M.; Heiselberg, P.K.; Yu, T. A review of integrated radiant heating/cooling with ventilation systems-Thermal comfort and indoor air quality. *Energy Build.* **2020**, *223*, 110094. [CrossRef]

Disclaimer/Publisher's Note: The statements, opinions and data contained in all publications are solely those of the individual author(s) and contributor(s) and not of MDPI and/or the editor(s). MDPI and/or the editor(s) disclaim responsibility for any injury to people or property resulting from any ideas, methods, instructions or products referred to in the content.

Article

Evaluation of the Smart Readiness Indicator for Educational Buildings

Gvidas Plienaitis [1], Mindaugas Daukšys [1], Evi Demetriou [2], Byron Ioannou [2], Paris A. Fokaides [1,2] and Lina Seduikyte [1,*]

[1] Faculty of Civil Engineering and Architecture, Kaunas University of Technology, Studentu Str. 48, LT-51367 Kaunas, Lithuania; gvidas.plienaitis73@gmail.com (G.P.); mindaugas.dauksys@ktu.lt (M.D.); eng.fp@frederick.ac.cy (P.A.F.)

[2] School of Engineering, Frederick University, 7, Frederickou Str., Nicosia 1036, Cyprus; res.de@frederick.ac.cy (E.D.); b.ioannou@frederick.ac.cy (B.I.)

* Correspondence: lina.seduikyte@ktu.lt

Abstract: The Smart Readiness Indicator (SRI) is an assessment scheme for the intelligence of buildings, which was introduced by the European Commission in the directive for the Energy Performance of Buildings in 2018. Since its introduction, many activities related to the maturation and employment of the SRI have been initiated. One of the adaptation needs of the SRI, revealed through public consultation with relevant stakeholders, is the requirement for a tailored SRI for different types of buildings. The aim of this study is to analyze possible scenarios to optimize the smartness performance, as addressed by the SRI score, in educational buildings. The subject of this study concerned campus buildings of the Kaunas University of Technology, in Lithuania. For the definition of the SRI, the calculation sheet developed by the European Commission was used. The effect of the improvements in the smartness performance of buildings on their energy efficiency was examined with the use of a whole-building, BIM-based energy assessment tool (IDA-ICE). The findings of this study revealed that despite the improvement in the automation and control levels of the building heating system, the maximum SRI values achieved deviate significantly by a high-smartness level. This study revealed the importance of services at a city level towards achieving the optimal smartness levels at a building unit level. It also delivered useful findings related to the linkage between energy and smartness performance of a building. The policy implication of the study findings also covers topics relevant to utilities management at a district level, as well as on the need for tailored SRI services catalogs for different types of buildings.

Keywords: SRI; energy efficiency; whole-building energy analysis; energy performance of buildings; educational buildings

Citation: Plienaitis, G.; Daukšys, M.; Demetriou, E.; Ioannou, B.; Fokaides, P.A.; Seduikyte, L. Evaluation of the Smart Readiness Indicator for Educational Buildings. *Buildings* 2023, 13, 888. https://doi.org/10.3390/buildings13040888

Academic Editor: Osama Abudayyeh

Received: 8 March 2023
Revised: 20 March 2023
Accepted: 26 March 2023
Published: 28 March 2023

Copyright: © 2023 by the authors. Licensee MDPI, Basel, Switzerland. This article is an open access article distributed under the terms and conditions of the Creative Commons Attribution (CC BY) license (https://creativecommons.org/licenses/by/4.0/).

1. Introduction

As humanity is transitioning to the era of smart buildings and smart cities, the requirement for the objective definition of the intelligence of building units arises [1]. When referring to the smartness of a building unit, this relates to the ability of a building to document, understand and adapt the performance of the building to user needs. These operations are usually addressed through the performance of the building automation and control systems and are aligned to the building technical systems, rather than the building shell.

The Technical Committee 247 of the European Standardisation Organization (CEN) [2], realizing this need, proceeded to the development of a series of standards which relate to the definition of the smartness of buildings. The EN 15232 standard, initially published in 2008 [3], introduced a method for classifying the smartness of building automation and control systems. This method introduced services and domains, based on which the different building automation and control systems are classified, as well as functionality levels, which mainly classify the smartness of different building technical systems.

This development was particularly significant for society, the market and the building automation and control industry. It was also expected to affect policy decisions, something which occurred a few years later, and particularly in 2018, with the adaptation of the Smart Readiness Indicator (SRI) by the European Commission, a methodology which proposes the definition of smartness of buildings, based on the grounds and principles of the 15232 standards. Since 2007, the 15232 standard was revised once in 2017 and was finally replaced with an ISO EN standard, the EN ISO 52120 standard of 2022 [4], which dominates the techniques and methods used Europe-wide, for the definition of the smartness levels of building technical systems.

In the Energy Performance of Buildings Directive (EPBD) recast of 2018 [5], the European Commission has introduced digitalization as one of the main factors for improving the energy efficiency of European Union (EU) Member States (MS). The digitalization of energy systems is expected to allow the integration of renewable energy into smart grids and smart buildings. It is expected that the efficiency of buildings will increase when electricity systems with their central operators—smart grids—will be connected to the energy systems (cooling and heating systems, gas grids) [6]. As a promotion of energy efficiency through smart building technologies, the EPBD recast introduced the Smart Readiness Indicator (SRI) concept. The SRI aims to provide a harmonized rating scheme for assessing the smartness of buildings across the EU MS and inform the tenants and landlords about the capacities of building automation and control systems (BACS). The SRI addresses both existing and new buildings, and it is anticipated to drive developments in the following years in the field of building upgrades. The SRI is a user-friendly, easy-to-understand tool that can be used by building owners, operators, designers, and other stakeholders. The tool is designed to provide an overview of the energy efficiency potential of a building and its components. Following the work conducted by technical consultants, the legal acts of the SRI scheme were entered into force in October 2020 by the European Commission's (EC) Regulation 2020/2156 [7] by detailing the technical modalities for its effective implementation with provisions for a non-committal test phase by MS.

The SRI services are evaluated based on seven impact categories, which are aligned to their readiness to adapt in response to the needs of the occupant, to facilitate maintenance and efficient operation and to adapt in response to the situation of the energy grid.

There are three SRI assessment methods described in the report on technical support for the development of the SRI for buildings:

- Simplified method: based on a checklist approach with a limited, simplified service list. Online self-assessment by end-user (no certification) or on-site inspection by a third-party qualified expert (formal certification). The duration is up to one hour. Used for residential buildings and small non-residential buildings (net surface floor area <500 m^2);
- Expert SRI assessment: based on a checklist approach, covering the full catalog of smart services. Online self-assessment by a technical expert (no certification) or on-site inspection by a third-party qualified expert (formal certification). The duration is half or one day, depending on the complexity. Used for non-residential buildings (and residential buildings if desired);
- In-use smart building performance: based on measured/metered data (potentially restricted set of domains) of in-use buildings. TBS self-reporting their actual performance. Gathering data over a long period. Used for residential and non-residential buildings. Restricted to occupied buildings (not in design phase).

In the SRI service catalogs, services are structured within the following domains: heating, cooling, domestic hot water, controlled ventilation, lighting, dynamic building envelope, electricity, electric vehicle charging and monitoring and control. The smart service impact criteria are energy savings on-site, maintenance and fault prediction, comfort, convenience, health and wellbeing, information for occupants and grid flexibility and storage.

EU MS may decide to implement the SRI in their territory for all buildings or only for certain categories of buildings. MS interested in the SRI scheme can start by launching a non-committal test phase. Feedback from national test phases will allow adjusting the implementation modalities of the scheme. MS must inform the European Commission prior to implementing the SRI test phase in their territory. The scheme is already being tested in Austria, Croatia, Czech Republic, Denmark, Finland and France. There are no specific guidelines from the European Commission for the SRI implementation according to regulation 2156/2020 [7]. This enables the national governing bodies of each MS to have the freedom and the ability to modify the SRI tool for their own testing phase.

2. Theoretical Background, Literature Overview

The field of SRI methodology research is relatively new and has limited research results so far. Nevertheless, some scientific literature is available. In 2020, Al Dakheel et al. [8] investigated various definitions of smartness used in the literature to describe smart buildings. The authors identified the basic features and technologies of smart buildings and highlighted that their minimum features must include the ability to respond to external and internal conditions. The authors proposed 36 KPIs and classified them based on their smartness features, after analyzing different reports, legislation, and research papers. In 2019, Janhunen et al. [6] conducted a study on the applicability of the SRI in Northern Europe, where heating accounts for a significant portion of building energy use due to the cold climate. The study explored three buildings and assessed district heating (DH) as a cold-climate solution to improve energy efficiency. However, the authors found that the SRI's system-oriented approach did not differentiate the unique features of cold-climate buildings, especially those with advanced DH systems. The authors also noted that SRI methodology allows too much subjectivity, which can manipulate scores to obtain more favorable results. A study by Ramezani et al. in 2021 [9] discussed challenges related to implementing SRI methodology and its use in the Mediterranean climate, specifically in Portugal. The study assessed two case buildings in Portugal, including an evaluation of indoor environment quality (IEQ) and energy savings. The authors concluded that the SRI framework is suitable for the Mediterranean climate, but when evaluating the improvement effect of energy and IEQ, SRI methodology did not fully recognize the influence of all implementations.

In 2020, Fokaides et al. presented a study on the impact of the SRI on building energy performance [10]. They evaluated an educational building using SRI methodology, which resulted in a total SRI score of 52%. The authors concluded that SRI methodology covers most aspects of buildings but should be integrated with other energy efficiency assessment processes and expanded to consider specific building types. Vigna et al. in 2020 [11] applied SRI methodology to a nearly zero-energy office building and evaluated it with two parallel groups of experts. They emphasized the importance of collecting data that directly affect the functionality of building services. The authors also highlighted that smartness should enhance buildings' energy efficiency and overall performance and that smartness should be evaluated in terms of how well buildings adapt their operation to the occupants' needs and the grid. In 2022, Apostolopoulos et al. [12], evaluated SRI methodology for residential buildings with different renovation scenarios. The authors assessed renovation costs to increase building smartness and evaluated the resulting SRI score. They concluded that relatively low expenditures are needed to increase the smartness of buildings constructed after EPBD implementation compared to older buildings.

In 2021, Canale et al. [13] applied SRI methodology to residential buildings using three different scenarios: the base scenario (building stock as it is), the energy scenario (simple energy retrofit), and the smart energy scenario (energy retrofit addressing the smartness of the building). The energy scenario resulted in a 15.7% SRI score, while the smart energy scenario resulted in a 27.5% SRI score.

Several studies analyzing building energy consumption and energy performance certificates (EPCs) have highlighted the importance of further developing EPCs and including the SRI in the system. These studies include Li et al. (2019) [14], Märzinger et al. (2019) [15], Koltsios et al. (2022) [16] and Seduikyte et al. (2022) [17]. The inclusion of the SRI in the system can motivate investment in smart technologies and energy savings. However, some SRI studies [9] have indicated that for non-residential buildings, amendments are still necessary to capture specific features, and the revision of weighting factors is required. Additionally, some studies have pointed out that the current SRI mainly focuses on a qualitative assessment of building smartness and does not consider the broader context of the district [18]. Sustainability and smartness are important in the context of smart cities (rapid adoption of emerging technologies, such as smart metering and sensors, load flexibility that will address current trends and challenges), nZEB buildings and renovation of the existing building stock [19–21].

The overview concerning the studies conducted in the field of SRI assessment reveals the need for the implementation of research activities related to the adaptation of the SRI to specific building types. Moreover, the research reveals that there is still a vague link between the SRI scores and the energy efficiency of buildings. Research work is still required to align the significance of the energy efficiency improvement with the optimization of the SRI score. Recognizing these gaps, this study aims to investigate the link between the improvement of the SRI score and energy efficiency.

The approach of this research was to investigate the influence of changes in the engineering systems on the SRI score and energy consumption of educational buildings.

The purpose of this study is to quantify the realistic limits that may be reached concerning the optimization of the SRI. The article is structured into four sections. The introductory section is followed by the materials and methods section, in which the research methods of the study are presented. The discussion and results section presents the findings of the SRI assessment for the pilot building, as well as the relation between the SRI improvement and the energy efficiency enhancement of the pilot building. The major findings of this study are summarized in the conclusions section.

3. Methods

In this section, the methods employed in this study are presented. In particular, the tools and conditions for implementing the SRI and the energy assessment of the investigated building unit are elaborated. In this section the case study building is also presented.

3.1. Smart Readiness Indicator (SRI) Assessment

The SRI assessment of the case study building was conducted in compliance with the content of the EU delegated Regulation 2020/2155, which establishes the definition of the SRI and a common methodology by which it is calculated. For the calculation of the SRI, the EU Commission Calculation Sheet Version 4 was employed. The calculation procedure was divided into three steps:

- Initially, general information concerning the building unit, its location and its employed systems was provided;
- In the second stage, the functionality levels of the selected building services were defined. For the definition of these services, field research was conducted, and input from the pilot's facility manager was received;
- The last stage of the calculation procedure comprised the implication of the findings, as well as a sensitivity analysis with the variation of the weighting factors of the assessment criteria.

In the SRI service catalogs, services are structured within the following domains: heating system, domestic hot water, cooling system, controlled ventilation, lighting, dynamic envelope, electricity, electric vehicle charging and monitoring and control. Mentioned domains cover fifty-two smart-ready services, and each service can be implemented with various degrees of smartness, that is, functionality levels. The functionality level 0 indicates a nonsmart service implementation. The highest functionality level leads to the highest smartness of the service. It means that this particular service offers more value-added impacts to building occupants or to the grid in comparison with services of lower functionality levels. Each domain has an impact on different categories: energy savings on site, flexibility for the grid and storage, comfort, convenience, wellbeing and health, maintenance and fault prediction and information available to occupants.

SRI methodology provides default weighting factors which depend on the building type (residential, non-residential) and the climate zone (e.g., Northern Europe, Western Europe, etc.).

The smart readiness score of a building is a percentage of how close (or far) the building is to the maximum smart readiness that it could reach.

The results are provided graphically, whereas the bar charts delivered by the calculation sheet are also used in this study to present the SRI assessment results.

3.2. Whole-Building Energy Assessment

For the calculation of the energy performance of the investigated building unit, the IDA ICE tool was used. IDA ICE allows the implementation of whole-building energy assessment. IDA ICE is a simulation application for the multi-zonal and dynamic study of indoor climate phenomena as well as energy use, supporting IFC BIM models. In particular, the building technical systems were simulated in detail, allowing for the implementation of parametric scenarios related to the upgrade of the investigated building's automation and control systems. In terms of this study, the use of thermostatic heads was investigated. The simulation delivered information related to building's energy consumption under diverse scenarios.

The geometrical model of the analyzed building is presented in Figure 1. Dynamic simulations were used to quantify the thermal energy used during the heating season under two scenarios: the business-as-usual scenario (BAU) and the scenario of installing thermostatic valves in the radiators. With regard to the calculation assumptions, it was considered that the amount of supply air was 18 $m^3/(hm^2)$ with an area per student of 2 m^2, and heat due to the lightning of 5 W/m^2. The heating was considered to start when the outdoor temperature was below 10 °C, and the temperature of the heat carrier was considered to be regulated according to the outdoor temperature sensor. For the operational schedule of the building, it was considered that students were present in the classrooms Monday to Friday, from 9 a.m. to 5 p.m.

Figure 1. Geometrical model of analyzed case study building.

3.3. Case Study Building

The investigated concept of this study was applied to a building of the Kaunas University of Technology (KTU), located in Kaunas, Lithuania. The building is used by the Faculty of Civil Engineering and Architecture of the university, and it also hosts the library of the institution.

The building (Figure 2), constructed in 1965, is of energy class C and has a total area of 14,824 m^2. Information related to the building technical systems are provided in Table 1.

Figure 2. Case study: an educational building of Kaunas University of Technology.

Table 1. Case study building: Technical systems information.

Building Technical System	Description
Heating	Energy carrier: District Heating Automation: Compensation sensor Circulation system: Variable speed Terminals: Radiators, without individual control
Sanitary hot water	Water heaters for toilets Individual heating system for canteen and gym
Lighting	Individual control at room level
Electricity	Grid-connected A roof-top grid-connected PV system is connected under net metering conditions

4. Results and Discussion

4.1. Calculation of Case Study Building SRI

Figure 3 presents the total score, as well as the impact scores and the domain scores of the SRI calculation of the test case building. In particular, the test case building was calculated to have a total SRI core of 26%, whereas the individual impact scores were calculated to be 54% for energy savings on-site, 5% for flexibility for the grid and storage, 36% for comfort, 19% for convenience, 14% for wellbeing and health, 19% for maintenance and fault prediction and 45% for the information for the occupants.

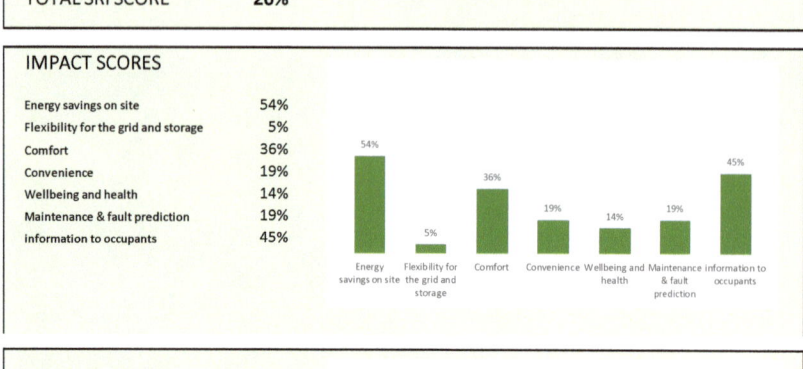

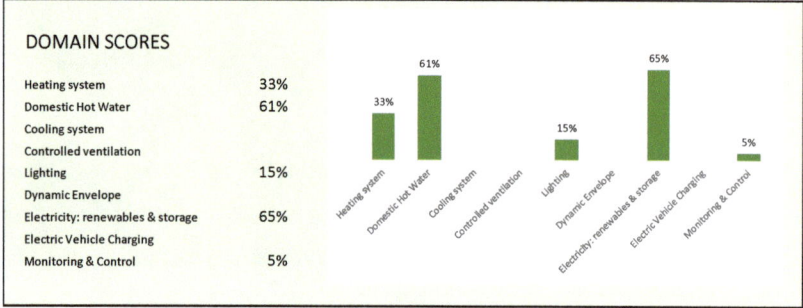

Figure 3. Results of a pilot building calculated SRI.

Concerning the domain scores, the following results were documented:
- The hot water system at the university building achieved a score of 61%, thanks to its ability to collect data on system performance and energy consumption;
- The electrical system score, with the inclusion of a smart solar power plant with an energy storage tank, was 65%, while it would have been only 13% without it;
- The technical areas of heating, lighting and monitoring and control were the lowest scoring areas, achieving 33%, 15%, and 5%, respectively;
- The flexibility of the grid and storage scored 5%, due to the lack of communication between the centralized heat and electricity networks and local building management systems;
- Scores for well-being and health, convenience, maintenance and fault prediction, comfort, and energy savings on-site were 14%, 19%, 19%, 36%, and 54%, respectively;
- The information-for-occupants criterion has the potential to reach up to 45%;
- Increasing the smartness of the heating and lighting areas could potentially increase the comfort of the rooms up to three times, and the building has the potential to save about half of the energy consumed.

4.2. Calculation of Whole-Building Energy Assessment

Figure 4 presents the results of IDA-ICE simulations for the amount of energy used for heating an educational building in two scenarios. The modernized system, which included the installation of radiator valves with thermostatic heads, showed a 12.2–22.2% reduction in energy consumption in the autumn and spring months and an 8.9–10% decrease in the winter months.

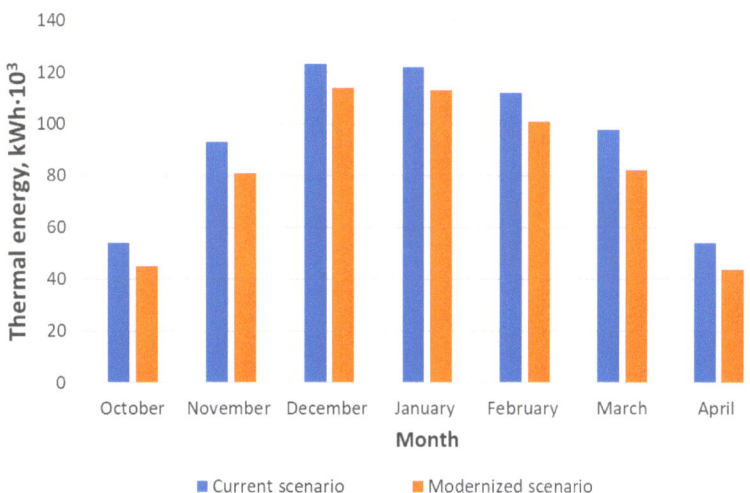

Figure 4. Heating energy of a pilot building under the two investigated scenarios.

4.3. Calculation of SRI Score with Building Energy Upgrade

By incorporating radiator valves with thermostatic heads into the educational building's heating system, an alternative scenario was created and analyzed using IDA-ICE. The SRI score of this modernized system was then computed and visualized in Figure 5.

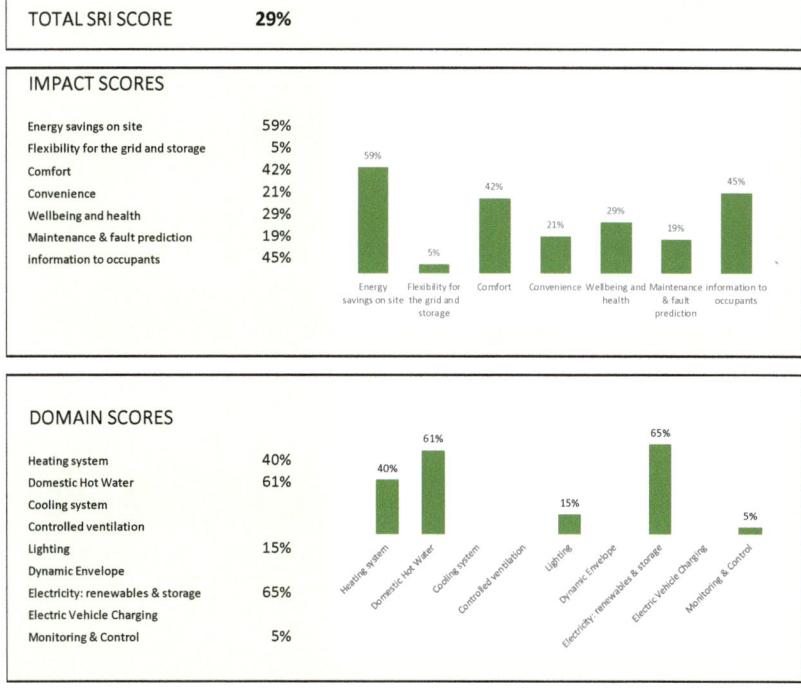

Figure 5. Results of a pilot building calculated SRI with upgraded heating system.

After performing SRI calculations to evaluate the existing educational building and a modernization scenario with installed radiator valves and thermostatic heads, the SRI score increased from 26% to 29%. Alternative solutions were then considered to further evaluate how they could influence the SRI score. The proposed alternative solutions for the engineering systems are presented in Table 2.

Table 2. Alternative solutions for the engineering systems of a pilot building.

Engineering System	Proposed Alternative Solutions
Heating	Heat source: City's centralized heat network Outdoor sensor: Regulates temperature of supplied heat carrier Variable speed circulation pumps: Connected to building management system Heating devices: Floor heating; each room controlled separately; connected to building management system
Sanitary hot water	Hot water is prepared at a heating point with smart automation.
Cooling	Variable refrigerant flow cooling system with indoor units in each room Controllers installed in each room separately System connected to building management system
Controlled ventilation	Ventilation system with heat recovery. Each room has a variable air volume ventilation system controlled by a room sensor based on CO_2.
Lighting	The lighting systems are controlled in each room separately by an automatic switch with the possibility of changing the light intensity.
Electricity system	No changes in the existing system.
Dynamic building envelope	External blinds are provided on the windows of the building, operating according to a solar illuminance sensor.
Electric vehicle charging	There are two charging points for electric cars in the faculty courtyard; you can choose the departure time there.

The calculated result of the SRI with an alternative solution for the investigated building is presented in Figure 6. Total SRI score is 67%.

When comparing the original SRI score of the educational building under investigation (Figure 3) to the score achieved with alternative solutions (Figure 6), it is evident that there were significant changes in the domain scores. Notably, the Monitoring and Control score increased from 5% to 67%, which could be attributed to the installation of fully automated engineering systems that offer better monitoring and control capabilities. Additionally, the Lighting score improved from 15% to 85%, following the provision of automatic switches with the ability to adjust light intensity. To achieve the maximum score in this domain, the lighting system should also be able to adjust the hue of the light.

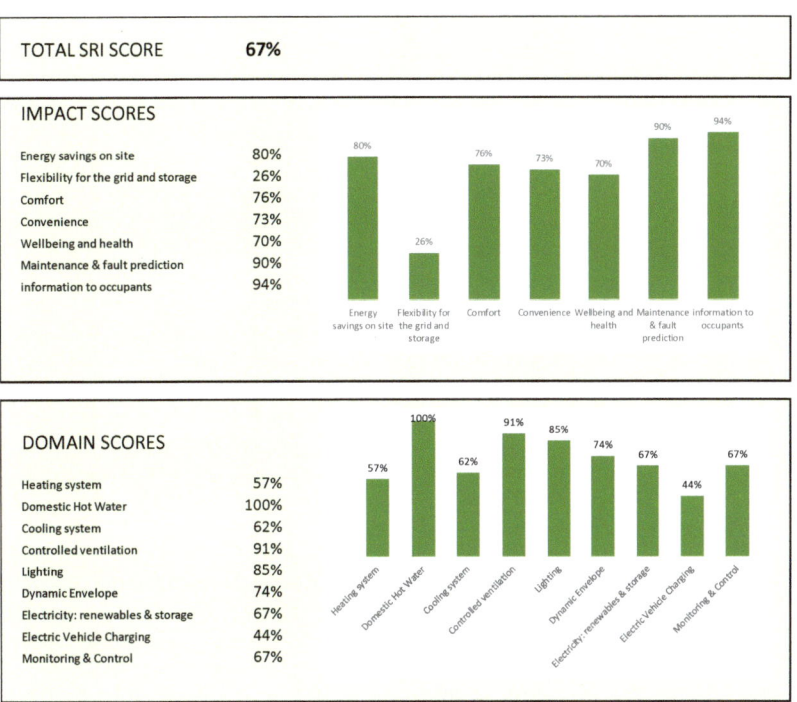

Figure 6. Results of a pilot building calculated SRI with upgraded building automation and control systems.

The Heating score also improved from 32% to 57%. However, to reach the maximum score like the Cooling system (which reached 62%), the heating/cooling should be based on occupancy sensors, and the system should be able to communicate with the city's centralized networks. The Electricity score remained relatively unchanged, while the Domestic Hot Water score reached the maximum score of 100%. However, the Electric Vehicle Charging score was the lowest among the alternative solutions investigated. To achieve the maximum score in this domain, electric car charging stations must make up more than 50% of the parking space, and the charging system should operate optimally based on a load of centralized electricity networks. Furthermore, it should be able to utilize electricity from electric cars.

5. Policy Implication of Study Findings

5.1. Need for Building-Type SRI Service Catalogs

The SRI takes into account various technical systems and indicators related to their automation and control levels. However, this study has shown that the methodology used to calculate the SRI needs to be adapted to different types of buildings. The study revealed that the technical systems installed in buildings differ depending on their intended use. For example, dwelling buildings have different technical systems compared to commercial or educational buildings. Therefore, the SRI methodology needs to be adapted to accurately reflect the smartness level of each type of building. The need for adapting the SRI methodology to different types of buildings has also been recognized at a national level. Several EU MS have already started working on the integration of the SRI into their building regulations and certification schemes. Adapting the SRI methodology to different types of buildings is significant because it allows for a more accurate measurement of the smartness level of buildings. This, in turn, can help identify areas where improvements can be made to increase energy efficiency and human comfort. It can also help promote the integration of renewable energy sources and smart technologies, leading to a more sustainable and comfortable living and working environment.

5.2. The Significance of Smartness Assessment in Regard to Utilities Management at City Level

One of the key pillars of the SRI is a building's ability to interact with smart grids, which allows for the transmission of information related to energy consumption and production. This interaction is particularly relevant in the case of quarantine or lockdown, where there is a need to manage the use of utilities such as energy, water, and waste production. The ability to interact with smart grids allows buildings to become prosumers, where they can both produce and consume energy. The SRI, therefore, can be a useful tool for managing building utilities at a city level. This is especially important during times when there is a strain on energy resources, such as during a lockdown. By accurately calculating the SRI, it is possible to identify buildings that can contribute to the generation of energy and manage the utilities in terms of generation, transmission and distribution. Public buildings, such as educational buildings, are particularly significant in this regard. The calculation of the SRI for public buildings is of great importance in managing the utilities during quarantine or lockdown. This allows for a smoother management of energy and other resources, ensuring that they are being used effectively and efficiently.

5.3. The Linkage between the SRI and the Energy Efficiency of Buildings

The main purpose of this study was to examine the relationship between the smartness levels of buildings and their energy performance. According to the findings of the study, the energy performance of buildings is related to the level of building automation and control systems for the building technical systems. The study revealed the need to consider both assessments in parallel to provide a more comprehensive evaluation of a building's overall performance. This aspect could have further policy implications, particularly in relation to the new design of energy performance certificates (EPCs). The study suggests that the SRI level of a building should be included in the EPC assessment to provide a more accurate picture of a building's energy performance. In the future, energy audits could also include smartness walk-through audits, and measures for energy upgrades could be adapted to include smartness upgrades. There is a need for further integration of the SRI into the energy efficiency assessment of buildings. This will require decision making at various levels, including the joint issuance of EPCs and the SRI and the integration of SRI findings into EPCs. By integrating the SRI into energy efficiency assessments, it will be possible to obtain a more comprehensive evaluation of building performance, which will aid in achieving energy efficiency goals and promoting sustainable development.

5.4. SRI Restrictions Related to District-Level Managed Services

The SRI scheme may not recognize the restrictions that district-level managed services, such as district heating, may introduce regarding the maximum SRI that can be achieved. This deficiency of the SRI scheme has also been identified in national testing phases. To address this issue, actions need to be taken to improve the conditions of the SRI calculation. It is not fair to blame a building if the level of service at a city level does not allow it to achieve its maximum smartness performance. Improving the SRI calculation conditions will enable a fairer evaluation of a building's smartness level and its ability to interact with city-level services, promoting sustainable development in cities.

6. Conclusions

The performed study did not only evaluate the SRI score of the educational building but also investigated the influence of changes in the engineering system on the SRI score and energy consumption.

The educational building in question was analyzed and found to have an SRI score of 26%. After implementing a modernized heating system, the SRI score increased to 29%, and with alternative engineering solutions, it reached 67%. While all impact scores for the alternative solutions scenario were above 70%, the flexibility for the grid and storage criteria remained the lowest, at 26%. It is important to note that each criterion is interrelated and can affect more than one impact criterion. However, achieving the maximum SRI score in Lithuania or other EU countries may not be possible due to centralized heating and cooling systems that lack smart networks. The assumption was made that a building with engineering systems that supports high thermal comfort classes would achieve the highest comfort criterion score. While modern engineering systems can achieve a high SRI score related to human comfort, the comfort criterion depends on the quality of design, installation work, and clothing allowed in the workplace. As a result, the calculated comfort score may not accurately reflect the comfort level maintained indoors. Additionally, differences may arise between the energy efficiency class of the building and the SRI score, as energy efficiency calculations do not take into account engineering system possibilities and characteristics. The IDA-ICE simulations indicated that minor modifications to the existing heating system could reduce the energy used for heating of the analyzed educational building.

The need for adapting the SRI methodology to different types of buildings is supported by both this study and the work conducted at a national level for the integration of the SRI in EU member states. By accurately measuring the smartness level of buildings, we can take steps towards a more sustainable and comfortable built environment. Additional indoor environment measurements are needed, to have accurate information about the comfort level in the premises. The study also revealed that the SRI is a valuable tool for managing building utilities at a city level. The ability of buildings to interact with smart grids and become prosumers is particularly relevant during quarantine or lockdown. By accurately calculating the SRI, it is possible to manage utilities effectively and ensure a sustainable use of resources. The study also highlights the need to consider building smartness levels alongside energy performance in the assessment of buildings. This will have policy implications, particularly in relation to the design of energy performance certificates. The integration of the SRI into energy efficiency assessments will require further action at various decision-making levels, and steps will need to be taken in the near future to achieve this. The study revealed that the SRI score is affected by the level of automation and control of services managed at a city level, such as district heating systems. These systems are managed at a central level, which can restrict a building's ability to achieve its maximum smartness level.

As digitalization is one of the main factors for improving the energy efficiency of the EU, the SRI rating system can help in assessing the smartness of buildings and inform responsible stakeholders about the capacities of building automation and control systems. From 2022, volunteer EU countries were able to launch the test phase or implement the SRI. The volunteering countries were Austria, Czech Republic, Croatia, Denmark, Finland and France. From the experience of the voluntary countries, more information or possible further development of the SRI methodology in research and the practical field is expected.

Author Contributions: Conceptualization, L.S.; data curation, M.D.; formal analysis, G.P.; funding acquisition P.A.F.; investigation, G.P. and E.D.; project administration, P.A.F.; supervision, L.S.; validation, B.I.; writing—original draft, G.P. and E.D.; writing—review and editing, L.S. and P.A.F. All authors have read and agreed to the published version of the manuscript.

Funding: This study gratefully acknowledges the financial support provided by the European Union's Horizon 2020 research and innovation program. Specifically, the "Development of Utilities Management Platform for the case of Quarantine and Lockdown" project funded under the Marie Sklodowska-Curie grant agreement No. 101007641.

Data Availability Statement: Not applicable.

Conflicts of Interest: The authors declare no conflict of interest.

References

1. Fokaides, P.; Apanaviciene, R.; Černeckiene, J.; Jurelionis, A.; Klumbyte, E.; Kriauciunaite-Neklejonoviene, V.; Pupeikis, D.; Rekus, D.; Sadauskiene, J.; Seduikyte, L.; et al. Research challenges and advancements in the field of sustainable energy technologies in the built environment. *Sustainability* **2020**, *12*, 8417. [CrossRef]
2. CEN/TC 247. (n.d.). Controls for Mechanical Building Services. Available online: https://standards.cen.eu/dyn/www/f?p=204:110:0::::FSP_ORG_ID:3836 (accessed on 10 January 2023).
3. EN 15232:2008; Industrial, Commercial and Residential Building—Impact of Building Automation, Control and Building Management on Energy Performance. European Committee for Standardization: Brussels, Belgium, 2008.
4. EN ISO 52120-1; Energy Performance of Buildings. Contribution of Building Automation, Controls and Building Management. General Framework and Procedures. International Organization for Standardization: Geneva, Switzerland, 2022.
5. Directive (EU) 2018/844 of the European Parliament and of the Council of 30 May 2018 Amending Directive 2010/31/EU on the Energy Performance of Buildings and Directive 2012/27/EU on Energy Efficiency. Available online: https://eur-lex.europa.eu/legal-content/EN/TXT/PDF/?uri=CELEX:32018L0844&from=IT (accessed on 2 March 2022).
6. Janhunen, E.; Pulkka, L.; Säynäjoki, A.; Junnila, S. Applicability of the Smart Readiness Indicator for Cold Climate Countries. *Buildings* **2019**, *9*, 102. [CrossRef]
7. Commission Implementing Regulation. (2020). (EU) 2020/2156 of 14 October 2020 Detailing the Technical Modalities for the Effective Implementation of an Optional Common Union Scheme for Rating the Smart Readiness of Buildings. OJ L, 431 (2020), 25–29. Available online: https://eur-lex.europa.eu/eli/reg_impl/2020/2156/oj (accessed on 10 January 2023).
8. Al Dakheel, J.; Del Pero, C.; Aste, N.; Leonforte, F. Smart buildings features and key performance indicators: A review. *Sustain. Cities Soc.* **2020**, *61*, 102328. [CrossRef]
9. Ramezani, B.; da Silva, M.G.; Simões, N. Application of smart readiness indicator for Mediterranean buildings in retrofitting actions. *Energy Build.* **2021**, *249*, 111173. [CrossRef]
10. Fokaides, P.; Panteli, C.; Panayidou, A. How Are the Smart Readiness Indicators Expected to Affect the Energy Performance of Buildings: First Evidence and Perspectives. *Sustainability* **2020**, *12*, 9496. [CrossRef]
11. Vigna, I.; Pernetti, R.; Pernigotto, G.; Gasparella, A. Analysis of the Building Smart Readiness Indicator Calculation: A Comparative Case-Study with Two Panels of Experts. *Energies* **2020**, *13*, 2796. [CrossRef]
12. Apostolopoulos, V.; Giourka, P.; Martinopoulos, G.; Angelakoglou, K.; Kourtzanidis, K.; Nikolopoulos, N. Smart readiness indicator evaluation and cost estimation of smart retrofitting scenarios—A comparative case-study in European residential buildings. *Sustain. Cities Soc.* **2022**, *82*, 103921. [CrossRef]
13. Canale, L.; De Monaco, M.; Di Pietra, B.; Puglisi, G.; Ficco, G.; Bertini, I.; Dell'Isola, M. Estimating the Smart Readiness Indicator in the Italian Residential Building Stock in Different Scenarios. *Energies* **2021**, *14*, 6442. [CrossRef]
14. Li, Y.; Kubicki, S.; Guerriero, A.; Rezgui, Y. Review of building energy performance certification schemes towards future improvement. *Renew. Sustain. Energy Rev.* **2019**, *113*, 109244. [CrossRef]
15. Märzinger, T.; Österreicher, D. Supporting the Smart Readiness Indicator—A Methodology to Integrate A Quantitative Assessment of the Load Shifting Potential of Smart Buildings. *Energies* **2019**, *12*, 1955. [CrossRef]
16. Koltsios, S.; Fokaides, P.; Georgali, P.; Tsolakis, A.C.; Chatzipanagiotidou, P.; Klumbytė, E.; Jurelionis, A.; Šeduikytė, L.; Kontopoulos, C.; Malavazos, C.; et al. An enhanced framework for next-generation operational buildings energy performance certificates. *Int. J. Energy Res.* **2022**, *46*, 20079–20095. [CrossRef]
17. Seduikyte, L.; Morsink-Georgali, P.-Z.; Panteli, C.; Chatzipanagiotidou, P.; Stavros, K.; Ioannidis, D.; Stasiulienė, L.; Spūdys, P.; Pupeikis, D.; Jurelionis, A.; et al. Next-generation energy performance certificates, what novel implementation do we need? In *Proceedings of the CLIMA 2022—14th REHVA HVAC World Congress, 22–25 May, Rotterdam, The Netherlands*; Delft University of Technology: Delft, The Netherlands, 2022; pp. 1–8. [CrossRef]
18. Märzinger, T.; Österreicher, D. Extending the Application of the Smart Readiness Indicator—A Methodology for the Quantitative Assessment of the Load Shifting Potential of Smart Districts. *Energies* **2020**, *13*, 3507. [CrossRef]
19. Gullbrekken, L.; Time, B. Towards Upgrading Strategies for nZEB-Dwellings in Norway. *J. Sustain. Arch. Civ. Eng.* **2019**, *25*, 35–42. [CrossRef]
20. Oprea, S.-V.; Bâra, A.; Marales, R.C.; Florescu, M.-S. Data Model for Residential and Commercial Buildings. Load Flexibility Assessment in Smart Cities. *Sustainability* **2021**, *13*, 1736. [CrossRef]
21. Tsirigoti, D.; Zenginis, D.; Bikas, D. Energy and Aesthetic Upgrading Interventions: Assessing Urban Block Renovation Scenarios. *J. Sustain. Arch. Civ. Eng.* **2021**, *29*, 62–82. [CrossRef]

Disclaimer/Publisher's Note: The statements, opinions and data contained in all publications are solely those of the individual author(s) and contributor(s) and not of MDPI and/or the editor(s). MDPI and/or the editor(s) disclaim responsibility for any injury to people or property resulting from any ideas, methods, instructions or products referred to in the content.

Article

Thermal Diagnosis of Ventilation and Cooling Systems in a Sports Hall—A Case Study

Maria Hurnik, Joanna Ferdyn-Grygierek, Jan Kaczmarczyk * and Piotr Koper

Faculty of Energy and Environmental Engineering, Silesian University of Technology, Konarskiego 20, 44-100 Gliwice, Poland; maria.hurnik@polsl.pl (M.H.); joanna.ferdyn-grygierek@polsl.pl (J.F.-G.); piotr.koper@polsl.pl (P.K.)
* Correspondence: jan.kaczmarczyk@polsl.pl; Tel.: +48-32-237-28-40

Abstract: Air conditioning systems in buildings consume a significant part of the world's energy, and yet there are cases wherein users are not satisfied with the quality of the thermal environment. Examples of such special cases are sports halls, which require different thermal conditions within a single zone. Thermal diagnostics for buildings can be used to diagnose problems. The aim of the paper was to analyse the effectiveness of the ventilation and cooling systems of a sports hall with a cubature of 16,300 m^3 and to check the possibility of managing the hall's cooling demands via the existing air conditioning system. Diagnostic measurements were performed, including in situ measurements of ventilation air flows from the diffusers and their temperatures, visualization of the supply air flows, and monthly registration of the indoor temperature in the hall at different set temperatures of the supply and exhaust air. Additionally, a numerical analysis, using EnergyPlus simulations, of cooling demand was performed with regard to the varying uses of the hall. The analysis based on measurement and simulation showed that it is not possible to remove heat gains from the hall with the current available ventilation air flow.

Keywords: sports hall; cooling; thermal diagnostic; thermal comfort; ventilation; building thermal simulation

1. Introduction

Maintaining appropriate thermal conditions and satisfactory air quality in buildings requires a significant amount of energy, with the literature stating that it accounts for about 40% of the total energy produced [1,2]. Despite high energy expenditure, it is not always possible in some existing buildings to obtain the required conditions of the internal environment. The reasons for such a situation may lie with the building itself, the way it is used, but also in the incorrect adoption of HVAC solutions at the design stage and/or their improper operation and regulation. Without identifying these causes, it is difficult to find a solution to the problem. Thermal diagnostic methods are a useful tool that allow for the systematic analysis and identification of observed problems.

Thermal diagnostics of a building includes the identification and assessment of all factors affecting its heat consumption in the building. These include heating, ventilation, air conditioning (HVAC) and hot water systems; building envelope; indoor environment; and building operation. Appropriately integrated partial diagnoses lead to the determination of the heat consumption of the entire building and provide data for the determination of improved solutions. When assessing energy consumption, attention should be paid to the quality of the indoor environment, as energy savings must not cause its deterioration. Two phases can be distinguished in the diagnostic procedure: inspections and diagnostic measurements. The main purpose of the inspection is to assess the technical condition of the systems and to compare the compliance of those systems with the designer's documentation. Diagnostic measurements are performed to evaluate the operation of the installations in situ under real conditions. Popiolek et al. [3] have proposed a comprehensive method of evaluating the physical envelope, HVAC systems and indoor environment [3]. Case studies of thermal diagnoses of building systems are presented in [4–7]. Comprehensive in situ measurements, including building envelope, ventilation with heat recovery, heating, and domestic hot water systems, were performed in two single-family houses in [4,5], while thermal diagnostics of ventilation systems and cooling sources for air conditioning systems carried out in office buildings are presented in [6] and [7], respectively.

Examples of particularly complex cases are large-volume facilities. These include sports halls in which thermal loads vary across a wide range. The ventilation and air-conditioning of sport halls requires a large number of air exchanges, especially during sports events attended by many spectators. The airflow pattern should be well planned and controlled to ensure acceptable indoor air quality and thermal comfort in the occupied zone for users with different activities, such as those of athletes and spectators. In addition, energy consumption should be taken into account.

Energy efficient ventilation of large enclosures, including sports halls, was the subject of the international research project IEA Annex 26 [8]. In the ventilation of these rooms, problems with undesirable thermal stratification, local overheating, drafts or spread of pollutants were observed. Ref. [8] provides an overview of techniques to measure and model air distribution in large, ventilated spaces.

Many research centres around the world have been dealing with energy efficient HVAC systems in sports halls. The indoor environment in sport halls has been evaluated according to air quality [8–17], microbiological pollution [18], dust pollution [19] and thermal comfort [10,14,20–22]. There has been an ongoing search for a proper ventilation and night cooling strategy [10,21,23] and various solar shading systems, which reduce solar heat gains [24], and lighting systems [25,26] have been analysed. For dynamic building simulation, the TRNSYS [10] and DesignBuilder [19] programmes were used. Air distribution inside sport halls was calculated using CFD codes: ANSYS [27] and IES Virtual Environment [28]. The temperature distributions for different HVAC systems in a training hall have been predicted using the FloVENT CFD code [14]. However, in the above-mentioned papers, no information on comprehensive thermal diagnosis of the cooling system in sports hall was found.

This paper presents thermal diagnostic tests performed in a sports hall for which users complained about overheating during the summer due to the ineffective operation of the air conditioning system. A retrofitting of an HVAC system cannot be undertaken without determining the reasons of such overheating. The aim of this paper was to analyse the effectiveness of the ventilation and cooling systems in a sports hall and to check the possibility of managing the cooling demands via its existing air conditioning system. The scope of the study included inspection of ventilation and cooling systems, in situ measurements of ventilation air volume flows from the diffusers and the temperature of the air supplied to the hall, visualisation of the supply air flows, and registration of the indoor temperature in the hall at different set temperatures of the supply and exhaust air flows. Temperatures in the different points of the hall and AHU were measured and registered continuously for about 30 days. On the basis of these measurements, the total and sensible power of the cooling system and the sensible power removed from the hall were determined. This method of cooling power determination can be useful in thermal diagnostics of HVAC systems in buildings. Additionally, a numerical analysis of cooling demand was performed for various operation scenarios of the hall. The effects of additional actions, such as the use of window roller shades and the introduction of additional sun control films on windows, were also examined.

2. Methods

2.1. Tested Building

The research was carried out in a sports hall with a cubature of 16,300 m^3, which includes a sports field with an area of 1387 m^2 and a height of 10 m, a fixed stands with an area of 152 m^2 for 288 people, and movable stands for 112 people (Figure 1a,b). The hall is located in southern Poland, which is characterized by a transitional temperate climate, between the continental and Atlantic climates, with relatively cold winters and warm summers (the Dfb class according to the Köppen–Geiger classification).

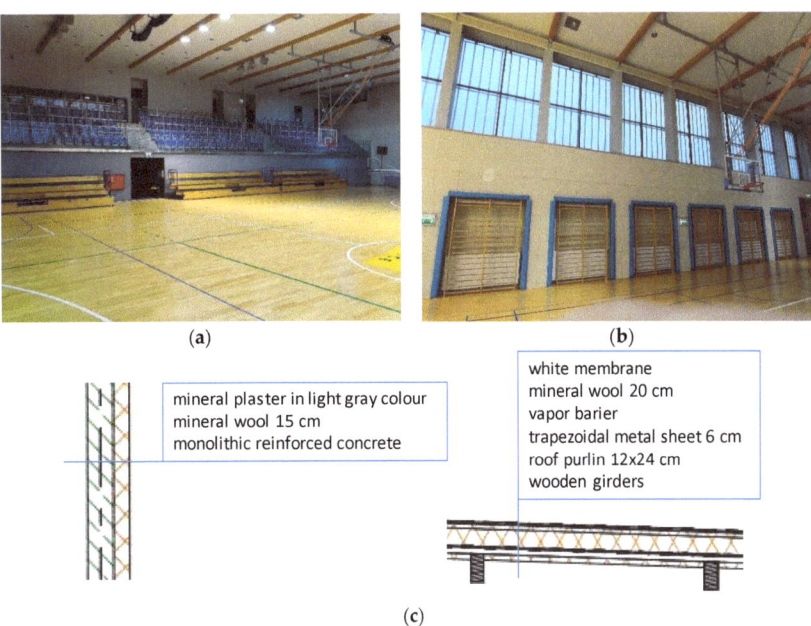

Figure 1. Hall: view on the stands (**a**); view on the windows (**b**); masonry and roof construction packages (**c**).

The walls of the hall have a monolithic reinforced concrete structure. The wooden roof is covered with a white membrane (U_{wall} = 0.20 W/(m²·K), U_{roof} = 0.15 W/(m²·K)) (Figure 1c). There are large windows in the external walls on the N, E and S sides (U_{glass} = 1.1 W/(m²·K)). The light-coloured internal roller shades have been installed on the windows, the positions of which are electrically controlled by a manually operated actuator. Some of the windows (N and S side) can be tilted.

The room is air-cooled with a supply-exhaust system. This system originally served only as ventilation and as additional heating of the hall above the standby temperature of 8 °C. In the second stage of the investment, the system was upgraded to cool the sports hall. The design air temperature in the hall is 24–26 °C. The air handling unit (AHU) is located on the roof and is equipped with a water heater and a water cooler. The constant air volume (CAV) system supplies air (11,500 m³/h) above the sports field with seven long-range air supply nozzles and to the stand area with seven swirl diffusers with a plenum box. According to the building documentation, the nozzles are set at angles of −29° to the axis of the ventilation duct (which was confirmed by inspection), and this is the recommended angle setting for the heating function. The angle was not changed when the system was upgraded to perform the cooling function. Air is removed from the ceiling of the hall through exhaust grilles. Thus, it is a classic mixing ventilation system.

A monoblock chiller with R-410A refrigerant is located on the roof of the building next to the AHU. The chiller is equipped with two compressors, one of which has an inverter to ensure the necessary adjustments to the cooling capacity. The temperatures of the coolant (40% water solution of ethylene glycol) at the inlet and outlet of the cooling coil are 6 °C and 12 °C, respectively.

According to technical data, for a ventilation airflow of 11,500 m³/h, the nominal total power of the cooling coil is 74.5 kW and the sensible power $Q_s|_{sup}^o$ is 54 kW (when outdoor air is cooled to 14 °C, from outdoor air temperature t_o = 32 °C to air temperature at the cooler outlet equal to the supply air temperature $t_c \cong t_{sup}$ = 18 °C). The sensible power $Q_s|_{sup}^o$ can be calculated from the formula:

$$Q_s|_c^o = V \cdot \rho \cdot c_p \cdot (t_o - t_c) = \frac{11500}{3600} \cdot 1.2 \cdot 1.005 \cdot 14 = 54 \text{ kW}. \tag{1}$$

Assuming that the temperature of exhausted air is equal to the average air temperature in the hall, t_i = 25 °C, the maximum sensible heat flux removed from the hall is:

$$Q_s|_{sup}^i = V \cdot \rho \cdot c_p \cdot (t_i - t_{sup}) = \frac{11500}{3600} \cdot 1.2 \cdot 1.005 \cdot 7 = 27 \text{ kW}. \tag{2}$$

The sensible heat gains assumed at the design stage were 69 kW. Thus, the sensible heat flux removed from the hall is approximately 2.6 times lower. This means that the ventilation airflow of 11,500 m³/h at a temperature of 18 °C will not ensure the air temperature in the hall at the level of 24 to 26 °C.

The controller maintains the temperature in the room at the set level. The system can operate in day or night mode. The operation of the automation system consists of regulating the supply temperature based on the readings of the temperature sensors of the outside air and the air exhausted from the room. The control system compares the set temperature with the current reading of the exhaust air temperature sensor.

2.2. Measurements

The tests performed included:

- measurement of airflows supplied to the hall with simultaneous measurement of the air temperature;
- visualisation of the supplied jets and determination of their throw length;
- continuous measurement and recording of temperature and humidity of the supplied air, exhausted air and air at the inlet and outlet of the cooling coil in AHU;
- continuous measurement and recording measurement of air temperature at the selected points in the occupied zone.

A list of measuring instruments with their range and measurement uncertainty is presented in Table 1. Additionally, meteorological data from a meteorological station located 6 km from the building were used for the analyses.

Table 1. List of measuring instruments.

Instrument	Type	Measurement Range and Uncertainty in Measurement	Purpose of Measurement
Balometer	ACCUBALANCE II	40–4000 m³/h ±3% of measured value	Measurement of volume flux of air supplied from diffusers
Thermometer	Testo 110	−50–150 °C ±0.2 °C (−25–75 °C)	Measurement of the temperature of air supplied from diffusers
Data logger APAR	AR235	Temperature: −30–80 °C, ±0.2 °C Relative humidity: 0–100%, ±3% (20–80%), ±3–5% (in the remaining range)	Measurement of air temperature and relative humidity: supply from the diffusers; in the occupied zone (7 meas. points); in the air-conditioning unit (2 meas. points)
Smoke generator	Antari 3000	–	Visualization of supplied jets

The measurement results were used to evaluate the thermal conditions in the room and to determine the total and sensible power of the cooling system at different temperatures, the humidity of the outside air, and the different values of the air temperature inside the hall, as set on the thermostat.

2.2.1. Measurements of Temperature and Volume Flux of the Supplied Air

Measurements of volume flux and temperature of the air supplied to the hall from all fourteen diffusers were performed once. The arrangement of the diffusers in the hall is shown in Figure 2. The long-range nozzles are marked with numbers 1 to 7, and the swirl diffusers are marked with numbers 8 to 14.

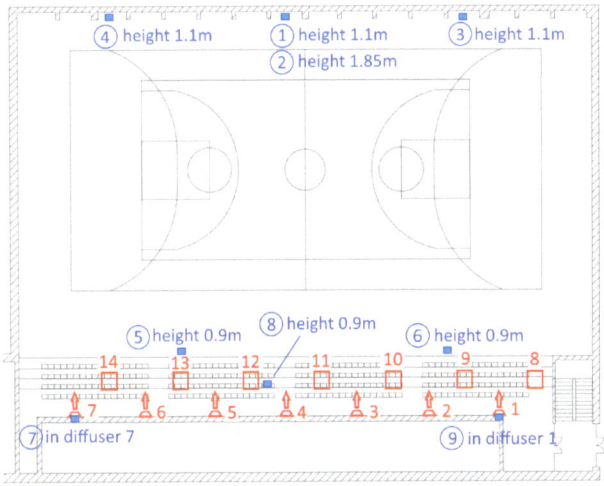

Figure 2. Sketch of the hall with diffusers (marked red, long-range jet nozzles 1–7, swirl diffusers 8–14) and the location of the temperature and humidity measurement points (marked blue in circles).

2.2.2. Visualisation of Supply Jets

In order to determine the throw length of the jets, they were visualised by adding smoke generated by the smoke generator to the supply air. Smoke dosing was carried out separately for one supply nozzle and one swirl diffuser. The air volume fluxes from the same type of diffusers were very similar; therefore, it can be assumed that the air velocity distributions are also similar.

Because the visualization may affect visibility, it was performed in an empty hall, late in the evening. The average outdoor air temperature of the 24 h preceding the measurements was 14.2 °C, and the daytime sun exposure did not exceed 300 W/m^2. Under these outdoor conditions the indoor air temperature in the hall was 21 °C.

2.2.3. Long-term Measurements of Air Temperature and Humidity

The tests of the efficiency of the cooling system were conducted in the warmest period of the year, i.e., in July and August. Continuous measurements and recording of air temperature and relative humidity were made at seven points located in the occupied zone. Data loggers 1, 2, 3 and 4 were placed near the external wall opposite the stands, data loggers 5 and 6 at the seats under the stands, and data logger 8 at the seat in the stands. Two data loggers placed in diffusers 1 and 7 (the data loggers marked 7 and 9 in Figure 2) measured the parameters of the air supplied to the hall. The locations of the data loggers in the hall are shown in Figure 2. The measurements in the occupied zone and in the diffusers were carried out from 27 July to 25 August 2021. In the period from 6 August to 25 August 2021, air temperature and humidity were measured at two points located in the AHU: the data logger located behind the cooler measured the parameters of the supplied air at the inlet of the supply fan, and the data logger placed at the inlet of the recuperator measured the parameters of the air removed from AHU. Due to technical reasons, the measurements of temperature and air humidity in the AHU started ten days later than measurements in the occupied zone and in the supplied jets. This period was sufficient to determine the cooling capacity of the HVAC system.

During the entire measurement period, the hall was used sporadically due to COVID restriction, so it can be assumed that internal heat gains were negligible. Measurement results were recorded with a time step of 5 min.

2.3. Thermal Simulation

To determine the cooling demand in the sports hall, a thermal simulation of the building was carried out using EnergyPlus 9.4 (US Department of Energy, Washington, DC, USA) [29]. The model included one thermal zone (Figure 3). The conditions in the unmodeled part of the building around the stands were assumed to be similar to the calculated zone. The following assumptions were made for the calculations:

- Simulation for the summer period 1.06 to 30.09 (2928 h) with a 15-min time step;
- Outdoor climate: typical climate data (TMY) for Katowice (typical meteorological years and statistical climatic data for the area of Poland for energy calculations of buildings [30]). The minimum temperature in the simulated period was 1.9 °C, the maximum temperature was 30.8 °C, and the average temperature was 16.2 °C;
- Indoor temperature: the lower limit was adopted according to the design assumptions—24 °C. In the model, the ideal loads air system was used, which supplies cooling or heating air to a zone in sufficient quantity to meet the zone load;
- Heat gains from lighting (400 W × 150 lamps = 60 kW): 15% convection (k), 85% radiation (r) (fluorescent lights built in the ceiling);
- Heat gains from people on the playing field [31]: sensible heat 145 W/person (k/r 50/50%), latent heat 280 W/person;
- Heat gains from people in stands [31] sensible heat 75 W/person (k/r 50/50%), latent heat 55 W/person;
- Non-transparent building partitions according to the current state;
- Transparent building partitions (windows) according to manufacturers' data for similar windows from the construction period of the facility. The optical properties of the glass were determined using a dedicated Window 7.8 program [32] (Table A1);
- Air infiltration: due to a mechanical ventilation system, external air infiltration was omitted. The windows in the building are relatively new (only ten years old) and tight and, with a small difference in temperature between the outside and the inside in summer, the infiltrating airflow is very small. Therefore, the error associated with this assumption is negligible;
- Internal shades (according to actual state): the solar radiation transmittance coefficient was adopted on the basis of catalogue data of similar roller shades available on the market. Two options were adopted: windows without shades or all windows covered.

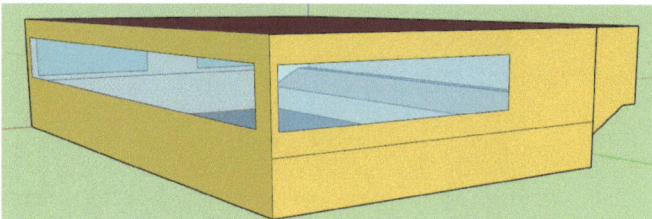

Figure 3. View of the geometry of the sports hall.

The calculations were carried out for three variants of use:

- A: empty hall;
- B: sports training on the playing field (20 people doing very hard work);

- C: hall with full occupancy of the audience (20 people doing very hard work and 400 people sitting).

3. Results

The ventilation and cooling systems were made according to the design of the building, but unfortunately the occupants complain of thermal discomfort in the summer. Retrofitting with a cooling system raises doubts and the question as to whether the air flow specified in the earlier project is sufficient to cover the cooling demand of the sports hall at the assumed supply air temperature has not been checked. Furthermore, the indoor temperature of 24 °C to 26 °C, assumed in the designed stage, is recommended for light work in the summer (for example, office work or spectating as part of an audience in the gym). The recommended temperature for very hard work, that is, high activity performed by players on the field, is 18–21 °C [33].

3.1. Ventilation Airflow Rate

The results of the measured volume flux and the temperature of the supplied air are presented in Table 2. The average temperature of the air supplied was 17.9 °C. The standard deviation of the air temperature measured in all diffusers was 0.5 °C.

Table 2. Supply airflow rate and air temperature.

Diffuser	Design Airflow Rate m³/h	Measured Airflow m³/h	Measured Air Temperature, °C
1	890	760	17.8
2	890	680	18.0
3	890	820	18.0
4	890	610	16.7
5	890	640	17.9
6	890	730	17.9
7	890	600	17.7
8	750	700	17.1
9	750	670	17.8
10	750	650	17.8
11	750	620	18.5
12	750	610	17.9
13	750	620	18.6
14	750	640	18.5

Total design airflow rate: 11,500 m³/h; total measured airflow rate: 9350 m³/h; average temperature: 17.9 °C.

The total measured ventilation air flow was 9350 m³/h ± 50 m³/h, and was lower by 8–33%, averaging 18%, than the design. From a hygienic point of view, with a maximum of 400 occupants, this flow is approximately 20% too low. When the hall is fully occupied, the air quality may be slightly unsatisfactory, but this is probably not often the case. Taking into account the hygiene conditions, it is not necessary to increase the ventilation airflow rate.

The average temperature of the air supplied was 17.9 °C. The standard deviation of the air temperature measured in all diffusers was 0.5 °C.

3.2. Evaluation of the Throw Length of Supply Jets

Visualization for the long-range nozzle was carried out twice, once for the nozzle directed downwards (setting found in the sport hall) and once upwards (setting recommended for a cold jet supply). The temperature of the supply air and the air at several points in the surrounding jet was measured. The supply air temperature measured in both diffusers was 16.6 °C. The air temperature in the hall was, on average, 21.7 °C, and its values differed very slightly within the uncertainty of its measurement. Thus, the temperature difference between the air supplied and the air in the room was −5 °C. Figures 4 and 5 show exemplary photos from tests of the jet throw length.

Figure 4. Visualization of the air stream from the nozzle directed downward (**a**) and upwards (**b**) in the 80th second.

Figure 5. Visualisation of the air stream from the swirl diffuser in the 30th second.

Figure 4 shows that, at the current setting of the long-range nozzles, the jets reach the level of the floor, covering the entire occupied zone. With a properly operating air conditioning system this setting may be unfavourable, as the jets reaching the occupied zone may cause a sensation of draught. However, in current conditions, when the temperature in the hall is too high, this setting may improve the thermal comfort of the athletes. When the nozzles are orientated up, the jets are attached to the ceiling and cover the entire width of the room.

In the case of the jet supplied from a swirl diffuser, and as shown in Figure 5, the cold air slowly descends downward and reaches the occupied zone. The conditions during the measurement were more conducive to such a development of the jet than the conditions when there are spectators in the audience. During a mass event, thermal plumes are formed above the spectators, which cause a vertical upward air movement. Thermal plumes work opposite to the supply jets, limiting their range. In this case, the throw length of the air jets may be smaller.

3.3. Temperature Measurement Results in the Hall

The time series of outdoor air temperature and indoor air temperature in the hall measured at seven points in the occupied zone, and the air temperature in the hall as set on the controller are shown in Figures 6a, 7a and 8a. Additionally, Figures 7a and 8a show the exhaust air temperatures measured in the air handling unit. The air temperature at all measurement points was continuously recorded, with the AHU turned on and off. In Figures 6–8, the periods in which the AHU was switched on are marked with a yellow background.

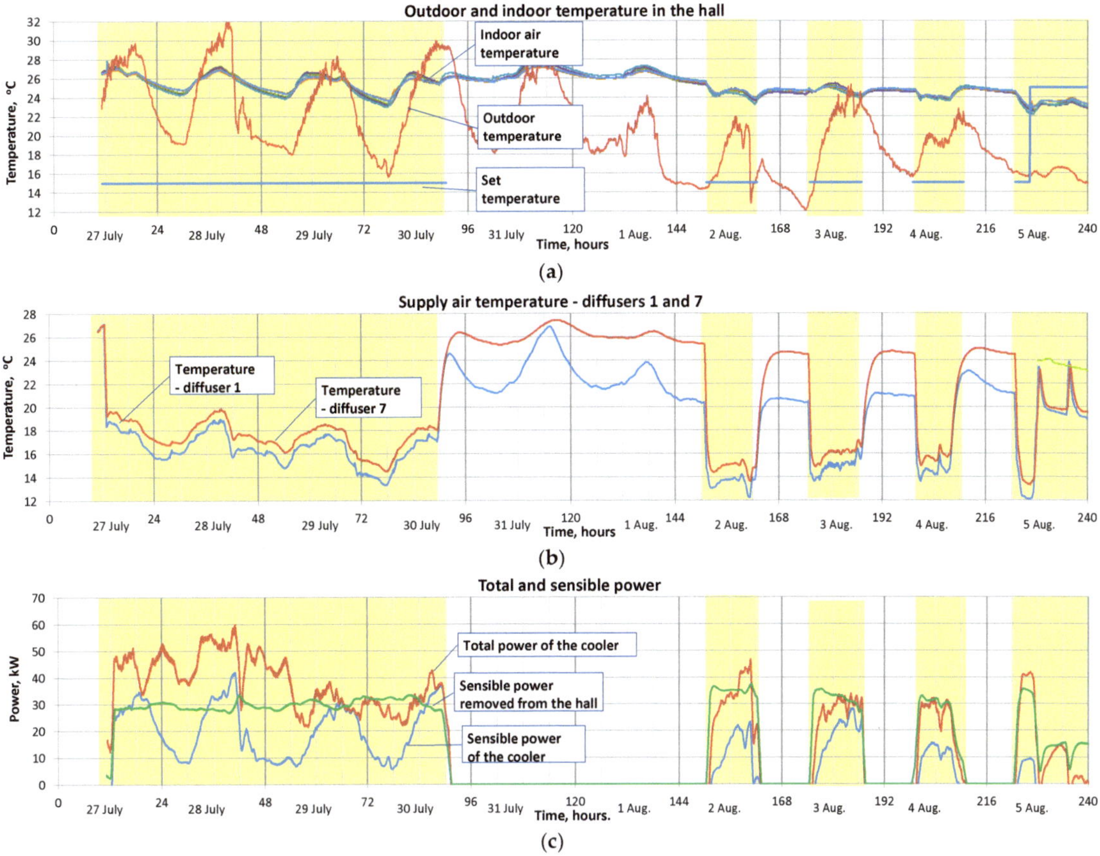

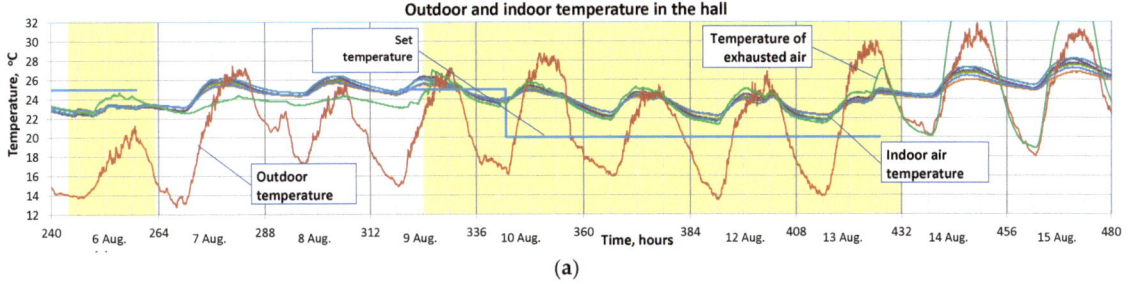

Figure 6. Time series of air temperature and the total and sensible power of the cooling system (27 July to 5 August): outdoor and indoor temperature in the hall (**a**); supply air temperature-diffusers 1 and 7 (**b**); total and sensible power (**c**).

Figure 7. *Cont.*

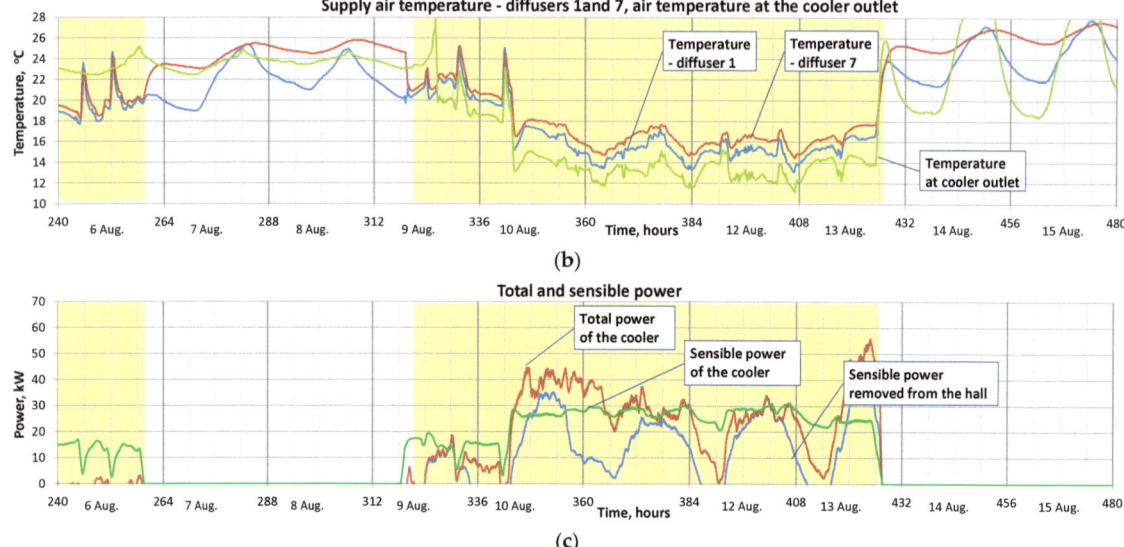

Figure 7. Time series of air temperature and the total and sensible power of the cooling system (6 August to 15 August): outdoor and indoor temperature in the hall (**a**); supply air temperature-diffusers 1 and 7, air temperature at the cooler outlet (**b**); total and sensible power (**c**).

The measurement results show that there are homogeneous thermal conditions in the occupied zone of the hall, that the temperature values recorded at various points in the hall are similar, and that the differences do not exceed 1 °C (Figures 6a, 7a and 8a). Furthermore, the temperature of the air exhausted from the hall is slightly (less than 1°C) higher than the average temperature measured in the occupied zone.

In the registered time series, three several-day periods of continuous operation of the control panel can be distinguished; these are hourly intervals: 11–91, 322–427 and 512–592. In the first period, the set temperature inside the hall was 15 °C (to determine the maximum cooling capacity), in the second period, in the range 322–335, the set temperature was 25 °C, and in the following hours of the second period and in the third period, it was 20 °C.

In the first period of continuous operation of the AHU (11–91 h) in the afternoon, the outdoor air temperature was high, at 28–32 °C. Despite the temperature setting in the hall being at 15 °C, the air-conditioning system did not provide the intended air temperature of 25 °C, and the temperature in the hall reached up to 27 °C. In the second period (322–427 h), the maximum outdoor air temperature was lower, at 24–30 °C. During most of the AHU operation time, when the outside air temperature was lower than 28 °C, the air temperature in the hall did not exceed 25 °C, but it was much higher than the set value of 20 °C. It was observed that, despite negligible internal gains and the continuous operation of the cooling system in the first and second periods, the average air temperature in the hall was about 2 °C higher than the average outside air temperature, which demonstrates the poor protection of the hall against heat gains from solar radiation.

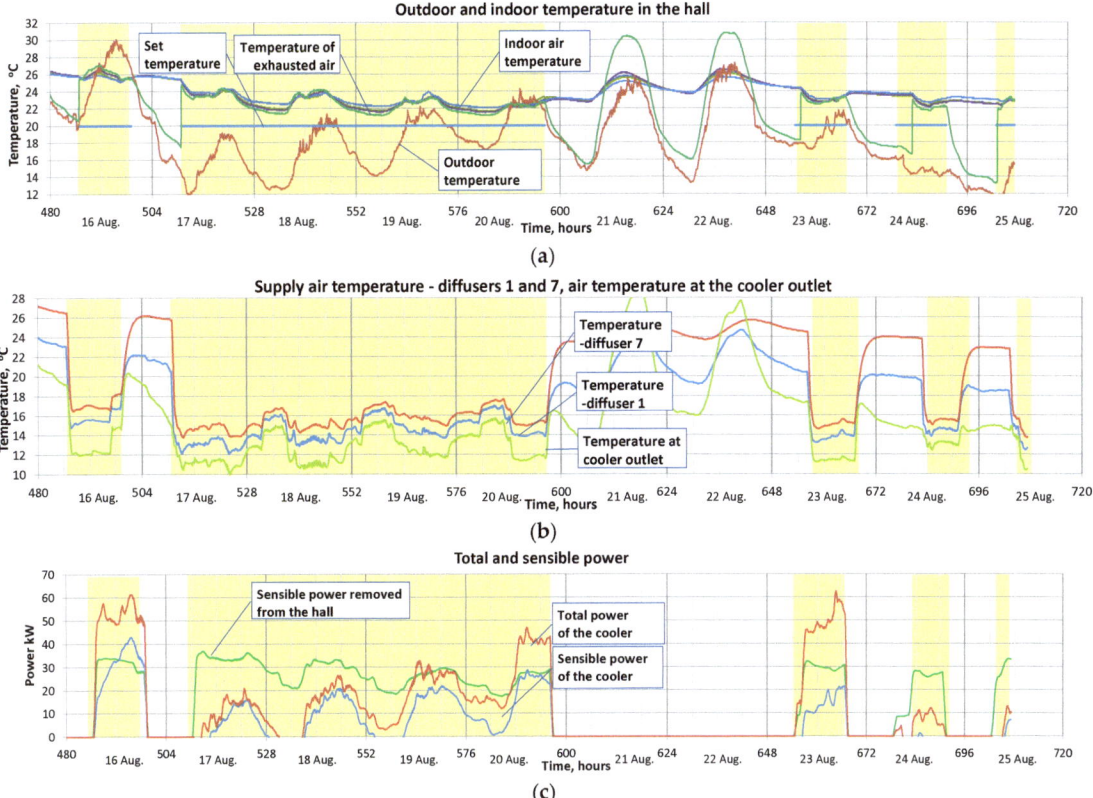

Figure 8. Time series of air temperature and the total and sensible power of the cooling system (16 August to 25 August): outdoor and indoor temperature in the hall (**a**); supply air temperature-diffusers 1 and 7, air temperature at the cooler outlet (**b**); total and sensible power (**c**).

In the third period (512–592 h), the maximum outdoor air temperature increased in the following days from 19 °C to 24 °C. The average daily temperature in the hall decreased slightly from 24 °C to 23 °C. Throughout the period, the air temperature in the room was higher than the outside air temperature and the set value of 20 °C.

Based on the temperature variations in the other, shorter, periods of the AHU operation (the AHU is turned on at 7 am and switched off around 7 pm), it can be observed that switching on the installation prevents the temperature from rising during the day, despite the increase in outdoor temperature.

3.4. Determination of the Total and Sensible Power of the Cooling System and the Sensible Power Removed from the Hall

3.4.1. Calculation Procedure

The sensible and total powers of the cooling system were determined on the measured temperature and relative humidity at the inlet and outlet of the cooling coil.

The sensible power of the cooling coil was determined from the equation:

$$Q_s|_c^o = V \cdot \rho \cdot c_p \cdot (t_o - t_c), \tag{3}$$

where:

V—airflow rate, m^3/h;
ρ—air density, kg/m^3;

c_p—specific heat of dry air, $c_p = 1.005$ kJ/(kg·K);
t_o—outdoor air temperature, °C;
t_c—temperature at the cooling coil outlet, °C.

Analysing the measurement results (Figures 7b and 8b), it was found that, during the AHU operation, the temperature of the air supplied to the hall was on average 2.3 °C higher than the temperature of the air measured directly at the cooling coil outlet. The reason for the increase in air temperature is because the air became heated in the supply fan and in the ducts that supply the air to the diffusers. For the calculation of the cooling coil capacity, it was assumed that the air temperature behind the cooler was 2.3 °C lower than the average temperature measured in diffusers 1 and 7.

The total power of the cooling coil was calculated from the equation:

$$Q_t|_c^o = V \cdot \rho \cdot (h_o - h_c), \tag{4}$$

where:
h_o—enthalpy of outdoor air, kJ/kg;
h_c—enthalpy of air at the cooling coil outlet, kJ/kg.

The enthalpy of outdoor air and enthalpy at the cooling coil outlet were calculated using the equation:

$$h_o = c_p \cdot t_o + x_o \cdot (r_o + c_{wp} \cdot t_o), \tag{5}$$

$$h_c = c_p \cdot t_{cc} + x_c \cdot (r_o + c_{wp} \cdot t_c), \tag{6}$$

where:
r_o—heat of vaporization, $r_o = 2501$ kJ/kg;
c_{wp}—specific heat of water vapor, $c_{wp} = 1.84$ kJ/kg.

The specific humidity x was determined from the relationship:

$$x = 0.622 \frac{\varphi \cdot p_s}{p_b - \varphi \cdot p_s}, \tag{7}$$

where:
φ—relative air humidity;
p_b—barometric pressure, Pa;
p_s—saturation pressure of water vapour calculated from the Clausius–Clapeyron equat

$$p_s = 611.213 \cdot \exp\left(\frac{17.5043 \cdot t}{241.2 + t}\right). \tag{8}$$

The sensible power (sensible heat flux) removed from the hall was calculated from the equation:

$$Q_s|_{sup}^{ex} = V_n \cdot \rho \cdot c_p \cdot (t_{ex} - t_{sup}), \tag{9}$$

where:
t_{ex}—temperature of the air exhausted from the hall, °C;
t_{sup}—supply air temperature (average of the values measured in diffusers 1 and 7), °C.

The air supply temperature measured at diffuser 7, which is located further from the AHU than diffuser 1, is systematically higher, by 0.5–1.0 °C, than the temperature measured at diffuser 1 (Figures 6b, 7b and 8b). A slightly higher air temperature at diffuser 7 may indicate insufficient thermal insulation of the supply air ducts.

Comparing the time series of the temperature of the air removed from the hall (measured at the inlet to the air handling unit in a shorter period, from 6 August to 25 August) and the average temperature measured inside the hall, it can be noticed that the temperature of the air removed from the hall is slightly, less than 1 °C, higher from the average temperature measured inside the hall. In the calculations for the entire measurement period from 27 July to 25 August, it was assumed that the temperature of the exhaust air was equal to the average temperature in the hall plus 1 °C.

3.4.2. Sensible and Total Power of the Cooling Coil

The maximum total power of the cooling coil, determined based on the measurement data recorded in the period from 27 July to 25 August 2021, was approximately 60 kW (Figure 8c). This value was approximately 20% lower than the assumed value of total power 74.5 kW when selecting the cooling coil. The reason for this difference is that the ventilation airflow rate is lower by approximately 18% than the design value. Due to the fact that the mean surface temperature of the cooling coil is approximately constant and equal to approximately 9 °C, the total power of the cooler decreases with decreasing temperature of the outdoor air. At an outdoor air temperature of approximately 16 °C, the total power of the cooling coil was approximately 25 kW (Figure 8c).

The sensible power is much lower than the total power of the cooler. The reason for this is that the water vapour condenses on the surface of the cooling coil and uses part of the total power for this phase change. The difference between total and sensible power depends on the moisture content in the outdoor air. The latent power (i.e., the difference between the total and sensible power of the cooler) as a function of the specific humidity in the outdoor air is shown in Figure 9. The data presented in Figure 9 were determined based on the temperature and humidity of the outdoor and supplied air recorded with a 5-min time interval during the AHU operation, in the period from 27 July to 25 August. The maximum sensible power of the cooling system, approximately 42 kW, occurred multiple times. The total and sensible power of the cooling coil varies widely with the temperature and moisture content of the outdoor air.

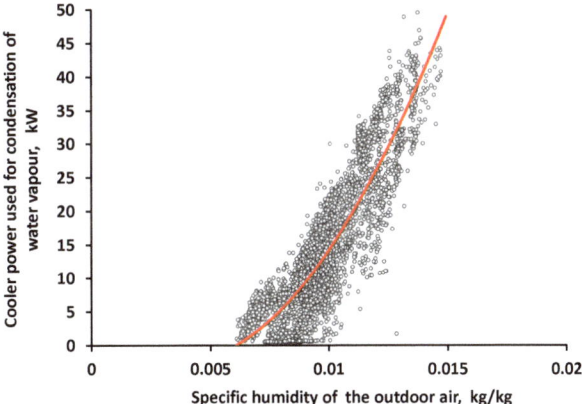

Figure 9. Cooler power used for condensation of water vapour as a function of the specific humidity of the outdoor air.

The sensible heat flux removed from the hall depends on the ventilation airflow and the temperature difference between the exhausted and supplied air. The air temperature during the measurement period varied from 13 °C to 19 °C with an average value of approximately 16 °C, and the exhausted air temperature was in the range from 23 to 28 °C with an average value of approximately 25.5 °C. The variability of the sensible heat flux removed from the room is shown in Figures 6c, 7c and 8c.

When the outdoor air temperature is high, the indoor air temperature exceeded the designed value of 25 °C, and most of the time the air temperature in the room was higher than the set value of 15 °C or 20 °C. Meanwhile, the sensible power removed from the hall has not changed significantly, from approximately 28 kW to approximately 32 kW. Therefore, it can be assumed that the maximum heat flux removed by the ventilation air from the hall is approximately 30 kW. The exceptions are two time periods: 240–258 and 318–340, in which the outside air temperature was lower than the set temperature of 25 °C. At that time, the average sensible heat flux removed from the hall was about 18 kW (Figure 7c).

3.5. Thermal Simulation Results

In order to determine the causes of the situation diagnosed during in situ measurements, thermal calculations of the sports hall were carried out in various states of its use. Table 3 presents the simulation results for the eight cases of internal and external heating load. The number of hours in which the cooling demand exceeds 27 kW is also given (see Section 2.1).

Table 3. Cooling demand calculation results.

Case	Occupants	Shades	Lighting	Cooling System Working Time	Max Cooling Demand Q_j, kW	Number of Hours with Cooling Demand > 27 kW (Percentage of Time)
A1	no	no	no	all day	45.8	126 (4.3%)
A2	no	yes	no	all day	42.9	35 (1.2%)
A3	no	yes	no	8 am to 8 pm	111.7	74 (2.5%)
B1	training (8 am to 8 pm)	yes	50%	all day	58.3	1221 (41.7%)
B2	training (8 am to 8 pm)	no	no	all day	48.4	300 (10.2%)
C1	full audience (8 am to 8 pm)	yes	100%	all day	122.3	1868 (63.8%)
C2	full audience (8 am to 8 pm)	yes	50%	all day	100.6	1663 (56.8%)
C3	full audience (8 am to 8 pm)	no	no	all day	90.5	1584 (54.1%)

Even in an empty room with shades on all windows and without using artificial lighting, the maximum cooling demand exceeded the available cooling power from the supplied air. Cooling power was insufficient in this case only for 1% of the summer season. Even without window shades, the excess was not greater than 5% of the time. These cases are similar to those analysed during the measurements. This confirms that the built thermal model is accurate. Simulations using TMY climate confirmed conclusions regarding the thermal conditions in the hall in the considered period. Case A2 can be treated as a comparative case with situations in which there are people and lighting in the room (it is the case with the lowest external and internal heat gains in the hall) and can be used to assess the impact of the use of shades on windows. In the hall, there are internal shades in a light colour. Visual evaluation showed that the shades are highly transparent and moved away from the window frame (on the sides and bottom of the shades, there are gaps that allow air to circulate between the window area and the room). In this case, a large part of the heat gains result from solar radiation entering the window space into the room. Figure 10 shows the variability of the cooling demand for cases without and with shades in July.

It should be noted that, due to the accumulation of heat in the building partitions, in the hot periods of the season, the cooling demand did not drop to zero even at night. Therefore, a very unfavourable case represents times when the cooling system is turned off at night (unfortunately, this is a typical situation in the hall). In the morning hours, a very large cooling power is therefore required to ensure 24 °C in the hall. Figure 11 shows the cooling demand on a selected day for such a case in comparison with the case wherein there is a 24-h operation of the system (empty-hall cases). After starting the system, the cooling demand was about three times higher than the average value in the following hours of the day and the maximum value when the system operated continuously around the clock. In practice, this means overheating of the room in the initial hours of its use, because the maximum cooling capacity of the system is limited. The time of overheating will depend on the hall load at the time of system start-up.

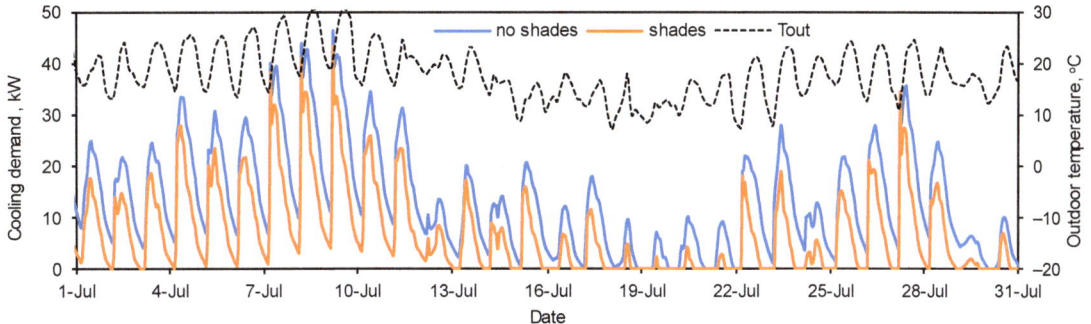

Figure 10. Cooling demand for cases A1 and A2.

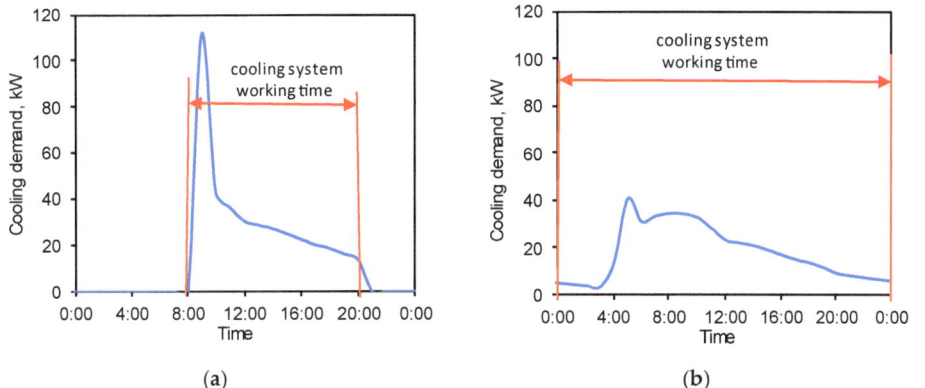

Figure 11. (**a**) Cooling demand on a selected day for the case when the cooling system is turned off at night—case A3—and (**b**) for the case when the system operates around the clock—case A2.

The next case considered was training in the hall without spectators (20 people doing very hard work). In cases where 50% of the lighting was used and without lighting and at the same time without shades on the windows, the maximum cooling demand exceeded two times the maximum cooling power of 27 kW. When using 50% the cooling power shortage occurred for 42% of the summer season; with lights off, it was for 10% of the season. However, it should be verified as to whether the level of natural lighting in the hall (without artificial lighting) is sufficient for performing sports activities.

In the case of a full audience, the cooling demand significantly exceeded the available cooling capacity (up to 4.5 times). With full occupancy and 100% lighting, the hall can be overheated for more than 60% of the summer season (June–September). Figure 12 shows the variability of cooling demand in one day at full load in the hall.

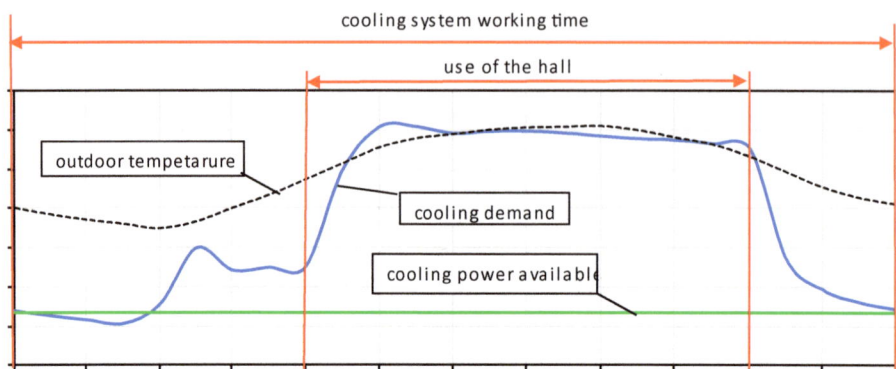

Figure 12. Cooling demand on the selected day (full load of the hall).

Due to the fact that the temperature in the sports hall in the summer should not exceed 21 °C, the question of how much the cooling demand increases in such a case was additionally evaluated. The results for the three selected cases are presented below:

- Empty hall (A2 case): 46.8 kW;
- Training (B1 case): 62.5 kW;
- Full audience (C1 case): 128.0 kW.

When meeting the indoor thermal conditions recommended by the standards (T_i = 21 °C), the cooling demand increased by an additional 5% to 10%, depending on the case of hall load. In summary, the calculated maximum cooling demand significantly exceeded the value specified in the design stage of the building; the value was underestimated by 77%. The efficiency of the designed system has been severely underestimated, as a result of which there are inadequate thermal conditions in the hall. When the hall is fully loaded, the gains from the sun (with windows covered with internal roller shades) account for almost 40% of the cooling demand, which is why the installation of much more effective external shields should be considered.

3.6. Evaluation of the Reduction of the Cooling Demand by the Use of Sun Control Window Films

Simulations of cooling demand in the summer season and heat demand in the winter season were performed to check the impact of the use of sun control window film. An empty hall was analysed without taking into account internal gains, represented by case A1 in Table 3. The assumed temperature set point was 20 °C for heating and 24 °C for cooling.

The main objective of solar control films is to reduce the amount of heat entering the building through the windows. Sunscreen films typically have a thin adhesive layer for optimal transparency and a metallic microlayer that evenly covers the film, reflecting infrared solar radiation and resulting in a significant reduction in solar heat gains. The calculations were based on 3M™ Sun Control window films from the Neutral series [34], which do not change the colour of the glass: RE35NEARL inner film and RE35NEARLXL outer film. Table A1 presents a comparison of the optical properties of the standard glazing unit assumed for the analysis and the same glazing unit with internal or external film. Figures 13 and 14 show the variability of the cooling demand in July and the heating demand in January. The results for case A2 are also presented to reflect the difference between placing shades versus sunscreens on the inside or outside of the windows.

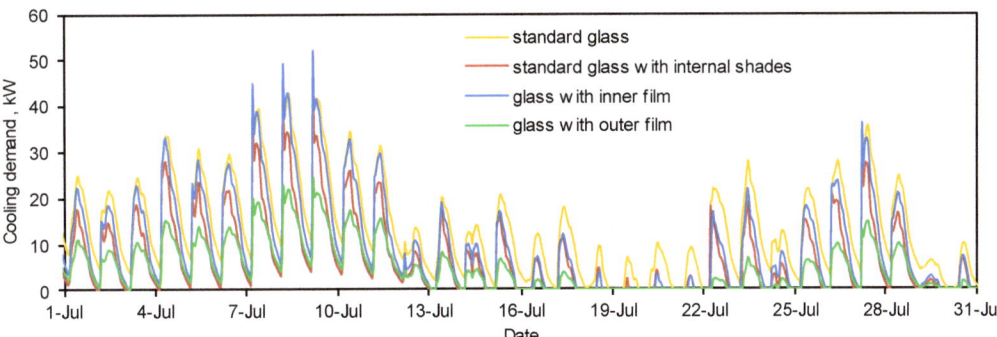

Figure 13. Cooling demand in July (empty hall).

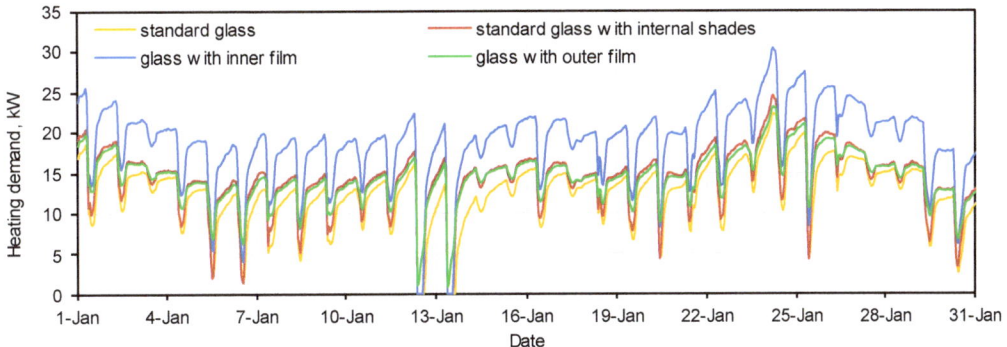

Figure 14. Heating demand in January (empty hall).

The inner film did not meet expectations. Cooling demand in summer was comparable to the "no film" case; the required maximum cooling power may be even higher (Table 4). Additionally, instantaneous heating demand in winter increased significantly in this case. The current solution with blinds on the inside of the window gave better results for both cooling and heating demands. The tested outer film reduced the maximum gains from the sun by almost half, additionally, it slightly affected the instantaneous heating demand in winter.

Table 4. Cooling and heating demand results for cases with film on windows.

Parameter	Standard Glass Assumed in Calculations	Standard Glass with Internal Shades	Glass with Inner Film RE35NEARL	Glass with Outer Film RE35NEARLXL
Maximum heating demand, kW	22.8	24.5	30.4	23.2
Maximum cooling demand, kW	45.8	42.9	51.4	24.1

The methodology for calculating the design heat load in accordance with the EN 12831-1:2017 standard [35] applicable in Poland does not take into account internal and external heat gains. Assuming that the calculations of design heat load in the tested hall were made correctly, it can be assumed that, regardless of the sun-control film used, the heating system should provide the required energy. However, this may affect the annual heat consumption of the building.

4. Conclusions

Based on the inspection and diagnosis performed on the air conditioning system in the sports hall, the following problems were identified:

- The performed measurements confirmed that, during periods when the outdoor air temperature exceeded 28 °C, the air temperature in the hall exceeded the design temperature of 24 °C to 26 °C even with no or negligible internal heat gains. The currently installed cooler used has insufficient cooling capacity. The sensible heat gains assumed at the design stage amounted to 69 kW, while the identified sensible heat flux removed from the hall was approximately 2.6 times lower. This means that the ventilation airflow of 11,500 m^3/h at the temperature of 18 °C cannot ensure the air temperature in the hall at the level of the design assumptions;
- The thermal conditions in all occupied zones in the hall were uniform. The temperature values recorded at various points in the hall were similar and the differences did not exceed 1 °C. This means that the system used does not ensure the required temperature in different zones of the hall. In this type of facility, in summer the temperature in the field zone should be 6–7 °C lower than in the spectator zone;
- The total and sensible power of the cooler varied greatly depending on the parameters (temperature and humidity) of the outdoor air. These changes, however, did not have a significant impact on the operation of the cooling system, i.e., the sensible heat flux removed from the hall by the ventilation air. The maximum values of the total and sensible power of the cooler (approximately 60 kW, and 42 kW respectively) were lower by approximately 4.2 kW, which is 20% of the values adopted for the selection of the cooler (74.5 kW and 55.0 kW). The reason for this difference is that the ventilation airflow rate is lower by approximately 18% than the design value;
- It is not possible to remove heat gains from the room with the current airflow of ventilation (which was confirmed based on measurement data and calculations). Air flow would have to be more than four times greater. It is not possible to supply this amount of air with the existing ventilation system. The expansion of the air system would be ineffective, and an additional (cooperating) cooling system is required;
- The cooling demand was underestimated at the design stage. The value calculated on the basis of computer simulation significantly exceeds (by 75%) the value estimated at the design stage. Simplified design methods can lead to significant calculation errors and thus problems later in the building operation. A very large part of the heat gains in the hall (approximately 40%) are solar gains due to the large windows. Unfortunately, the internal roller blinds used do not significantly reduce solar gains;
- The cooling system is turned off at night, adding to the problem of cooling during the day.

In order to solve the problems with cooling the main hall, the following actions are proposed:

- In the hottest periods of the summer season, it is recommended to operate the cooling system continuously or at least 24 h before using the room;
- Additionally, the introduction of night cooling on these nights, when there is a significant drop in the outdoor air temperature, is recommended. High-efficiency exhaust fans should ensure intensive air exchange, with outside air inflow through opening hatches or windows. For this purpose, additional exhaust fans (e.g., roof fans) must be installed. Fans should be installed at a certain distance from the opened windows, to ensure intensive mixing and air exchange throughout the hall. The operation of the night cooling system should be automatically controlled. The opening of windows at night should be coupled with the operation of the fans;
- The introduction of external sun shields controlled by a solar radiation sensor. Consideration should also be given to the use of more effective sunscreens on the windows, for example, the window films tested in this study that were glued to the outside of the glass;

- The replacement of light sources with energy-saving sources, e.g., LED ones, to reduce internal heat gains;
- The use of an additional system based on cooling the internal air in recirculation mode should be considered. This is because, in buildings with high internal heat gains, all-air cooling systems are ineffective, as they require very large volumes of ventilation air. Air distribution in the room is likely to be problematic in this case due to the risk of drafts. The existing ventilation system in the hall ensures the required amount of fresh air defined by hygiene requirements. For this purpose, it is possible to use, for example, fan coil units supplied with coolant from an additional cooling unit, or multi-split units with direct evaporation of the refrigerant. The selection of devices should be based on simulation calculations of the hall's heat loads. The additional air conditioning system should cooperate with the existing air system, e.g., during periods of high external air temperature and/or with high internal heat gains.

Author Contributions: Conceptualization, M.H., J.F.-G. and J.K.; methodology, M.H., J.F.-G., J.K. and P.K.; software, J.F.-G.; formal analysis, M.H. and J.F.-G.; data curation, P.K. and M.H.; writing—original draft preparation, M.H. and J.F.-G.; writing—review and editing, J.K. and P.K.; supervision, J.F.-G. All authors have read and agreed to the published version of the manuscript.

Funding: This research received no external funding.

Data Availability Statement: Not applicable.

Acknowledgments: The work supported by the Polish Ministry of Education and Science within the research subsidy.

Conflicts of Interest: The authors declare no conflict of interest.

Appendix A

Table A1. Optical properties of glazing units.

Parameter	Description	Standard Glass Assumed in Calculations	Glass with Inner Film RE35NEARL	Glass with Outer Film RE35NEARLXL
SHGC	Solar heat gain coefficient	0.623	0.552	0.362
Tvis	Visible light transmission of the glazing system	0.803	0.333	0.370
Rfvis	Front surface reflectance of the glazing system	0.124	0.247	0.145
Rbvis	Back surface reflectivity of the glazing system	0.128	0.190	0.195
Tsol	Solar transmission of the glazing system	0.539	0.268	0.254
Rfsol	Front surface solar reflectance of the glazing system	0.280	0.216	0.181
Rbsol	Back surface solar reflectance of the glazing system	0.267	0.203	0.305
Abs1	Solar absorptance for layer 1	0.093	0.081	0.522
Abs2	Solar absorptance for layer 2	0.089	0.436	0.043

References

1. Ngarambe, J.; Yun, G.Y.; Santamouris, M. The use of artificial intelligence (AI) methods in the prediction of thermal comfort in buildings: Energy implications of AI-based thermal comfort controls. *Energy Build.* **2020**, *211*, 109807. [CrossRef]
2. Moran, P.; Goggins, J.; Hajdukiewicz, M. Super-insulate or use renewable technology? Life cycle cost, energy and global warming potential analysis of nearly zero energy buildings (NZEB) in a temperate oceanic climate. *Energy Build.* **2017**, *139*, 590–607. [CrossRef]
3. Popiolek, Z. Comprehensive Onsite Thermal Diagnostics of Buildings in Practice, Vol. 5 of the Guidebook of Thermal Diagnostics of Buildings, Gliwice, Politechnika Śląska. *Wydział Inżynierii Sr. Energetyki* **2013**, *447*. (In Polish).
4. Hurnik, M.; Specjal, A.; Popiolek, Z.; Kierat, W. Assessment of Single-Family House Thermal Renovation Based on Comprehensive on-Site Diagnostics. *Energy Build.* **2018**, *158*, 162–171. [CrossRef]
5. Specjal, A.; Lipczyńska, A.; Hurnik, M.; Król, M.; Palmowska, A.; Popiolek, Z. Case Study of Thermal Diagnostics of Single-Family House in Temperate Climate. *Energies* **2019**, *12*, 4549. [CrossRef]
6. Blaszczok, M.; Król, M.; Hurnik, M. On-Site Diagnostics of the Mechanical Ventilation in Office Buildings. *Archit. Civ. Eng. Environ.* **2017**, *10*, 145–156. [CrossRef]
7. Hurnik, M.; Blaszczok, M.; Król, M. On-Site Thermal Diagnostics of Cooling Sources for Air Conditioning Systems in Office Buildings. *Archit. Civ. Eng. Environ.* **2017**, *10*, 157–163. [CrossRef]
8. Heiselberg, P.K. *Ventilation of Large Spaces in Buildings: Analysis and Prediction Techniques*; Department of Mechanical Engineering, Aalborg University: Aalborg, Denmark, 1998.
9. Tolis, E.I.; Panaras, G.; Douklias, E.; Ouranos, N.; Bartzis, J.G. Air Quality Measurements in a Medium Scale Athletic Hall: Diurnal and I/O Ratio Analysis. *FEB-Fresenius Environ. Bull.* **2019**, *28*, 658.
10. Accili, A.; Ortiz, J.; Salom, J. *Energy Strategies to NZEB Sports Hall*; Palma de Mallorca: Barcelona, Spain, 2016.
11. Salonen, H.; Salthammer, T.; Morawska, L. Human Exposure to Air Contaminants in Sports Environments. *Indoor Air* **2020**, *30*, 1109–1129. [CrossRef]
12. Szczepanik-Ścislo, N. Indoor Air Quality and Control Methods for Mechanical Ventilation Systems inside Large Passive Objects. *Tech. Trans.* **2020**, *117*, e2020001. [CrossRef]
13. Ilieș, D.C.; Buhaș, R.; Ilieș, A.; Gaceu, O.; Onet, A.; Buhaș, S.; Rahotă, D.; Dragoș, P.; Baiaș, Ș.; Marcu, F. Indoor Air Quality Issues. Case Study: The Multipurpose Sports Hall of the University of Oradea. *Environ. Eng. Manag. J.* **2018**, *17*, 2999–3005.
14. Seduikyte, L.; Stasiulienė, L.; Prasauskas, T.; Martuzevičius, D.; Černeckienė, J.; Ždankus, T.; Dobravalskis, M.; Fokaides, P. Field Measurements and Numerical Simulation for the Definition of the Thermal Stratification and Ventilation Performance in a Mechanically Ventilated Sports Hall. *Energies* **2019**, *12*, 2243. [CrossRef]
15. Szulc, J.; Cichowicz, R.; Gutarowski, M.; Okrasa, M.; Gutarowska, B. Assessment of Dust, Chemical, Microbiological Pollutions and Microclimatic Parameters of Indoor Air in Sports Facilities. *Int. J. Environ. Res. Public Health* **2023**, *20*, 1551. [CrossRef] [PubMed]
16. Ma, Y.; Lin, S.; Liu, L.; Pan, Z.; Chen, H.; Peng, Q. Investigation on Indoor Air Quality in the Badminton Hall of Wuhan Sports University in winter Based on Subjective Questionnaire Survey and Field Test. *Proceedings* **2020**, *49*, 148.
17. Huang, X.; Chen, G.; Zhao, C.; Peng, Y.; Guo, W. Post Occupancy Evaluation of Indoor Environmental Quality of Sports Buildings at Hot and Humid Climate from the Perspective of Exercisers. *Build. Environ.* **2022**, *226*, 109760. [CrossRef]
18. Lenart-Boroń, A.; Drab, D.; Chrobak, J. Microbiological Aerosol, Particulate Matter Concentrations and Antibiotic Resistant Staphylococcus spp. in the Premises of Poland's Oldest Agricultural School. *Atmosphere* **2021**, *12*, 93. [CrossRef]
19. Bralewska, K.; Rogula-Kozłowska, W.; Bralewski, A. Size-Segregated Particulate Matter in a Selected Sports Facility in Poland. *Sustainability* **2019**, *11*, 6911. [CrossRef]
20. Fantozzi, F.; Lamberti, G. Determination of Thermal Comfort in Indoor Sport Facilities Located in Moderate Environments: An Overview. *Atmosphere* **2019**, *10*, 769. [CrossRef]
21. Dudzińska, A.; Kisilewicz, T. Efficiency of Night Ventilation in Limiting the Overheating of Passive Sports Hall. In *MATEC Web of Conferences*; EDP Sciences: Les Ulis, France, 2020; Volume 322, p. 01031.
22. Kisilewicz, T.; Dudzińska, A. Summer Overheating of a Passive Sports Hall Building. *Arch. Civ. Mech. Eng.* **2015**, *15*, 1193–1201. [CrossRef]
23. Seppᵃnen, O. Ventilation Strategies for Good Indoor Air Quality and Energy Efficiency. *Int. J. Vent.* **2008**, *6*, 297–306.
24. Dudzińska, A. Efficiency of Solar Shading Devices to Improve Thermal Comfort in a Sports Hall. *Energies* **2021**, *14*, 3535. [CrossRef]
25. Gürlich, D.; Reber, A.; Biesinger, A.; Eicker, U. Daylight Performance of a Translucent Textile Membrane Roof with Thermal Insulation. *Buildings* **2018**, *8*, 118. [CrossRef]
26. Spunei, E.; Frumușanu, N.-M.; Măran, G.; Martin, M. Technical–Economic Analysis of the Solutions for the Modernization of Lighting Systems. *Sustainability* **2022**, *14*, 5252. [CrossRef]
27. Koper, P. Performance Assessment of Air Conditioning Installation in Multifunctional Sports Hall Using CFD Simulations. *Archit. Civ. Eng. Environ.* **2016**, *9*, 123–134. [CrossRef]
28. Rajagopalan, P.; Elkadi, H. Thermal and Ventilation Performance of a Multifunctional Sports Hall within an Aquatic Centre. In Proceedings of the Building Simulation, Sydney, Australia, 14–16 November 2011.

29. Mitchell, R.; Kohler, C.; Curcija, D.; Zhu, L.; Vidanovic, S.; Czarnecki, S.; US Department of Energy. *Engineering Reference, EnergyPlusTM Version 9.4.0 Documentation*; US Department of Energy: Washington, DC, USA, 2020. Available online: https://energyplus.net/Sites/All/Modules/Custom/Nrel_custom/Pdfs/Pdfs_v9.4.0/EngineeringReference.Pdf (accessed on 15 January 2023).
30. Typical Meteorological and Statistical Climatic Data for Energy Calculations of Buildings. Available online: https://dane.gov.pl/en/dataset/797,typowe-lata-meteorologiczne-i-statystyczne-dane-klimatyczne-dla-obszaru-polski-do-obliczen-energetycznych-budynkow (accessed on 19 February 2023).
31. *ANSI/ASHRAE Standard 55*; Thermal Environmental Conditions for Human Occupancy. American Society of Heating, Refrigerating and Air-Conditioning Engineers (ASHRAE): Atlanta, GA, USA, 2017.
32. Mitchell, R.; Kohler, C.; Curcija, D.; Zhu, L.; Vidanovic, S.; Czarnecki, S.; Arasteh, D. WINDOW 7 User Manual. Lawrence Berkeley National Laboratory, March 2019. Available online: https://Windows.Lbl.Gov/Tools/Window/Documentation (accessed on 14 December 2022).
33. *ISO 7730:2005*; Ergonomics of the Thermal Environment—Analytical Determination and Interpretation of Thermal Comfort using Calculation of the PMV and PPD Indices and Local Thermal Comfort Criteria. International Organization for Standardization: Geneva, Switzerland, 2005.
34. 3MTM Sun Control Window Film Neutral Series. Available online: https://www.3m.com/3M/en_US/p/d/b00016672/ (accessed on 19 February 2023).
35. *EN 12831-1:2017*; Energy Performance of Buildings—Method for Calculation of the Design Heat Load—Part 1: Space Heating Load, Module M3-3. European Committee for Standardization: Brussels, Belgium, 2017.

Disclaimer/Publisher's Note: The statements, opinions and data contained in all publications are solely those of the individual author(s) and contributor(s) and not of MDPI and/or the editor(s). MDPI and/or the editor(s) disclaim responsibility for any injury to people or property resulting from any ideas, methods, instructions or products referred to in the content.

Article

A Comparison of the Project Management Methodologies PRINCE2 and PMBOK in Managing Repetitive Construction Projects

Aivaras Simonaitis [1], Mindaugas Daukšys [1,*] and Jūratė Mockienė [1,2]

[1] Faculty of Civil Engineering and Architecture, Kaunas University of Technology, 44249 Kaunas, Lithuania; aivaras.simonaitis@gmail.com (A.S.); jurate.mockiene@ktu.lt (J.M.)
[2] Faculty of Engineering Sciences, Kaunas University of Applied Engineering Sciences, 50155 Kaunas, Lithuania
* Correspondence: mindaugas.dauksys@ktu.lt

Abstract: Nowadays, companies employ various project management (PM) methodologies to ensure that their projects are effective and successful. It is worth knowing that differences in principles and processes of PM methodologies influence the use of different PMs in managing non-repetitive and repetitive construction projects. This paper presents the selection and application of a rational construction PM methodology to a repetitive construction project after a comparison of two PM methodologies, namely Project Management Body of Knowledge (PMBOK) and Projects IN Controlled Environments (PRINCE2). The object of this study is a repetitive anti-corrosion works project for steel structures conducted at Company X. The research was carried out in two steps. First, a quantitative survey of the respondents from companies involved in the management and execution of construction projects was conducted with the aim to identify a rational approach to construction PM. The questionnaire consisted of fourteen closed-ended questions, six of which were generic and eight were PMBOK- and PRINCE2-specific questions. Companies that took part in the quantitative study identified the PRINCE2 project management approach as the most suitable for managing a repetitive construction project. Using the PRINCE2 PM methodology, the repetitive construction project would aim to provide as much information as possible to the project participants, form a team and assign team leaders responsible for the phases, establish a financial plan, a detailed timetable for the execution of the works, a quality control plan, and a plan of responsible persons, and detail the technological sequencing of the works. Second, a quantitative study on the selection of a rational construction project management approach for a repetitive construction project was pursued, and a qualitative assessment of construction project monitoring trends and actions was conducted. The qualitative research was performed using a structured interview method and asking the representatives of different companies X, Y, and Z the same 15 questions. The results of the qualitative research showed that a successful PM depends on the size of the project team, the PM tools and methodologies used, the PM philosophy, and the frequency of monitoring and discussing the project progress.

Keywords: project management methodologies; PRINCE2; PMBOK; repetitive construction projects

Citation: Simonaitis, A.; Daukšys, M.; Mockienė, J. A Comparison of the Project Management Methodologies PRINCE2 and PMBOK in Managing Repetitive Construction Projects. *Buildings* **2023**, *13*, 1796. https://doi.org/10.3390/buildings13071796

Academic Editor: Pramen P. Shrestha

Received: 31 May 2023
Revised: 5 July 2023
Accepted: 12 July 2023
Published: 14 July 2023

Copyright: © 2023 by the authors. Licensee MDPI, Basel, Switzerland. This article is an open access article distributed under the terms and conditions of the Creative Commons Attribution (CC BY) license (https://creativecommons.org/licenses/by/4.0/).

1. Introduction

Why do construction projects take longer to implement than they should? Why do they fail to deliver the desired results on time and to ensure sustainability? The analysis of project management (PM) experience from different sectors has revealed such aspects as mistakes made by project team members, changes in law, unclear roles and responsibilities, ineffective communication, problems with suppliers, and the sustainability of the results. These factors do not really answer the questions raised above. Apparently, there are general laws of PM and challenges that project managers (PJMs) have to handle.

In the digital economy, business and governmental organisations design and implement projects of all types and sizes. However, not all project activities run smoothly. Despite the availability of financial resources and competent PM teams, projects are not always completed on time, they often exceed the budget, they do not meet quality and time requirements, and they fail to meet sustainability goals. What does this mean? If a PJM sets a deadline, the staff usually report the work carried out when the deadline comes, even if the work was finished earlier. What are the factors behind this? The project team may try to avoid additional tasks because next time new tasks may have shorter deadlines and less-inspired staff members may be late in completing the planned goals. Poor planning and high urgency can lead to project failure or the need for additional resources due to the changes in scope, lack of quality, and weak sustainability. Project planning is a crucial stage. It is inappropriate to start implementing ideas without a good-quality plan.

The benefits of a good PM methodology include the easy handling and distribution of project reports, transparent management practices, successful risk management, effective problem-solving, easy measurement of the completion percentage, improved control and command of the project, regulated stakeholder inventories, the measurement of accomplishments against plans, improved estimation for future planning, and identifying objectives that cannot be met or will be exceeded [1]. PM is employed to ensure the efficient and effective organisation of a project from its initiation to its completion [2–5]. It also encompasses all strategies and activities that ensure the project's success based on the quality, cost, and timeliness triangle. In this case, standard PM frameworks that include PMBOK (Project Management Body of Knowledge), PRINCE2 (Projects IN Controlled Environments), and the international standard ISO 21500:2012 are used for project management [6,7]. Both the PMBOK and the PRINCE2 PM methodologies have positive and improvable areas. According to popularity in regions, the PMBOK PM methodology is more widespread in North America, while the PRINCE2 management methodology is more popular in Europe, as it originated in the United Kingdom [4,5]. The PMBOK methodology represents the collective knowledge that is widely accepted as best practice in the field of PM, while PRINCE2 is focused on the business aspects of the project and emphasises a structured organisational approach to PM with an emphasis on dividing the project into manageable stages and a product-oriented approach to planning [8,9]. The above-mentioned standard PM frameworks are popular with scientists who explore non-repetitive construction projects.

Repetitive construction projects refer to projects that involve the construction of recurring units, where each unit consists of the same group of sequential activities. These projects often occur in the construction industry, such as building multiple identical houses, constructing similar sections of a highway, or conducting similar anti-corrosion works for steel structures. The repetitive nature of these projects offers several advantages including significant time and cost savings. By working on similar units with the same sequence of activities, construction crews and resources can maintain continuity and efficiency. Once they become familiar with the process, they can work more quickly and effectively, reducing the overall construction time [10–12]. Maintaining work continuity in repetitive construction projects is crucial for achieving time and cost savings, and it can be achieved through various means:

- By keeping a consistent workforce throughout the project, there is no need for the frequent hiring and firing of labour. This reduces recruitment and training costs as well as the time required to onboard new workers. Skilled workers who are familiar with the project's requirements can continue working on subsequent units, ensuring higher productivity and efficiency.
- In repetitive projects, skilled labour becomes more proficient and experienced over time. By retaining these skilled workers, their expertise and knowledge are preserved, leading to improved performance and quality. They become familiar with the project's specific requirements and can work more efficiently, minimising errors and rework.
- In repetitive projects, the use of equipment can be optimised since the same activities are repeated. By maintaining work continuity, equipment idle time is minimised as

well. Equipment can be kept operational and utilised efficiently without long periods of downtime between units. This reduces equipment-related costs and increases overall productivity.
- As workers become more experienced with the sequential activities involved in the project, they can perform their tasks more quickly and accurately. This leads to improved productivity and reduced construction time with each successive unit.

On the other hand, maintaining work continuity in repetitive projects indeed presents an additional constraint that traditional scheduling and planning tools and techniques struggle to address effectively [13–15]. The following negative aspects can be mentioned:
- Traditional scheduling tools are often rigid and not well-suited to adapt to the specific requirements of repetitive projects. These methods typically assume a high level of task variability, making it difficult to account for the repetitive nature of the project and the streamlined workflow it entails.
- Repetitive projects demand efficient resource allocation and management to maintain work continuity. However, traditional tools may not adequately account for the optimisation of crew allocation, equipment utilisation, and material flow. This can lead to suboptimal resource allocation, increased idle time, and reduced productivity.
- Traditional planning and scheduling techniques often overlook or underestimate the learning effect, resulting in unrealistic timelines and cost projections.
- While repetitive projects consist of recurring units, there may still be variations in design, site conditions, or other factors. Traditional tools may struggle to handle these variations effectively, leading to challenges in maintaining the desired work continuity and achieving accurate project planning.

The purpose of standard PM frameworks is to establish clear project objectives, prioritise quality, enhance communication, and provide professionals with the necessary project management tools. As a result, it is not clear which PM methodology—PRINCE2 or PMBOK—would be the most rational and efficient for implementing repetitive construction projects. Therefore, the purpose of this study is to determine which PM methodology is more effective for repetitive construction projects. In order to identify a rational methodology for construction PM, an analytical quantitative survey of respondents from companies involved in the management and execution of construction projects was conducted. In addition, a descriptive qualitative study was performed in order to identify the trends and actions of monitoring the progress of construction projects.

2. Background

It is known that PM combines knowledge, skills, tools, and techniques to efficiently manage project activities and meet stakeholders' expectations [1]. It involves nine knowledge areas and enables project PJMs to ensure that projects are conducted rationally and efficiently. PM techniques are primarily applied in the planning and control of time, cost, and quality to achieve success in different projects. However, there is a lack of precision in distinguishing between the P and PM. The overlapping definitions of the two can potentially impact their relationship. Therefore, understanding the difference between the P and the PM can increase the likelihood of a project's success, as concluded by the authors of [9].

The rising stakeholder expectations demand for integrating all PM activities within an organisation. This integration is known as organisational project management (OPM), which combines organisation and PM. The foundation of organisation theory is based on organisational structure, forms, and the concept of integration. PM is linked to organisations through the use of projects as platforms for improving business, implementing changes, fostering innovation, and gaining a competitive edge, as explained by the authors of [16]. In their research, Aubry et al. [17] highlighted a growing interest in social perspectives that take politics, organisational dynamics, paradoxes, and pluralism into consideration. This presents an opportunity for PM scholars to contribute to management and organisational theory. The premise of the authors of [18] is that the project management office (PMO) is an integral part of a complex network that connects strategy, projects, and structures. Thus, it provides a point of entry to the organisation for studying the fundamentals of OPM. The proposed theoretical framework refers to three complementary fields, namely innovation, sociology, and organisational theory, to provide a novel understanding of the PMO and OPM. Other researchers [19] suggest that robust management structures, particularly PMOs, can enhance the oversight capabilities of contractors when implementing sustainable procurement management (SPM) processes.

Numerous projects may encounter significant delays that exceed their initial time and cost estimates. Construction delays can be attributed to various parties including the owner, contractor, and other involved parties. Thus, it is crucial to identify the responsibility for delays among these parties, and it is necessary to understand the causes and types of delays [20–26]. In this case, different approaches have been offered by scientists to decrease the likelihood of delays in construction [24,25,27]. An important aspect of the construction industry is the implementation of PM success methods. The authors of [28–32] in their research identified factors that influence a construction project's success. A summary of studies that analyse the causes of delays in construction projects, approaches to decrease the likelihood of delays in construction, and the success of PM implementation methodologies is presented in Table 1.

Table 1. Summary of studies that analyse the causes of delays in construction projects, approaches to decrease the likelihood of delays in construction, and the success of PM implementation methodologies.

Authors	Major Aspects
	Causes of delays in construction projects
Aubry et al. [21]	Relationship between organisational culture and the extent of delays.
Sweis et al. [22]	Financial difficulties experienced by the contractor and excessive change orders requested by the owner.
Arditi and Pattanakitchamroon [23]	Availability of scheduling data, analyst's familiarity with the project software capabilities, clear specifications in the contract regarding concurrent delays, and float ownership.
Kim et al. [24]	Insufficient consideration of concurrent delays and inadequate consideration of time-compressed activities.
Gunduz et al. [25]	Quantification of the likelihood of delays in construction projects before the bidding stage.
Mahamid [26]	Payment delays, poor labour productivity, lack of skilled personnel, frequent change or-ders, and rework.
Arantes and Ferreira [33]	Late progress payments by the owner to the contractor, slow decision-making by the owner, owner interference, increase in the scope of the works, modifications of orders, inappropriate planning and scheduling, errors and discrepancies in drawings, contractors' financial difficulties, late delivery of materials, changes to the specifications of materials during construction, late procurement of materials, bidding and contract award process, impracticable schedule and specifications in the contract, deficient communication between parties, disputes and negotiations between the parties, and late permits from authorities.
	Approaches to decrease the likelihood of delay in construction
Kim et al. [24]	Delay Analysis Method Using Delay Section (DAMUDS).
Gunduz et al. [25]	Relative Importance Index (RII) methodology with fuzzy logic integrated.
Shi et al. [27]	The methodology involves using a series of equations that can be quickly implemented into a computer program and provide rapid access to project delay data and activity contributions.
Arantes and Ferreira [33]	ISM-MICMAC analysis methodology to support the development of delay mitigation measures (DMMs) in construction projects.
	Causes of PM implementation methodologies success
Gudienė et al. [28]	Factors influencing a construction project's success: external factors, institutional factors, project-related factors, PM-/team member-related factors, project-manager-related factors, client-related factors, and contractor-related factors
Radujkovič and Sjekavica [29]	Competent project manager (PJM), a competent team, good coordination between the manager and the team, an adequate organisational structure, culture, atmosphere, and competence, as well as a high usage of PM methodologies, methods, tools, and techniques
Radujkovič and Sjekavica [30]	Continuous development of competencies and improvement of management methodologies
Greenwood and Miller [31]	The organisational components of management activities have to include meeting the initially set deadlines and costs, making more efficient use of resources, adopting an appropriate management style, facilitating communication among the participants, and ensuring stakeholder satisfaction, with a particular focus on the project owner
Ingle and Mahesh [32]	Customer relations, safety, schedule, cost, quality, productivity, finance, communication and collaboration, environment, and stakeholder satisfaction.

The PM process typically involves establishing a business plan for the project, preparing an opportunity statement that aligns with the management's strategy, defining a business model for the project, and identifying potential risks in advance [2–5]. The Project Management Institute has developed the PMBOK methodology as a guide to ensure a standardised set of principles and knowledge in the field of PM [8]. The PMBOK methodology comprises a framework of nine knowledge areas, which are divided into activities across five stages of the project life cycle [8]. It was approved by the American National Standards Institute (ANSI) as early as 1998, and the sixth standard version was released in 2017 [4,5]. Another widely used methodology is PRINCE2, which was developed by the Central Computer and Telecommunications Agency in 1989. PRINCE2 is focused on the business aspects of a project and emphasises a structured organisational approach to PM [8,9]. At the request of the UK government, the PRINCE2 PM methodology was released in 1996, and the sixth version of the methodology was released in 2017, which is applicable to projects to this day [4,5]. The release of the seventh edition of PMBOK in 2021 brought about a significant transformation in the approach to project management [7]. In this latest edition, the previously detailed process and group-based view of project management have been replaced by a comprehensive principle-based perspective. This shift has broadened the applicability of the PMBOK document, making it relevant and adaptable to all types of projects.

Matos et al. [34] explained that while PMBOK defines a project as a temporary endeavour aimed at creating a unique product, service, or result, PRINCE2 defines a project as a management environment created to deliver one or more business products according to a specified business case. In order to provide an integrative approach in respect to business stakeholders and openness on the international scale, the authors of [6] combined the two PM standards (PMBOK and PRINCE2) and proposed a new hybrid approach to PM. A comparative analysis of the controls of PRINCE2 and PMBOK is presented in Table 2.

Table 2. Comparative analysis of controls of PRINCE2 and PMBOK.

PRINCE2	PMBOK
Seven principles (continued business justification, learning from experience, defined roles and responsibilities, manage by stages, manage by exception, focus on products, and tailored to the project environment) [3,4,8].	There is no comparative analysis of controls with PRINCE2 [4].
Seven themes (business case, organisation, quality, planning, risk, change, and progress) [3,4,8].	Ten knowledge areas (project integration, scope, time, cost, quality, resources, communication, risks, procurement, and stakeholder management) [2,4].
Seven processes (project initiation, project planning, project control, stage boundaries, product delivery management, project closure, and project monitoring) [3,4,8].	Five process groups (initiating, planning, executing, monitoring and controlling, and closing) [2,4].
Forty-one activities (for example, in the project initiation process, work on the project summary and business case refinement is a repetitive activity, and discussions and document refinement should occur continuously [3,4,8].	Forty-nine processes (grouped by process groups and knowledge areas, such as project integration management and, creating a PM plan in the planning process group) [2,4].
Forty specified tools and techniques [3,4,8].	One hundred and thirty-two specified tools and methodologies [2,4].

While PMBOK and PRINCE2 have similar control definitions, there are many differences between the two PM methodologies. One PM methodology may have a more-detailed process, while another may have a less-detailed one. PMBOK can be identified as a methodology that comprehensively covers such knowledge areas: project integration management, project scope management, project time management, project cost management, project quality management, project resource management, project communication management, project risk management, project procurement management, and project stakeholder management [4]. Meanwhile, the PRINCE2 PM themes cover the following topics: business case, changes, and progress; planning and progress; quality; planning, organization; risk; undefined; organisation [4]. It is seen that the PRINCE2 PM themes cover all the same topics apart from one topic—procurement management, which is not covered by PRINCE2 PM.

Another advantage of the PRINCE2 methodology is the integration of processes with the seven themes. The integration of processes and themes creates a methodology that can be applied widely to manage projects of any type. The author [4] analysed the seven PRINCE2 PM processes and the process model that includes the PMBOK PM methodology. It was stated that the PRINCE2 PM processes cover the following four management levels: corporate or programme management, directing (project board), managing (PJM), and delivering (team manager). Meanwhile, the PMBOK methodology prioritises the level of managing for which the PJM is responsible.

The drawbacks of the PMBOK PM methodology include a lack of responsibility for PM team members and overly detailed descriptions of certain aspects. The PJM plays the main role in the PMBOK methodology, with the primary responsibilities being assigned to them, but the responsibilities of other project team members have been defined ambiguously [4]. The main drawback of the PRINCE2 PM methodology is the limited selection, only 40, of tools and methodologies. Although the PRINCE2 PM methodology does not limit the use of best practice tools and methodologies from external sources, in comparison, PMBOK has 132 integrated tools and methodologies. Additionally, each project team member must know their tasks and responsibilities precisely in the PRINCE2 methodology, and if at least one team member does not follow the PRINCE2 methodology process and flow, the project can become uncontrollable. According to the PBMOK methodology, the schedule baseline serves as a fundamental component of delay analysis in projects [35]. The schedule baseline refers to the authorised project timeline, against which actual dates and modifications must be compared to assess schedule delays in the project model. When updating the project schedule, it is crucial to retain accurate data on project time performance. Any alteration to the critical path within the schedule baseline results in delays. In their study, the authors of [36] put forward a risk management framework that refers to the PMBOK standards. This framework aims to assist in the selection of appropriate risk response strategies for addressing a specific case study within a construction company. To handle the complexity of the proposed model, different state-of-the-art metaheuristic algorithms were employed. Applying the principles of the PMBOK methodology to identify sources of health, safety, and environmental (HSE) risks and utilising a fuzzy analytic hierarchy process can enhance the accuracy of risk assessments for hazards in construction projects [37]. This comprehensive approach enables a more realistic estimation of the risk index, thereby improving the overall understanding and management of HSE risks in construction projects.

Hence, while project management standards may have their unique features, they generally encompass elements such as terminology, areas of knowledge, an administration system, the project life cycle, and alignment with organisational objectives. These components provide guidance and a common framework for professionals to effectively manage projects [7].

3. Research Object and Methodology

This part presents the project under analysis on anti-corrosion works for steel frames of gas distribution stations at Company X, the description project management model at Company X, the identification of problems related to project management at Company X by conducting repetitive construction projects, and the aim and progress of execution of quantitative and qualitative research.

3.1. The Repetitive Construction Project

The project under analysis was an anti-corrosion works project for steel frames of gas distribution stations. The company that carried out the anti-corrosion work on gas distribution stations and allowed to use of the obtained data for the research was called

Company X. The project consisted of abrasive cleaning by sandblasting and anti-corrosion painting of nine gas distribution stations. The total value of the anti-corrosion works for the gas distribution stations was EUR 92.565 including value-added tax.

The planned duration of the project was 23 working days, and the maximum number of staff required during the project was 17. The number of staff depended on the number of gas distribution stations running in parallel. After the welding of the gas distribution stations, the metal structures were transported to the sandblasting chamber, and after the sandblasting, the structures were transported to the anti-corrosive painting chamber. According to the size of the metal structures, the sandblasting work was planned to be carried out within a maximum of 2 working days (4 shifts), and the anti-corrosive painting work was planned to be carried out within 5 working days (10 shifts) in order to meet the specified deadlines. The two-shift front end was more than half of the planned project duration.

Depending on the type of gas distribution station (three different types in total), the metal structures were painted with three to five different anti-corrosive painting systems. The surfaces of the metal structures were painted with the paint systems appropriate for the operating characteristics of the distribution station. Not only did each painting system have different paint dry film thicknesses (ranging from 50 to 150 μm) and several intermediate layers (ranging from 2 to 4) but also colour variants (grey, yellow, or red) depending on the type of gas distribution station (Figure 1).

(a)

(b)

Figure 1. Examples of a steel frame of gas distribution stations: (**a**) station type no. 1; (**b**) station type no. 3.

The anti-corrosion coating was subject to high-quality requirements. Before the transportation of the structures, the work carried out for the installation of the equipment was handed over to the FROSIO (nor. Faglig Råd for Opplæring og Sertifisering av Inspektører innen Overflatebehandling) inspector responsible for the condition of the welded metal structures, the surface preparation, and the anti-corrosion coating. At the time of the intermediate acceptance, the FROSIO inspector had to be provided with the gas distribution station's performance documentation, such as measurements of the sub-surface preparation of the metallic structures, dustiness, salt content, and adhesion of the paint coating. In addition, intermediate-temperature and dry-coating measurements had to be carried out in accordance with the requirements of the project's technical specifications.

The sandblasting and anti-corrosive painting works were performed according to technical documents and standards that governed the installation technology. During the sandblasting and painting operations, it was necessary to ensure intermediate quality control by measuring the number of soluble contaminants present on the surface to be sandblasted, ensuring the quality of the abrasive used, ensuring the correct air temperature during sandblasting, anti-corrosive painting, etc. Thus, the project specified the quality requirements, the criteria to be met, and their values. The quality control plan for the anti-corrosion work is presented in Table 3.

Table 3. Quality control plan for anti-corrosion work.

No.	Name of Quality-Related Activity	Relevant Standard	Satisfactory Criterion/Value
1. Before sandblasting			
1.1	Steel surface preparation (welds and imperfections).	ISO 8501-3:2006 [38]	P2
1.2	Determination of the presence of soluble contaminants on the surface.	ISO 8502-6:2020 [39]/ISO 8503-5:2017 [40]	3 µg/cm
2. During sandblasting			
2.1	Quality of the surface to be cleaned before sandblasting.	SSPC-SP1 [41]	Oil- and grease-free surface.
2.2	Blast cleaning abrasives control: conductivity and lubricant contamination.	ASTM D4940-15:2020 [42]	<250 µS/cm at 20 °C temp.
2.3	Quality of supply air.	-	Oil-, water-, and moisture-free.
2.4	Environmental conditions during sandblasting: air temperature; relative humidity; dew point; and surface temperature.	ISO 8502-4:2017 [43]	Air and surface temperatures according to the material's technical data sheet; relative humidity not exceeding 85%; and difference from the dew point greater than 3 °C.
2.5	Air compressor blotter test.	ASTM D4285:2018 [44]	Free of oil and moisture.
3. Before anti-corrosive painting			
3.1	Visual inspection of the sandblasted carbon steel surface.	ISO 8501-1:2007 [45]	SA 2.5
3.2	Determination of dustiness on the surface.	ISO 8502-3:2017 [46]	Maximum quantity—2, size—2.
3.3	Roughness check on the surface of carbon steel.	ISO 8503-2:2012 [47]/ISO 8503-5:2017 [40]	Medium (45–75 µm).
4. During anti-corrosive painting			
4.1	Environmental conditions during painting: air temperature; relative humidity; dew point; and steel surface temperature.	ISO 8502-4:2017 [43]	Air and surface temperatures according to the material's technical data sheet; relative humidity not exceeding 85%; and difference from the dew point greater than 3 °C.
4.2	Dry film thickness check.	-	Before each layer, according to the specified dry film thickness of each layer.
4.3	Layered painting.	-	Before every layer of paint.
5. After anti-corrosive painting			
5.1	Adhesion measurement (adhesion to coating) test.	ISO 4624:2016 [48]	A value greater than 6 MPa; Performed on SDS >200 µm

If the values did not meet the intended criteria, e.g., the SA 2.5 cleanliness class was not achieved during the sandblasting according to the ISO 8501-1:2007 standard [45], these areas were re-sandblasted to ensure the cleanliness class specified in the design.

3.2. Project Management Model at Company X

The PM in Company X was based on a customised model developed internally. The responsibilities and duties of the project team were defined in the project implementation process so that each member of the project team had a clear understanding of his/her tasks. The project initiation phase involved the appointment of a PJM, an understanding of the project purpose and expectations, the phasing and scheduling of the project, a review of contractual requirements for additional documentation (e.g., site access permits), a risk assessment, and the creation of the project initiation documents, namely the appointment letter, the order, and the site file.

The standard PM organisational structure in Company X included the following personnel: project director, PJM, works manager, works organisation manager, engineer, delivery manager, works supervisor, and workers. As one of the key criteria of the project was to complete the work on time, the technical director was also involved in the PM organisational structure. The responsibilities of the project team are broken down by function below (Table 4).

Table 4. Responsibilities of the project team by function at Company X.

Function	Project Responsibilities
Technical director	Advises the project team on technological matters and assists the PJM in making decisions; participates in project discussions and provides support in resolving project-related issues.
Project director	Supervises the PJM, organises regular project meetings, and assists in managing project issues; may also arrange additional meetings with the technical director, project director, works organisation manager, and PJM as needed; implements PMT and signs and approves project contracts.
Works organisation manager	Ensures an adequate number of contractors and workers, plans human resources and manages their employment and allocation in projects; organises construction documents based on client needs, including assignments, permits, orders, and instructions, which are provided alongside the site file; is responsible for monitoring compliance with safety requirements.
Project manager	Estimates and submits commercial proposals to clients, clarifies technical and commercial issues, and negotiates contracts; organises project team meetings, provides project briefings to contractors, monitors work execution, and hands over the site file containing relevant documents; plans initial equipment requirements, facilitates material procurement and equipment delivery, and coordinates construction permits with the client; establishes project control principles, oversees contractor and subcontractor work, monitors project progress, and makes necessary adjustments; ensures quality execution through periodic checks, approves statements and invoices, manages project costs, and participates in project discussions.
Supply manager	Ensures the availability of necessary tools as per the list provided by the PJM and contractor; arranges for equipment hire or purchase and its delivery to the site; is responsible for purchasing services, seeking and reserving accommodation, assembling and inspecting the specified equipment, and ensuring its readiness for use; also handles transport arrangements and reports any discrepancies as needed.
Warehouseman	Carries out delegated tasks from the delivery manager, arranges necessary equipment based on the provided list by the PJM and contractor, and ensures that the issued equipment for the project is in good working condition.
Engineer	Performs assigned tasks by the PJM, prepares construction documents such as assignment permits and orders, and compiles the site file; registers reports submitted by the contractor in the system, and provides plan-invoice outputs to the PJM, works organisation manager, and contractor weekly; also handles forms F2/F3 and internal acts.
Works supervisor	Formulates tasks for employees, monitors the progress and quality of work throughout the project; updates the equipment list as needed, signs construction documents including assignment permits, orders, and briefings based on customer requirements, and maintains object files; also submits the completed work to the PJM.

3.3. Problems Related to Project Management at Company X

The problems that related to project management at Company X were identified by the project team conducting repetitive anti-corrosion work projects for steel frames of gas distribution stations. The following problems were identified during the repetitive construction PM:

- Increased design quantities that were not paid for by the contract;
- The technological solutions for the project were only adjusted during the project;
- Lack of quality control of the project and poor quality of the anti-corrosion work carried out on certain steel structures;
- Failure to assign the responsible works managers to the relevant work operations;
- Works not foreseen in the project (installation of elevation aids, coating of steel structures before painting, and repeating the same operation twice);
- Delays in the agreed work schedule;
- Exceeded material resources.

The project foresaw the execution of 1360 m² of anti-corrosive painting, but the actual quantities were higher, i.e., 1653 m² (21.54% increase), and the contract did not provide payment for additional works. As the actual quantities were not in line with the projected quantities, the cost of the project increased, requiring more man-hours, materials, and machinery to complete the project. During the project, the observed over-quantities led to additional orders for materials, which were transported directly from the paint supplier, resulting in higher project costs. The following cost increases were identified: the wage bill increased by 37.68%, the number of materials used increased by 38.69%, and the cost of machinery increased by 68.47%. But the direct and other costs decreased by 7.66%.

Increased design quantities, technological complexity, and continuous work in several shifts were identified as PM weaknesses, which had a major impact on the overall project's result. This shows that the above-mentioned management areas have to be improved to seek good project results. In this case, it was decided to apply two PM methodologies, PMBOK and PRINCE2, to improve the management areas in conducting repetitive construction projects at Company X.

3.4. Methodologies

Analytical quantitative research in the form of a survey and qualitative research using a structured interview method were carried out by the researchers. To enhance the validity and reliability of research results, it is crucial to integrate quantitative and qualitative approaches in the research design and data collection. By adopting a mixed-methods research approach, researchers can strengthen their ability to derive trustworthy and compelling conclusions from empirical research [49].

3.4.1. The Analytical Quantitative Study

In order to help Company X to solve the problems related to the project management of their repetitive anti-corrosion work projects, analytical quantitative research was carried out in the form of a survey. By employing this technique, researchers can create an appealing and well-structured questionnaire with a suitable introduction, clear instructions, and a thoughtfully arranged set of questions with aligned response alternatives that aim to facilitate ease of response for the respondents [50]. The researchers developed a survey questionnaire to assess the perceptions of the respondents of construction PM and execution companies of the application importance of two PM methodologies, PMBOK and PRINCE2, for repetitive construction projects. Not only did it aim to extract general information about the respondents (age, experience in the construction sector, etc.) but also their knowledge of PM methodologies. The survey analysed two PM methodologies, PMBOK and PRINCE2, which were discussed in the literature review section above. The survey results were used to determine which of the two PM approaches would be more suitable for a repetitive construction project.

The questionnaire for the quantitative study on the selection of a rational construction PM approach for the repetitive process consisted of fourteen closed-ended questions, six of which were generic and eight were PMBOK- and PRINCE2-specific questions. The six general questions were related to gender, age group, educational background, work experience, position in the company, and scope of the company activities. Each of the eight questions on project management had two possible answer options, one related to PMBOK and the other to the PRINCE2 methodology. The respondents, by choosing the answer options, could consider what methodology would be more suitable for managing repetitive construction projects. The questions given to the respondents are discussed in Section 4.1. A repetitive construction project involving the abrasive cleaning and anti-corrosive coating of steel structures of gas distribution stations was presented for the case study. In addition, it was highlighted that the project was subject to stringent requirements in terms of the budget, the quantities of materials, and relatively short lead times.

The quantitative research was carried out over two weeks, from 22 November to 5 December 2021, targeting the respondents from construction PM and execution companies. The survey was conducted online using the Google Forms platform. Construction companies with recurring construction projects in Lithuania were surveyed. The survey link was distributed via email and the LinkedIn platform. The respondents were randomly selected if their job title was related to project management in the company. Based on the findings of the researchers [22], the selection of a simple random sampling method ensured that every element within the population had an equal opportunity to be included in the sample. A total of 104 respondents took part in the quantitative study over a two-week period. Responses to the questionnaire were then collected and analysed.

3.4.2. The Descriptive Qualitative Study

To identify trends and actions related to monitoring the progress of construction projects, a descriptive qualitative study was conducted. In qualitative research, structured interviews serve as a prevalent method for data collection. It is worth noting that the effectiveness of the study results is fundamentally influenced by the quality of the interview guide [51]. To collect the needed information, the qualitative research was carried out using a structured interview method, asking representatives of different companies, X, Y, and Z, the same 15 questions. The qualitative study involved three different-sized companies dealing with construction PM and implementations. In the analysis of the replies received, the companies were identified as Company X, Company Y, and Company Z.

The descriptive qualitative study aimed to obtain information about the structure of the company, the tools, methods, and methodologies used to improve PM, project progress monitoring, progress indicators monitored in the report, additional report information, and other relevant issues, as listed in Table 5. Author [52] in his work provided a working knowledge of the whole building industry, i.e. the technical skills required to manage a construction project from conception through occupancy. The useful information provided helped to prepare the questionnaire for the descriptive qualitative study.

Table 5. Company representatives were asked to answer the following questions.

No.	Questions
1.	How are projects organised and managed within the company? Please describe the PM structure and responsibilities.
2.	What tools/methods, PM methodologies, or approaches do you use to manage projects?
3.	How do these tools, techniques, PM methodologies, or approaches contribute to improving project delivery?
4.	How do you monitor the progress of construction projects (e.g., human resources, materials and machinery)?
5.	What indicators do you track in project progress?
6.	How often is the project progress monitored?
7.	Does the existing project progress report provide clear and relevant information?
8.	What would you improve in the project progress report?
9.	What decisions does monitoring project progress help you make?
10.	At what frequency is the project progress discussed with the contractors and how is the discussion organised?
11.	Do staff easily assimilate project progress information?
12.	Who in your company takes appropriate action to ensure that the targets are met?
13.	If you see that a project is not going according to the plan, what action do you take? Which units of the company are involved in changing the project plan?
14.	What are the success factors for PM in your opinion?
15.	What are the reasons for project delays in your opinion? What should be avoided in order to deliver the projects on time?

The interviews in the qualitative research were carried out in person and online. The results obtained are analysed in Section 4.3.

4. Results, Discussion and Recommendations

This part presents the results of the quantitative results, gives recommendations for improvement of the implemented repetitive project management at Company X, presents the results of a descriptive qualitative study, and discusses the sustainability aspect of project management for future studies.

4.1. The Results of the Analytical Quantitative Study

A survey was conducted in order to determine which of the two PM approaches would be more suitable for a repetitive construction project; 104 respondents took part in the quantitative study over a two-week period. The respondents' answers to the questions and the choice of the most appropriate option are presented in Tables 6 and 7. The first six questions were generic and related to gender, age, educational background, working experience, and current position in a company. The answers showed that the majority of the respondents were male (77.0%), aged 25–30 (33.7%), held a Bachelor's degree (61.5%), and had 5–10 years of working experience (39.4%). By position in the company, the majority of the respondents were PJMs (25.0%) and came from companies mainly specialised in construction works (45.2%) (see Table 6).

Table 6. Respondents' characterisation.

No.	Questions	Number of Respondents Who Selected the Answer	Answers as a Percentage
1.	Please indicate your gender.	Men: 80	~77.0%
		Women: 24	~23.0%
2.	Please indicate your age group choosing the appropriate option.	18–25 age group: 25	~24.0%
		25–30 age group: 35	~33.7%
		30–40 age group: 31	~29.8%
		40–50 age group: 8	~7.7%
		50+ age group: 5	~4.8%
3.	What is your educational background? Please select the appropriate option.	Secondary education: 5	~4.8%
		Professional education: 5	~4.8%
		Bachelor's degree: 64	~61.5%
		Master's degree: 28	~26.9%
		PhD degree: 2	~1.9%
4.	What is your work experience in the construction sector by year? Please select the appropriate option.	Up to 1 year: 9	~8.7%
		1–5 years: 34	~32.7%
		5–10 years: 41	~39.4%
		10–20 years: 14	~13.5%
		20 years and more: 6	~5.7%
5.	What is your position in the company? Please select the appropriate option.	Assistant manager of projects and works: 20	~19.2%
		Project engineer: 25	~24.0%
		Work manager: 12	~11.5%
		Project manager: 26	~25.0%
		Construction manager: 9	~8.7%
		Head of department: 11	~10.6%
		Head of the company: 1	~1.0%
6.	What is the scope of your company activities? Please select the appropriate option.	General construction contractor: 42	~40.4%
		Specialised construction works: 47	~45.2%
		Real estate development: 15	~14.4%

Table 7. Survey responses.

No.	Questions	Number of Respondents Who Selected the Answer	Answers as a Percentage
1.	Which definition of a project do you think is more appropriate for the repetitive construction project? Option 1. Project—a temporary activity designed to create a unique product, service or result. Option 2. Project—a temporary organisation set up to implement one or more products according to a defined business plan.	Option 1 (related to PMBOK): 48 Option 2 (related to PRINCE2): 56	~46.2% ~53.8%
2.	Which PM model do you think is more appropriate for the repetitive construction project? Option 1. The PMBOK PM model consists of 5 groups of management processes: initiation, planning, execution, monitoring and control, and closure. Each stage of each process group is followed by a specific deliverable or feedback. Option 2. The PRINCE2 PM model consists of 7 groups of management processes: project supervision, project inception, project planning, stages boundary management, stages control, product development management and project closure. Each process is reviewed by the PJM and approved by the project board.	Option 1 (PMBOK): 23 Option 2 (PRINCE2)—81	~22.1% ~77.9%
3.	Which definition of the project team's responsibilities do you think is more appropriate for the repetitive construction project? Option 1. PJM is the person responsible for leading a project team to achieve the project objectives. Responsible for completing the tasks assigned to the project team. Project team is a group of people working towards common project goals, under the authority of a PJM. Option 2. PJM is the person whose day-to-day focus is on the project, liaising with the project board throughout the project. Delegates tasks to a team leader (e.g., the works manager or works supervisor). Team Leader—responsible for the execution of the tasks assigned by the PJM and the work performed. Regularly delegates completed work for the PJM. Project staff is subordinate to team leaders to carry out assigned tasks.	Option 1 (related to PMBOK)—40 Option 2 (related to PRINCE2)—64	~38.5% ~61.5%
4.	Which definition do you think is more appropriate for the repetitive construction project? Option 1. The roles and responsibilities of project team members should be discussed and may be specified during the project. Option 2. The roles and responsibilities of the project members must be described, and each participant must have a clear understanding of their roles and responsibilities before the project starts.	Option 1 (related to PMBOK)—18 Option 2 (related to PRINCE2)—86	~17.3% ~82.7%
5.	Do you agree with the statement that the project in question must adhere strictly to the chosen PM methodology?	Agree (related to PRINCE2)—65 Disagree (related to PMBOK)—39	~62.5% ~37.5%
6.	Please choose the statement you think is most appropriate for the repetitive construction project. Option 1. The PM methodology must be descriptive, i.e., it describes processes and knowledge areas but does not specify how they are to be used. Option 2. The PM methodology must be prescriptive, i.e., describing what is to be done and when.	Option 1 (related to PMBOK)—18 Option 2 (related to PRINCE2)—86	~17.3% ~82.7%
7.	Please choose the statement you think is most appropriate for the repetitive construction project. Option 1. The PM methodology must anticipate the tools and techniques that can be applied to the project. Option 2. The PM methodology has less defined tools and techniques but is not limited to the use of best practice tools and techniques from outside the PM methodology.	Option 1 (related to PMBOK)—41 Option 2 (related to PRINCE2)—63	~39.4% ~60.6%
8.	Please choose the statement you think is most appropriate for the repetitive construction project. Option 1. Depending on the progress of the project, a PMO is organised at the request of the PJM or instruction from the project board, involving the project board (e.g., heads of department and company) and the PJM. Option 2. The project board (e.g., head of department and head of the company) must control the work of the PJM and his/her team from project initation to project closure, regardless of the progress of the project.	Option 1 (related to PMBOK)—47 Option 2 (related to PRINCE2)—57	~45.2% ~54.8%

The remaining eight out of fourteen questions were related to PMBOK and PRINCE2 PM methodologies, and the respondents had to choose an appropriate option from the two presented. The majority of the respondents (see Table 7) chose the answer options related to the PRINCE2 PM methodology, i.e., they considered this methodology to be more suitable for managing repetitive construction projects.

To summarise the results of the quantitative study based on the responses of the 104 respondents, the PRINCE2 PM methodology was more rational for the project under consideration. The answers to the eight questions related to PM methodologies in the quantitative survey showed that the PRINCE2 PM methodology was given a priority. After the conversion of the answers of the one hundred and four respondents to eight questions into points, PMBOK scored 274 points, whereas PRINCE2 scored 558 points (see Figure 2). As a percentage, the PRINCE2 PM methodology accounted for 67% of the total points and the PMBOK methodology for 33%.

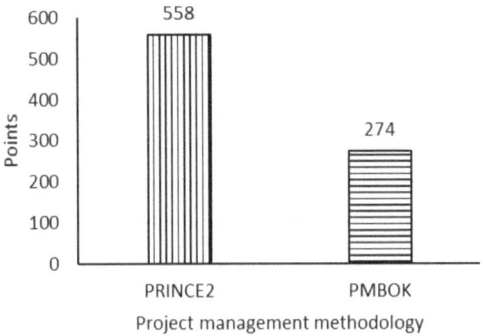

Figure 2. PRINCE2 vs. PMBOK according to quantitative study.

Thus, based on the results of the quantitative study, the PRINCE2 PM methodology was chosen for further development of the repetitive construction project.

One of the principles of the PRINCE2 PM methodology is to learn from experience, so the lessons learnt must be applied to the repetitive project under analysis. The project was integrated through three main elements of PRINCE2: principles, themes, and processes [5]. The business case, organisation, quality, planning, risks, change, and progress are some of the key elements of PRINCE2 PM. In this case, the processes and tasks of the PRINCE2 PM methodology for the repetitive construction project were outlined and compared with the structure of the project in question and the tasks carried out [5]. A comparison of the PM processes using the PRINCE2 project management methodology and standard project management is presented in Table 8.

Table 8 shows that the fundamental differences between the objectives of the PRINCE2 PM methodology for a repetitive construction project and a completed project are the information flow and different PM processes. For example, a repetitive construction project would introduce phases and phase boundary management, whereas an implemented project is not divided into phases. According to the PRINCE2 PM methodology, a repetitive construction project would aim to provide as much information as possible to the project participants, form a team and assign team leaders responsible for the phases, establish a financial plan, a detailed timetable for the execution of the works, a quality control plan, and a plan of responsible persons, and detail the technological sequencing of the works. These things were not carried out in the project under consideration. In addition, a repetitive project based on the PRINCE2 PM methodology would include a problem register, a risk and action plan, more frequent monitoring of the project progress, and more frequent meetings with the project' team leaders and/or team members.

Table 8. Comparison of the PM processes using the PRINCE2 project management methodology and standard project management.

No.	Project Management Processes	The PRINCE2 Project Management Methodology	Tasks Completed in the Project under Review/Completed by Standard Project Management
1	Project supervision	Regular project meetings with the project board two times a week from the start of the project to the end.	Project discussions/meetings in critical situations separately with managers.
2	Beginning of the project	Preparation of information and plans in accordance with the principles and theorems of PM methodology. Establishment of a site file, an assignment, and a project order with responsible persons.	Preparation of an object file with technological guidelines (painting technology, drawings, and project requirements) and creation of an assignment authorisation and a project order with responsible persons.
3	Planning the project	The PJM approves the documentation set out at the start of the project and adds or assigns a task to add information if the information is missing.	Human resources, equipment, and materials are approved by the PJM.
4	Phase boundary management	The PJM, together with the project board, carefully analyse the phases in progress before providing the PJM with relevant information. If necessary, the project plan is updated.	The project is not phased.
5	Phase control	The PJM designates the persons responsible for the execution and quality of the works for the intermediate control of each phase. In the event of deviations, they inform the PJM, who takes the initiative for project changes and takes decisions. Meetings are organised by the PJM with the project team leaders and/or members.	The project is not phased.
6	Product development management	The PJM carries out a quality control check before the work is delivered to the client. The project board controls communication between the PJM and the project team.	Quality control of the PJM is introduced during the project following the comments made.
7	Closure of the project	Final project documentation is produced, a register of learning from experience is kept, and the results of the project are summarised and evaluated in the company.	Final project documentation, summarising, project results and progress, is presented at the company.

4.2. Recommendations for Improvement of the Implemented Repetitive Project Management at Company X

Given the results of the quantitative study and the comparison presented in Table 9, the PRINCE2 PM methodology was chosen for improvement of the implemented repetitive project management at Company X. The problems related to project management at Company X are presented in Section 3.3. Some recommendations about how to improve the identified problem areas using the PRINCE2 methodology are given below.

Table 9. The answers of the Company X representative.

Question No.	Summarised Answers to the Questions
1.	PM starts with the signing of the contract. The PJM is fully responsible for the success of the project and manages the team to ensure that the outcome of the project meets the terms of the contract, the detailed and technical designs and the budget. For standard, small-scale projects, the project team is quite simple: a PJM, a works manager, a foreman and an engineer. However, the PJM also uses other people in the organisation, such as the supply manager and the warehouse manager. The PM reports in weekly meetings to his/her line manager (head of the department), who controls basic criteria, i.e., time, scope and budget. The unit manager also approves the essential PM tools (PMT), the project plan, and the project schedule. If the project has a deviation from the plan that affects the budget, the unit manager has to approve the costs in the project change committee, which can include everyone, from the accountant to the company shareholders. Thus, the PJM makes decisions that do not affect the budget and plan and do not impact the business. The role of the unit manager is to oversee the execution of the process at various stages, monitor adherence to the plan, approve essential PMT, and organise changes in the event of a major deviation. Clearly, non-standard projects require additional team members, although the roles of the main actors in the project remain almost unchanged. The PJM is supported by an engineer, a supplier, a works manager, a foreman or more, and the control function is taken over by the PMC (project management committee) rather than the department manager.
2.	Tools need to be user-friendly and standardised so that the PJM can manage the project efficiently and those responsible can monitor the project outcome or deviations. The main tools for PM are the risk management software (which we currently use in MS Excel spreadsheet), the project plan, and the project schedule (MS Project). We also use project progress tracking software, which was specifically developed for our company, since in small-scale projects the most important impact on the budget is the man-hours and the material yield. These indicators need to be monitored on an ongoing basis to be able to manage the result and make the necessary decisions. To monitor the budget, we use a database (Power BI) which allows us to monitor data at different cross-sections, outliers, or comparisons with the plan. Project progress and deviations need to be clearly and quickly understood, which is why the traffic light principle is used so that we can quickly see if there is a deviation and if we need to make decisions as soon as the tool is opened.
3.	It is difficult to compete in the market with other organisations if you do not have key advantages (technology you have developed, machinery that your competitors do not have, or the materials you produce yourself), so you have an advantage in PM. The key is to have a standardised project procedure that clearly defines roles, who is responsible for what, who makes decisions and specifies the process and the tools used. As a result, the PJM no longer has to think about what to do or where to go in case of deviations and can therefore be more productive and focused on the outcome of the project.
4.	We can monitor the working hours and materials spent in the previous period for the next day on the traffic light basis. We can see the deviation of one hour and the information is very accurate. The data is also stored to be used for new estimates. We have a unique in-house software for this. We use a tracking program to monitor the machinery where we can see its working hours and fuel consumption rates.
5.	The tools we use are quite effective allowing us to compare the plan of working hours with the fact and the project budget. We can also monitor the result, but it is difficult to understand what decisions are needed in the event of a deviation, so it is most important to monitor the indicators affecting the project result. In our company, the indicators that influence the result the most are man-hours, material yield, and project resources, so we monitor these indicators the most.
6.	It is sufficient to monitor the project result once a month for our projects, as we do not have the possibility to do it more frequently, but the PM has to monitor the indicators that influence the project result on a daily basis to achieve the best project result. And because the tools are user friendly, it takes only 5–10 min a day to review the key indicators.
7.	The traffic-light-based progress tracking software is very user-friendly and time-saving if there is no need for a PJM and a unit manager to go into the figures in detail. Also, it should be standardised. If the traffic light is yellow, the PJM has to make decisions; if it is red, the head of the unit has to find out the reason, and we intend to standardise it in the future.
8.	At the moment, we cannot monitor the direct costs of projects in the current period and compare them with the plan. This can only be conducted at the beginning of the month when the PJM does the material and cost write-off and the accounting department enters the data. If we could monitor this at least every week, we would be able to manage projects more accurately. Clearly, like other indicators, it should give a quick and clear indication of whether there is a deviation and additional time is required.

Table 9. *Cont.*

Question No.	Summarised Answers to the Questions
9.	We do not monitor indicators for the sake of it but to make decisions that will help us manage the outcome of the project. Decisions can take many forms, such as changing the technology or equipment, adjusting the schedule, negotiating extensions, increasing resources, subcontracting, etc., if productivity is not being achieved. If material yields are exceeded, it may be necessary to change materials, order additional quantities, renegotiate with the customer, and arrange a meeting with material and equipment suppliers. It may even be the case that if the indicators show that we will not only be over the budget but also under the cost, the project change committee may decide to cancel the contract.
10.	Usually, if there is no deviation, progress is not discussed with the workers, but this is a bad example, and in any case, progress should be discussed at least once a week. Workers are motivated when they know where they stand. We have tried in the company to send project progress by e-mail, but the figures are not manageable for everyone, so our e-mails get ignored. For larger projects, we use a model where we present the previous day's progress to the lower management, which helps us communicate the problems that are preventing us from achieving the desired result. The project progress meeting should be no longer than 20–30 min, during which indicators are reviewed and problems are mapped. Another meeting is organised to address the problems.
11.	It is important to show the indicators for which they are directly responsible, i.e., labour productivity and material yield. Although we do not hide the financial indicators of the project in our company, the latter would be redundant. The information conveyed must be simple and clear, and the employees must understand what is going on with the project within 5 min of looking at it, which is why coloured boxes and arrows are used.
12.	The PM is always responsible for the outcome of the project, so he is naturally responsible for monitoring and controlling progress, although he may delegate this task to an engineer to inform him of deviations. The PJM takes action when he sees a deviation and delegates tasks to other team members to clarify or correct the deviation depending on the problem. Alternatives should always be applied, taking the costs and foreseeable risks into account to achieve the best or most appropriate result.
13.	The PM is the person who directly influences the outcome of the project and is solely responsible for communication, both internally and externally. The PJM has the ability to decide which line managers he/she needs to support if the redesign results in a change in the final outcome that is either unchanged or insignificant to the project budget. But if, due to deviations or alternatives chosen, the planned budget is insufficient, the PJM has to approach the division manager with the estimated additional budget needed, who (un)allocates the additional budget as far as possible, or summons a project change committee, which decides on the impact on the company's results or on the future business. Modification of the project plan and schedule is mandatory in the case of major deviations.
14.	There are three success factors for PM: Firstly, a standardised and user-friendly project implementation procedure, which is where it all starts. Secondly, the behaviour of the PJM in the preparation phase of the project. The better you prepare the project before it starts, the fewer small problems you have to tackle and can concentrate on the execution phase. The primary tools for project preparation are the project plan, the schedule, and the risk management plan. Thirdly, the PJM's attitude to problems and deviations. If he or she does not feel responsible for the outcome of the project, the outcome will never be good.
15.	The reasons for project delay can be many and varied, e.g., inappropriate choice of materials, insufficient team expertise, inappropriate team structure, poor project preparation, unreliable equipment or suppliers, etc. We can anticipate and prepare for problems in advance and anticipate what we will do if there is a problem. This is a risk management plan.

Increased design quantities. The customer calculated the design quantities using software. However, the actual demand for human resources and the yield of materials was higher than the customer's design calculations. The recalculation of the quantities showed an increase of 21.54% over the projected quantities. Using the PRINCE2 PM methodology, the project initiation process, i.e., the pre-project phase, must clarify and justify the client's expectations and enable a smooth start of the planned project. Therefore, the PJM has to give the task to a team member to check the project quantities internally against the customer's drawings to avoid any misunderstandings during the execution of the project. The implementation of embedded sensors for automated tracking of construction equipment and materials on-site can lead to cost reductions in material expenses [53]. By leveraging this technology, construction projects can achieve improved efficiency and accuracy in managing their resources, resulting in potential savings in material costs.

Technological solutions. The technological painting solutions were refined during the project. Initially, the painting process was experimental in order to determine the most rational sequence of coat application. The lack of clarity in the tasks and the allocation of responsibilities in the teams before the start of the project resulted in poor-quality painting and inefficiency, as the work was conducted on the basis of impulsive decision-making. According to PRINCE2 PM principles related to roles and responsibilities and the nature of the organisation, before the start of a project, everyone involved in the project must know their responsibilities and tasks. Also, the PRINCE2 methodology calls for a clearly defined structure and sequence to guide the entire project team. Although the timeframe for the project preparation was shortened, a joint meeting of the project board (technical director, project director, works supervisor, and PJM), with the participation of the works supervisor, was able to find compromises to improve the project.

Lack of quality control. The poor quality of the anti-corrosive painting during the project execution indicated a lack of project control. Prior to approving the next stage of the project, the PJM must ensure that the work is carried out to the specified requirements and high standards. Failure to ensure the quality led to repeated operations that take as much time as the initial operation. Several factors contributed to this shortcoming, such as the inadequate induction of new works managers and their teams into the project, a lack of feedback from the works managers, and the absence of a coordinating person. A project team is formed with respect to the size and complexity of a project. The standard project team in Company X consisted of a project director, a PJM, a works organisation manager, an engineer, a supply manager, a works manager, and workers. The PRINCE2 methodology places an exceptional emphasis on quality control. Each member of the project team must have a clear understanding of the quality requirements, and the responsible persons (PJM, works manager) must be able to ensure that the work is carried out to a high standard. For example, to manage the project properly, the sandblasting work had to be divided into phases to ensure the quality of the project and to assign the responsible works managers for the sandblasting work. According to the authors of ref. [7], the analysis of stakeholders in a construction project should serve as a foundation for supporting and strengthening team-building activities. These activities encompass various aspects such as strategy development, aligning with project goals, defining members' roles, and managing interfaces between different sections. By leveraging the outputs of stakeholder analysis, team-building efforts can be enhanced and aligned with the needs and expectations of relevant stakeholders involved in the project.

Failure to assign responsible works managers. The work was carried out in two shifts (morning and evening), with information being passed from the morning shift to the evening shift by the works managers. However, the evening shift and the morning shift did not meet, and thus the quality of the project was compromised because hard-to-reach areas were inadequately prepared and/or unpainted due to the lack of communication. To avoid quality control errors, it is necessary to assign a person responsible for certain operations, such as the works manager responsible for the quality of the sandblasting work, person X, who identifies and defines the shortcomings of the evening shift and communicates the backlog, highlighting the hard-to-reach spots, and the works manager Y who is responsible for the sandblasting. The PRINCE2 PM methodology helps to avoid such errors by breaking the project down into phases, thereby increasing control over certain operations, such as the first coat of paint. Once the project has been phased, it is necessary to assign responsible persons to ensure the quality of the phase. No further work is to be undertaken until the first phase, the first coat of paint in this case, has been approved. The PJM is responsible for the approval of the phase and may delegate the intermediate control of operations to team leaders.

Works not foreseen in the project. The project in question provided that the customer would supply scaffolding, but due to the poor mobility of the scaffolding, the work was carried out inefficiently, and it was not possible to reach a part of the steel structures. The contractor, therefore, decided to provide a mobile aluminium scaffolding system. Another problem encountered was the need for additional polyethilene film cover for the structures while using different painting systems. The change in the painting sequence necessitated a change in the pre-design solutions and an additional cover of the steel structures to achieve the highest quality. In this case, a detailed project preparation plan would have helped to avoid unforeseen activities or to plan them most efficiently, as emphasised by the PRINCE2 PM principles. Time and resource commitment must be clear, and everyone involved in the project must know in advance what is expected from the product. According to the authors of [54], the increasing adoption of prefabrication, modular construction, and additive manufacturing offers significant benefits in terms of reducing fabrication costs and enabling the timely completion of construction projects. The construction industry can streamline the production process, enhance cost-effectiveness, and improve project delivery timelines.

Delays in the agreed work schedule. Although the project timelines were clearly defined and the timeframe for each gas distribution station was known, the technological process of the project was only revised during the project, and errors were unavoidable. The painting of the steel structures in different sequences presented technological challenges, resulting in a deteriorated quality of the painting. The increase in project volumes also had an impact on the duration of the work, as the number of workers had to be increased to complete the work on time. The PRINCE2 PM model can be used to overcome such problems by checking the contractual quantities at the start of the project, preparing a detailed project plan, and phasing the work to be carried out in the project, thus eliminating the risks associated with the project delays. According to the authors of [35], by utilising project control software systems, specifically those professionally designed for project time and cost management, PJMs can effectively plan the construction sequence, monitor the progress of project activities, and update the project's overall progress. These software systems enable PJMs to identify and track project delays efficiently.

Exceeding foreseen material resources. The overruns consisted of poor workmanship and work not foreseen in the design. For example, the demand for materials due to substandard work almost doubled as a repainting operation was required. Also, without knowing the exact quantities of the project, it was difficult to predict the exact material resources. This proves once again that in order to manage a project according to the business plan, it is essential to follow PRINCE2 PM principles and processes to avoid mistakes during the project. Firstly, the contractual quantities of the project have to be checked before the project starts. Also, the phases of work to be carried out are to be identified not only for quality control but also for intermediate control of material utilisation.

Thus, by following the PRINCE2 PM methodology, it is possible to avoid the problems that occur in certain phases and processes of PM.

4.3. The Results of a Descriptive Qualitative Study

Three companies of different sizes agreed to take part in the qualitative study. These companies were asked a series of questions (see Table 5), which were answered by company representatives. To ensure the confidentiality of information, the companies participating in the qualitative research were labeled as Company X, Company Y, and Company Z.

Company X was a specialised construction company providing services in constructing and repairing infrastructure, industrial, energy, commercial, and other building projects. Its main activities included anti-corrosion protection (blasting, shot blasting, anti-corrosion painting. and thermal coating), concrete repair and protection, installation of cast-in-place floors, fireproofing, and supply and installation of fire protection products. The company had more than 120 employees. The answers of the Company X representative are given in Table 9.

The second company participating in the qualitative study, labelled as Company Y, was also involved in specialised construction work. The portfolio of work included the design, installation, and assembly of water boreholes, the installation of water supply using hydrophores and frequency converters, the construction of domestic wastewater treatment plants and pumping stations, the restoration of borehole performance, deep grounding, etc. The company had around 90 employees. The answers of the Company Y representative are given in Table 10.

Table 10. The answers of the Company Y representative.

Question No.	Summarised Answers to the Questions
1.	New projects are allocated according to the PJM's workload and the number of projects he manages. From that point on, he/she is responsible for the implementation of the project and for taking decisions to implement it, i.e., from the drafting and modification of the draft contract to the delivery of the project. Surely, the PJM is not responsible for the progress of the work assigned to the work manager, with whom the possible solutions are discussed regularly. The PJM has to keep an eye on the progress of the work, as he takes decisions that the work manager will have to implement. In comparison, private-sector projects are much simpler and require less time input compared to legal entity projects, so 'bigger' projects also require extra attention, such as attending the meetings of site construction contractors, constant communication with the construction manager, adjustments to the working design, constant site visits, etc., whereas private-sector projects can be visited up to three or four times if no problems arise. If the project is problematic, if the PJM is unable to resolve the problem or if advice is needed from someone with more expertise, general management meetings are usually held with the participation of the directors of different divisions. A standard project team consists of a PJM, a works manager, a foreman, and workers.
2.	Various tools are used, such as Dalux software and the Electronic Construction Work Log (ECWL). Our company does not have one specific PMT; usually we adapt to the client's requirements (apps, ECWL, etc.) for larger works. Project allocations are usually based on a hierarchical model, i.e., division directors allocate work to lower-level managers according to the staff availability, who in turn allocate it to work managers, who, in turn, allocate it to workers. A project is usually allocated by the PJM according to priorities, contractual obligations, and deadlines.
3.	New technologies are always good, but they also take time to adapt to. Nevertheless, any digitisation or systematisation of mechanical work is appreciated. For example, the ECWL greatly facilitates the description of the work and thus saves precious professional time. Also, the documentation is not lost and is archived immediately. The app is quite new, so mistakes can happen.
4.	We are currently testing trial versions of various applications to find out what the company needs and which will work best. Recently, MS Excel has been used to track human resources, materials, and machinery, and is also used for materials and warehouses.
5.	Project monitoring consists of many components, such as monitoring whether the project is on track, as most estimates are calculated on a project-by-project basis, worker time is calculated based on past projects of a similar nature, and the time taken to complete various tasks. The progress of the work must also be mentioned, as it is rare for everything to run smoothly in construction, and decisions need to be taken as soon as possible if a problem arises. The quantities of materials used and the wear and tear of machinery are also monitored. The financial results of projects are monitored every month.
6.	Project monitoring depends on the complexity of the site, with private-sector sites requiring much less attention as the solutions are not complex, with some exceptions though. Special structures require much more supervision, as they involve more contractors with whom regular communication is required, and the technical solutions are much more complex and need to be coordinated with a greater number of responsible persons. For example, in the private sector, it is usually sufficient to visit a site up to two to three times (depending on the scope of the work), whereas, in larger sites, meetings are usually held two to three times a week until the project is delivered.
7.	The information is really clear, but we would always like to see improvements, as most of the company staff is older in age and it is quite difficult to do so. Our observation is that each PJM uses his methodologies and principles, and of course, it would be ideal if everyone used the same principle. It would be much simpler, which is why we are trying to innovate in the company.
8.	System optimisation, software upgrades, and deployment are required, as with the better flow of information about the facility, the profit estimation is more accurate. It is also worth mentioning that the company employs about 60% older-age workers who would find all this problematic in their day-to-day work, and we believe that additional funding is needed to improve their computer literacy.

Table 10. *Cont.*

Question No.	Summarised Answers to the Questions
9.	Monitoring project progress helps you keep up with on-site activities so that when problems arise, decisions are made more efficiently and quickly. It also shows the number of staff needed to run the site. Time costs are assessed, which also helps to assess the scope and timing of future works, i.e., more accurate pricing in proposals and more precise schedules for the execution of works.
10.	The frequency of project progress tracking depends on the complexity of the project. If the project is simple and on track, the project will be monitored by liaising with the works manager or the executor of the work. For larger projects (lasting more than one month), such as general contracts, meetings with the works managers, and contractors, are held at least once a week to discuss the progress of the project while at the same time detailing the work fronts.
11.	Workers working on site are given a technical brief, both digitally and on paper (for older people), which contains only relevant and necessary information about the site so they are not overloaded with excessive information. Financial indicators are not provided to the on-site workers, as this information would be superfluous for them (except for supervisors). Each PJM uses a different methodology to assess the progress of the project.
12.	The appropriate action to ensure the achievement of the targets is taken by the PJM or, in exceptional cases, the project director. All decisions relating to the works are to be taken by the PJM. In the case of delays, the PJM will hold a meeting in the company, listen to the proposals, evaluate them, and take an appropriate or alternative decision.
13.	If the implementation is not going according to the plan, the first step is for the PJM to try to find out the reasons. This is followed by a meeting or simply discussing with the works managers how to optimise the work at the same time as updating the work schedules. The PJM submits proposals to the project director on how to optimise the work or how to allocate additional funds to the project. The PJM is always in contact with the client. Changes to the plan affect all levels, from department heads, who have to approve the changes, to the workers, who recieve new technical tasks included in the new work plan.
14.	A constant interest in the site, and monitoring any problems on site, which may not necessarily be our company's, but may also affect the work we do, so that we can predict and prepare for future problems. Moreover, close cooperation with other companies working on the site, whose help is sometimes very valuable. A good relationship with the client or its representative is essential for the success of the project.
15.	In our case, project delays are mostly caused by production processes, machinery breakdowns, or delays in materials supply. Often it is also the lack of competence of PM or works managers, which leads to unforeseen work. Designers' decisions and adjustments can also contribute to project delays. Although such risks are quite difficult to avoid, they should be managed by choosing reliable design companies (with extensive experience in specialised work). To avoid project delays, the competence of managers must be continuously upgraded and mechanisms updated.

Company Z was a general construction contractor engaged in industrial, residential, public, administrative, and cultural heritage construction and reconstruction, offering advanced, innovative, and sustainable construction solutions to the public and private sectors. The company applied advanced certification systems and international building standards—LEED, BREEAM, BIM, and ISO. The company employed more than 230 people. The answers of the Company Z representative are given in Table 11.

The summary of responses provided in Tables 9–11 reveal that construction PJMs are responsible for managing the entire construction project, from the planning and design to execution and completion. Therefore, they have the power to make decisions that can affect the successes and effectiveness of the project.

Comparing the answers given by the representatives of the different companies, some similarities and differences were identified. These are summarised in Table 12.

Table 11. The answers of the Company Z representative.

Question No.	Summarised Answers to the Questions
1.	First and foremost, the life of a project starts with a successful sales process, which is the responsibility of the commercial department. Later, as the project moves into the execution phase, a construction execution team is appointed. A typical construction execution team consists of a PJM, a construction manager, a project engineer, and foremen. Depending on the scope or complexity of the project, the project team may be larger, as one project may have several construction managers, project engineers, or construction managers who divide the work among themselves. The PJM is responsible for the success of the project and makes decisions and gives approvals to the issues during the project.
2.	Successful and effective PM consists of applying tools, techniques, and management philosophies to achieve the project's objectives. The company's PM uses Building Information Modelling (BIM), which helps to monitor and manage the performance of a building throughout its lifetime, avoiding errors that lead to additional time, cost, and material costs. An intelligent cost forecasting and management software has been implemented. By assessing the purpose of the building, its materials, and other parameters relevant to the project, it allows accurate forecasting of the project costs, opening up the possibility of optimising costs during the project implementation phases. Dalux software is used to record defects and observations, Trimble is used to disseminate information, and MS Project software is used to draw up the project plan, monitor and update the schedule, and draw up the activation plan. In addition, LEAN PM philosophies are widely used in the PM. Visual aids, such as information and accident prevention posters, and a site plan are displayed next to the project to inform the employees. Five Sigma techniques, which are part of the LEAN PM principles, help to organise and control the work in progress, ensure and increase safety on the construction site, avoid repeating mistakes, and promote progress and continuous improvement. These are some of the PM philosophies, tools, and techniques that are applied in PM on different levels of the project team.
3.	The above tools, methodologies, and PM philosophy help to avoid mistakes, save time, costs, and materials, optimise costs during the project phases, plan and control the project, organise and control the project work, prevent accidents, increase safety on site, and promote continuous progress during the project.
4.	The progress of the project in terms of monitoring human resources, materials, and machinery is carried out in several ways—by applying the Asaichi methodology of the LEAN philosophy, which records the objectives and the results of the previous week. The breakdown by work items and the results are recorded in a table. Tracking software is used to monitor the machinery, which indicates the time the equipment is running and the fuel consumed. The progress is also tracked and recorded on the job board on the construction site, and meetings are organised to discuss the project progress.
5.	The PJM monitors labour costs, material, and machinery plans and facts, forecasts project duration, and tracks financial results to derive the overall project outcome.
6.	Construction managers monitor the project progress daily and take appropriate actions if they see that a project is not on track. The financial performance of the project is monitored every month after the write-offs have been made and the subcontractors' invoices have been approved, and the results are discussed with the project board during the meetings.
7.	The project progress report is presented in an informative manner, with relevant information and results. To facilitate the day-to-day monitoring of the project progress, deviations are highlighted in different colours, with positive deviations in green, minor deviations requiring attention in yellow, and negative positions requiring improvement decisions in red. Material yield and machinery deviations are indicated by arrows comparing the plan with the fact.
8.	There is enough information in the daily project report, and there is room for improvement on the financial side of the project results, as accurate conclusions and assumptions can only be drawn one month after the approval of the payroll, the write-offs of materials and machinery, and the confirmation of subcontractors' invoices. But a more detailed tracking of the financial progress should be relevant for projects that are not on track and that deviate significantly from the targets, as the additional processing of the information would put an additional burden on the responsible parties. In such cases, intermediate actuation and cost write-offs should be made.
9.	Tracking the project progress helps in decision making. For example, if the work is behind the schedule, additional staff is involved in the project; if work is slowing down but it is not possible to speed up the work process, an additional agreement is negotiated with the client to extend the deadlines. If, due to technology changes, some works cannot be completed on time, additional funds are agreed upon with the client; if materials or machinery used for the works carried out in the project exceed the planned resources, the use of materials and machinery is reviewed for proper application.

Table 11. *Cont.*

Question No.	Summarised Answers to the Questions
10.	The progress of the project is discussed at a meeting attended by the construction manager, the works managers, and the contractors. The meeting will discuss current outputs, issues and problems, and measures to improve performance. Usually, the meeting is held once a week, but the frequency may be increased or reduced depending on the project. The discussion ensures the transfer of information within the project.
11.	Project progress information is passed on to the contractors, who share the information with their staff, so the last link in the analysis of the project progress information is the contractor. The project progress reports show the most relevant indicators for the contractors, such as deviations in outputs and materials, highlighted by colours and arrows.
12.	The construction manager and the works manager take appropriate action to ensure that the outputs foreseen in the design are achieved and monitor the progress of the works and take measures where necessary. Deviations are communicated to the contractors by the supervisors or, in the absence of such supervisors in the design, by the construction manager. The PM, the construction manager, and the works manager are involved in monitoring the progress of the projects, and the works manager is involved in implementing changes. The progress of the projects is continuously monitored by the PJM, as he is responsible for the overall success and financial performance of the project.
13.	A project plan is a roadmap to help steer a project in the right direction. If the project is not going according to the plan, additional meetings are organised to solve problems and to clarify the issues that need to be addressed here and now. To know what changes are needed, it is necessary to have a risk plan in place so that problems are identified in advance and possible solutions are known. Depending on the situation, the project change plan may involve different people from the construction manager to the commercial director. If the problems are routine and ordinary, they are dealt with within the project team by the PJM, but if the PJM is unable to make decisions and the problems are extraordinary, then the higher levels, such as the construction director or his deputy, the commercial director, are called in.
14.	The success factors for PM consist of external and internal factors. The external success factors include detailed and correct design decisions and the timely delivery of materials and machinery. We cannot change the external circumstances, but we can control the internal circumstances, such as a prompt reaction to changes in the project, good teamwork, timely decisions, communication, and continuous planning (updating the project schedule and plan).
15	The causes of project delays, like success factors, can be both external and internal. External causes of delays include inconsistencies in designers' drawings, delays in additional solutions, lack of communication with builders, etc. Internal causes of delays include lack of communication, miscommunication, human error, chaotic decisions, lack of skills of the project team, and inappropriate structure of the project team. The external causes of delays cannot be directly influenced, but the internal problems can be solved, and communication is the most important tool to eliminate them. To deliver projects on time, it is necessary to keep the project under constant control, to be aware of the problems and to solve them, and if the problems cannot be solved by the PJM, to raise them to a higher level and take the necessary decisions.

Table 12. Comparison of responses of the representatives of Companies X, Y and Z.

Criterion	Company X	Company Y	Company Z
Number of employees	100–200 employees.	Up to 100 employees.	More than 200 employees.
Type of construction activity	Specialised construction work.	Specialised construction work.	A general construction contractor.
Standard project team	PJM, works manager, engineer, works executor, and employees.	PJM, works manager, works executor, and employees.	PJM, building construction manager, construction works manager, project engineer, executors, and employees.
Tools, techniques, methodologies, and philosophies used for PM	Risk management using MS Excel, MS Project, Power BI, and bespoke project progress monitoring software.	Dalux and MS Excel software, and Electronic Construction Work Log.	BIM, Dalux, Trimble, MS Project, MS Excel software, smart cost forecasting, and management software. LEAN PM philosophy tools—5S and Asaichi.

Table 12. Cont.

Criterion	Company X	Company Y	Company Z
Key indicators on project progress	Man-hours, materials and machinery, and human resources.	Duration of work, wear and tear on materials, and machinery.	Man-hours, materials and machinery, and project duration.
Frequency of monitoring project progress	Indicators for ongoing work projects—every day. Financial indicators—every month.	For small-scale projects—depending on the progress of the project. For medium and larger projects—two to three times a week. Monthly discussion of financial results.	Indicators of work in progress—every day. Financial indicators—every month.
Frequency of project progress discussions with the implementers	If there are no deviations, progress is not discussed with the contractors. For larger projects—daily meetings, up to 30 min.	Undefined for small projects (up to 1 month) and at least once a week for larger projects.	Usually once a week, with building construction manager, the construction works manager, and the contractors attending the meeting.
Areas for improvement in the project progress report	Comparison of direct project costs in the current period with the plan every week.	System optimisation, software upgrades, and installation.	Bi-weekly monitoring of the financial performance of projects not on track.
What changes can be made as a result of project progress monitoring?	Changes to work technology or equipment, negotiating work extensions, negotiating unforeseen work, and terminating contracts.	Changes to work schedules, project staffing levels, and cost estimates for subsequent projects.	The number of workers on the project, the agreement with the client on the extension of deadlines, the agreement on additional funds with the client, and the use of materials and machinery.
Action if the project does not go according to the plan	A project change committee is organised, and the PJM contacts the head of the unit with an estimate of the additional budget needed.	The reasons for deviations are identified, a meeting is held to optimise the work, and the project plan and schedule are updated. If an additional budget is needed for a project, the PJM makes proposals to the project director.	Priority problems are addressed, meetings are organised with the project team, and, in the case of extraordinary problems, heads of departments attend the meetings to help take appropriate decisions.
The following PM success factors were identified	Standardised and user-friendly implementation procedure, detailed project preparation, and PJM's handling of problems and deviations.	Focus on the project, tracking and optimising malfunction, communicating closely with collaborating companies, and maintaining good relations with the client.	External factors—detailed and correct design decisions and timely delivery of materials and machinery. Internal factors—self-sustained communication of the project team, good teamwork, and continuous planning.
Identified reasons for project delays	Inadequate materials, lack of team expertise, inappropriate team structure, and poor project preparation.	Machinery failures, delays in material supply, lack of competence of project teams, and designers' decisions.	External factors include inconsistencies and delays in design solutions and a lack of communication with contractors. Internal factors include chaotic decisions, lack of skills of the project team, and inappropriate structure of the project team.

According to Table 12, the standard project teams varied according to the size of the company. In Company Y, the standard project team consisted of a PJM, a works manager, a works supervisor, and workers, while in Company X, an engineer was added, and in Company Z, in comparison with Company Y, the standard project team included a construction manager, a project engineer, and several works supervisors. This shows that the companies implemented projects of different sizes. Of course, the composition of a project team may vary for certain projects.

The comparison of companies X, Y, and Z in terms of the tools and techniques used for PM, PM approaches, and philosophies revealed significant differences. General contracting Company Z used six different applications for PM such as Building Information Modeling, an intelligent cost forecasting and management program, Dalux, Trimble, and the MS

Project software, and it applied LEAN PM philosophies, which helped it to stay on track with project plans. Company X used four different applications for PM such as a unique software that tracks labour hours and material output based on work carried out, MS Excel, MS Project, and Power BI, but it did not identify the PM methodologies and philosophies used. Company Y employed the least number of tools, only three, which were Dalux, an electronic journal of construction works, and MS Excel, and it did not indicate any specific PM methodologies or philosophies but plans to implement additional PM tools shortly. In the future, various technologies such as machine learning, automation, artificial intelligence, the Internet of Things (IoT), data analytics, drone technologies, and robotics will present exciting opportunities for the workplace and, particularly, for the intelligent management of projects [55]. As project managers play a crucial role in implementing these transformative changes, the profession needs to be agile and adaptable to embrace the emerging wave of digitalisation. By embracing these advancements, project managers can effectively leverage the potential benefits and drive successful project outcomes in this rapidly evolving digital landscape.

Among the three companies (X, Y, and Z), the relevant project progress indicators were similar: monitoring man-hours and deviations in terms of materials and machinery used. Company X also focused on the number of human resources in the project, while Company Y additionally monitored the wear and tear of machinery, and Company Z monitored the duration of projects.

The comparison of the monitoring frequency and discussion of the progress of corporate projects revealed both differences and similarities. The companies participating in the qualitative study discussed financial results in monthly meetings. Companies X and Z monitored the progress of ongoing work projects daily, whereas in Company Y, the progress was monitored depending on the progress or size of the project. The notable differences between the companies could be seen in the discussion of the progress of the projects with the contractors. In Company X, no discussion was organised if the project was implemented according to the plan, except for larger projects, in which case the meetings were held every day. In Company Y, there was no defined periodicity of project progress discussion for small projects, but for larger projects, the meetings were held at least once a week. In Company Z, a meeting was held once a week regardless of the project results. Currently, virtual project teams have the advantage of utilising a wide range of technologies and communication tools to facilitate collaboration, even when team members are geographically dispersed, both domestically and internationally [55]. Tools such as Zoom meetings, Slack, FaceTime, Periscope, video conferences, and chat rooms have become invaluable assets for virtual teams. By harnessing these technologies, project teams can effectively bridge the distance and create seamless communication channels enabling efficient collaboration and coordination regardless of physical location.

The areas for improvement in the project progress reports of companies X and Z were the same, with representatives of both companies stating that they would like to see the financial results more often, while the representative of Company Z said that the company would like to systematise everything and introduce additional software, as there was a lack of supporting tools. The monitoring of project progress in companies X, Y, and Z helped to deal with the changes in project schedules and changes in the number of employees. In addition, the decision to terminate a contract may be taken in Company X. The general contracting Company Z monitored the sustainable use of materials and machinery to ensure their efficient use and to avoid downtime and keep within the planned quantities of materials. By leveraging the power of data analytics, project managers gain the capability to analyse and dissect intricate project data providing valuable insights and real-time predictions [56]. These analytical reports enable project managers to make informed decisions by drawing upon historical data, allowing them to effectively maintain project schedules and adhere to budgetary constraints [57]. This predictive information empowers project managers to proactively address potential challenges, optimise resource allocation, and ensure successful project outcomes.

The companies participating in the qualitative study had different perceptions of the project change plan. Company X organised a project change committee, where problems were identified and solved, while in Company Y, the causes of deviations were first identified, and then a meeting was organised to optimise the project plan, and finally, in Company Z, project changes were solved by the project team with the participation of department managers in meetings if extraordinary problems were encountered.

Successful PM in Company X consisted of a standardised implementation procedure, priority attention to the pre-project phase, and responsiveness of PJMs to problems and deviations. In Company Y, the focus was placed on the project, the optimisation of bottlenecks, and close communication with the customer. Company Z distinguished internal and external success factors in PM. The external success factors were detailed, and correct project decisions and the internal success factors involved timely communication of the project teams and constant planning. To enhance the PM success and improve the overall project results, continuous development of competencies and improvement of management methodologies are highly significant [30]. The authors of [58] examined organisation and management processes, empirically focusing on ontological and epistemological issues related to process research. They present effective methodological strategies for conducting empirical process studies and highlight some unique forms of insight that process research can provide.

It can be argued that the reasons for project delays identified by the companies in the qualitative study were similar. A lack of staff competence, inadequate project team structure, delays in material and machinery supply, breakdowns, and inconsistencies in design solutions were the main reasons for not meeting project deadlines. Leveraging Building Information Modeling and other simulation technologies can result in reduced project delivery time and enhanced cost management, ensuring projects stay within the allocated budget [59]. By utilising these advanced tools, project teams can optimise project planning, streamline coordination among stakeholders, and identify potential issues before they arise. This proactive approach facilitates efficient decision-making, mitigates risks, and improves overall project performance, leading to timely project completion and adherence to budgetary constraints. According to the authors of [60], incorporating design thinking into PM education is necessary to produce more effective PJMs and reduce the occurrence of project failures in the future.

4.4. Sustainability Aspect in Project Management for Future Studies

Construction projects have a significant impact on the environment and the society; therefore, construction project management and sustainability are strongly linked. This means that sustainability in construction project management requires the consideration of the environmental, social, and economic impact of a project throughout its life cycle. A variety of project management methodologies must be used to ensure that the desired sustainability outcomes are achieved. According to the authors of [61], integrating sustainability concepts into construction processes at both the strategic and operational levels has a profound impact on employees, the community, and the environment. Not only does this incorporation of sustainable practices benefit the well-being of employees, it also contributes to the enhancement of the community and the preservation of the environment. By prioritising sustainability in construction, organisations can drive positive changes and promote a more responsible and environmentally conscious approach to their operations. Embracing sustainable development enables project managers to shift their focus towards various aspects such as value creation, performance enhancement, efficiency improvement, business agility, project excellence, operational quality, paradigm shifts in thinking, flexibility, and more. New questionnaires can be designed based on the sustainable factors that include environmental, social, and economic impacts.

Future studies should also be linked with sustainability in construction PM through energy-efficient building development. One of the things which sustainability in construction PM could adopt is green building practices. These practices involve designing and constructing energy-efficient buildings using renewable resources and minimising waste. Green buildings are designed to reduce the carbon footprint of the building and ensure that the building operates efficiently throughout its life cycle. Given the increasing environmental concerns, reverse logistics also plays a crucial role in promoting sustainability within the construction industry. According to the authors of [62], reverse logistics has the potential to address and mitigate some of the adverse environmental impacts associated with construction activities, but its implementation level in the construction industry is still low. The root barriers to adopting reverse logistics in construction are as follows: the lack of financial incentives to incorporate recycled materials, the lack of knowledge about RL, the lack of technical support, standard codes, and regulations in favour of using recycled materials, the lack of information sharing, cooperation, and coordination among entities of the supply chain, the fact that current buildings have not been designed for deconstruction, and the lack of construction and demolition waste management and recycling infrastructures and markets for the materials resulting from construction and demolition waste.

In the case of PJMs, they have a critical role to play in promoting sustainability in the future. PJMs could also adopt sustainable procurement practices to ensure that the materials used in a construction project are sustainable. Sustainable procurement involves selecting materials that are environmentally friendly, socially responsible, and economically viable. Sustainable procurement practices can reduce the carbon footprint of a construction project, improve working conditions for workers, and support the local economy. In addition, they could promote sustainability through waste reduction and management. Waste reduction involves minimising the amount of waste generated during a construction project, while waste management involves the responsible disposal of waste. By adopting these practices, construction PJMs can minimise the environmental impact of a construction project. According to the authors of [63], inadequate stakeholder engagement, ineffective scope management, subpar schedule management, and insufficient resource management have detrimental effects on project sustainability over time. Moreover, these factors adversely impact employee wellbeing, resulting in decreased productivity, and they undermine the effectiveness of organisational management. The authors of [64] in their study indicated that obstacles such as 'insufficient sub-contractor cooperation,' 'resistance to changes in existing company structure and policy,' and 'the need for additional employee training' pose significant barriers to the successful implementation of environmental management systems in the construction industry. These identified barriers highlight that the primary challenge faced by construction professionals regarding environmental management systems implementation revolves around the level of emphasis placed on effectively communicating environmental concerns.

5. Conclusions

Based on the research conducted, the following conclusions were drawn:

1. The respondents identified the PRINCE2 project management approach as the most suitable for managing a repetitive construction project. According to the PRINCE2 project management methodology, a repetitive construction project would aim to provide as much information as possible to the project participants, to form a team and assign team leaders responsible for the phases, to establish a financial plan, a detailed timetable for the execution of the works, a quality control plan, and a plan of responsible persons, and to detail the technological sequencing of the works.
2. The rational project management approach PRINCE2 should be integrated into the management of a project under study by applying the seven principles, seven themes, and seven processes. Regular project meetings with the project board should be organised, information and plans should be prepared according to the principles and themes of the project management methodology, the project manager should appoint the persons responsible for the execution of the work, quality, and material control, and the phases should be analysed responsibly with the project board. Before the works are handed over to the client, the project manager should carry out a quality

3. control check, and the project board should control the communication between the project manager and the project team. In the project closure process, a learning-from-experience re-list should be completed, and the project results should be summarised and evaluated.
3. The analysis of the anti-corrosion works project under the PRINCE2 methodology suggested the following areas for improvement: checking the project quantities and describing the technological process before the start of the project to eliminate the risks associated with project delays; periodic control of the quality of the intermediate, understanding of quality requirements, and ensuring that work is carried out to a high standard; clarifying responsibilities and tasks in the teams before the start of the project; breaking the project down into phases to increase control over certain operations (such as the first coat of paint); and preparing a detailed project plan to avoid unforeseen activities (such as an additional cover of the steel structures).
4. The results of the qualitative research and the comparison of the participating companies revealed the following key similarities in project management: monitoring of relevant project progress indicators, project changes based on the project progress report, and the monitoring of project delays. Meanwhile, the main differences in project management among the surveyed companies were the project team size, the tools and methodologies used for project management, the project management philosophy, and the frequency of monitoring and discussing project progress.
5. According to the studied companies, a successful project management should consist of a standardised implementation procedure, priority attention to the pre-project phase and responsiveness of project managers to problems and deviations, focus on the project, optimisation of mistakes and close communication with the customer, detailed and correct project decisions, timely communication of the project teams, and constant planning. Meanwhile, the reasons for project delays were mainly related to the lack of staff competence, inadequate project team structure, delays in materials and machinery supply, breakdowns, and inconsistencies in design solutions.
6. Sustainable construction projects can minimise a project's environmental impact, improve working conditions, and support the local economy by adopting green building practices, sustainable procurement practices, and waste reduction and management practices. By prioritising sustainability development in construction, project managers can shift their focus towards various aspects such as value creation, performance enhancement, efficiency improvement, business agility, project excellence, operational quality, paradigm shifts in thinking, flexibility, and more. Future studies should also be linked with sustainability in construction project management through energy-efficient building development. One of the things which sustainability in construction project management could adopt is green building practices.

Author Contributions: Conceptualisation, A.S. and M.D.; methodology, A.S., M.D. and J.M.; software, A.S.; validation, A.S.; formal analysis, A.S.; investigation, A.S. and M.D.; resources, A.S. and J.M.; data curation, A.S. and M.D.; writing—original draft preparation, A.S., M.D. and J.M.; writing—review and editing, A.S., M.D. and J.M.; visualisation, A.S.; supervision, M.D. All authors have read and agreed to the published version of the manuscript.

Funding: This research received no external funding.

Data Availability Statement: Data sharing is not applicable.

Acknowledgments: The authors are grateful to the company LLC 'Švykai' for the technical support and data used in the research.

Conflicts of Interest: The authors declare no conflict of interest.

References

1. Al-Zwainy, F.M.S.; Mhammed, I.A.; Raheem, S.H. Application Project Management Methodology in Construction Sector: Review. *Int. J. Sci. Eng. Res. IJSER* **2016**, *7*, 244–253.
2. Project Management Institute. *A Guide to the Project Management Body of Knowledge (Pmbok® Guide)*; Project Management Institute: Newtown Square, PA, USA, 2017; ISBN 9781628251845.
3. PRINCE2 Training: Construction Industry. 2021. Available online: https://www.prince2training.co.uk/blog/prince2-for-the-construction-industry/ (accessed on 10 December 2021).
4. Knowledge Train®: PRINCE2® vs the PMBOK® Guide: A Comparison. 2021. Available online: https://www.knowledgetrain.co.uk/project-management/pmi/prince2-and-pmbok-guide-comparison (accessed on 10 December 2021).
5. Murray, A.; Bennett, N.; Bentley, C. *Managing Successful Projects with PRINCE2*; TSO: London, UK, 2015; ISBN 0113310595.
6. Jaziri, R.; El-Mahjoub, O.; Boussaffa, A. Proposition of a hybrid methodology of project management. *Am. J. Eng. Res. AJER* **2018**, *7*, 113–127.
7. Faraji, A.; Rashidi, M.; Perera, S.; Samali, B. Applicability-Compatibility Analysis of PMBOK Seventh Edition from the Perspective of the Construction Industry Distinctive Peculiarities. *Buildings* **2022**, *12*, 210. [CrossRef]
8. APM Group. *PRINCE2 Case Study. PRINCE2 and PMI/PIMBOK®. A Combined Approach at Getronics*; The APM Group Limited: Bucks, UK, 2002. Available online: https://silo.tips/download/contents-4-current-perceptions-of-relative-positioning-of-prince2-and-pmbok-appe (accessed on 13 June 2023).
9. Munns, A.K.; Bjeirmi, B.F. The role of project management in achieving project success. *Int. J. Proj. Manag.* **1996**, *14*, 81–87. [CrossRef]
10. Bakry, I.; Moselhi, O.; Zayed, T. Optimized acceleration of repetitive construction projects. *Autom. Constr.* **2014**, *39*, 145–151. [CrossRef]
11. Long, L.D.; Ohsato, A. A genetic algorithm-based method for scheduling repetitive construction projects. *Autom. Constr* **2009**, *18*, 499–511. [CrossRef]
12. Vanhoucke, M. Work continuity constraints in project scheduling. *J. Constr. Eng. Manag. ASCE* **2006**, *132*, 14–25. [CrossRef]
13. Russell, A.D.; Wong, W.C.M. New generation of planning structures. *J Constr Eng Manag. ASCE* **1993**, *119*, 196–214. [CrossRef]
14. Reda, R.M. RPM: Repetitive project modelling. *J. Constr. Eng. Manag. ASCE* **1990**, *116*, 316–330. [CrossRef]
15. Hyari, K.; El-Rayes, K.; Asce, M. Optimal Planning and Scheduling for Repetitive Construction Projects. *J. Manag. Eng.* **2006**, *22*, 11–18. [CrossRef]
16. Drouin, N.; Muller, R.; Sankaran, S. The nature of organizational project management through the lens of integration. In *Cambridge Handbook of Organizational Project Management*; Cambridge University Press: Cambridge, MA, USA, 2017; pp. 9–18. ISBN 978131666224-3. [CrossRef]
17. Aubry, M.; Lavoie-Tremblay, M. Rethinking organizational design for managing multiple projects. *Int. J. Proj. Manag.* **2018**, *36*, 12–26. [CrossRef]
18. Arditi, D.; Nayak, S.; Damci, A. Effect of organizational culture on delay in construction. *Int. J. Proj. Manag.* **2017**, *35*, 136–147. [CrossRef]
19. Ershadi, M.; Jefferies, M.; Davis, P.; Mojtahedi, M. Achieving Sustainable Procurement in Construction Projects: The Pivotal Role of a Project Management Office. *Constr. Econ. Build.* **2021**, *21*, 45–64. [CrossRef]
20. Arditi, D.; Robinson, M.A. Concurrent delays in construction litigation. *Cost Eng.* **1995**, *37*, 20–28.
21. Aubry, M.; Hobbs, B.; Thuillier, D. A new framework for understanding organisational project management through the PMO. *Int. J. Proj. Manag.* **2007**, *25*, 328–336. [CrossRef]
22. Sweis, G.; Sweis, R.; Abu Hammad, A.; Shboul, A. Delays in construction projects: The case of Jordan. *Int. J. Proj. Manag.* **2008**, *26*, 665–674. [CrossRef]
23. Arditi, D.; Pattanakitchamroon, T. Selecting a delay analysis method in resolving construction claims. *Int. J. Proj. Manag.* **2006**, *24*, 145–155. [CrossRef]
24. Kim, Y.; Kim, K.; Shin, D. Delay analysis method using delay section. *J. Constr. Eng. Manag.* **2005**, *131*, 1155–1164. [CrossRef]
25. Gunduz, M.; Nielsen, Y.; Ozdemir, M. Fuzzy assessment model to estimate the probability of delay in Turkish construction projects. *J. Constr. Eng. Manag.* **2013**, *31*, 04014055. [CrossRef]
26. Mahamid, I. Effects of Design Quality on Delay in Residential Construction Projects. *J. Sustain. Archit. Civ. Eng.* **2021**, *1*, 118–129. Available online: https://sace.ktu.lt/index.php/DAS/article/view/20531 (accessed on 10 December 2021). [CrossRef]
27. Shi, J.; Cheung, S.O.; Arditi, D. Construction delay computation method. *J. Constr. Eng. Manag.* **2001**, *127*, 60–65. [CrossRef]
28. Gudienė, N.; Banaitis, A.; Banaitienė, N.; Lopes, J. Development of a conceptual critical success factors model for construction projects: A case of Lithuania. *Procedia Eng.* **2013**, *57*, 392–397. [CrossRef]
29. Radujkovič, M.; Sjekavica, M. Project management success factors. *Procedia Eng.* **2017**, *196*, 607–615. [CrossRef]
30. Radujkovic, M.; Sjekavica, M. Development of a project management performance enhancement model by analysing risks, changes, and management. *Građevinar* **2017**, *69*, 105–120. [CrossRef]
31. Greenwood, R.; Miller, D. Tackling design anew: Getting back to the heart of organizational theory. *Acad. Manag. Perspect.* **2010**, *24*, 78–88. [CrossRef]
32. Ingle, P.V.; Mahesh, G. Construction project performance areas for Indian construction projects. *Int. J. Constr. Manag.* **2020**, *22*, 1443–1454. [CrossRef]

33. Arantes, A.; Ferreira, L.M.D.F. Development of delay mitigation measures in construction projects: A combined interpretative structural modeling and MICMAC analysis approach. *Prod. Plan. Control* **2023**, 1–16. [CrossRef]
34. Matos, S.; Lopes, E. Prince2 or PMBOK—A question of choice. *Procedia Technol.* **2013**, *9*, 787–794. [CrossRef]
35. Sepasgozar, S.M.E.; Karimi, R.; Shirowzhan, S.; Mojtahedi, M.; Ebrahimzadeh, S.; McCarthy, D. Delay Causes and Emerging Digital Tools: A Novel Model of Delay Analysis, Including Integrated Project Delivery and PMBOK. *Buildings* **2019**, *9*, 191. [CrossRef]
36. Safaeian, M.; Fathollahi-Fard, A.M.; Kabirifar, K.; Yazdani, M.; Shapouri, M. Selecting Appropriate Risk Response Strategies Considering Utility Function and Budget Constraints: A Case Study of a Construction Company in Iran. *Buildings* **2022**, *12*, 98. [CrossRef]
37. Soltanzadeh, A.; Mahdinia, M.; Omidi Oskouei, A.; Jafarinia, E.; Zarei, E.; Sadeghi-Yarandi, M. Analyzing Health, Safety, and Environmental Risks of Construction Projects Using the Fuzzy Analytic Hierarchy Process: A Field Study Based on a Project Management Body of Knowledge. *Sustainability* **2022**, *14*, 16555. [CrossRef]
38. ISO 8501-3:2006; Preparation of Steel Substrates before Application of Paints and Related Products—Visual Assessment of Surface Cleanliness Preparation Grades of Welds, Edges and other Areas with Surface Imperfections. International Standard Organization: Geneva, Switzerland, 2006.
39. ISO 8502-6:2020; Preparation of Steel Substrates before Application of Paints and Related Products—Tests for the Assessment of Surface Cleanliness—Part 6: Extraction of Water Soluble Contaminants for Analysis (Bresle Method). International Standard Organization: Geneva, Switzerland, 2020.
40. ISO 8503-5:2017; Preparation of Steel Substrates before Application of Paints and Related Products—Surface Roughness Characteristics of Blast-Cleaned Steel Substrates—Part 5: Replica Tape Method for the Determination of the Surface Profile. International Standard Organization: Geneva, Switzerland, 2017.
41. SSPC-SP1:2015; Solvent Cleaning. The Society for Protective Coatings: Pittsburgh, PA, USA, 2015.
42. ASTM D4940-15(2020); Standard Test Method for Conductimetric Analysis of Water-Soluble Ionic Contamination of Blast Cleaning Abrasives. American Society for Testing and Materials: West Conshohocken, PA, USA, 2020.
43. ISO 8502-4:2017; Preparation of Steel Substrates before Application of Paints and Related Products—Tests for the Assessment of Surface Cleanliness—Part 4: Guidance on the Estimation of the Probability of Condensation Prior to Paint Application. International Standard Organization: Geneva, Switzerland, 2017.
44. ASTM D4285-83(2018); Standard Test Method for Indicating Oil or Water in Compressed Air. American Society for Testing and Materials: West Conshohocken, PA, USA, 2018.
45. ISO 8501-1:2007; Preparation of Steel Substrates before Application of Paints and Related Products—Visual Assessment of Surface Cleanliness Rust Grades and Preparation Grades of Uncoated Steel Substrates and of Steel Substrates after Overall Removal of Previous Coatings. International Standard Organization: Geneva, Switzerland, 2007.
46. ISO 8502-3:2017; Preparation of Steel Substrates before Application of Paints and Related Products—Tests for the Assessment of Surface Cleanliness—Part 3: Assessment of Dust on Steel Surfaces Prepared for Painting (Pressure-Sensitive Tape Method). International Standard Organization: Geneva, Switzerland, 2017.
47. ISO 8503-2:2012; Preparation of Steel Substrates before Application of Paints and Related Products—Surface Roughness Characteristics of Blast-Cleaned Steel Substrates Method for the Grading of Surface Profile of Abrasive Blast-Cleaned Steel—Comparator Procedure. International Standard Organization: Geneva, Switzerland, 2012.
48. ISO 4624:2016; Paints and Varnishes—Pull-Off Test for Adhesion. International Standard Organization: Geneva, Switzerland, 2016.
49. Ribeiro, A.M.; Arantes, A.; Cruz, C.O. Barriers to the Adoption of Modular Construction in Portugal: An Interpretive Structural Modeling Approach. *Buildings* **2022**, *12*, 1509. [CrossRef]
50. Forza, C. Survey research in operations management: A process-based perspective. *Int. J. Oper. Prod. Man.* **2002**, *22*, 152–194. [CrossRef]
51. Kallio, H.; Pietila, A.M.; Johnson, M.; Kangasniemi, M. Systematic methodological review: Developing a framework for a qualitative semi-structured interview guide. *J. Adv. Nurs.* **2016**, *72*, 2954–2965. [CrossRef]
52. Gould, F.E. *Managing the Construction Process: Estimating, Scheduling, and Project Control*; Pearson Longman: London, UK, 2012; ISBN 978-0138135966.
53. Majrouhi, S.J. Influence of RFID technology on automated management of construction materials and components. *Sci. Iran.* **2012**, *19*, 381–392. [CrossRef]
54. Tay, Y.W.D.; Panda, B.; Paul, S.C.; Noor Mohamed, N.A.; Tan, M.J.; Leong, K.F. 3D printing trends in building and construction industry: A review. *Virtual Phys. Prototyp.* **2017**, *12*, 261–276. [CrossRef]
55. Aliu, J.; Oke, A.E.; Kineber, A.F.; Ebekozien, A.; Aigbavboa, C.O.; Alaboud, N.S.; Daoud, A.O. Towards a New Paradigm of Project Management: A Bibliometric Review. *Sustainability* **2023**, *15*, 9967. [CrossRef]
56. Tiwari, S.; Wee, H.M.; Daryanto, Y. Big data analytics in supply chain management between 2010 and 2016: Insights to industries. *Comput. Ind. Eng.* **2018**, *115*, 319–330. [CrossRef]
57. Ram, J.; Afridi, N.K.; Khan, K.A. Adoption of Big Data analytics in construction: Development of a conceptual model. *Built Environ. Proj. Asset Manag.* **2019**, *9*, 564–579. [CrossRef]
58. Langley, A.; Smallman, C.; Tsoukas, H.; Van De Ven, A.H. Process studies of change in organization and management: Unveiling temporality, activity, and flow. *Acad. Manag. J.* **2013**, *56*, 1–13. [CrossRef]

59. Chu, M.; Matthews, J.; Love, P.E.D. Integrating mobile Building Information Modelling and Augmented Reality systems: An experimental study. *Autom. Constr.* **2018**, *85*, 305–331. [CrossRef]
60. Ewin, N.; Luck, J.; Chugh, R.; Jarvis, J. Rethinking Project Management Education: A Humanistic Approach based on Design Thinking. *Procedia Comput. Sci.* **2017**, *121*, 503–510. [CrossRef]
61. Toljaga-Nikolić, D.; Todorović, M.; Dobrota, M.; Obradović, T.; Obradović, V. Project Management and Sustainability: Playing Trick or Treat with the Planet. *Sustainability* **2020**, *12*, 8619. [CrossRef]
62. Pimentel, M.; Arantes, A.; Cruz, C.O. Barriers to the Adoption of Reverse Logistics in the Construction Industry: A Combined ISM and MICMAC Approach. *Sustainability* **2022**, *14*, 15786. [CrossRef]
63. Govindaras, B.; Wern, T.S.; Kaur, S.; Haslin, I.A.; Ramasamy, R.K. Sustainable Environment to Prevent Burnout and Attrition in Project Management. *Sustainability* **2023**, *15*, 2364. [CrossRef]
64. Beck Schildt, J.C.; Booth, C.A.; Horry, R.E.; Wiejak-Roy, G. Stakeholder Opinions of Implementing Environmental Management Systems in the Construction Sector of the U.S. *Buildings* **2023**, *13*, 1241. [CrossRef]

Disclaimer/Publisher's Note: The statements, opinions and data contained in all publications are solely those of the individual author(s) and contributor(s) and not of MDPI and/or the editor(s). MDPI and/or the editor(s) disclaim responsibility for any injury to people or property resulting from any ideas, methods, instructions or products referred to in the content.

Article

Investigation of Sensible Cooling Performance in the Case of an Air Handling Unit System with Indirect Evaporative Cooling: Indirect Evaporative Cooling Effects for the Additional Cooling System of Buildings

Attila Kostyák, Szabolcs Szekeres and Imre Csáky *

Department of Building Services and Building Engineering, Faculty of Engineering, University of Debrecen, Ótemető Str. 2-4, 4028 Debrecen, Hungary; kostyak.attila@eng.unideb.hu (A.K.); szekeres@eng.unideb.hu (S.S.)
* Correspondence: imrecsaky@eng.unideb.hu; Tel.: +36-52-415-155 (ext. 77772)

Abstract: Previous studies have shown that the amount of energy consumed by mechanical cooling can be significantly reduced by the indirect evaporative cooling (IEC) process. By increasing the heat recovery efficiency of air handling units (AHUs), sensible cooling performance can be achieved with the IEC process for a significant part of the cooling season. This study determined the sensible cooling performance under which outdoor air conditions can be achieved. With IEC, the indoor humidity load cannot be adequately managed and must be solved by a supplementary cooling system, which may require additional cooling energy. This study shows the effect of the set indoor humidity on the amount of cooling energy required. The increase in energy consumption of the supplementary cooling system has been determined by simulation and for which indoor air conditions the amount of cooling energy used can be optimized if only IEC cooling is used in the air handling unit.

Keywords: indirect evaporative cooling; evaporative cooling; air handling unit; cooling system; cooling energy; heat recovery

Citation: Kostyák, A.; Szekeres, S.; Csáky, I. Investigation of Sensible Cooling Performance in the Case of an Air Handling Unit System with Indirect Evaporative Cooling: Indirect Evaporative Cooling Effects for the Additional Cooling System of Buildings. *Buildings* **2023**, *13*, 1800. https://doi.org/10.3390/buildings13071800

Academic Editors: Lina Šeduikytė and Jakub Kolarik

Received: 26 June 2023
Revised: 7 July 2023
Accepted: 10 July 2023
Published: 14 July 2023

Copyright: © 2023 by the authors. Licensee MDPI, Basel, Switzerland. This article is an open access article distributed under the terms and conditions of the Creative Commons Attribution (CC BY) license (https://creativecommons.org/licenses/by/4.0/).

1. Introduction

With ever-increasing outdoor peak temperatures and energy prices, system solutions that increase cooling efficiency are becoming essential. Direct evaporative cooling (DEC) has been used for a long time in industrial refrigeration systems [1]. The cooling process is also often used to improve the operating conditions of the cooling circuit [2,3]. For outdoor, or semi-open areas, DEC is also ideal [4]. The direct evaporative cooling process in dry hot climates is also suitable for treating higher-quality comfort areas [5]. Higher comfort requirements are difficult to achieve in regions with a humid climate with DEC [6]. DEC cooling solutions have a limitation in comfort areas as the evaporated water vapor influences the sense of comfort [7]. Keeping the relative humidity in the comfort space within a given zone limits the maximum cooling capacity of DEC systems compared to the cooling capacity that can potentially be achieved [8]. In recent years, several studies have investigated the potential of indirect evaporative air cooling [9]. With indirect evaporative cooling (IEC), the treated air is not directly supplied to the comfort space and therefore does not affect it [10]. One of the most common applications of indirect evaporative cooling is to cool the air exhausted by air handling units [11]. Using a heat recovery unit, it is possible to pre-cool the outdoor air without mass transfer [9]. Due to the efficiency of the IEC process, the cooling demand of air handling units produced by the refrigeration circuit can be significantly reduced, saving a significant amount of energy [12]. With the increase in heat recovery efficiencies, it is now possible to achieve isothermal inflow or sensible cooling power input using only IEC cooling methods, which creates an opportunity to review the technical design of building mechanical systems. As the efficiency of the indirect cooling process depends on several parameters, a complex assessment is required to determine the energy savings that can be expected using IEC alone and the operating range at which sensible cooling performance or isothermal inflow can be achieved [13]. Determination of isothermal inflow boundary parameters helps to design predictive control of supplementary cooling systems. If the outdoor weather conditions do not reach the isothermal inflow threshold for the whole cooling season, it is possible to treat the air with IEC only [14]. In this case, however, it must be considered that the humidity of the external air cannot always be treated properly by the IEC procedure. The humidity level of the external air may exceed the level that should be maintained indoors, so the management of the humidity level falls to the low-temperature supplementary cooling system. The humidity level of the external air may exceed the level that should be maintained indoors, leaving the task of managing the humidity level to the low-temperature supplementary cooling system. A good example of this is where the air handling unit supplies 100% fresh air to the occupied zone and the premises are cooled by low-temperature supplementary cooling systems (VRF, fan-coil, etc.). The additional cooling energy demand due to humidity load should be considered when designing the supplementary cooling system [15].

This study presents the theoretical basis of the DEC and IEC processes. It will show how to define the operating zone in which the sensible cooling performance of the IEC process can be achieved. The performance of an air handling unit using only IEC cooling was investigated under laboratory conditions when the supplementary cooling energy was provided by a fan-coil system for high-quality comfort space. Based on several years of meteorological data, it was simulated how much cooling energy can be saved by using the IEC procedure and how much the additional cooling system varies the amount of cooling energy depending on the amount of humidity load from the external environment.

2. Materials and Methods

When designing office comfort areas, the primary aim is to create ideal working conditions. Employee well-being is necessary for effective work [16]. Many factors affect human comfort, a major part of which is influenced by the ventilation system. The air handling unit, which is part of the air handling system, influences the air temperature, humidity, and air quality in the comfort space. A significant amount of energy is used to set and maintain these comfort parameters [17]. The air distribution system associated with air handling units influences the movement of air in the occupied zone and the variability of air movement [18]. The sizing of air handling systems in practice is carried out according to several standards (ISO 7730:2006; CR 1752:2000; ASHRAE 55) [19–21]. The standards contain several methods for determining the amount of fresh air. The required volumetric flow rate of fresh air can be determined, for example, based on the headcount method, where the amount of necessary fresh air is determined by the performed activities and the number of workers [20]. Another method is the coverage method, where the minimum air volume required to flush the room is determined [20]. Calculating the amount of fresh air needed to dilute pollutant concentration is another option for determining the minimum fresh air requirement (in offices, this usually means measuring CO_2 concentration) [22]. According to standard recommendations, by selecting the highest value from the minimum fresh air requirements determined using the methods for determining the amount of fresh air, we can serve the treated area with an adequate amount of fresh air from all aspects. Treating fresh air during the cooling season results in significant cooling energy consumption, which is usually provided by chillers and heat pumps [23]. The cooling solutions employed have a significant impact on the extent of peak summer electricity consumption [24]. By employing direct and indirect evaporative cooling solutions, we can significantly reduce the consumption of cooling systems [25]. In sizing the air handling system, optimizing between energy and comfort considerations is of immense importance. Increasing the amount of fresh air is advantageous from a comfort perspective but requires additional energy investment from an energy perspective. Additionally, the desired indoor air conditions significantly influence the cooling energy requirements of the room. When designing internal air conditions, it is important to examine what level of comfort provides adequate comfort for people while keeping the cooling energy demand low [26]. In our study, we present our calculations for an indoor target temperature of 25 °C, which, according to standards, can provide high-quality comfort during the summer season [21]. The sizing steps presented can be applied to other indoor design air conditions as well.

2.1. Calculation Methods

Heat recovery units have been used in air handlers for a long time for the pre-treatment of outdoor air. With the help of heat recovery units, the relative cooling energy of the exhaust air can be used to pre-cool the outdoor air. The efficiency of heat recovery can be described by the ratio of the enthalpy change between the inlet and outlet points of the heat exchanger. If the temperature of the exhaust air from the indoor space is higher than the dew point temperature of the outdoor air, no condensation will occur during heat recovery. In this case, the efficiency of heat recovery can be determined from the temperature change (assuming that the mass flow rates on both sides are equal) [27].

$$\eta_{rec.} = \frac{h_{rs} - h_i}{h_x - h_0} = \frac{h_x - h_{re}}{h_x - h_0} \simeq \frac{T_x - T_{re}}{T_x - T_o} \quad (1)$$

In direct evaporative cooling, the air is forced to flow through a wetting pad. Heat and mass transfer occur simultaneously between the moving air mass and the large water film. The energy required for evaporation comes from the medium adjacent to the water surface (air), so the temperature of the air decreases depending on the amount of water evaporated. The energy taken from the medium returns to the gaseous mixture along with the evaporating material, so during the process, the enthalpy of the water vapor–air mixture changes only slightly [28]. If no external heat is added or removed from the system during the process, meaning that the process occurs adiabatically, the enthalpy of the medium increases with the enthalpy of the evaporated water vapor [29]:

$$\Delta h = \Delta x \times c_{pv} \times t_v [kJ/kg] \quad (2)$$

Practical calculation methods disregard the change in the enthalpy of the medium since its magnitude and effect are negligible compared to the absolute enthalpy value of the system. In the following, we calculate taking into account the above simplification. If we consider the enthalpy of the air–water vapor mixture to be constant ($h \approx$ constant) during the process, the resulting air temperature after the procedure can be expressed as a function of the initial air state (temperature, humidity) and the amount of evaporated water using Equation (3):

$$t_e = \frac{-r \times \Delta x}{c_{pa} + c_{pv} \times \Delta x} + t_x [°C] \quad (3)$$

The amount of evaporable water depends on the state of the treated air and the efficiency of the evaporative cooling system.

$$\varepsilon = \frac{T_{in} - T_{out}}{T_{in} - T_{wb\ out}} [-] \quad (4)$$

If the initial air state is known, it is possible to determine the required evaporative cooling efficiency to achieve a given amount of moisture uptake. The table shows that, in addition to the initial temperature, the relative humidity greatly influences the amount of water that can be evaporated.

The achievable evaporative efficiency greatly depends on the method of humidification. The formation of a large free water surface required for adiabatic humidification can be achieved in several ways: water spray, cooling pad technology, etc. [30,31]. Our measurements were performed using cellulose cooling pads. Previous research has shown that the efficiency of evaporative cooling with cellulose-based filters ranges from around 60 to 90% depending on the thickness of the filter and the flow rate [32]. Taking this into account, the amount of evaporated water corresponding to the highlighted efficiency value in Table 1 is expected to depend on the internal air condition.

Table 1. Required evaporative efficiency in function of added water vapor (in several relative humidity cases).

Handled Air Starting Parameters	Δx Added Evaporated Water (g/kg)													
	0.25	0.50	0.75	1.00	1.25	1.50	1.75	2.00	2.25	2.50	2.75	3.00	3.25	3.50
25 °C														
30 RH%	6%	12%	18%	23%	29%	35%	41%	47%	52%	58%	64%	70%	76%	81%
40 RH%	7%	14%	21%	28%	35%	42%	49%	56%	63%	70%	77%	84%	91%	98%
50 RH%	9%	17%	26%	34%	43%	52%	60%	69%	77%	86%	95%	-	-	-
60 RH%	11%	22%	33%	44%	55%	66%	77%	88%	99%	-	-	-	-	-
70 RH%	15%	30%	45%	60%	75%	90%	-	-	-	-	-	-	-	-
80 RH%	23%	46%	68%	91%	-	-	-	-	-	-	-	-	-	-

As a function of the evaporating water, the treated air cools and becomes more humid, which results in an increase in the cooling performance of the heat recovery system (see Figure 1). With the help of this process, the treated outdoor air can be cooled below room temperature. The boundary value of the external temperature for isothermal supply air can be derived from (5)–(7).

$$t_i - t_s = 0 \; [°C] \qquad (5)$$

$$t_i - [t_{o,limit} - \eta(t_{o,limit} - t_e)] = 0 \; [°C] \qquad (6)$$

$$t_{o,limit} = \frac{t_i - \eta t_e}{1-\eta} = \frac{t_i - \eta \left[\frac{-r\Delta x}{c_{pa}+c_{pv}\Delta x} + t_i\right]}{1-\eta} = t_i - \frac{\eta \left[\frac{-r\Delta x}{c_{pa}+c_{pv}\Delta x}\right]}{1-\eta} \; [°C] \qquad (7)$$

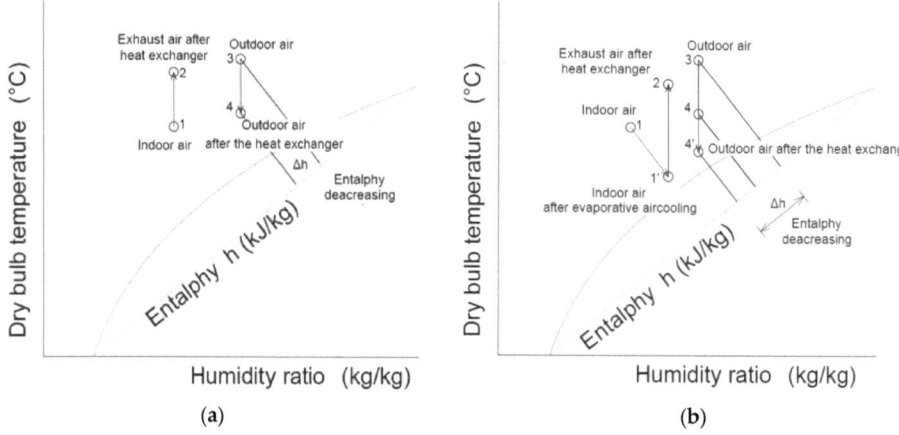

Figure 1. Heat recovery process presented in Mollier h-x diagram (**a**) AHU without IEC (**b**) AHU with IEC.

If the outdoor temperature is lower than the calculated limit, then with indirect evaporative cooling, we can achieve a supply air temperature lower than the indoor temperature, which means that we can introduce a sensible cooling performance into the room with the cooling process, provided that the absolute humidity of the external air does not reach the condensation limit. Based on the above, we also need to determine an absolute humidity limit in addition to the temperature limit to write the operating range of indirect evaporative cooling. The humidity limit is the saturated humidity corresponding to the air state after the evaporative cooling. If the air at the temperature limit contains more humidity than this limit, then the condensation heat of this humidity will be higher than that of the air cooled by the evaporative path involved in the heat exchange. This results in condensation, and its energy requirement reduces the temperature difference created. To determine the humidity limit, we need to determine the saturation vapor pressure of the air after evaporative cooling, which can be determined using the Antoine equation [33]:

$$p_{ws} = e^{\left(23.752 + \frac{-4134.9088}{238.5104 + t_e}\right)} \; [\text{Pa}] \tag{8}$$

Given the knowledge of saturation vapor pressure and atmospheric pressure, the limit of absolute humidity can be determined as a function of the temperature that occurs after evaporative cooling.

$$x_{o,limit} = 0.622 \frac{p_{ws}}{p_0 - p_{ws}} \; [\text{kg/kg}] \tag{9}$$

If we assume that p_0 is equal to the normal atmospheric pressure (101,325 Pa), and the relationship derived from the Antoine equation is inserted, then:

$$x_{o,limit} = 0.622 \frac{e^{\left(23.752 + \frac{-4134.9088}{238.5104 + t_e}\right)}}{101325 - e^{\left(23.752 + \frac{-4134.9088}{238.5104 + t_e}\right)}} \; [\text{kg/kg}] \tag{10}$$

The equation can be used to determine the air temperature after evaporative cooling as a function of the initial air temperature and the amount of evaporated water, according to Equation (3), which can be inserted into Equation (10) to obtain the following relationship:

$$x_{o,limit} = 0.622 \frac{e^{\left(23.752 + \frac{-4134.9088}{238.5104 + \left[\frac{-r\Delta x}{c_{pa} + c_{pv}\Delta x} + t_i\right]}\right)}}{101325 - e^{\left(23.752 + \frac{-4134.9088}{238.5104 + \left[\frac{-r\Delta x}{c_{pa} + c_{pv}\Delta x} + t_i\right]}\right)}} \; [\text{kg/kg}] \tag{11}$$

If the humidity level of the external air is higher than the critical humidity level, condensation occurs during heat exchange. The enthalpy line of the air state determined by the critical temperature and critical humidity represents the limit where the air can be cooled down to the indoor air state.

$$h_{o,limit} = c_{pa} t_{o,limit} + x_{o,limit} \left(r + c_{pv} t_{o,limit}\right) \; [\text{kJ/kg}] \tag{12}$$

$$h_{o,limit} = c_{pa}\left\{t_i - \frac{\eta\left[\frac{-r_0\Delta x}{c_{pa}+c_{pv}\Delta x}\right]}{1-\eta}\right\} + 0.622 \frac{e^{\left(23.752 + \frac{-4134.9088}{238.5104 + \left[\frac{-r\Delta x}{c_{pa}+c_{pv}\Delta x}+t_i\right]}\right)}}{101325 - e^{\left(23.752 + \frac{-4134.9088}{238.5104 + \left[\frac{-r\Delta x}{c_{pa}+c_{pv}\Delta x}+t_i\right]}\right)}} \left(r + c_{pv}\left\{t_i - \frac{\eta\left[\frac{-r\Delta x}{c_{pa}+c_{pv}\Delta x}\right]}{1-\eta}\right\}\right) \tag{13}$$

The limit values $t_{0,limit}$, $x_{0,limit}$, and $h_{0,limit}$ determine the operating range using Equations (7), (11) and (13), where the treated air has a dry bulb temperature equal to or lower than the indoor air temperature after indirect evaporative air cooling. The boundaries of the operating zone can be determined based on the indoor air temperature, the amount of evaporated water, and the heat recovery efficiency. For example, assuming a heat recovery efficiency of 80% and an indoor air state of 25 °C and 40% RH, the operating range varies depending on the amount of evaporated water, as shown in Figure 2. Using the above formulas, it can be analyzed whether the indirect cooling process can be applied independently for air treatment based on the desired internal comfort level and the degree of evaporation.

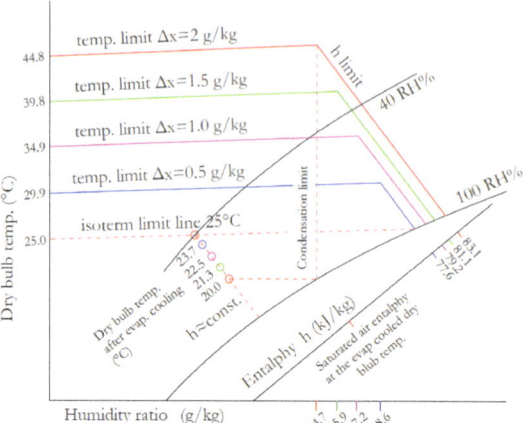

Figure 2. Operating range depends on the additional water vapor by IEC to reach sensible cooling performance.

If the outdoor air condition does not cross the determined boundary during the summer season, the perceptible cooling effect can be achieved with the supplied air. If the outdoor air condition is at the boundary, isothermic supply can be realized. For example, by adiabatic evaporation of 1.5 g/kg water from the air extracted from the room during the IEC process, at 80% heat recovery efficiency, it is possible to reach a supplied air temperature below 25 °C up to 39.8 °C and an absolute humidity of 15.9 g/kg. If the outdoor air humidity exceeds the value of 15.9 g/kg, then the boundary is defined by the enthalpy line corresponding to the air condition of 39.8 °C and 15.9 g/kg humidity (83.1 kJ/kg). Taking into account the indoor air condition and the meteorological data of the installation area, it can be assessed whether the IEC process can be applied independently in air treatment equipment. This question is particularly relevant in the case of building renovations. Numerous office buildings are constructed so that an air handling system takes care of the treatment of fresh air and its introduction into the occupied zones, while a separate heat emitter system takes care of the supply of cooling and heating energy (e.g., fan-coil, VRF systems). These cooling systems generally have low–medium temperatures, which allow for handling of moisture load in the rooms.

2.2. Measuring Methods

A unique measuring system was set up in the air conditioning laboratory of the Building Services and Building Engineering Department at the University of Debrecen to investigate the effect of indirect evaporative cooling on the energy consumption of the air conditioning system. A unique direct evaporative cooling unit (DEC) was installed in the laboratory's air conditioning system as can be seen in Figure 3, creating a dry IEC air handling unit.

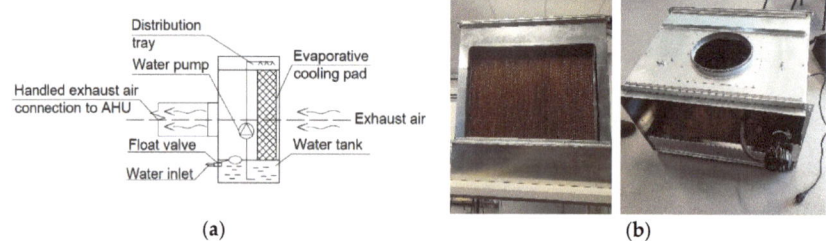

Figure 3. (**a**) Scheme of the evaporative air cooler (**b**) Built evaporative air cooler unit.

The evaporative cooling unit was connected to the drinking water network. The water level in the unit is regulated by a float valve. A distribution network with a pump is responsible for wetting the evaporative cooling pad. The size of the evaporative cooling pad is 0.89 × 0.72 m with a thickness of 100 mm. During the operation of the system, the evaporative efficiency was calculated according to Equation (4). The DEC unit is connected to the exhaust branch of the air handling unit. The air conditioning system was designed according to Figure 4.

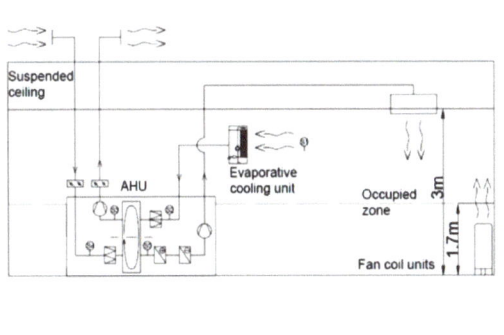
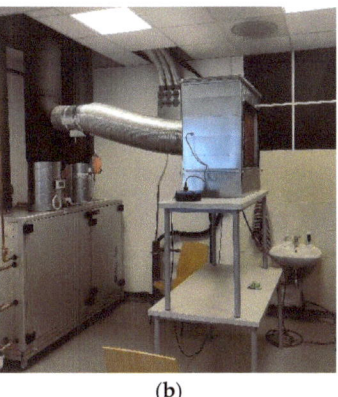

(a) (b)

Figure 4. (a) Scheme of the laboratory measurement (b) Direct evaporative air cooler unit applied to the AHU exhaust branch (without insulation).

We closed off the exhaust branches connected to the air handling system, and air was only extracted from the interior through the evaporative cooling unit. The air handling system delivers 100% fresh air to the laboratory and does not have a recirculation branch. During the measurements, balanced ventilation was implemented. A volumetric flow rate of 1000 m^{3}h^{-1} was set during the measurements. A rotary heat recovery unit was installed in the air handling system.

The measurements were carried out between 26 July and 17 September 2021 and were conducted for 19 days (2592 observations). During the measurement period, close attention was paid to time intervals lasting from 12 to 16 h, which were the most critical periods considering the outdoor temperature. The measurements started with a room temperature of 25 °C, and the room temperature was maintained using the laboratory fan-coil system. During the measurement period, there were summer [Figure A1], hot [Figure A2], and torrid days [Figure A3] (Table 2) at the installation site, which therefore enabled us to analyze the system's operation for all outdoor air conditions that occurred during the cooling season.

Table 2. Categories of days during the measurements.

	Summer Days 30 °C > t_{max} > 25 °C	Hot Days 35 °C > t_{max} ≥ 30 °C	Torrid Days t_{max} ≥ 35 °C
Number of days (pc)	10	7	2

The measurements were carried out using the Testo Saversis Monitoring system (Figure 5). The central unit of the system communicates with the sensors via radio frequency. If the sensors go out of range or become shaded during data collection, their own memory is capable of collecting data, and when the radio frequency connection is restored, the data stored in the sensors are downloaded to the central unit.

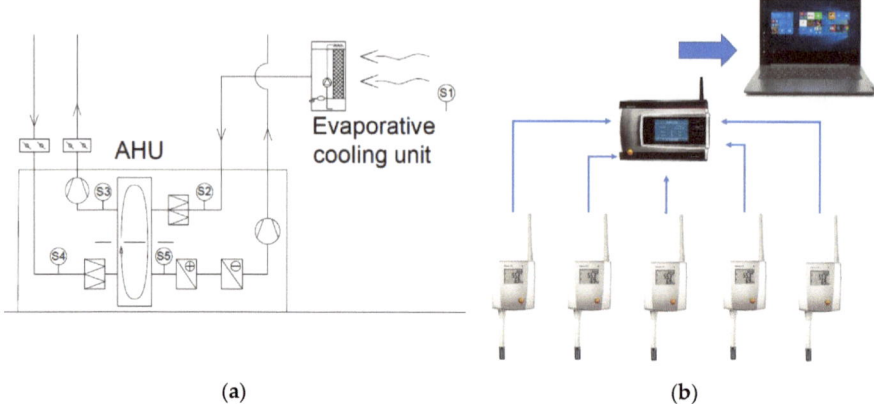

Figure 5. (**a**) The positions of the instruments (**b**) Scheme of the Testo Saveris measurements system. S1—Testo Saveris H3 D-2 sensor (T.S.s.); Measured area: air extracted from the room. S2—T.S.s.; air extracted from the room treated with evaporative cooling. S3—T.S.s.; air extracted from the room after passing through the heat recovery unit. S4—T.S.s.; external air. S5—T.S.s.; external air after passing through the heat recovery unit.

The capabilities of the measurement system allowed us to place the sensors inside the air handling unit. At the end of each measurement cycle, the sensors that were shaded by the unit body were able to connect to the central unit by opening the service doors, so it could summarize the measured results at the end of each measurement cycle.

Control measurements were conducted to verify if there is a significant transfer of substances that would modify the moisture content of the treated fresh air from outside while maintaining appropriate indoor air conditions for the external and office areas during summer. As none of the media involved in heat exchange reached the dew point of the other medium during the entire measurement series, the observed changes in moisture content could be a consequence of measurement error and substance transfer occurring through the regenerative heat exchanger.

Based on the measured data, as can be seen in Figure 6, a slight material transfer was observed; however, considering the margin of error of the instruments (Table 3), the result was not significant.

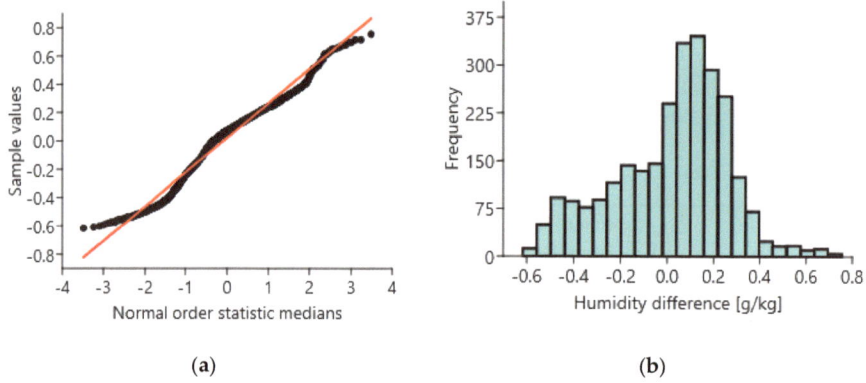

Figure 6. (**a**) Normal probability test of the air leakage (**b**) Histogram of the absolute air humidity difference (before and after the heat recovery unit).

Table 3. Main properties of the instruments.

Measuring Instruments	Measured Parameter	Range	Accuracy	Resolution
Hygro-thermometer (T.S.s. H3 D-2)	Temperature Relative humidity	−20 … + 50 °C 0 … 100% RH	+/−0.5 °C +/−3% RH	0.1 °C 0.1%
Digital differential pressure manometer (Testo 400)	Pressure drop	−100 … + 200 hPa	+/−0.3 Pa	0.001 hPa

2.3. Presentation of Daily Measurements

The daily measurements were started simultaneously with the operation of the air handling unit and the evaporative cooling unit. The results of the measurement method are presented for the day of 26 July 2021, and the general data of that day are included in the Appendix A [Figure A2]. The measurements were carried out in the same way every day of measurement. The figure shows the changes in temperature and humidity during the measurement period. Since the indoor air state changed only slightly, its treatment with DEC resulted in almost the same air temperature, with little variability. The fresh air treated with the IEC procedure resulted in a lower supply air temperature than the indoor air temperature (Figure 7), so the HVAC system provided sensible cooling performance in the room throughout the measurement period.

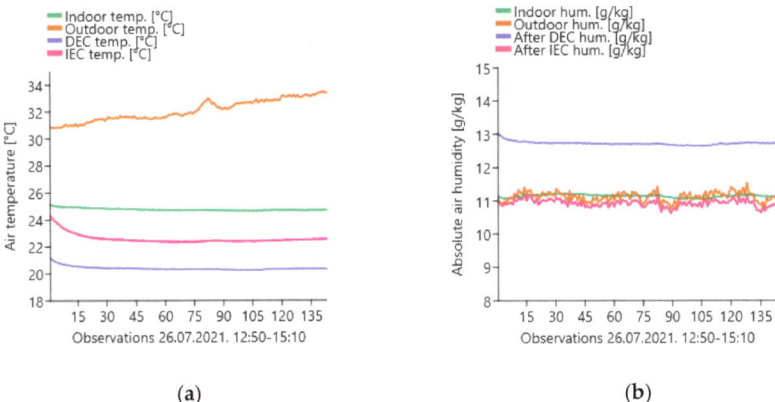

Figure 7. (a) Measured temperatures (b) Measured absolute humidity on 26 July 2021.

Figure 8 demonstrates how the heat recovery and evaporative cooling efficiency (1) build up after the start of the equipment until the quasi-stationary state of the system components is reached. The outlier data points shown in the box plot were also generated during the start-up and ramp-up phases. On the presented measurement day (26 July 2021), the median value of the heat recovery efficiency of the air handling unit was 82.3%, and the median value of the efficiency of the DEC process (4) was 75.1%.

The total and sensible cooling performance values achievable by the IEC procedure can be determined with the help of the measured results. Based on the results of the measurements taken on a given day, the median value of the total cooling performance was 3.18 kW, while the median value of the sensible cooling performance was 0.76 kW (Figure 9).

2.4. The Conclusions of the Measurement Series

The above measurements were carried out in the same way on every measurement day. Based on the combined results of the measurement days (Figure 10), the median value of the evaporative efficiency (4) was 0.737 throughout the entire measurement series

(19 measurement days). The measured results are consistent with the expected efficiency value according to the literature examined when using cellulose cooling pads [32]. Since the outdoor air conditions have no influence on the DEC, and only the indoor air conditions affect the DEC, these can be combined into one figure (see Figure 10). Similarly, the heat recovery efficiency primarily depends on the technical design of the air handling unit (AHU) and is only slightly affected by the external and internal air conditions. The outlier data points shown in the box plot as mentioned earlier in this paper were generated during the start-up and ramp-up phases of the measurements.

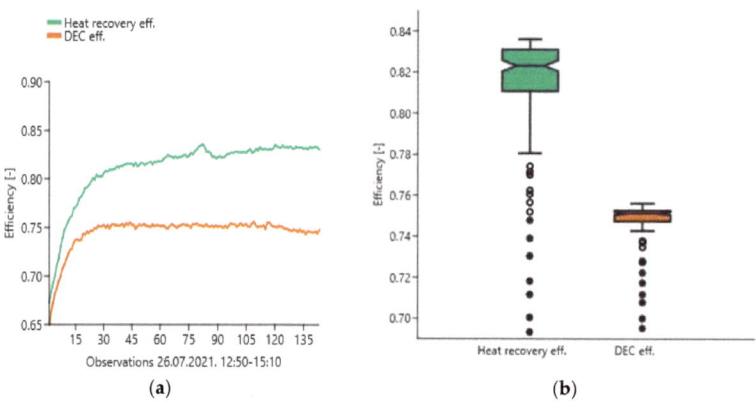

Figure 8. (**a**) Measured efficiency of heat recovery and direct evaporative air cooler units (**b**) Box plots of measured values on 26 July 2021.

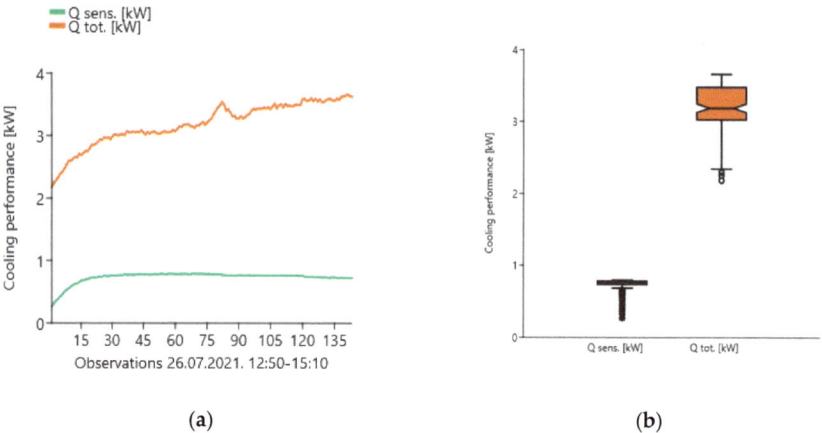

Figure 9. (**a**) Measured total and sensible cooling performance of the AHU by IEC (**b**) Box plots of measured values on 26 July 2021.

The heat recovery efficiency of the air handling unit was determined using Equation (1). Throughout the entire test period, the median value of the heat recovery efficiency was 0.824. The measured heat recovery efficiency meets the expectations of modern times, so the results of the laboratory tests can be extended to real air handling units. During the measurements, the aim was to examine whether the air handling unit is capable of properly treating the fresh air using only indirect evaporative air cooling. During the measurement period, the total and sensible cooling performance of the heat recovery was examined under the assumption of the same heat recovery efficiency, with an airflow rate of 1000 $m^3 h^{-1}$, in both conventional and IEC applications.

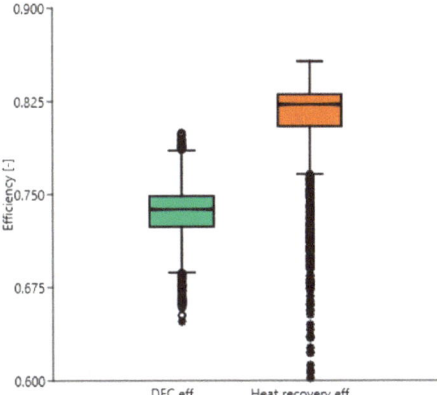

Figure 10. Box plots of measured efficiency of the heat recovery and direct evaporative air cooler units 26 July–17 September 2021. Measurements were conducted for 19 days (2592 observations).

Based on the results, significant total cooling performance can be achieved in both cases, but significant sensible cooling performance could only be achieved with the use of the IEC as can be seen in Figure 11. It is important to note that the outdoor absolute humidity load affects the room if the outdoor air humidity is higher than the desired indoor value. The IEC procedure is not capable of handling this humidity load. It is necessary to handle the extra humidity load, which can be achieved either with a built-in condensation temperature below the surface temperature of the cooling coil or with a low-temperature cooling system in the room (fan-coil, VRF systems). With the use of the IEC, the total cooling energy demand of the building is significantly reduced. However, when used alone, the cooling energy required to handle the humidity load must be provided by the supplementary cooling system, which can increase the energy consumption of the supplementary cooling system. Based on the above, there are both reducing and increasing effects on the energy consumption of the supplementary cooling system, so how much cooling energy the supplementary cooling system needs to provide with the use of the IEC cannot be determined simply. This can be examined by using real meteorological data and regulating indoor air conditions.

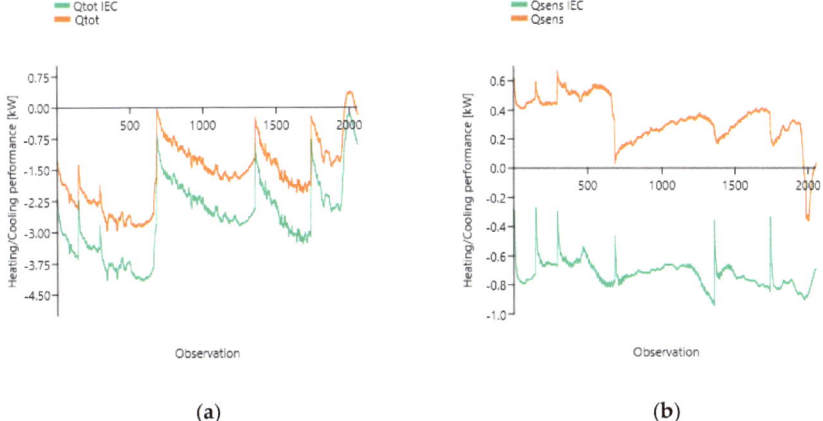

Figure 11. (**a**) Measured total cooling performance by IEC and calculated total cooling performance by original heat recovery unit (without IEC) (**b**) Measured sensible cooling performance by IEC and calculated sensible cooling performance by original heat recovery unit (without IEC).

3. Simulation

The aim of this study was to determine at what internal moisture content unchanged supplementary cooling energy consumption with the application of IEC can be achieved. The results largely depend on the meteorological conditions of the installation area. The investigation was carried out using data from the accredited Agricultural Meteorological Station of the University of Debrecen for the period between 2019 and 2021, from 1 May to 30 September. During the examined periods, the operation of IEC was simulated taking into account the measured data, determining the amount of extra energy required if the indoor target temperature is 25 °C and the allowable maximum indoor relative humidity is 30, 40, 50, 60, 70, 80 RH%. In the calculations, a 0.737 evaporative efficiency was assumed, which determines the amount of evaporated water, and a 0.824 heat recovery efficiency, which can determine the achievable supply air temperature and the felt cooling capacity inside the building.

During the simulation, the calculations were performed while the air handling unit was providing a volume flow rate of 1000 m^{3}h^{-1}, ensuring balanced ventilation. The total cooling energy generated by the IEC during the investigated period ($E_{IEC\,tot.}$) was determined, as well as the sensible cooling energy ($E_{IEC\,sens.}$). The difference between the extra drying energy demand caused by the supplied air ($E_{dehum.}$) and the sensible cooling performance determines how much the cooling energy requirement of the additional cooling system changes ($\Delta E_{add.c}$).

$$\Delta E_{add.c.} = E_{IEC\,sens.} - E_{dehum.} \quad (14)$$

The change in the cooling energy used by the supplementary cooling system and the difference between the total cooling energy generated by the IEC determine the total cooling energy that does not need to be covered by any other mechanical cooling when using the IEC (E_{tot}).

$$E_{tot} = E_{IEC\,tot.} - \Delta E_{add.c.} \quad (15)$$

The simulated results are summarized in Tables 4–6. The highlighted values in the table represent the most favorable values for each column, while the bolded row shows the results corresponding to the indoor air state with the smallest difference between the sensible cooling and dehumidification energy generated by the IEC.

Table 4. 1 May 2019–30 September 2019, Debrecen, Hungary simulation values.

Indoor Air Condition	$E_{IEC\,tot.}$ [kWh]	$E_{IEC\,sens.}$ [kWh]	$E_{dehum.}$ [kWh]	$\Delta E_{add.c.}$ [kWh]	E_{tot} [kWh]
25 °C 30 RH%	1492.16	954.03	1605.26	651.23	840.93
25 °C 40 RH%	**1337.44**	**794.67**	**845.43**	**50.76**	**1286.68**
25 °C 50 RH%	1185.62	638.30	305.33	−332.97	1518.59
25 °C 60 RH%	1036.79	485.00	47.77	−437.23	1474.02
25 °C 70 RH%	891.01	334.84	0.81	−334.03	1225.04
25 °C 80 RH%	748.36	187.92	0.00	−187.92	936.28

Table 5. 1 May 2020–30 September 2020, Debrecen, Hungary simulation values.

Indoor Air Condition	$E_{IEC\,tot.}$ [kWh]	$E_{IEC\,sens.}$ [kWh]	$E_{dehum.}$ [kWh]	$\Delta E_{add.c.}$ [kWh]	E_{tot} [kWh]
25 °C 30 RH%	1831.45	1202.00	3587.29	2385.29	−553.84
25 °C 40 RH%	1637.80	1002.55	2515.63	1513.08	124.72
25 °C 50 RH%	1447.79	806.84	1507.97	701.13	746.66
25 °C 60 RH%	**1261.51**	**614.97**	**672.28**	**57.31**	**1204.20**
25 °C 70 RH%	1079.05	427.04	166.95	−260.09	1339.14
25 °C 80 RH%	900.52	243.15	16.69	−226.46	1126.98

Table 6. 1 May 2021–30 September 2021, Debrecen, Hungary simulation values.

Indoor Air Condition	$E_{IEC\ tot.}$ [kWh]	$E_{IEC\ sens.}$ [kWh]	$E_{dehum.}$ [kWh]	$\Delta E_{add.c.}$ [kWh]	E_{tot} [kWh]
25 °C 30 RH%	2449.92	1418.84	4026.32	2608.32	−158.4
25 °C 40 RH%	2213.70	1175.53	2724.89	1549.36	664.34
25 °C 50 RH%	1981.90	936.79	1558.38	621.59	1360.31
25 °C 60 RH%	**1754.66**	**702.72**	675.18	−81.12	1835.78
25 °C 70 RH%	1532.08	473.47	169.39	−303.61	1835.69
25 °C 80 RH%	1314.30	249.15	13.17	−235.98	1550.28

If the set relative humidity is low in indoor environments, the cooling energy that can be extracted by the IEC process increases because the indoor air, at a given evaporative efficiency, can absorb more moisture. Therefore, in the examined cases, $E_{IEC\ tot.}$ and $E_{IEC\ sens.}$ are at their maximum at 25 °C and 30% RH. However, the lower the maximum relative humidity set in the indoor environment, the higher the moisture load introduced by the outdoor air, which increases the cooling energy demand of the additional cooling systems due to the increased energy required for dehumidification. In terms of minimizing the energy devoted to managing the moisture load ($E_{dehum.}$), the most favorable indoor air condition among the examined cases is at a maximum allowable condition of 25 °C and 80% RH for the purpose of minimizing the energy devoted to managing the moisture load.

Since the energy devoted to managing the moisture load is at a maximum at 25 °C and 30% RH, the largest decrease in cooling energy with the application of the IEC (E_{tot}) does not necessarily coincide with the maximum value of $E_{IEC\ tot.}$. Based on the simulation results, it can be seen that the maximum E_{tot} is at 25 °C and 50% RH in 2019, 25 °C and 70% RH in 2020, and 25 °C and 60% RH in 2021.

According to the simulated results, based on the results of the years 2019, 2020, and 2021, if the acceptable maximum humidity inside is 60RH%, the excess dehumidification energy and the IEC-induced sensible cooling energy input almost balance each other during the cooling season. In this study, at the examined 25 °C indoor temperature, the 60RH% air condition is at the limit and falls into the acceptable range under office conditions (0.5 clo; 1.1 met, 0.1 m/s) [21]. The maximum relative humidity inside was determined by the above simulation, so the simulated limit air condition can only be considered as an upper limit.

Based on the above simulation, the acceptable hottest and most humid indoor air condition in this case is 25 °C 60 RH%. For this indoor air condition, it is necessary to examine which outdoor air conditions can be efficiently handled only using the IEC procedure in the air handling unit (Figure 12). Knowing the indoor air condition (25 °C, 60 RH%), the evaporative efficiency (0.737), and the heat recovery efficiency (0.824), the isotherm or the sub-isotherm inlet temperature working field can be determined using Equations (6), (10) and (12), as can be seen in Figure 12.

working field calculations	
Calculated air conditions	values
Indoor air temperature	25 °C
Indoor air relative humidity	60 RH%
Addition absolute humidity by DEC	1.62 g/kg
Outside air temperature limit by IEC (6)	43.89 °C
Outside relative humidity limit by IEC (10)	15.58 g/kg
Outside enthalpy limit by IEC (12)	84.33 kJ/kg

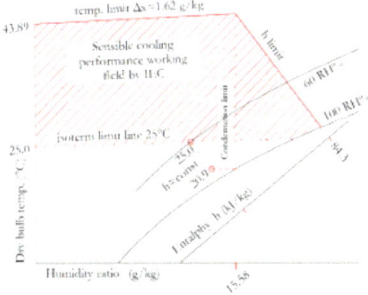

Figure 12. Estimated working field of sensible cooling performance by IEC.

Based on the determined limit values and operational parameters, it can be concluded that, considering the examined meteorological data, the air handling unit is capable of achieving sensible cooling performance solely by utilizing the IEC throughout the entire cooling season at the installation site.

4. Conclusions

By using IEC (Indirect Evaporative Cooling), the cooling demand of air handling units can be significantly reduced. With the increase in heat recovery efficiency of air handling units, it has become possible to achieve supply air temperatures lower than the indoor air temperature and achieve sensible cooling performance during a significant part or even the entire cooling season, exclusively through the application of IEC.

With the help of the presented connections in this study, it is possible to determine the operating range where sensible cooling performance can be achieved exclusively through the application of IEC, considering the specific equipment characteristics and desired indoor air conditions. Based on the measurements and taking into account meteorological data, it can be stated that in the investigated area with a continental European climate, significant cooling performance can be achieved throughout the entire cooling season. The presented diagnostic and computational method can be extended and implemented in any installation location.

The fact that sensible cooling performance can be achieved over a wide operational range with the IEC procedure does not necessarily mean that the cooling energy generated by supplementary cooling systems decreases, as the moisture load from the outdoor environment cannot be fully or partially managed with the IEC procedure. It is recommended to combine the IEC procedure with supplementary cooling systems operating below the dew point temperature to serve high-quality comfort spaces.

Under conditions of equilibrium absolute humidity corresponding to the given supply air volume flow rate, the HVAC system does not cause changes in cooling energy from the perspective of supplementary cooling systems serving the indoor space (the cooling energy demand of the air handling unit decreases significantly).

If we want to maintain absolute humidity in the indoor environment lower than the equilibrium state, efforts should be made to minimize the quantity of fresh air and the associated moisture load for reducing the cooling energy demand. However, if the acceptable absolute humidity in the indoor environment is above the equilibrium state, increasing the airflow managed by the IEC can reduce the energy demand of supplementary cooling systems.

Based on the examined meteorological data, it has been determined at what indoor air conditions the additional energy required for dehumidification and the decreased cooling energy demand due to the sensible cooling performance of the IEC procedure balance each other, considering a given airflow rate. In the studied case, the near-equilibrium state can be achieved with a maximum allowable indoor air condition of 25 °C and 60% relative humidity, resulting in acceptable comfort conditions in an office environment.

Author Contributions: Conceptualization, A.K., S.S. and I.C.; methodology, A.K., S.S. and I.C.; software, A.K., S.S. and I.C.; validation, A.K., S.S. and I.C.; formal analysis, A.K., S.S. and I.C.; investigation, A.K., S.S. and I.C.; resources, A.K., S.S. and I.C.; data curation, A.K., S.S. and I.C.; writing—original draft preparation, A.K., S.S. and I.C.; writing—review and editing, A.K., S.S. and I.C.; visualization, A.K., S.S. and I.C.; supervision, I.C.; project administration, A.K., S.S. and I.C.; funding acquisition, A.K., S.S. and I.C. All authors have read and agreed to the published version of the manuscript.

Funding: Project no. TKP2021-NKTA-34 has been implemented with support provided by the National Research, Development and Innovation Fund of Hungary, financed under the TKP2021-NKTA funding scheme.

Institutional Review Board Statement: Not applicable.

Informed Consent Statement: Not applicable.

Data Availability Statement: Not applicable.

Conflicts of Interest: The authors declare no conflict of interest.

Abbreviations

List of symbols

c_{pa}	specific heat of the air at constant pressure [kJ kg^{-1}K^{-1}]
c_{pv}	specific heat of the water vapor at constant pressure [kJ kg^{-1}K^{-1}]
E	cooling energy [kJ; kWh]
h	enthalpy [kJ kg^{-1}]
T	temperature [°C]
T_{wb}	wet-bulb temperature [°C]
x	humidity ratio of air [kg$_v$/kg$_a$]
r	latent heat of vaporization [kJ kg^{-1}]
Q	cooling performance [kW]

Greek symbols

η	effectiveness of heat recovery unit [%]
ε	effectiveness of direct evaporative cooling [%]
Δ	the sum of differences [-]

Abbreviations

DEC	direct evaporative air cooling
IEC	indirect evaporative air cooling
RH	relative humidity

Subscripts

add. c	value of the additional cooling system
dehum	dehumidification value
e	after evaporative cooling
i	indoor
in	input value
o	outdoor
o, limit	limit of the outdoor value
out	output value
rs	recovery unit, supply side
re	recovery unit, exhaust side
s	supply
sens	sensible
x	extract
out wb	wet-bulb value of the output value
v	water vapor
tot	total

Appendix A

Summary statistics	values
Date:	04.09.2021
Category:	Summer day
Min. air temperature:	10.2 °C
Max. air temperature:	26.5 °C
Mean air temperature:	17.9 °C
Min. relative air humidity:	28.9 RH%
Max. relative air humidity:	96.9 RH%
Mean relative air humidity:	70.4 RH%
Min. air enthalpy	29.0 kJ/kg
Max. air enthalpy	49.6 kJ/kg
Mean air enthalpy	39.4 kJ/kg

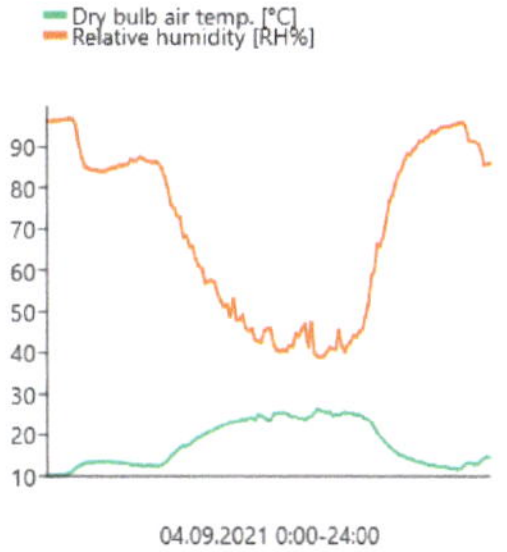

Figure A1. Main information about a typical "Summer day" during the measurement.

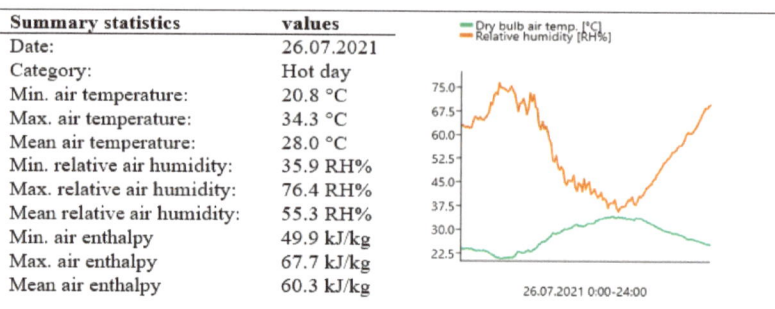

Summary statistics	values
Date:	26.07.2021
Category:	Hot day
Min. air temperature:	20.8 °C
Max. air temperature:	34.3 °C
Mean air temperature:	28.0 °C
Min. relative air humidity:	35.9 RH%
Max. relative air humidity:	76.4 RH%
Mean relative air humidity:	55.3 RH%
Min. air enthalpy	49.9 kJ/kg
Max. air enthalpy	67.7 kJ/kg
Mean air enthalpy	60.3 kJ/kg

Figure A2. Main information about a typical "Hot day" during the measurement.

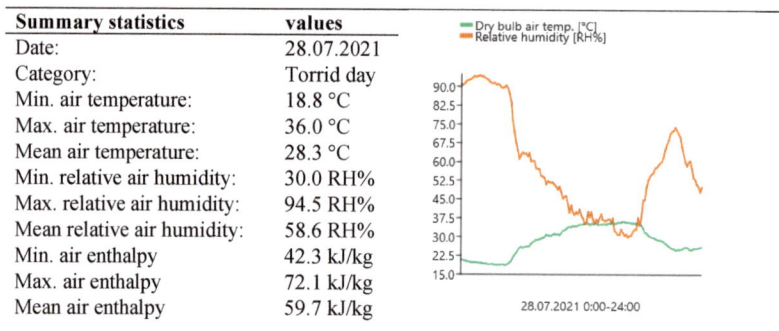

Summary statistics	values
Date:	28.07.2021
Category:	Torrid day
Min. air temperature:	18.8 °C
Max. air temperature:	36.0 °C
Mean air temperature:	28.3 °C
Min. relative air humidity:	30.0 RH%
Max. relative air humidity:	94.5 RH%
Mean relative air humidity:	58.6 RH%
Min. air enthalpy	42.3 kJ/kg
Max. air enthalpy	72.1 kJ/kg
Mean air enthalpy	59.7 kJ/kg

Figure A3. Main information about a typical "Torrid day" during the measurement.

References

1. Fisenko, S.; Brin, A.; Petruchik, A. Evaporative cooling of water in a mechanical draft cooling tower. *Int. J. Heat Mass Transf.* **2004**, *47*, 165–177. [CrossRef]
2. Harby, K.; Gebaly, D.R.; Koura, N.S.; Hassan, M.S. Performance improvement of vapor compression cooling systems using evaporative condenser: An overview. *Renew. Sustain. Energy Rev.* **2016**, *58*, 347–360. [CrossRef]
3. Ketwong, W.; Deethayat, T.; Kiatsiriroat, T. Performance enhancement of air conditioner in hot climate by condenser cooling with cool air generated by direct evaporative cooling. *Case Stud. Therm. Eng.* **2021**, *26*, 101127. [CrossRef]
4. He, J. A design supporting simulation system for predicting and evaluating the cool microclimate creating effect of passive evaporative cooling walls. *Build. Environ.* **2011**, *46*, 584–596. [CrossRef]
5. Bishoyi, D.; Sudhakar, K. Experimental performance of a direct evaporative cooler in composite climate of India. *Energy Build.* **2017**, *153*, 190–200. [CrossRef]
6. Xuan, Y.; Xiao, F.; Niu, X.; Huang, X.; Wang, S. Research and applications of evaporative cooling in China: A review (II)—Systems and equipment. *Renew. Sustain. Energy Rev.* **2012**, *16*, 3523–3534. [CrossRef]
7. Tewari, P.; Mathur, S.; Mathur, J.; Loftness, V.; Abdul-Aziz, A. Advancing building bioclimatic design charts for the use of evaporative cooling in the composite climate of India. *Energy Build.* **2019**, *184*, 177–192. [CrossRef]
8. Sonawane, T.; Patil, S.; Dube, A.; Chaudhari, B.D.; Sonawane, T.R.; Patil, S.M. A Review on Evaporative Cooling Technology Performance assesment of green supply chain practice View project Design of off grid solar roof top system View project A Review on Evaporative Cooling Technology. *Int. J. Res. Advent Technol.* **2015**, *3*, 88–96. Available online: https://www.researchgate.net/publication/295448903 (accessed on 9 July 2023).
9. Duan, Z.; Zhan, C.; Zhang, X.; Mustafa, M.; Zhao, X.; Alimohammadisagvand, B.; Hasan, A. Indirect evaporative cooling: Past, present and future potentials. *Renew. Sustain. Energy Rev.* **2012**, *16*, 6823–6850. [CrossRef]
10. Xuan, Y.; Xiao, F.; Niu, X.; Huang, X.; Wang, S. Research and application of evaporative cooling in China: A review (I)—Research. *Renew. Sustain. Energy Rev.* **2012**, *16*, 3535–3546. [CrossRef]
11. Min, Y.; Chen, Y.; Shi, W.; Yang, H. Applicability of indirect evaporative cooler for energy recovery in hot and humid areas: Comparison with heat recovery wheel. *Appl. Energy* **2021**, *287*, 116607. [CrossRef]
12. Delfani, S.; Esmaeelian, J.; Pasdarshahri, H.; Karami, M. Energy saving potential of an indirect evaporative cooler as a pre-cooling unit for mechanical cooling systems in Iran. *Energy Build.* **2010**, *42*, 2169–2176. [CrossRef]

13. Min, Y.; Chen, Y.; Yang, H. A statistical modeling approach on the performance prediction of indirect evaporative cooling energy recovery systems. *Appl. Energy* **2019**, *255*, 113832. [CrossRef]
14. Noor, S.; Ashraf, H.; Sultan, M.; Khan, Z.M. Evaporative Cooling Options for Building Air-Conditioning: A Comprehensive Study for Climatic Conditions of Multan (Pakistan). *Energies* **2020**, *13*, 3061. [CrossRef]
15. Shi, W.; Ma, X.; Gu, Y.; Min, Y.; Yang, H. Indirect evaporative cooling maps of China: Optimal and quick performance identification based on a data-driven model. *Energy Convers. Manag.* **2022**, *268*, 116047. [CrossRef]
16. Al Horr, Y.; Arif, M.; Katafygiotou, M.; Mazroei, A.; Kaushik, A.; Elsarrag, E. Impact of indoor environmental quality on occupant well-being and comfort: A review of the literature. *Int. J. Sustain. Built Environ.* **2016**, *5*, 1–11. [CrossRef]
17. Kusiak, A.; Xu, G.; Tang, F. Optimization of an HVAC system with a strength multi-objective particle-swarm algorithm. *Energy* **2011**, *36*, 5935–5943. [CrossRef]
18. Cheng, Y.; Niu, J.; Du, Z.; Lei, Y. Investigation on the thermal comfort and energy efficiency of stratified air distribution systems. *Energy Sustain. Dev.* **2015**, *28*, 1–9. [CrossRef]
19. *MSZ EN ISO 7730:2005*; Ergonomics of the Thermal Environment. Analytical Determination and Interpretation of Thermal Comfort Using Calculation of the PMV and PPD Indices and Local Thermal Comfort Criteria. Hungarian Standard Commitment: Budapest, Hungary, 2000.
20. *MSZ CR 1752:2000*; Ventilation for Buildings. Hungarian Standard Commitment: Budapest, Hungary, 2000.
21. *ANSI/ASHRAE Standard 55-2013*; Thermal Environmental Conditions for Human Occupancy. ASHRAE Standard: Atlanta, GA, USA, 1992.
22. Sun, Z.; Wang, S.; Ma, Z. In-situ implementation and validation of a CO_2-based adaptive demand-controlled ventilation strategy in a multi-zone office building. *Build. Environ.* **2011**, *46*, 124–133. [CrossRef]
23. Kusiak, A.; Zeng, Y.; Xu, G. Minimizing energy consumption of an air handling unit with a computational intelligence approach. *Energy Build.* **2013**, *60*, 355–363. [CrossRef]
24. Ürge-Vorsatz, D.; Cabeza, L.F.; Serrano, S.; Barreneche, C.; Petrichenko, K. Heating and cooling energy trends and drivers in buildings. *Renew. Sustain. Energy Rev.* **2015**, *41*, 85–98. [CrossRef]
25. Al Horr, Y.; Tashtoush, B.; Chilengwe, N.; Musthafa, M. Operational mode optimization of indirect evaporative cooling in hot climates. *Case Stud. Therm. Eng.* **2020**, *18*, 100574. [CrossRef]
26. Yang, L.; Yan, H.; Lam, J.C. Thermal comfort and building energy consumption implications—A review. *Appl. Energy* **2014**, *115*, 164–173. [CrossRef]
27. Roulet, C.-A.; Heidt, F.; Foradini, F.; Pibiri, M.-C. Real heat recovery with air handling units. *Energy Build.* **2001**, *33*, 495–502. [CrossRef]
28. Fouda, A.; Melikyan, Z. A simplified model for analysis of heat and mass transfer in a direct evaporative cooler. *Appl. Therm. Eng.* **2011**, *31*, 932–936. [CrossRef]
29. Kovačević, I.; Sourbron, M. The numerical model for direct evaporative cooler. *Appl. Therm. Eng.* **2017**, *113*, 8–19. [CrossRef]
30. Montazeri, H.; Blocken, B.; Hensen, J. Evaporative cooling by water spray systems: CFD simulation, experimental validation and sensitivity analysis. *Build. Environ.* **2015**, *83*, 129–141. [CrossRef]
31. Wu, J.; Huang, X.; Zhang, H. Theoretical analysis on heat and mass transfer in a direct evaporative cooler. *Appl. Therm. Eng.* **2009**, *29*, 980–984. [CrossRef]
32. Malli, A.; Seyf, H.R.; Layeghi, M.; Sharifian, S.; Behravesh, H. Investigating the performance of cellulosic evaporative cooling pads. *Energy Convers. Manag.* **2011**, *52*, 2598–2603. [CrossRef]
33. Alklaibi, A. Experimental and theoretical investigation of internal two-stage evaporative cooler. *Energy Convers. Manag.* **2015**, *95*, 140–148. [CrossRef]

Disclaimer/Publisher's Note: The statements, opinions and data contained in all publications are solely those of the individual author(s) and contributor(s) and not of MDPI and/or the editor(s). MDPI and/or the editor(s) disclaim responsibility for any injury to people or property resulting from any ideas, methods, instructions or products referred to in the content.

Article

Investigation of Microclimate Parameter Assurance in Schools with Natural Ventilation Systems

Tomas Makaveckas [1,*], Raimondas Bliūdžius [2], Sigita Alavočienė [1], Valdas Paukštys [2] and Ingrida Brazionienė [1]

[1] Alytus College, Studentu St. 17, LT-62252 Alytus, Lithuania; sigita.alavociene@akolegija.lt (S.A.); ingrida.brazioniene@akolegija.lt (I.B.)
[2] Institute of Architecture and Construction, Kaunas University of Technology, Tunelio 60, LT-44405 Kaunas, Lithuania; raimondas.bliudzius@ktu.lt (R.B.); valdas.paukstys@ktu.lt (V.P.)
* Correspondence: tomas.makaveckas@akolegija.lt

Abstract: Slow population growth has limited the construction of new schools, leading to the renovation of existing buildings to achieve energy efficiency goals. While improvements are made to thermal insulation, heating and ventilation systems often remain outdated, presenting challenges in maintaining indoor air quality (IAQ) in schools, where children spend a significant amount of time in densely populated classrooms, and whose health is more affected by IAQ than that of adults. Therefore, this study assessed the possibilities to achieve IAQ requirements in schools ventilated by opening the windows by monitoring the carbon dioxide (CO_2) concentration, temperature, and relative humidity (RH) fluctuation. The results of the study have shown that it is not feasible to achieve the defined IAQ parameters in classrooms through window opening alone. The measured CO_2 concentration during lessons in many cases exceeded the limit value of 1000 ppm and did not decrease to ambient levels when the windows were opened during the break. Additionally, the internal air temperature dropped below the normative when lessons started, and RH was significantly below the recommended minimum value on all days. It was also found that the use of thermal energy decreases evenly because of inefficient air change leaving no direct economic leverage for the installation of efficient ventilation systems.

Keywords: indoor air quality; indoor air exchange; school ventilation; CO_2 monitoring; temperature/humidity fluctuations

1. Introduction

In many European countries, due to the lack of population growth, not a lot of new schools are built, and school buildings built 30 or more years ago are mostly adapted to today's needs. In Eastern Europe, these are low-energy-efficiency buildings with powerful central heating systems and a natural ventilation system, where air enters classrooms through leaky windows or inefficient air vents and is removed through ventilation ducts installed in internal partitions [1,2]. According to Šadauskienė et al. [3], the thermal resistance of existing school building roofs and walls in Lithuania was about 1 m²·K/W, the annual heat loss was up to 180 kWh/m² per year, and the air tightness of the building n_{50} ranged from three to seven times per hour. It is obvious that it is impossible to ensure modern energy efficiency and air quality requirements in such school buildings, so school buildings are being renovated. Since these are public buildings, renovation decisions are usually motivated by compliance with the energy efficiency requirements of buildings specified in building regulations. Investments to improve energy efficiency are easily justified by the saved thermal energy during the corresponding payback period [4]. It is much more difficult to justify investments in an efficient ventilation system, especially for ensuring indoor air quality (IAQ), because there are no clear and approved economic assessment methodologies, so after renovation, the natural ventilation system is often left.

Even worse, old leaky windows with frost vents are being replaced with new energy-efficient and airtight windows, restricting airflow to the premises. In order to maintain at least the minimum IAQ requirements, all that remains is to open the windows when the classrooms are empty. However, there is no detailed analysis of how window opening affects IAQ.

Nowadays, people spend more than 80% of their time indoors [5,6]. Moreover, children spend almost 12% of this indoor time in classrooms; the occupancy density of many classrooms is about four times higher than that of office buildings [7], so poorly ventilated school buildings are a key source of the potential for health problems from indoor air pollutants [8]. In terms of IAQ, multiple pollutants: particulate matter concentrations, volatile organic compounds, microorganisms, excessive concentrations of carbon dioxide (CO_2), etc., which can cause health problems, are often measured in schools [9]. The effect of variations in indoor air temperature, relative air humidity, and air movement speed is less often discussed, as it is more associated with well-being than with illness.

To improve IAQ, ventilation systems are used to ensure the necessary air exchange in the premises. According to Sadrizadeh et al. [8], the CO_2 concentration is an indicator of the sufficiency of the rate of outside air supply per person; the higher the CO_2 concentration, the poorer the ventilation and the higher concentrations of other pollutants.

Excessive exposure to CO_2 might cause headaches, fatigue, and lower performance at work. According to different standards, the concentration of CO_2 in the premises should not exceed 2500 parts per million (ppm), and the recommended upper limit is 1000 ppm [10–12]. Indoor CO_2 concentrations above 1000 ppm are generally considered indicative of unacceptable ventilation rates. However, concentrations of CO_2 below 1000 ppm do not always guarantee that the ventilation rate is adequate for the removal of air pollutants from other indoor sources [13]. There is enough evidence to state that CO_2 has a direct health effect on humans, but only at concentrations much higher than those found in normal indoor applications [7,14]. As stated by Talarosha et al. [15] children are more susceptible to air pollutants and require more oxygen for their growth. Indoor CO_2 concentrations greater than 1000 ppm will affect student health, concentration levels, and academic performance.

According to Petersen et al. [16], low ventilation rates are especially critical for children as they inhale more air in relation to their body weight compared to adults because their tissues and organs are actively growing. This study showed that the performance of the students improved significantly in four of four performance tests when the outdoor air supply rate increased from an average of 1.7 to 6.6 L/s per person. Another study showed that a higher ventilation rate (up to 10 L/s per person) could improve school performance by 14.5% [17].

Sundell et al. [18], based on Klauss et al., provide the following historical chronology of ventilation requirements: in 1836, Tregold made one of the earliest estimates of ventilation requirements. He estimated a minimum required ventilation rate of about 2 L/s per person; later, Billings (1893) calculated a ventilation requirement of about 15 L/s per person, and this value was valid until 1931 when the New York State Commission found that 5–7.5 L/s per person was adequate in schoolrooms. Since 2001, the ventilation rate used is typically 7.5 L/s per person in densely occupied spaces. According to the hygiene standards of the Republic of Lithuania, HN 21:2017 'School carrying out general education programs. General requirements for health safety', the average concentration of CO_2 in classrooms and teaching rooms, measured between the beginning and end of the lessons, must not exceed 2745 mg/m^3 (1500 ppm), and the maximum acceptable concentration of CO_2 in the air of these rooms is 9000 mg/m^3 (5000 ppm) [19].

The main aim of ventilation is to create an IAQ that is more suitable for people and processes than that occurring in an unventilated building, thereby diluting and removing the pollutants produced by inhabitants, their activities, and the indoor surroundings [20]. However, the operation of any ventilation system to reduce the concentration of CO_2 indoors is inseparable from the change in the temperature and relative humidity (RH) of the indoor air. Temperature and RH are important physical factors that affect the sense of well-being [21]. Studies showed that reducing temperature by 1 °C could improve school performance by 3.5% [17]. Air temperature also affects thermal comfort, which can negatively affect student health and educational performance [9]. Ma et al. [22] noticed that children worldwide have different thermal comfort temperatures in different seasons and different climatic conditions. They provide several examples of other research: in England, the comfort temperature for children (11–16 years of age) is 16.5 °C in winter and 19.1 °C in summer; in South Korea, the comfort temperature for children (4–6 years of age) is 22.1 °C; in Australia, the comfort temperature is 24.5 °C in winter and 22.5 °C in summer for students in primary and secondary school. The HN 21:2017 hygiene norms of the Republic of Lithuania recommend that the air temperature in classrooms, teaching rooms, and the assembly hall should be at least 18 °C and not higher than 22 °C in the cold season, and not higher than 26 °C in the warm season [19].

One of the functions of ventilation is to remove moisture from the building, which is emitted by the occupants, generated through their activities (cooking, showering, cleaning, etc.) and produced through chronic leaks from piping, roofs, or through the walls of basements [21]. Controlling RH is important for the comfort of children and for the prevention of moisture accumulation, which can lead to mould growth [21]. Low levels of humidity are unacceptable due to the decrease in mucous available to the eyes and nose, leading to the drying of these organs and thus irritation [23]. On the other hand, high levels of RH may support the growth of pathogenic or allergenic microorganisms (e.g., mould on the inner side of the elements of the building envelope) [5]. However, when moisture emissions are low, ventilation greatly reduces the relative humidity. The hygiene norms HN 21:2017 of the Republic of Lithuania recommend that RH in school premises be maintained at a level of 35–60% in the cold season and 35–65% in the warm season [19].

The ventilation of the premises is always related to the loss of heat in the building. When evaluating the energy efficiency of buildings, heating the amount of air necessary for ventilation is considered a mandatory part of the energy consumption of buildings. However, the energy efficiency of buildings is reduced by the energy consumption to heat the air with excessive infiltration, so the aim is to avoid this by increasing the level of air tightness of buildings and installing adjustable mechanical ventilation systems. Natural ventilation systems are considered to be poorly regulated, so the energy consumption for ventilation is considered higher. It is even more difficult to estimate the energy consumption related to ventilation, when ventilation takes place by opening the windows, and even such a methodology does not exist, although the assessment of the increase in energy consumption due to opening the windows would add economic motivation to install mechanical ventilation systems in schools. Even worse, old room heating systems that have not been renovated and do not have automatic regulation mean that the air temperature in heated schools is often higher than the norm, also increasing the total heating costs of the building [24].

According to a World Health Organization (WHO) report, most schools (86%) in Europe use natural ventilation, only 7% use assisted ventilation, and only 7% of schools use mechanical ventilation [21]. The problems related to ventilation also exist in many Lithuanian school buildings constructed in the Soviet period [1]. According to data from the Lithuanian Statistics Department, in 2022, there were 957 general education schools in Lithuania with 330,262 students [25]. A significant number of Lithuanian schools are renovated; however, only a few of them have a mechanical ventilation system. According to Zemitis et al. [26], the lack of modern ventilation systems in schools means that most students are taking classes in buildings with the only possible air exchange through open windows. Kapalo et al. [2] stated that even a short 10-min class ventilation has a significant impact on classroom air quality, which means that even brief ventilation has a positive impact on classroom IAQ. However, there is no detailed analysis performed on how window opening affects indoor air quality parameters: CO_2 concentration, indoor air temperature, and relative air humidity, nor is there any energetic or economic assessment method for this type of ventilation. The WHO survey suggested that technical and operational requirements must be established to ensure adequate ventilation: minimum number and surface area of vents in natural ventilation systems; minimum number of windows/window surface area; functioning heating systems and temperature controls; specifications of mechanical systems; protocol for opening windows during classes and breaks; CO_2-based demand-controlled mechanical systems; and increase in teacher awareness [21]. Therefore, this research aimed to analyse the impact of opening windows as part of the natural ventilation of school rooms on the parameters of the internal microclimate (CO_2 concentration, temperature, and RH) in renovated naturally ventilated schools in Lithuania. Three schools of different sizes and years of construction were selected for the study. Monitoring of the CO_2 concentration, air temperature and RH was carried out in classrooms of various sizes and occupancy. During the study, the outside air temperature, cloudiness, classroom occupancy, and duration of opening the windows were recorded. The air change rate was measured in all classrooms using the blower door method before starting the measurements. The results of the study are presented as a dependence of CO_2 concentrations in classrooms on window opening patterns, considering the classroom occupancy, measured standardised air exchange rate, and outside climate parameters. In addition, the impact of window opening on the heat consumption of the building was also analysed.

2. Materials and Methods

2.1. Investigated Schools and Classrooms

Three schools (School A, School B and School C) located in Alytus, Lithuania, were selected for the study. All schools were built between 1960 and 1980 and have undergone renovations. The renovation works included the replacement of windows and external doors, thermal insulation of the facades and roofs, and modernization of the heating system. Ventilation during renovation has not been improved in either of these schools and is still only natural with openable windows. All schools are in the city centre, close to busy streets. The classrooms selected for the study had similar volumes and window areas in all schools and were located on the eastern or southern facade. The floor area of these classrooms varied from 49.2 to 54.25 m^2. According to the hygiene norms HN 21:2017, each student must be provided with at least 1.7 m^2 of learning space per classroom [19]. The occupancy density in selected classrooms varied from 2.05 to 3.51 m^2 per person, thus meeting the requirements of HN 21:2017. Data for schools are presented in Table 1.

Table 1. Data on schools and surveyed premises.

School	Year of Construction	Renovation Completed	Area of School, m^2	Details of the Test Facilities			
				Floor Area, m^2	Occupancy Density Person/m^2	Area of Windows, m^2	Window Orientation
A	1963	2009	3751.11	49.2	0.30	8.1	East
B	1977	2005	7292.11	54.25	0.46	16.96	South-East
C	1970	2001	5458.40	49.28	0.38	16.07	South-East

The premises were in continuous classroom use at the time of the survey. Heating in the classrooms is provided with water-filled radiators with thermostatic valves. The three study classrooms have typical school furniture and the floor is covered with linoleum. The density of people who occupy the selected rooms during the research is given in Table 2.

During the study, teachers were allowed to ventilate the classroom as normal: when students leave the classroom, all windows are opened at an angle of about 90° and kept until the end of the break. There are no people in the classroom during the break. At the end of the break, the windows are closed, and only then do the students enter the classroom.

Table 2. Density of persons in the classroom, person/m^2.

		First Lesson	Second Lesson	Third Lesson	Fourth Lesson	Fifth Lesson	Sixth Lesson
		8:00–8:45	8:55–9:40	9:50–10:35	10:50–11:35	12:05–12:50	13:00–13:45
School A	8 March 2022	0.26	0.26	0.02	0.26	0.02	0.12
	9 March 2022	0.22	0.22	0.04	0.02	0.22	0.22
	10 March 2022	0.28	0.02	0.28	0.28	-	-
School B	22 March 2022	0.46	0.46	0.46	0.46	0.46	-
	23 March 2022	0.46	0.46	0.46	0.46	0.46	-
	24 March 2022	0.46	0.46	0.46	0.46	0.46	-
School C	25 April 2022	0.37	0.02	0.34	0.37	0.37	0.28
	26 April 2022	0.37	0.37	0.34	0.37	0.37	-
	27 April 2022	0.37	0.37	0.37	0.37	-	-

2.2. Measurement Equipment Used in Research

The indoor air quality meter used for the study was a Combo IAQ Meter 77,597 portable indoor air quality (IAQ) meter (Figure 1), compliant with EN 50270:2015, which measures:

- Air temperature from −20 °C to 60 °C (air temperature resolution 0.1 °C; air temperature accuracy ±0.6 °C).
- Relative humidity (RH) from 0 to 99.9% (RH resolution 0.1% RH; RH accuracy ±3% RH (at 25 °C, 10~90% RH), others ±5% RH).
- Carbon dioxide (CO_2) concentration from 0 to 9999 ppm (CO_2 resolution 1 ppm; CO_2 accuracy ±30 ppm ± 5% of reading (0~5000 ppm), other ranges are not specified).

Figure 1. Combo IAQ Meter 77597.

Levels of CO_2, internal temperature and RH were monitored at 5-min intervals throughout the occupied day, in locations close to the occupied zone at seated head height, at 1.1 m height in the centre of the classroom (Figure 2). The meter was placed on a table that was unoccupied throughout the measurement.

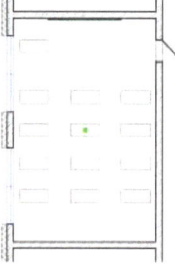

Figure 2. Location of the meter in the standard classroom layout (green dot).

The air exchange rate per hour n_{50} of each classroom was measured with the standardised fan pressurization method, using the Retrotec Blower Door System (Figure 3) with the following technical specifications: measurement accuracy ±3.0%, measurement uncertainty 8.3%. Tests were carried out once before the monitoring of IAQ with sealed ventilation exhaust ducts under negative indoor pressure conditions.

Figure 3. Blower door test.

2.3. Monitoring Sequence of Window Opening, CO_2 Concentration, Indoor Air Temperature, and RH

The periods of approximately −10–+10 °C external air temperatures were chosen for the study. It is the longest period in the school year and, thus, the opportunities for opening windows to ensure indoor microclimate parameters in classrooms are the most relevant and cannot be ignored. Hot summer is less relevant for general education schools, because they are not used during the school holidays, and on warm spring days, when the outside air temperature is similar to the inside, the windows can be kept open during classes as well. During the summer, overheating problems arise, not ventilation-related, which are solved by other means. In the cold winter period, when the outside air temperature is below −10 °C, and without testing, it is clear that the temperature conditions will not be suitable.

Firstly, prior to the three-day IAQ monitoring, a blower door test was carried out once to determine air exchange in every classroom at a pressure difference of 50 Pa. Monitoring was carried out for three consecutive days during the heating season in one representative classroom in each school. Measurements were made during all teaching hours, including breaks. The school day varied depending on the classroom schedule from 8:00 to 13:45. The survey recorded the start and end time of each lesson, the number of pupils in the lesson, and the opening and closing times of the windows. Measurement dates, window opening times, and climatic conditions during the measurement period are presented in Table 3.

Table 3. Climatic data determined during the measurement period.

School	Date	Opening Time of the Window	Outside Temperature When the Window Is Open	Information about Clouds
A	8 March 2022	8:45–9:00 9:40–9:50 10:35–10:50 12:00–12:15 12:55–13:00	−2 °C −3 °C −2 °C −1 °C 0 °C	Cloudy
	9 March 2022	7:45–8:00 8:50–9:00 10:45–11:05 11:35–12:05	−4 °C −3 °C −1 °C 0 °C	Cloudy with sunshine
	10 March 2022	7:45–8:00 8:45–9:00 9:40–9:50 10:35–10:50 12:00–13:20	−9 °C −8 °C −6 °C −5 °C −3 °C	Overcast

Table 3. Cont.

School	Date	Opening Time of the Window	Outside Temperature When the Window Is Open	Information about Clouds
B	22 March 2022	8:45–9:00 9:20–9:45 10:35–10:45 11:35–12:00	+ 6 °C + 8 °C + 10 °C + 11 °C	Sunny
	23 March 2022	8:40–8:50 9:25–9:45 10:30–11:20 12:00–12:50	+ 4 °C + 7 °C +11 °C + 14 °C	Sunny
	24 March 2022	8:50–9:35 10:35–10:45 11:30–11:45	+2 °C +3 °C + 5 °C	Sunny
C	25 April 2022	8:55–9:05 9:35–9:40 10:35–10:55 11:40–12:00	+6 °C +6 °C +7 °C +8 °C	Overcast Overcast Cloudy with sunshine Cloudy
	26 April 2022	8:50–9:00 9:40–9:55 10:35–10:55 11:20–11:35 11:45–12:00	+7 °C +8 °C +9 °C +10 °C +11 °C	Overcast Cloudy Cloudy Cloudy Cloudy
	27 April 2022	8:50–9:00 9:40–9:55 10:35–10:55	+6 °C +6 °C +7 °C	Sunny

Wind speeds were similar on all monitoring days, ranging from 1 to 5 m/s.

2.4. Assessment of Air Exchange and Heat Consumption for Class Ventilation

The volume of air changing through the open windows during one break $V_{air,exc.op.}$ is calculated using the equation:

$$V_{air,\ exc.\ op.} = \frac{V_{room}}{N}, \text{m}^3; \quad (1)$$

where: V_{room}—volume of the room, m^3.

N—the proportion of indoor air that has changed.

The volume of air that changes in the room through leaks in one lesson $V_{air,exc.leak}$ is calculated using the equation:

$$V_{air,exc.\ leak} = 0.5 \cdot n_{50} \cdot \frac{z_c}{60} \left(0.75 \cdot \frac{\rho_{air}}{2 \cdot 50} \cdot \left(0.9 \cdot v_{wind,m} \right)^2 \right)^{\frac{2}{3}} \cdot V_{room}, \text{m}^3; \quad (2)$$

where 0.5—a coefficient to account for the fact that air infiltration into the building does not occur throughout the envelope area.

n_{50}—the value of the air change in the building as determined by tests (h^{-1}).

z_c—duration of the lesson, min.

60—number of seconds per minute.

0.75—aerodynamic coefficient of the building.

ρ_{air}—air density (kg/m^3); ρ = 1.21 kg/m^3.

0.9—a coefficient that takes into account the reduction in wind speed due to the presence of various barriers in the vicinity of buildings.

$v_{wind,m}$—wind speed (m/s).

50—the pressure at which the building's airtightness is measured (Pa).

2/3—the value of the grade indicator determined by blower door tests.

The amount of air required for ventilation during a lesson, $V_{air,exc.req.}$, is calculated using equation:

$$V_{air,exc.req.} = \frac{V_{req} \cdot z_c \cdot 60 \cdot NS}{1000}, \text{ m}^3; \quad (3)$$

where: V_{req}—fresh air flow required by hygiene standards, L/s.

z_c—duration of the lesson, min.

60—number of seconds per minute.

NS—number of students in the class during the lesson.

1000—number of litres per cubic metre.

The amount of air to be additionally heated in case of required ventilation achieved is calculated using the equation:

$$V_{air,\,exc.} = V_{air,exc.req.} - V_{air,exc.op.} - V_{air,exc.leak.}, \text{ m}^3; \quad (4)$$

The amount of thermal energy used to heat the changed air, Q_h, is calculated using the equation:

$$Q_h = \frac{V_{air,\,exc.} \cdot \rho_{air} \cdot c_{air} \cdot (t_i - t_e)}{3600}, \text{ kWh}; \quad (5)$$

where ρ_{air}—density of air at a temperature of 0 °C, kg/m^3.

c_{air}—specific heat capacity of the air, kJ/(kg·K).

$(t_i - t_e)$—internal and external air temperatures, °C.

3600—conversion to kWh.

3. Results

3.1. Monitoring Results for School A

The monitoring of School A was carried out in the presence of freezing outside temperatures and high cloud cover. The air change rate per hour (n_{50}) in this classroom was measured at 1.55 1/h. On the first day of monitoring (8 March 2022), the outside air temperature was −3–0 °C. The carbon dioxide (CO_2) concentration, indoor air temperature, and occupant density during lessons on 8 March 2022 in School A are shown in Figure 4.

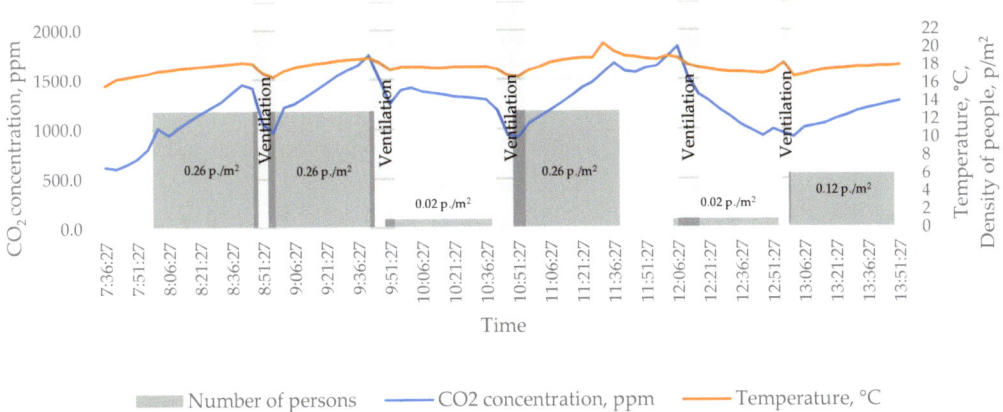

Figure 4. CO_2 concentrations, air temperature, and density of people during class on 8 March 2022 in School A.

As can be seen in the CO_2 concentration graph in Figure 4, the room was well-ventilated at the beginning of the class (CO_2 concentration around 600 parts per million (ppm)). Classroom occupancy was average for all lessons, with 0.26 people/m^2. The CO_2

concentration started to increase before the lesson started, as the students gathered in the classroom earlier. During the first 45 min of the first lesson, the CO_2 concentration in the classroom increased from 800 ppm to 1450 ppm (+650 ppm). Opening the windows for 15 min reduced the CO_2 concentration from 1450 ppm to approximately 970 ppm (−480 ppm), which did not cover the accumulation of CO_2 during the lesson. During the second lesson, the CO_2 concentration increased from 970 ppm to 1646 ppm (+676 ppm), which is very similar to the increase during the first lesson. During the break, the window was open only for 10 min, so the CO_2 concentration decreased from 1646 ppm to 1250 ppm (−396 ppm), roughly proportional to the decrease in the first break. During the third lesson, there was only one person in the classroom (approximately 50 m^2 of classroom space) and the CO_2 concentration decreased by only about 100 ppm. This shows that the change in indoor air without opening the window is quite low. During the third 15-min break, the CO_2 concentration dropped from 1189 ppm to 920 ppm (−269 ppm), which is similar to the drop during the first break. During the fourth lesson, when the same number of students reassembled, the CO_2 concentration increased by approximately the same amount as during the first lesson (+742 ppm). During the sixth lesson, when the class was about twice as small as during the first lessons, the CO_2 concentration increased from 916 ppm to 1254 ppm (+338 ppm) in 45 min, that is, approximately 2 times less.

The temperature curve is smooth, with temperatures below the norm at the beginning of the lessons but close to 18 °C during the other lessons. During ventilation, the indoor temperature dropped by about 1.5 °C when the windows were open, but quickly rose to the norm when the windows were closed.

The relative humidity (RH) also decreased by about 5% but remained close to the norm of 35%.

On the second monitoring day (9 March 2022), the outside air temperature was close to the first day's temperature, with lower cloud cover. Classroom occupancy was slightly lower than on the first day, with 0.22 person/m^2. This time, the classroom was ventilated before the first lesson; the first lesson started at a CO_2 concentration of 574 ppm and reached a concentration of 1228 ppm (+654 ppm) at the end of the lesson (Figure 5). During the first break, the windows were only opened for 10 min, but the decrease in CO_2 concentration was −395 ppm, like the window opening of 15 min on the first monitoring day. During the third lesson, there were two people in the classroom and the classroom was not ventilated before and during the lesson, which shows that without opening the windows there is insufficient air change in the room and the CO_2 concentration is not reduced.

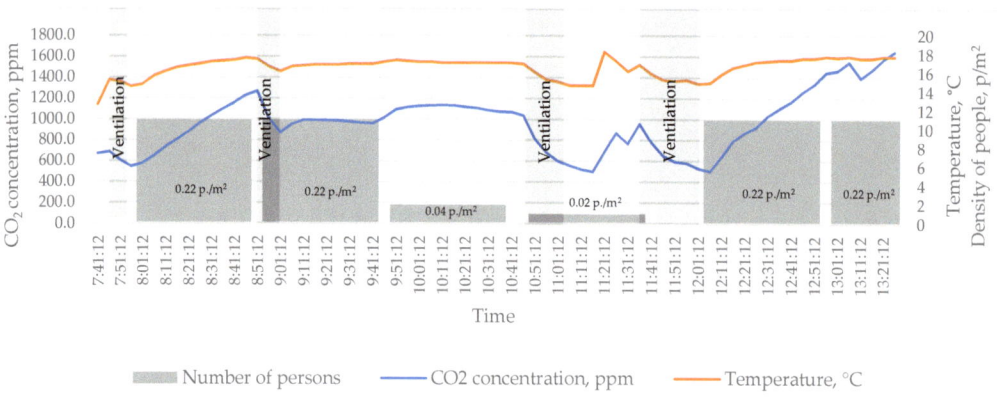

Figure 5. CO_2 concentrations, temperature, and density of people during class on 9 March 2022 in School A.

The increase in CO_2 concentration during lessons five and six shows that during the first break without ventilating the classroom, the CO_2 concentration reaches a high value of 1650 ppm.

The indoor air temperature curve for the day shows that the temperature was kept lower during the night, with a drop of about 1.5 °C when the windows are open and stabilised between lessons five and six without ventilation. The RH of the indoor air is lower than on the first day, between 30% and 35%, which is due to the lower number of people in the room, especially during the third and fourth lessons.

On the third day of monitoring (10 March 2022), the outside air temperature was −9 to −3 °C with high cloud cover. Classroom occupancy was higher than before, with 0.28 person/m² during the first, third and fourth lessons. The classroom was ventilated before the first lesson and the lesson started at a concentration of 602 ppm, which increased to 1370 ppm (+768 ppm) at the end of the lesson. This shows that increasing the occupancy of the classroom quickly leads to an increase in CO_2 concentration (Figure 6). From the monitoring data for this day, opening a window at a lower outside air temperature leads to a greater air change and, consequently, to a greater reduction in the CO_2 concentration in the room air. During the 15 min window opening between the third and fourth lessons, the CO_2 concentration in the room decreased from 1354 ppm to 680 ppm (−674 ppm). The trends in CO_2 regulation when opening the windows at school A were similar on all three days of monitoring.

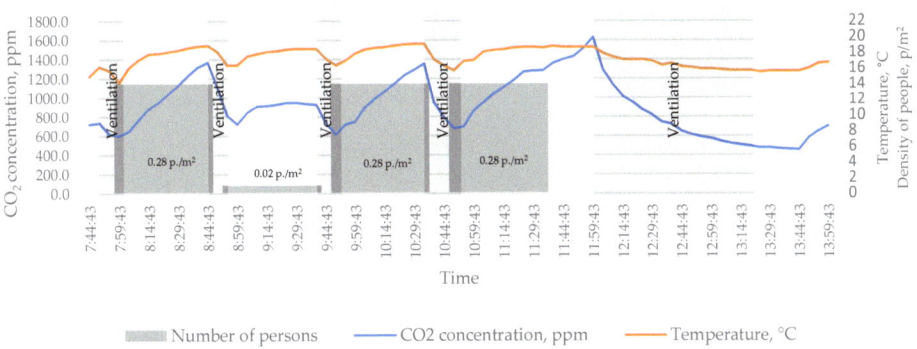

Figure 6. CO_2 concentrations, temperature, and density of people during class on 10 March 2022 in School A.

The room air temperature decreased more during the opening of the windows than during the first lessons, by around 2 °C. This is a significant drop, but the speed with which the room air temperature recovers when the windows are closed suggests that air changes during ventilation do not have time to cool the room surfaces.

Indoor RH is much lower than on previous days, with 11% RH measured at one point. This is below the regulatory limit. Cold air entering the room intensively absorbs moisture, so ventilation at lower outside air temperatures removes moisture from the room more quickly.

3.2. Monitoring Results for School B

The monitoring of School B took place at positive outside temperatures on sunny days. The air change rate per hour (n_{50}) in this classroom was measured at 2.26 1/h. On the first day of monitoring (22 March 2022), the outside air temperature varied between 6 and 11 °C, resulting in a high temperature of +25 °C in the classroom at the beginning of lessons (Figure 7).

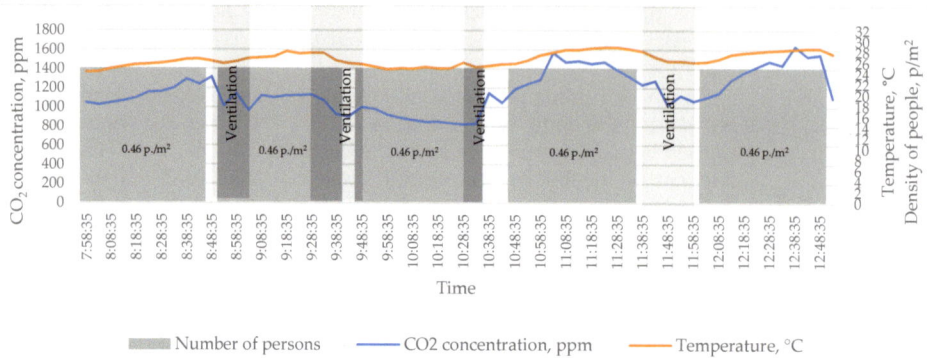

Figure 7. CO_2 concentrations, temperature, and density of people during class on 22 March 2022 in School B.

The situation was similar on the other monitoring days: on the second day (23 March 2022), the outside air temperature varied between 4 and 14 °C, with a high inside air temperature of +23 °C in the morning and a steady rise during the day, and on the third monitoring day (24 March 2022), the outside air temperature varied between 2 and 5 °C, but the classroom temperature was +22 °C in the morning and rose to almost +28 °C during the day (Figure 8). During all lessons, the density of people in the room was 0.46 people per m².

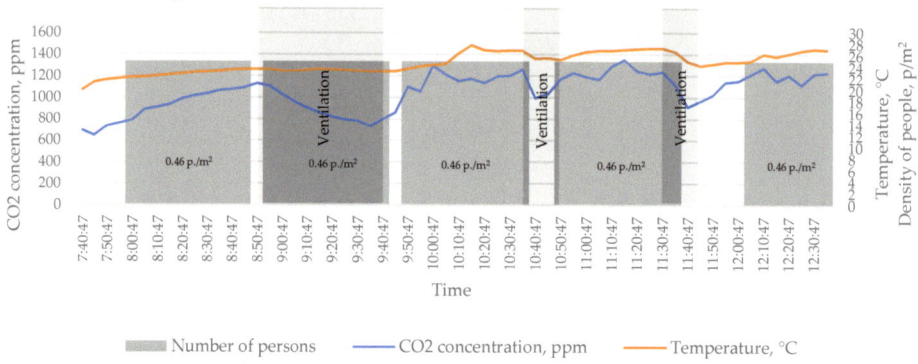

Figure 8. CO_2 concentrations, temperature, and density of people during class on 24 March 2022 in School B.

The CO_2 concentration in this school varied very unevenly during the school day, with frequent increases and decreases during the school day, regardless of ventilation. Passive sunscreens should be installed on windows as the room becomes too hot and ventilation by opening the windows does not reduce the indoor air temperature.

3.3. Monitoring Results for School C

The monitoring of School C was carried out in the presence of positive outside temperatures and high cloud cover. The air change rate per hour in this classroom was measured at 2.50 1/h. On the first day of monitoring (25 March 2022), the outside air temperature was 6–8 °C. The concentration of CO_2, the temperature of the indoor air and the density of the residents during the lessons on 04.25 at School C are shown in Figure 9.

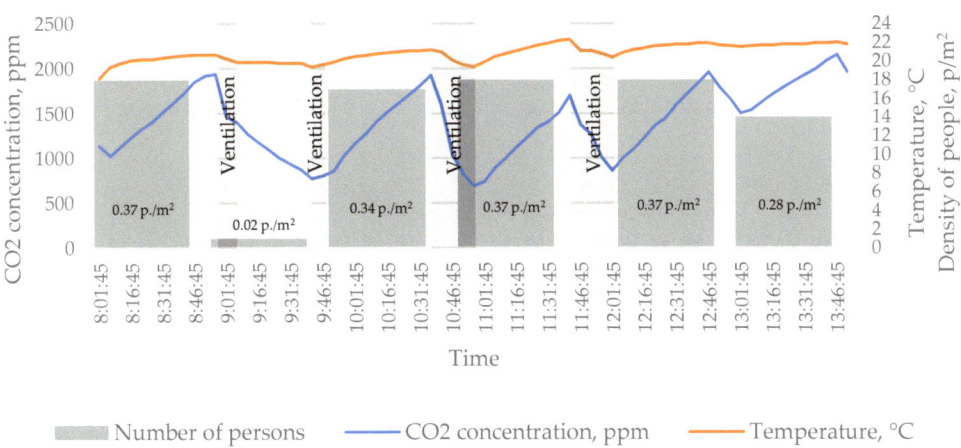

Figure 9. CO_2 concentrations, temperature, and density of people during class on 25 April 2022 in School C.

As can be seen in the CO_2 concentration graph in Figure 9, the CO_2 concentration at the beginning of the class was 1026 ppm. Classroom occupancy was average for all lessons, with 0.30 person/m². The CO_2 concentration started to increase before the lesson started, as the students gathered in the classroom earlier. During the first 45 min of the first lesson, the CO_2 concentration in the classroom increased from 1026 ppm to 1940 ppm (+914 ppm). Opening the windows for 10 min reduced the CO_2 concentration from 1940 ppm to approximately 1400 ppm (−540 ppm), which did not cover the accumulation of CO_2 during the lesson. During the second lesson, the classroom was empty, so the concentration of CO_2 decreased from 1400 ppm to 872 ppm (−528 ppm). During the second break, the window was open for 5 min, decreasing the CO_2 concentration from 872 ppm to 773 ppm (−99 ppm), recognizing that it was empty before. During the third lesson, the density of the classroom rose to 0.34 person/m², thus increasing the CO_2 concentration from 855 ppm to 1927 ppm (+1072 ppm). During the third 20-min break, the CO_2 concentration dropped from 1927 ppm to 740 ppm (−1187 ppm), which is like the drop during the two-times shorter first break. During the fourth lesson, when the density of people in the classroom was a bit higher (0.37 person/m²), the CO_2 concentration increased by approximately the same amount as during the first lesson (+1011 ppm). The fourth 20-min break decreased the CO_2 concentration from 1701 ppm to 861 ppm (−840 ppm), as before. The fifth lesson was attended by the same number of pupils as before, the CO_2 concentration increased from 861 ppm to 1959 ppm (+1098 ppm) in 45 min, i.e., the same as before. Between the fifth and sixth lessons, no window opening occurred, so on this 15-min break, the CO_2 concentration decreased from 1959 ppm to 1499 ppm (−460 ppm), i.e., around twice less compared to opening of the window.

The temperature curve is smooth, with temperatures close to 18 °C at the beginning of the lessons, and not higher than given in the norms during the other lessons. During ventilation, the indoor temperature dropped by about 1.5 °C when the windows were open, but quickly rose to the norm when the windows were closed.

The RH also decreased by about 8% but remained within the limits of the norms.

On the second monitoring day (26 April 2022), the outside air temperature was close to the first day's temperature, with lower cloud cover. Classroom occupancy was slightly higher than on the first day, with 0.36 person/m². This time, the first lesson started at a CO_2 concentration of 941 ppm and reached a concentration of 1961 ppm (+1020 ppm) at the end of the lesson (Figure 10). During the first break, the windows were only opened for 10 min, but the decrease in CO_2 concentration was −694 ppm, like the 10 min window opening on the first monitoring day. The classroom was ventilated by opening windows after every lesson. The ventilation period was usually 15 min, so we can see the same tendencies in Figure 10, as described in the first lesson, and the results of the second day are similar to the results of the first day.

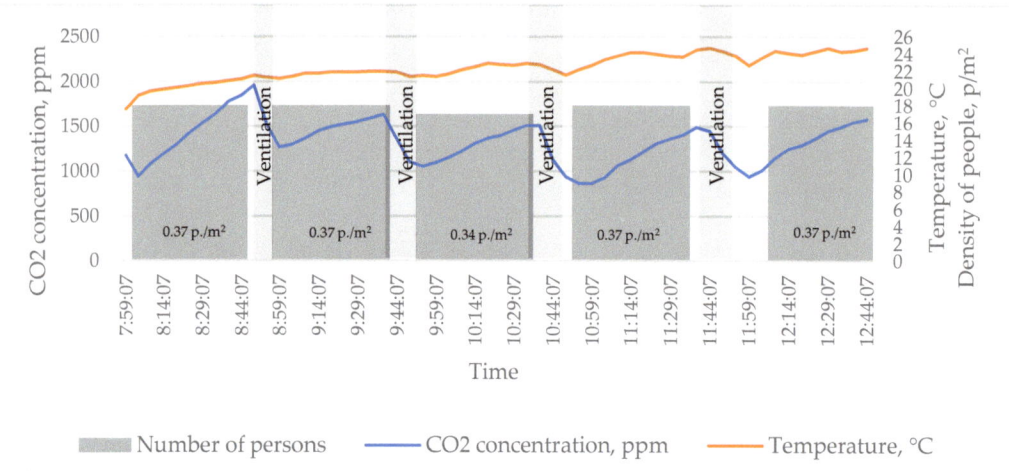

Figure 10. CO_2 concentrations, temperature, and density of people during class on 26 April 2022, in School C.

The indoor air temperature curve for the day shows that the temperature was kept lower during the night, with a drop of about 1.5 °C when the windows are open and constantly increasing during the day. The RH of the indoor air is like the first day, between 35% and 50%.

On the third day of monitoring (27 April 2022), the outside air temperature was 6–7 °C, and the weather was sunny. Classroom occupancy was the same as the day before. The room air temperature dropped more during the opening of the windows than during the first two days, by around 3–3.5 °C. This is a significant drop, but the speed with which the room air temperature recovers when the windows are closed suggests that the air changes during ventilation do not have time to cool the room surfaces. The lowest indoor humidity was recorded on the third day. After the third ventilation, indoor humidity dropped significantly, from 24 to 16%. This is below the regulatory limit.

The trends in CO_2 regulation when opening the windows in School C were similar on all three days of monitoring. The longer the window is open for ventilation, the bigger the drop in CO_2 concentration. The results of the monitoring of School C showed that the airtightness of the classroom is the highest of all the schools studied, with the highest increases in CO_2 concentrations for similar occupancy levels. This is confirmed by the clear dependence of the decrease in CO_2 concentration on the time of window opening.

4. Discussion

4.1. Dependence of the Drop in CO_2 Concentration When the Windows Are Open on the Duration of Window Opening at Different Outside Temperatures

Table 4 shows the data for all ventilations: duration of ventilation, change in carbon dioxide (CO_2) concentration, change in indoor temperature during ventilation, outside air temperature during ventilation, and occupant density before and after ventilation.

Table 4. Ventilation summary data.

School	Day	Ventilation Duration	Occupant Density before Ventilation, p./m²	Change in CO_2 Concentration, Parts per Million (ppm)		Indoor Temperature Change (Outdoor Temperature)
				Before	After	
School A	I	15 min	0.26	1450	970	−0.9 °C (−2 °C)
		10 min	0.26	1646	1250	−1.3 °C (−3 °C)
		15 min	0.02	1189	920	−1.1 °C (−2 °C)
		15 min	0.26	1831	1348	−1.3 °C (−1 °C)
		5 min	0.02	993	916	−1.5 °C (0 °C)
	II	15 min	0	669	543	−0.5 °C (−4 °C)
		10 min	0.22	1266	871	−1.3 °C (−3 °C)
		20 min	0.04	1035	555	−2.3 °C (−1 °C)
		30 min	0.02	955	499	−1.9 °C (0 °C)
	III	15 min	0	735	602	−2 °C (−9 °C)
		15 min	0.28	1370	809	−2.5 °C (−8 °C)
		10 min	0.02	931	617	−2.1 °C (−6 °C)
		15 min	0.28	1354	677	−3.4 °C (−5 °C)
		80 min	0.28	1628	493	−3 °C (−3 °C)
School B	I	15 min	0.46	1316	965	+0.5 °C (+6 °C)
		25 min	0.46	1129	1000	−1.8 °C (+8 °C)
		15 min	0.46	821	829	+0.3 °C (+10 °C)
		25 min	0.46	1236	1065	−2.5 °C (+11 °C)
	II	10 min	0.46	1067	906	−0.5 °C (+4 °C)
		20 min	0.46	1076	986	+0.3 °C (+7 °C)
		50 min	0.46	1443	756	−0.6 °C (+11 °C)
		50 min	0.46	1503	780	−(+14 °C)
	III	45 min	0.46	1129	859	−(+2 °C)
		10 min	0.46	1262	1034	−1.3 °C (+3 °C)
		15 min	0.46	1238	914	−3 °C (+5 °C)
School C	I	10 min	0.37	1940	1400	−0.8 °C (+6 °C)
		5 min	0.02	872	773	−0.4 °C (+6 °C)
		20 min	0.34	1927	740	−1.8 °C (+7 °C)
		20 min	0.37	1701	861	−1.9 °C (+8 °C)
	II	10 min	0.37	1529	1267	−0.3 °C (+7 °C)
		15 min	0.37	1631	1053	−0.5 °C (+8 °C)
		20 min	0.34	1510	864	−0.7 °C (+9 °C)
		15 min	0.37	1449	940	−0.5 °C (+10 °C)
	III	10 min	0.37	1408	1178	−0.5 °C (+6 °C)
		15 min	0.37	1566	675	−3.6 °C (+6 °C)
		20 min	0.37	1389	456	−2.8 °C (+7 °C)

As can be seen in Table 4, the CO_2 concentration is highly dependent on the timing of window opening, the duration of which was usually between 10 and 20 min. It has been observed that such a short period of time leads to a reduction in CO_2 concentrations between 15% and, in some cases, 65%. The presence of people during ventilation does not affect the slower decrease in CO_2 concentration. Opening the windows for ventilation results in changes in temperature in the room. Because the studies were carried out during the cold season, the negative temperature outside also influenced faster cooling of the indoor air, with a 10 to 20 min ventilation that generally cools the room by 1 to 3 degrees, resulting in an indoor temperature that did not meet hygiene standards, and also changed the relative humidity (RH) value to a lower value as the air entered the room.

Opening the windows for ventilation makes it very difficult to keep the air quality parameters constant, as it is only when the window is closed that the CO_2 concentration starts to rise again and usually reaches undesirable levels in the middle of the lesson. As the study did not specify when to open the windows, it can be seen that the window was not always opened between lessons. Normally, CO_2 concentrations in a well-ventilated classroom start to rise from a minimum and reach the maximum recommended value (1000 ppm) in the middle of the lesson. In an unventilated classroom, the CO_2 concentration drops very little during breaks and always remains above 1000 ppm.

As can be seen in the graphs above (Figures 4–10), Schools A and C show a clear decrease in CO_2 concentrations in proportion to the time the windows are open, depending on the magnitude of the initial concentration. Opening the window (depending on the duration of the opening) returned the CO_2 concentration to its initial state, that is, before each lesson, the CO_2 concentration dropped to a level similar to at the beginning of the previous lesson. This did not happen in School B because the internal temperature was always quite high.

4.2. Variation in Indoor Air Temperature with Window Opening Time and Outside Air Temperature

The variation in indoor air temperature with window opening time and outside air temperature at the time of window opening is analysed below. Based on the data in Table 4, each window opening period can be divided into different temperature ranges.

We first analyse the cases where the outside temperature ranged from -6 °C to -9 °C. At these temperatures, there were two ventilation cycles of 15 min and one of 10 min. In all cases, the indoor temperature dropped by about 2 degrees.

The next case to be analysed was the outdoor temperature of 0 to -5 °C. At these temperatures, there was one ventilation cycle of 5 min, two ventilation cycles of 10 min, five ventilation cycles of 15 min, and two ventilation cycles of 20 min or more. The 5-min ventilation cycle took place at an outdoor temperature of 0 °C and resulted in a 1.5-degree decrease in room temperature (from 18.3 °C to 16.8 °C). The two 10-min ventilation sessions took place when the outside temperature was -3 °C and the room temperature dropped by 1.3 degrees. During 15-min ventilation, the indoor temperature decreased by approximately 1 degree, except for one case where the indoor temperature decreased by 3.4 degrees during ventilation and the outside air temperature was -5 °C at the time of ventilation. Ventilation cycles of more than 20 min cooled the room temperature by 2 degrees.

Another case to be analysed was an outdoor temperature of 1 to 5 °C. At these outdoor temperatures, two ventilation cycles of 10 min and one of 15 min were carried out. One 10-min ventilation after the first lesson reduced the indoor temperature by half a degree and the other after the third lesson the following day reduced the indoor temperature by 1.3 degrees. Both stints of ventilation reduced the CO_2 concentration by the same amount from the initial value, but the room temperatures recorded before the start of the ventilation were 25.2 °C in the first case and 26.9 °C in the second case. During 15-min ventilation, the room temperature was reduced by 3 °C, while the room temperature before ventilation was 27.3 °C.

The last case analysed was the case where the outside air temperature was between 6 and 10 °C. At these outside temperatures, one ventilation of 5 min was carried out, during which the room temperature dropped by 0.4 degrees. The 10 min ventilation sessions resulted in a drop in room temperature of about 0.5 to 0.8 degrees. A 15-min ventilation in one school raised the room temperature by 0.5 degrees; in another school, it reduced the room temperature by half a degree. In both cases, the outside air temperature was similar, as was the classroom occupancy. An exceptional case was a 15-min ventilation that reduced the room temperature by more than 3.6 degrees (the room temperature before the ventilation started was 24 °C, the outside air temperature during ventilation was 6 °C). There were four 20-min ventilation cycles, two of which were like the 15-min ventilation cycles: in one school, the room temperature rose, and in the other school, it fell. However, despite these occurrences, the room temperature dropped by about 2 degrees during these ventilations. The last three cases are ventilation cycles of 25 min, during which the room temperature dropped by about 2 to 2.5 degrees.

4.3. Variation in RH with Window Opening Time and Outside Air Temperature

Table 5 shows the data for all ventilation events: duration of ventilation, change in RH, outside air temperature at the time of ventilation.

Table 5. RH summary data.

School	Day	Ventilation Duration	RH Before the Start of Ventilation	RH After Ventilation	Outside Temperature
School A	I	15 min	40%	36.3%	−2 °C
		10 min	39.4%	33.9%	−3 °C
		15 min	36.2%	31.1%	−2 °C
		15 min	41.9%	36.8%	−1 °C
		5 min	38.1%	31.2%	0 °C
	II	15 min	42.1%	35%	−4 °C
		10 min	40.7%	34.2%	−3 °C
		20 min	36.2%	29%	−1 °C
		30 min	41.7%	29.1%	0 °C
	III	15 min	36.4%	22.5%	−9 °C
		15 min	32.9%	18.1%	−8 °C
		10 min	26.5%	15%	−6 °C
		15 min	29.7%	14.1%	−5 °C
		80 min	35.1%	11.5%	−3 °C
School B	I	15 min	20.7%	18.9%	+6 °C
		25 min	21.6%	19.5%	+8 °C
		15 min	21.4%	22.7%	+10 °C
		25 min	21.8%	22.6%	+11 °C
	II	10 min	22.2%	21.3%	+4 °C
		20 min	21.2%	19.3%	+7 °C
		50 min	23.9%	17.2%	+11 °C
		50 min	28.2%	17%	+14 °C
	III	45 min	27%	23%	+2 °C
		10 min	24.3%	23.2%	+3 °C
		15 min	23.9%	26.4%	+5 °C
School C	I	10 min	48.3%	44.7%	+6 °C
		5 min	40.7%	41.2%	+6 °C
		20 min	49%	41.9%	+7 °C
		25 min	44.6%	40.4%	+8 °C
	II	10 min	50.6%	45.1%	+7 °C
		15 min	45.6%	40.8%	+8 °C
		20 min	41.8%	36.9%	+9 °C
		15 min	37.6%	39.8%	+10 °C
		15 min	36.2%	34%	+11 °C
	III	10 min	31.7%	29.9%	+6 °C
		15 min	27.4%	21.5%	+6 °C
		20 min	23.8%	16.3%	+7 °C

In School A, the outdoor temperature on the third day was lower than on the other days, which affected the RH. With each opening of the window for ventilation, the cold air entering the room dehumidified the indoor air, resulting in a drop in RH on this day, with a particularly significant drop in RH during lesson 5 of 1.7 times. When the outside temperature is below −5 °C (Table 5), it is more difficult to control the RH, which drops below the norm when the window is opened and, with each ventilation session, the cold air entering the room makes the air even drier, which can cause dry eyes and other similar ailments in pupils. Without opening a window to ventilate the room, RH increases, but so does the room temperature and CO_2 concentration.

In School B, the RH was significantly below the recommended minimum value (35%) on all days. RH is influenced by indoor temperature, which was significantly higher than hygiene standards on all days in this school, which also led to air drying. In School B, unlike School A, it is not possible to control the temperature and RH by opening windows for ventilation, although the CO_2 concentration can be adjusted in this way. It is easier to control the microclimatic parameters when the outside temperature is below 5 °C, which was recorded on the last day of monitoring. At an outdoor temperature of around 10 °C, it is more difficult to control the RH, and the school has a constant tendency to exceed the comfort temperature, resulting in a feeling of dryness in the rooms, as shown by the RH measurements. At outdoor temperatures between 0 and 10 °C, classroom ventilation decreases CO_2 concentration, RH, and temperature. If the classroom is not ventilated at these outdoor temperatures, an increase in CO_2 and RH is recorded, but the room temperature remains almost unchanged.

At School C, the RH was within the normal range for the first two days of the study, but on the third day, the RH steadily decreased because the classroom temperature on that day was the highest of the three days of the study and it was rising all day, while the RH fell at the same time. In School C, a similar situation to School B is observed: at outdoor temperatures between 0 and 10 °C, the ventilation of the classroom decreases the CO_2 concentration, RH, and temperature. When the classroom is empty, the RH decreases, whereas when the classroom is later filled with pupils, the RH increases. The same trend is observed for outdoor temperatures between 10 and 15 °C.

4.4. Assessment of Air Exchange and Heat Consumption for Class Ventilation

To assess the intensity of air change in the classes and associated heat losses, three monitoring cases were selected for analysis (Table 6).

Table 6. Results for air change, ventilation demand, and associated energy consumption.

	School A	School B	School C
Monitoring date	10 March 2022	24 March 2022	25 April 2022
Volume of the room, V_{room}, m³	167.3	154.82	142.59
Measured air change rate, n_{50} (h⁻¹)	1.55	2.26	2.50
Outside air temperature, °C	−3–−9	+3–+5	+6–+8
The volume of air changing through the open windows during one break, $V_{air,exc \cdot op.}$, m³	167	93	143
The volume of air that changes in the room through leaks in one lesson, $V_{air,exc \cdot leak.}$, m³	41	56	58
The amount of air required for ventilation during a lesson, $V_{air,exc \cdot req.}$, m³	263	506	365
The amount of thermal energy used to heat the changed air, Q_h, kWh	−0.48 *	−2.73 *	−0.72 *

* "−" indicates that the class ventilation consumes less thermal energy due to low ventilation rates.

The analysis of the CO_2 monitoring carried out in School A on 10 March 2022 (Figure 6) shows that the classroom was well-ventilated before the lesson, but during the lesson, the CO_2 concentration curve was increasing intensively, clearly indicating insufficient ventilation. When the window was opened during the first break, the CO_2 concentration approached the baseline level, indicating that almost all the air in the room had changed. The change in CO_2 concentration during the third lesson and the break afterward also indicates that the air in the room changes when the windows are open during the break.

Analysis of the monitoring of CO_2 concentrations in School B on 24 March 2022 (Figure 8) shows that opening the windows during breaks after the third lesson results in a drop of only about 50% in CO_2 concentrations compared to the increase during the lesson and after the fourth lesson in a decrease of about 70%. In this case, not all of the air has changed, so it is assumed that on average 60% of the indoor air volume changes during the break when the windows are opened.

Analysis of CO_2 concentration monitoring (Figure 9) carried out in School C on 25 April 2022 shows that a longer window opening period ensures that CO_2 concentrations that have increased during the lesson return to their initial level, but that even in a less airtight room with a higher outside air temperature, the increase in concentration during the lesson is sudden. These experimental results confirm that indoor ventilation conditions deteriorate as the temperature of the outside air increases.

The results of the calculation of air change, ventilation demand, and associated energy consumption using Equations (1)–(5) in the Methodology section are shown in Table 6. The analysis of air change and associated heating energy usage has shown that even moderate-sized leakages in rooms have several times lower air change compared to the ventilation demand of the classes, so the required air change cannot be achieved without supply and extract air devices in the classes. The intensity of the air change when windows are open decreases as the outside air temperature increases, so in warmer periods, the duration of the breaks is insufficient to fully ventilate the rooms. The results of the analysis also show that natural ventilation of school classrooms by opening windows even reduces the real energy consumption of heating compared to properly ventilated school premises, as indoor air change does not meet the hygiene standards.

The dependencies and trends identified during this research are relevant for general education schools, where the number of children in classes is large, there are many lessons in a row and breaks are short, when the unoccupied room duration is not enough to ventilate them and stabilise their air temperature. In other schools, the conditions of use of the premises will be different, less intensive, so their ventilation by opening the windows may give better air quality results. Air temperature fluctuations after opening the windows can be smaller in schools with heating systems with automated regulation.

5. Conclusions

It was noticed that a periodic window opening during each break significantly reduces the carbon dioxide (CO_2) concentration in the classroom but may not be sufficient to reach the continuous comfort limit in an airtight room, especially during warmer periods. Monitoring of classrooms with different airtightness has shown that the control of CO_2 concentration and indoor air temperature will be more effective when the indoor airtightness can be controlled, e.g., by installing a window micro-ventilation function. When rooms are less airtight, lower peak values of air parameters are obtained because the room is partially ventilated through the leaks all the time. However, this is only relevant at lower outside temperatures. When it reaches +10 °C, air filtration through leaks is greatly reduced, and its influence on indoor air parameters becomes insignificant.

The air in the entire room can only change during a 15-min break at low outside temperatures. At positive temperatures, opening the windows for 15 min is not sufficient to remove the CO_2 concentration accumulated during the lesson. The higher air leakage from classrooms helps flatten the CO_2 concentration curves only at lower temperatures. When the outside temperature is higher, only forced ventilation can produce adequate indoor air quality (IAQ). A drop in indoor air temperature when the window is open is only acceptable at positive temperatures. When the outside air temperature is negative, the indoor temperature before the start of the lesson may be 1–2 °C below the normative temperature and does not meet the hygiene requirements. The opening of windows dries out the indoor air, especially at low outside temperatures. The study found that indoor air becomes drier in leaky classrooms with more constant air changes. The results also showed that opening a window during a break only significantly changes some of the indoor air, as the CO_2 concentration and air temperature immediately begin to increase when the window is closed. The use of a fan to move air during the break would increase the intensity of the air change in the room.

It is not possible to directly apply economic weights to the installation of an efficient ventilation system in schools under renovation, as poorly ventilated classrooms consume less heating energy than classrooms that meet the ventilation requirements, and the requirements for classroom ventilation systems and IAQ in schools should therefore be formalised in the building regulations.

Author Contributions: Conceptualization, T.M., R.B., S.A. and I.B.; Methodology, T.M., R.B. and S.A.; Validation, T.M. and R.B.; Formal Analysis, T.M. and R.B.; Investigation, S.A., V.P. and I.B.; Data Curation, T.M., S.A. and V.P.; Writing—Original Draft Preparation, T.M.; Writing—Review and Editing, R.B. All authors have read and agreed to the published version of the manuscript.

Funding: This research received no external funding.

Data Availability Statement: Not applicable.

Conflicts of Interest: The authors declare no conflict of interest.

References

1. Pikutis, R.; Šeduikytė, L. Estimation of the effectiveness of renovation work in Lithuanian schools. *J. Civ. Eng. Manag.* **2006**, *12*, 163–168. [CrossRef]
2. Kapalo, P.; Klymenko, H.; Zhelykh, V.; Adamski, M. Investigation of indoor air quality in the selected Ukraine classroom–Case study. Proceedings of CEE 2019, Lviv, Ukraine, 11–13 September 2019; Blikharskyy, Z., Koszelnik, P., Mesaros, P., Eds.; Springer: Cham, Switzerland, 2020; Volume 47. [CrossRef]
3. Šadauskienė, J.; Šeduikytė, L.; Paukštys, V.; Banionis, K.; Gailius, A. The role of air tightness in assessment of building energy performance: Case study of Lithuania. *Energy Sustain. Dev.* **2016**, *32*, 31–39. [CrossRef]
4. Jensen, P.A.; Thuvander, L.; Femenias, P.; Visscher, H. Sustainable building renovation–strategies and processes. *Constr. Manag. Econ.* **2022**, *40*, 157–160. [CrossRef]
5. Kapalo, P.; Domnița, F.; Bacoțiu, C.; Spodyniuk, N. The impact of carbon dioxide concentration on the human health–case study. *J. Appl. Eng. Sci.* **2018**, *8*, 61–66. [CrossRef]
6. Moya, T.A.; van den Dobbelsteen, A.; Ottelé, M.; Bluyssen, P.M. A review of green systems within the indoor environment. *Indoor Built Environ.* **2019**, *28*, 298–309. [CrossRef]
7. Luther, M.B.; Horan, P.; Tokede, O. Investigating CO_2 concentration and occupancy in school classrooms at different stages in their life cycle. *Arch. Sci. Rev.* **2018**, *61*, 83–95. [CrossRef]
8. Sadrizadeh, S.; Yao, R.; Yuan, F.; Awbi, H.; Bahnfleth, W.; Bi, Y.; Cao, G.; Croitoru, C.; de Dear, R.; Haghighat, F.; et al. Indoor air quality and health in schools: A critical review for developing the roadmap for the future school environment. *J. Build. Eng.* **2022**, *57*, 104908. [CrossRef]
9. Kapalo, P.; Mečiarová, Ľ.; Vilčeková, S.; Burdová, E.K.; Domnița, F.; Bacoțiu, C.; Péterfi, K.A. Investigation of CO_2 production depending on physical activity of students. *Int. J. Environ. Health Res.* **2019**, *29*, 31–44. [CrossRef] [PubMed]
10. Zhang, D.; Ortiz, M.A.; Bluyssen, P.M. A review on indoor environmental quality in sports facilities: Indoor air quality and ventilation during a pandemic. *Indoor Built Environ.* **2022**, *32*, 1–21. [CrossRef]
11. Krawczyk, D.A.; Zielinko, P.; Rodero, A. Measurements of carbon dioxide concentration and temperature in dormitory rooms in Poland and Spain—A case study. *IOP Conf. Ser. Earth Environ. Sci.* **2019**, *214*, 012056. [CrossRef]
12. Song, G.; Ai, Z.; Liu, Z.; Zhang, G. A systematic literature review on smart and personalized ventilation using CO_2 concentration monitoring and control. *Energy Rep.* **2022**, *8*, 7523–7536. [CrossRef]
13. Daisey, J.M.; Angell, W.J.; Apte, M.G. Indoor air quality, ventilation, and health symptoms in schools: An analysis of existing information. *Indoor Air* **2003**, *13*, 53–64. [CrossRef]
14. Satish, U.; Mendell, M.J.; Shekhar, K.; Hotchi, T.; Sullivan, D.; Streufert, S.; Fisk, W.J. Is CO_2 an indoor pollutant? Direct effects of low-to-moderate CO_2 concentrations on human decision-making performance. *Environ. Health Perspect.* **2012**, *120*, 1671–1677. [CrossRef] [PubMed]
15. Talarosha, B.; Satwiko, P.; Aulia, D.N. Air temperature and CO_2 concentration in naturally ventilated classrooms in hot and humid tropical climate. *IOP Conf. Ser. Earth Environ. Sci.* **2020**, *402*, 012008. [CrossRef]
16. Petersen, S.; Jensen, K.L.; Pedersen, A.L.S.; Rasmussen, H.S. The effect of increased classroom ventilation rate indicated by reduced CO_2 concentration on the performance of schoolwork by children. *Indoor Air* **2016**, *26*, 366–379. [CrossRef] [PubMed]
17. Shendell, D.G.; Prill, R.; Fisk, W.J.; Apte, M.G.; Blake, D.; Faulkner, D. Associations between classroom CO_2 concentrations and student attendance in Washington and Idaho. *Indoor Air* **2004**, *14*, 333–334. [CrossRef]
18. Sundell, J.; Levin, H.; Nazaroff, W.W.; Cain, W.S.; Fisk, W.J.; Grismund, D.T.; Gyntelberg, F.; Li, Y.; Persily, A.K.; Pickering, A.C.; et al. Ventilation rates and health: Multidisciplinary review of the scientific literature. *Indoor Air* **2011**, *21*, 191–204. [CrossRef]
19. *HN 21:2017*; Mokykla, Vykdanti Bendrojo Ugdymo Programas. Bendrieji Sveikatos Saugos Reikalavimai: Vilnius, Lithuania, 2017.

20. Sundell, J. On the history of indoor air quality and health. *Indoor Air* **2004**, *14*, 51–58. [CrossRef]
21. World Health Organization. *School Environment: Policies and Current Status*; Report; WHO Regional Office for Europe: Copenhagen, Denmark, 2015.
22. Ma, F.; Zhan, C.; Xu, X.; Li, G. Winter thermal comfort and perceived air quality: A case study of primary schools in severe cold regions in China. *Energies* **2020**, *13*, 5958. [CrossRef]
23. Abdel-Salam, M.M.M. Investigation of indoor air quality at urban schools in Qatar. *Indoor Built Environ.* **2019**, *28*, 278–288. [CrossRef]
24. Ramos, N.M.M.; Almeida, R.M.S.F.; Curado, A.; Pereira, P.F.; Manuel, S.; Maia, J. Airtightness and ventilation in a mild climate country rehabilitated social housing buildings—What users want and what they get. *Build. Environ.* **2015**, *92*, 97–110. [CrossRef]
25. Statistics Lithuania, Official Statistics Portal. General Schools. 2021–2022. Available online: https://osp.stat.gov.lt/statistiniu-rodikliu-analize?hash=96f65e9b-794b-4f3b-845d-c7576edf57cf#/ (accessed on 4 July 2022).
26. Zemitis, J.; Bogdanovics, R.; Bogdanovica, S. The study of CO_2 concentration in a classroom during the COVID-19 safety measures. *E3S Web Conf.* **2021**, *246*, 01004. [CrossRef]

Disclaimer/Publisher's Note: The statements, opinions and data contained in all publications are solely those of the individual author(s) and contributor(s) and not of MDPI and/or the editor(s). MDPI and/or the editor(s) disclaim responsibility for any injury to people or property resulting from any ideas, methods, instructions or products referred to in the content.

Article

Evaluating Reduction in Thermal Energy Consumption across Renovated Buildings in Latvia and Lithuania

Aleksejs Prozuments [1,*], Anatolijs Borodinecs [1,*], Sergejs Zaharovs [1], Karolis Banionis [2], Edmundas Monstvilas [2] and Rosita Norvaišienė [2]

[1] Department of Heat Engineering and Technology, Faculty of Civil Engineering, Riga Technical University, LV-1048 Riga, Latvia

[2] Institute of Architecture and Construction, Kaunas University of Technology, 44405 Kaunas, Lithuania; rosita.norvaisiene@ktu.lt (R.N.)

* Correspondence: aleksejs.prozuments@rtu.lv (A.P.); anatolijs.borodinecs@rtu.lv (A.B.)

Abstract: Currently, the optimization of thermal energy consumption in buildings is considered a suitable alternative in the construction of new buildings, as a result of which the overall energy efficiency of the building increases. Thus, this study examined the efficiency and efficacy of different building renovation packages conducted across several buildings in Latvia and in Lithuania (across a larger building stock). In the first section of this study, 13 multi-apartment residential houses with 3 building renovation packages have been investigated in the city of Daugavpils, Latvia, in order to determine the actual reduction in heat energy consumption across each of the renovation implementation packages. The study findings indicate that changes in Latvian building regulations regarding insulation thickness did not significantly impact thermal energy consumption in fully renovated buildings. However, the combination of facade renovations, upgraded heating systems, and improved ventilation systems resulted in substantial energy savings, with an average reduction of 50.59% in thermal energy consumption for space heating across the reviewed multi-apartment residential building stock. In the following section of this study, the impact of the Energy Performance of Buildings Directive (EPBD) on building energy efficiency in Lithuania has been examined. The results show that over a 10-year period in the 2000s, Lithuanian building stock experienced a 20% increase in energy efficiency, followed by an additional 6.3% increase between 2010 and 2016. The mandatory requirement for renovated buildings to achieve a minimum energy efficiency class has resulted in significant reductions in energy consumption for heating purposes. The findings underscore the effectiveness of building renovation packages and the EPBD regulations in enhancing energy efficiency and promoting sustainable building practices. The importance of heat metering, consideration of indoor air temperature, and the need to address indoor air quality during renovations were also highlighted.

Keywords: thermal energy consumption; building renovation; multi-apartment buildings; energy savings

Citation: Prozuments, A.; Borodinecs, A.; Zaharovs, S.; Banionis, K.; Monstvilas, E.; Norvaišienė, R. Evaluating Reduction in Thermal Energy Consumption across Renovated Buildings in Latvia and Lithuania. *Buildings* **2023**, *13*, 1916. https://doi.org/10.3390/buildings13081916

Academic Editor: Gerardo Maria Mauro

Received: 21 June 2023
Revised: 20 July 2023
Accepted: 26 July 2023
Published: 27 July 2023

Copyright: © 2023 by the authors. Licensee MDPI, Basel, Switzerland. This article is an open access article distributed under the terms and conditions of the Creative Commons Attribution (CC BY) license (https://creativecommons.org/licenses/by/4.0/).

1. Introduction

With the rising demand for energy and the diminishing availability of fossil fuel-based energy sources, which are both environmentally unsustainable and costly, enhancing the energy efficiency of building stock is a matter of concern worldwide. Building stock accounts for approximately 40% of total final energy use across developed countries, constituting up to one-third of the worldwide greenhouse gas emissions. Thus, upgrading existing buildings to be more energy efficient and implementing sustainable design principles are crucial. This necessitates the adoption of advanced technologies, innovative building materials, and energy-efficient systems to reduce energy consumption and promote a transition toward renewable and low-carbon energy sources. By upgrading the energy efficiency of the existing building stock, governments can mitigate energy demand, enhance energy security, and contribute to the overall sustainability and resilience of the built environment, as well as meet carbon emission reduction targets stipulated in their agenda [1,2].

Building energy efficiency is a dynamically and rapidly growing field and has certainly become a separate industry and a research area over recent decades, as it requires an involvement of highly skilled professionals and continuous research and development activities. In line with the industry's growth, the market availability and promotion of sustainable and energy-efficient products and solutions has increased. This development is in large part driven by national and regional energy and environmental building codes and regulations. Regulatory building codes have proven to be an effective way to promote energy efficiency in buildings. Many governments across the world have put forward nationwide long-term energy use reduction goals for newly constructed and existing building stock that are reinforced by stringent UN regulations aimed at addressing the environmental impact and climate change [3–7].

Residential heating is a compelling issue requiring immediate attention, particularly in regions characterized by mild and cold climates, where low outdoor temperatures persist for extended periods and the heating season lasts approximately 5–6 months. These conditions necessitate efficient and effective heating systems to maintain comfortable indoor environments and reduce building energy consumption. The duration of the heating period is of particular importance as it influences energy demand, costs, and has a substantial environmental impact. Therefore, in these regions, careful consideration must be given to the selection and optimization of building energy efficiency, including heating technology, such as centralized heating systems, heat pumps, or alternative energy sources, to ensure sustainable and cost-effective solutions for residential heating needs [8,9].

Latvia and Lithuania, as Northern European countries, have been implementing and continue to implement energy efficiency programs for existing and future building stock, which help reduce energy consumption and GHG emissions. One of the feasible approaches to reduce the energy consumption of a building is a partial or full renovation. A partial renovation of a residential building is primarily related to insulating the building envelope or retrofitting the heating system, while the full renovation of the building includes both the insulation of the building envelope and the modernization of the heating system [10–12]. It should be noted that before implementing measures to upgrade a building's energy efficiency, what measures stipulate higher feasibility in energy consumption reduction and economic terms have to be thoroughly evaluated [13–15].

Buildings in the EU-27 account for approximately 55% of total electricity consumption and roughly 40% of total final energy consumption on average. Followed by transport and industry, the building industry is the largest end-use energy sector in Europe [16–18]. In Latvia and Lithuania, buildings' energy use share is higher than the EU-27 average (45%) due to the poor energy performance of the existing building stock that features a high share of structures constructed between 1945 and 1990, and is now obsolete with regards to meeting stringent energy performance criteria. Up until the late 1990s and early 2000s, buildings in Latvia and Lithuania were constructed in accordance with regulatory codes, which were insufficiently rigorous and thorough with regards to thermal performance stipulations. As a result, the bulk of the existing building stock that has not undergone deep renovation features poor thermal insulation, excessive outdoor air infiltration, and condensation occurrence within the external wall structures. Moreover, the absolute majority of the building stock constructed between 1945 and 1990 lacks mechanical ventilation systems, and thus the air exchange occurs due to natural ventilation and/or outdoor air infiltration through the external elements (walls and roofs), which entails major thermal energy losses, especially during critical cold season months [19–21].

As building energy consumption continues to rise, it is expected to constitute an even larger proportion of the overall energy consumption. Consequently, it becomes imperative to devote more stringent and thorough attention to building energy efficiency. Enhancing energy efficiency in buildings is crucial for mitigating the environmental impact, reducing energy demand, and ensuring long-term sustainability [22–24]. This entails the implementation of rigorous building energy upgrade measures such as advanced building design, energy-efficient technologies, and effective energy management systems. By prioritizing building energy efficiency, governments can strive toward achieving energy conservation goals, reducing greenhouse gas emissions, and promoting a sustainable future [25–27].

Furthermore, it is important to note that there is a notable research gap in the field of renovation measure efficiency within the Baltic countries. While several studies have been conducted in this field in various regions worldwide, a limited number of studies have been conducted in the Baltic countries' context.

2. Methodology

2.1. Building Selection Criteria: Case Study of Latvia

The research methodology pertained to analyzing the available thermal energy consumption data (kWh) of 13 renovated multi-apartment residential houses in Daugavpils (city in south-eastern Latvia) throughout the period from 2012 to 2021. This was aimed at identifying common guidelines that would help to classify various buildings that had undergone deep renovation in certain time periods, as well as understanding the effectiveness of energy efficiency measures. Taking into account the fact that the share of renovated residential houses in Daugavpils is very low (around 1% of the total multi-apartment residential stock), this study included multi-apartment residential buildings that meet the following criteria:

- the multi-apartment residential house put into operation before the end of 2000;
- the total combined floor area of the residential premises is greater than 300 m^2;
- the total floor area of uninhabitable premises (shop, office, etc.) does not exceed 50% of the total area of the residential building;
- the technical documentation of the planned renovation package has been developed for energy efficiency improvement measures;
- the multi-apartment building has received municipal or state co-financing support for the renovation project implementation.

The climate data for the city of Daugavpils are added in the table below (Table 1).

Table 1. Climate characteristics of the city of Daugavpils.

City	The Average Air Temperature of the Coldest Five Days	The Average Air Temperature of the Coldest Five Days, the Probability of Exceeding Which Is		Coldest Monthly Temperatures					
		0.02	0.1	I	II	III	X	XI	XII
Daugavpils	−23.3	−26.4	−22.3	−42.7	−43.2	−32.0	−14.7	−24.1	−38.7

Daugavpils experiences a cold climate characterized by long, harsh winters. The city's geographical position in the eastern part of the country exposes it to continental influences, resulting in significantly cold temperatures. The average winter temperatures in Daugavpils can often drop below freezing, requiring adequate heating measures to ensure comfort and energy efficiency in buildings. As is seen in Figure 1, the temperature range in the city of Daugavpils in the cold season of the year (October thru March) drops below 0 °C very frequently, falling even below −15 °C.

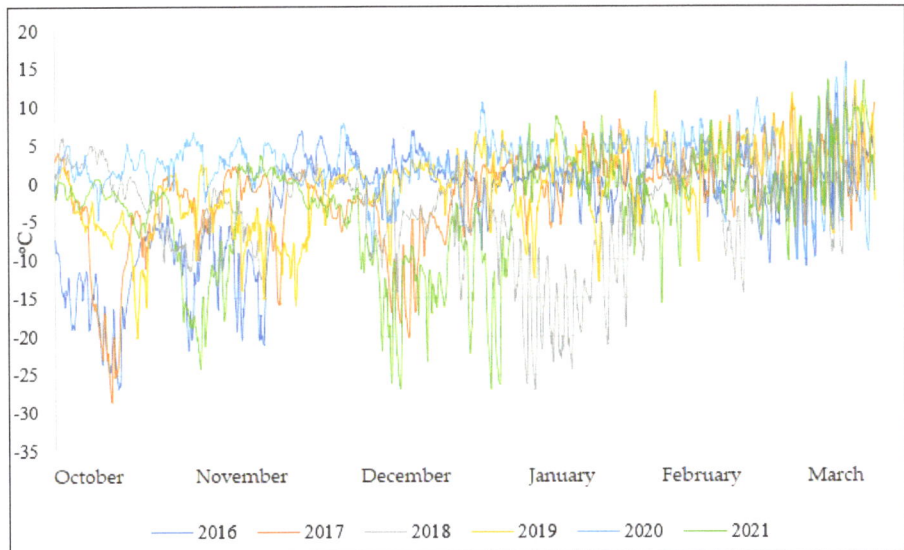

Figure 1. Recorded outdoor air temperatures between the months of October and March (heating season) over the period of 2016–2021 in city of Daugavpils.

The datalogger used to measure the temperature, RH, and CO_2 level was Extech Industries Datalogger SD800 (Table 2)

Table 2. Technical parameters of the datalogger.

Parameter	Value
Humidity measuring accuracy	±4%
Humidity measuring range	10…90% RH
Humidity measuring resolution	0.1% RH
Measurement accuracy	±5%
Measuring instrument features	automatic temperature compensation
Measuring range	0…4000 ppm
Temperature measurement accuracy	±0.8 °C
Temperature measurement resolution	0.1 °C
Temperature measuring range	0…50 °C
Type of meter	datalogger

2.2. Profile of the Examined Buildings

All of the selected multi-apartment buildings were constructed before 1980, while the renovation measures to improve the energy efficiency of those buildings were implemented in the period from 2012 to 2020 (pertaining to the year of completion). The selected buildings were assigned a sequence number, as well as the building group category describing the renovation package with regards to the implemented renovation measures.

To attain the most impartial findings regarding accomplished energy efficiency advancements, the analyzed buildings were categorized into separate groups based on specific criteria (such as characteristics of structural elements, serial type of the multi-apartment building characterizing their total floor count, layout features and materials used in their building envelope, etc.). The photos of the analyzed building types are added in Figure 2. Considering the aforementioned criteria, the categorization of the renovation multi-apartment buildings is as follows:

- The A-group comprises buildings #1, #3, and #12, where the classification of buildings is based on the reference serial-type standard for residential houses (series 316), while building #9 belongs to an enhanced series of prior residential structures (series 318) constructed using similar materials as series 316.
- The B-group consists of buildings #5 and #6, which were categorized based on a specific serial-type standard of residential houses commonly referred to as "Stalinka". These buildings share similar architectural characteristics, including the same number of floors, apartment layout, and a distinctive design element known as a "turn".
- The C-group comprises buildings #2, #8, and #10, which are two-story structures; however, it should be noted that building #8 currently has three floors, despite originally being constructed with only two floors in 1974.
- The D-group residential houses are quite different (#4—building of series 602; #7—small-family residential house; #11—building of series 467; #13—building of series 103), but all residential houses have one thing in common which are elements of external enclosing structures, i.e., hollow reinforced concrete panels.

Figure 2. Photos of reference buildings employed in the study (from left: group-A, group-B, group-D, no group-C picture is available).

The D-group of the examined residential buildings exhibits notable variations (#4 is 602 serial-type buildings, #7 and #11 are serial-type 467 buildings, and #13 is serial-type 103 building). Nonetheless, a shared characteristic among all residential houses within this group is the 9 to 10 story floor count, the utilization of external enclosing structures comprising hollow reinforced concrete panels.

Despite the fact that residential houses are divided into groups according to one or more characteristics, it is necessary to divide the groups of buildings into smaller groups (subgroups) in order to evaluate the effectiveness of the implemented measures. For the comparison of the objects under study, we assume a division into three subgroups, i.e., buildings with an insulated facade and renovated heating system (first subgroup); renovated heating system (second subgroup); insulated building facade (third subgroup).

In order to assess the effectiveness of the implemented measures across the dispersed building stock, it was necessary to further subdivide the groups of residential houses into smaller subgroups. Thus, it was proposed to divide the implemented renovation measures into three subgroups: the first subgroup comprises buildings with both an insulated facade and a renovated heating system, the second subgroup includes buildings with a renovated heating system only, and the third subgroup consists of buildings where only facade insulation was implemented. This subdivision allows for a more detailed evaluation of the outcomes achieved within each subgroup and enables a comprehensive analysis of the impact of specific interventions on energy efficiency and thermal performance of the residential buildings.

Table 3 compiles the building profile information and directory used for the evaluation.

Table 3. The characteristics of the examined multi-apartment buildings.

	Building Characteristics					Implemented Renovation Measures			Heating System Description	
No.	Building ID #	Building Group	Floor Area, m²	Year of Construction	Renovation Completion Year	Facade Renovation	Ventilation System Renovation	Upgrade/ Modernization of Heating System	Heating System *	Distribution *
1	1	A1	2840.4	1960	2012	+	-	+	one-pipe	top
2	9	A1	1920.07	1971	2015	+	-	+	two-pipe	bottom
3	12	A1	2820.4	1963	2019	+	+	+	one-pipe	top
4	3	A3	2803.5	1960	2013	+	-	-	one-pipe	top
5	5	B1	1468.1	1957	2012	+	-	+	two-pipe	top
6	6	B1	1806.1	1955	2012	+	-	+	two-pipe	bottom
7	2	C1	522.3	1949	2012	+	-	+	two-pipe	bottom
8	10	C2	317.5	1957	2017	-	-	+	two-pipe	bottom
9	8	C3	1004.25	1974	2013	+	-	-	one-pipe	top
10	7	D1	2681.5	1980	2013	+	-	+	one-pipe	top
11	11	D1	2656.42	1977	2018	+	+	+	one-pipe	bottom
12	13	D2	2069.97	1973	2020	-	-	+	two-pipe	bottom
13	4	D3	1957.92	1980	2013	+	-	-	one-pipe	top

* A one-pipe or single-pipe heating system uses a single pipe to distribute both the supply and return of hot water or steam in one circuit, while a two-pipe system has separate pipes for supply and return circuit. "Top" and "bottom" refer to the vertical positioning of heating system components, with "top" typically denoting higher floor levels and "bottom" referring to lower floor levels.

2.3. Building Selection Criteria: Case Study of Lithuania

To investigate the impact of the Energy Performance of Buildings Directive (EPBD) on building energy efficiency in Lithuania, this study employed a quantitative methodology. Data from 5558 multi-apartment buildings, registered between 2014 and 2020, were analyzed. These buildings were classified into energy efficiency classes ranging from A++ to G. The primary focus was on the certification calculations of renovated multi-apartment buildings, specifically those classified as energy performance classes C and D. Energy efficiency certificates (EPCs) issued and registered in Lithuania between 2007 and 2021, totaling 257,196, were also considered. This study compared the average final energy consumption and primary energy consumption for heating purposes across different energy efficiency classes, highlighting the changes resulting from building renovation efforts.

2.4. Result Evaluation

Energy consumption data for the renovated buildings were collected using energy meters installed within each building. These energy meters recorded the thermal energy consumption (kWh) on a regular basis, allowing for a detailed analysis of energy usage patterns before and after the implementation of the renovation measures. The data collected included information on thermal energy consumption for space heating, providing a comprehensive understanding of the energy performance of the buildings. By comparing the energy consumption data (kWh) before and after the renovations, this study was able to quantify the actual energy savings achieved through the implemented measures and assess the effectiveness of the renovation strategies in reducing thermal energy consumption.

To determine the actual thermal energy savings (Δ) as a result of the building renovation measures, the following equation was used:

$$\Delta = \frac{Q_i^0 - Q_j^1}{Q_i^0} * 100 \ [\%]$$

where:

- Q_i^0—the actual thermal energy consumption of the building in the year i before renovation [kWh];
- Q_j^1—actual thermal energy consumption of the building in year j after renovation [kWh].

3. Results

3.1. Case Study of Latvia

Within building group-A (Figure 3), the pre-renovation thermal energy consumption for space heating ranged from 142.77 to 159.42 kWh/m², while the thermal energy consumption during the most recent heating season varied from 73.06 to 103.47 kWh/m². Due to the absence of the A2 subgroup within the group-A building sample, it was not possible to estimate the energy savings achieved solely by renovating the heating system without insulating the facade.

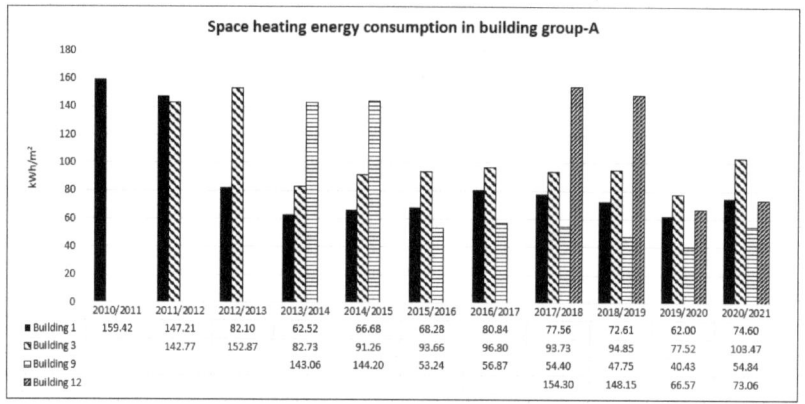

Figure 3. Comparison of thermal energy consumption for space heating within building group-A [kWh/m²].

However, upon comparing the average heat energy consumption before and after renovation, the average savings are as follows: building #1—53.09%, building #3—37.92%, building #9—64.31%, and building #12—53.84%. Analyzing the results of building #3 (subgroup A3), it is evident that the savings significantly differ from the A building group belonging to subgroup A1. This finding indicates that the facade insulation upgrade alone, without adjustments to the heating system in a specific series-type house, is not sufficiently effective. Buildings #1 and #12 exhibit rather similar results to building #9 in terms of the thermal energy savings. Despite variations in the thickness of the facade insulation (#1 and #9—100 mm stone wool, #12—150 mm stone wool), the distribution type of the heating system plays a more significant role than the insulation thickness itself. Buildings #1 and #12 have a common heating distribution pattern (single-pipe distribution from the attic), while building #9 features a different heating system distribution (two-pipe bottom distribution from the basement).

Building group-B (Figure 4) constitutes the smallest group in the study, containing only one subgroup, B1. Both buildings share similar layouts and external building envelope materials. However, despite these similarities, the average thermal energy consumption for space heating per square meter is higher for building #5 compared to building #6, both before renovation (#5: 187.55; #6: 143.12 kWh/m²) and after renovation (#5: 79.61; #6: 55.06 kWh/m²).

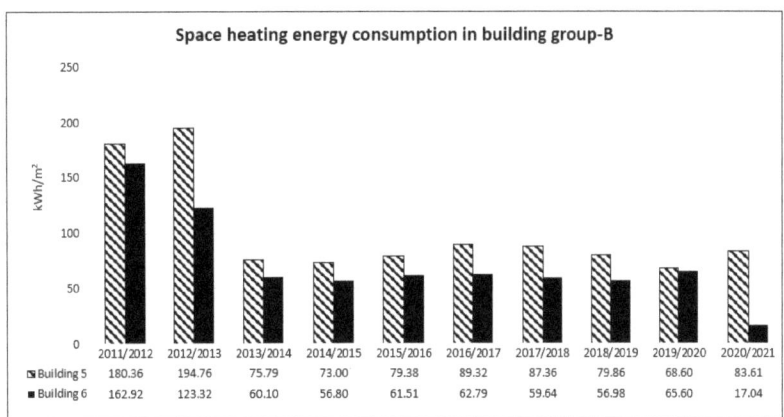

Figure 4. Comparison of thermal energy consumption for space heating within building group-B [kWh/m²].

The average thermal energy savings are as follows: building #5—57.55%, building #6—55.05%. Despite having different heating distribution systems (building #5: two-pipe top distribution from the attic; building #6: horizontal distribution with individual heat meters), the thermal energy savings are quite similar. When comparing buildings with the same heating distribution system (upper distribution from the attic) from different groups/subgroups, such as A1 and B1 (buildings #1, #12, and #5), the thermal energy saving for building #5 is on average 4% higher compared to buildings #1 and #12. This implies that the reduction in thermal energy consumption in this case is influenced by the technical characteristics of the building's fundamental enclosing structures.

Building group-C (Figure 5) encompasses all individual subgroups for the detailed result comparison, consisting of buildings that had undergone facade and heating system renovations (C1), only the heating system upgrade/renovation (C2), or only the facade insulation (C3).

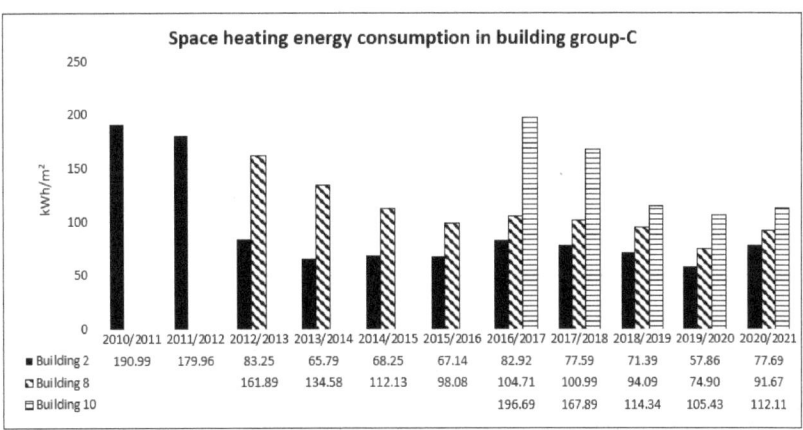

Figure 5. Comparison of thermal energy consumption for space heating within building group-C [kWh/m²].

Notably, buildings #2 and #8 feature partial insulation due to their status as cultural and historical monuments. Additionally, buildings #2 and #10 share the same heating system distribution (horizontal distribution with individual heat meters). The implementation of energy efficiency measures enables the evaluation of each building under equivalent conditions. The average heat energy savings are as follows: building #2—60.92%, building #8—31.21%, and building #10—33.40%. These results indicate that heating system modernizations (C2) yield similar savings to facade insulation (C3), and, when combined (C2 + C3), the energy savings are lower than those achieved through a comprehensive renovation (C1).

Comparing buildings with the same heating distribution system and insulation thickness from different groups/subgroups, such as B1 and C1 (buildings #6 and #2), the thermal energy savings are notably close (<2%). This finding suggests that the thermal system distribution pattern is a significant factor for buildings with a smaller total floor space for living areas (up to 522.3 m^2). Conversely, buildings with a similar insulation thickness from different groups/subgroups, such as A3 and C3 (buildings #3 and #8), demonstrate nearly identical results in thermal energy savings (<1%), indicating an average thermal energy resource saving of 31% for brick-type buildings.

Building group-D (Figure 6) encompasses all subgroups for result comparison, including buildings within the same subgroup (D1) with varying thermal insulation thicknesses but using the same insulation material (mineral wool). The average thermal energy savings are as follows: building #4—42.17%, building #7—57.19%, building #11—67.22%, and building #13—36.17%. These results indicate that the lowest savings were achieved in buildings where only the facade was insulated (third subgroup). A similar outcome was observed for building #8 (C3—31.21%) and building #3 (A3—30.69%).

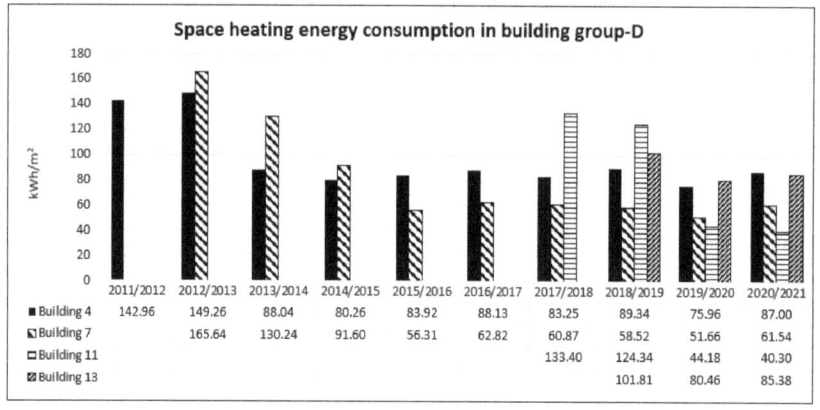

Figure 6. Comparison of thermal energy consumption for space heating within building group-D [kWh/m^2].

Furthermore, the results demonstrate that, similar to building group-C, the renovation of the heating system (second subgroup) yields nearly identical thermal energy savings in cases where the facade is not insulated (third subgroup). It is important to note that building #13 had its heating system replaced during renovation, similar to building #9 (bottom distribution with two-pipe system from the basement), which falls within the first subgroup (A1) with a thermal energy saving of 65.02%. Theoretical calculations suggest that insulating building #13 with a 100 mm thick thermal insulation could result in very close savings compared to building #9.

Additionally, building #7 and building #11 feature the same subgroup (D1), sharing the similar heating distribution system (single-pipe top/bottom distribution), but differing in thermal insulation thickness—100 mm (#7) and 150 mm (#11). In this case, the 10% difference in thermal energy savings for panel-type residential houses is directly influenced by the insulation thickness.

Table 4 compiles averaged thermal energy consumption (kWh/m^2) and thermal energy savings (%) before and after renovation implementation. The data provided in the table allow for investigating the correlation between different retrofit packages and the relative humidity (RH) and carbon dioxide (CO_2) levels in the buildings after renovation, where RH and CO_2 were monitored. Buildings underwent various retrofit packages, including facade upgrades, heating system upgrades, and full retrofit packages (facade, ventilation, heating system upgrades). These buildings exhibited notable energy savings ranging from 34.80% to 67.22%.

Table 4. The comparison of measured parameters in the examined multi-apartment buildings.

No.	Building ID #	Building Group	Average Thermal Energy Consumption Q for Space Heating [kWh/m^2]			Average Measured Parameters		Comments
			Before Renovation, Q_i per Floor Area	After Renovation, Q_j per Floor Area	Savings [%]	RH [%]	CO_2 [ppm]	
1	1	A1	153.32	71.91	53.10%	37.5	1079	Facade + heating system upgrade
2	9	A1	143.63	51.25	64.31%	25.5	457	Facade + heating system upgrade
3	12	A1	151.23	69.81	53.84%	42.6	1314	Full retrofit package
4	3	A3	147.82	91.75	37.93%	N/A	N/A	RH, CO_2 were not monitored
5	5	B1	187.56	79.62	57.55%	44.3	1096	Facade + heating system upgrade
6	6	B1	143.12	55.06	61.53%	40.0	1399	Facade + heating system upgrade
7	2	C1	185.48	72.43	60.95%	N/A	N/A	RH, CO_2 were not monitored
8	10	C2	182.29	110.62	39.31%	62.0	2796	Only heating system upgrade
9	8	C3	148.23	96.65	34.80%	N/A	N/A	RH, CO_2 were not monitored
10	7	D1	147.94	63.33	57.19%	41.9	747	Facade + heating system upgrade
11	11	D1	128.87	42.24	67.22%	29.9	438	Full retrofit package
12	13	D2	127.81	82.92	36.17%	34.2	1131	Only heating system upgrade
13	4	D3	146.11	84.49	42.17%	35.4	1496	Only facade renovation

When examining the average RH levels, it is important to note that the measured RH levels were not excessively high and remained within the comfort range of 30 to 70%. This suggests that the retrofit packages implemented in these buildings, which included improvements to the building envelope and heating systems, potentially contributed to better moisture control and healthier indoor environments. However, in buildings 9 and 11, the average RH level was <30% which suggests that the humidity level is rather low. Although the cause behind the low RH levels in buildings 9 and 11 may require further investigation, it is highly likely that the retrofit measures implemented in these buildings, such as improved insulation or sealing, inadvertently resulted in reduced moisture infiltration from outside. Inadequate humidity control systems or insufficient moisture sources within the buildings may have also contributed to the low RH levels.

Similarly, when examining the CO_2 levels in these buildings, it was observed that all buildings had CO_2 concentrations below 1500 ppm, except for building #10 where the average CO_2 concentration was 2796 ppm, critically exceeding the DIN 1946 Part 2 stipulated limit for a healthy IEQ of 1500 ppm. The CO_2 level <1500 ppm in other buildings indicates improved ventilation and indoor air quality, which can be attributed to the retrofit measures implemented. These measures likely enhanced air circulation and facilitated the removal of indoor air pollutants. In building #10, only the heating system upgrade was carried out, suggesting that the overly high CO_2 concentration might have been an underlying issue, and further retrofit measures are suggested (improved air circulation, mechanical ventilation system) to control the CO_2 concentration within the premises. The measured humidity level is also higher in building #10, suggesting that insufficient air exchange and poor ventilation lead to both moisture and indoor pollutant build-up.

Further research and monitoring are necessary to obtain more conclusive evidence regarding the correlation between specific retrofit strategies and the RH and CO_2 levels in buildings, as these parameters are position-dependent, implying that the measurement and data might be influenced by the specific position or location where they are taken, and therefore the readings may fluctuate substantially depending on the sensor position with regards to the building layout and potential sources of pollution.

3.2. Case Study of Lithuania

Most buildings by area (83%) in Lithuania were built before 1993 (Table 5). Insulation materials were not used for the better thermal insulation of these buildings and only the specific thermal resistance of the building materials (such as bricks, blocks, or panels) determined the thermal resistance of the building. In addition, a significant part of these buildings has not been renovated either by participating in renovation programs. As a result, a large part of the building stock is in poor technical condition (especially in the apartment segment). A total of 58% of the area of the building fund consists of buildings built between 1961 and 1992, the architectural and structural diversity of which is probably not great [28–31]. Accordingly, there is a potential for implementing repeated (standard) renovation solutions, especially in the apartment building segment, where ~72% buildings were built between 1961 and 1992.

Table 5. Building stock by year of construction completion (thousand m^2).

Type of Building	Year of Construction								Total	Total %
	<1900	1901–1960	1961–1992	1993–2005	2006–2013	2014–2016	2017–2018	2019		
1. Residential	1.765	23.105	72.038	11.067	10.461	4.841	3.768	1.958	129.004	64%
1–2 apartment buildings	1.212	17.095	29.160	6.628	7.231	3.912	2.861	1.441	69.540	34%
Multi-apartment buildings	553	6.010	42.878	4.439	3.230	929	907	517	59.464	29%
2. Non-residential	840	8.384	44.337	7.405	6.477	2.360	1.954	913	72.670	36%
Industrial	235	3.416	23.537	3.382	2.627	968	891	433	35.490	18%
Administrative	169	1.554	5.706	924	844	351	332	217	10.097	5%
Educational	118	1.257	6.367	386	220	125	16	14	8.503	4%
Trade	47	495	2.375	1.627	1.631	491	316	83	7.064	4%
Treatment	27	467	1.973	218	178	50	37	2	2.952	1.46%
Accommodation	40	261	987	257	424	207	206	116	2.497	1.24%
Culture	145	467	1.449	122	59	14	23	0	2.279	1.13%
Service	31	227	1.231	291	203	87	83	45	2.199	1.09%
Other	28	239	711	199	291	66	49	5	1.589	0.79%
Total	2.605	31.489	116.375	18.472	16.938	7.201	5.722	2.872	201.674	100%
Total in %	1%	16%	58%	9%	8%	4%	3%	1%	100%	

The typical energy consumption designed for in these houses is for 160–180 kWh/m^2 per year. In terms of heating systems, according to data published by the Lithuanian Heating Association [32], 46.97% of multi-apartment buildings are supplied by a district heating system, with the remaining share of multi-apartment buildings using individual boiler modules present within the building or inside of individual apartments. A smaller percentage of multi-story residential buildings are heated via the use of electric radiators [33]. Notwithstanding the high percentage of multi-apartment buildings connected to a central district heating system, by area, this is significantly smaller, with only 26% of the total area comprising the entire Lithuanian multi-apartment building stock currently connected to a centralized heating system. To address this, the Lithuanian long-term strategy is to transform the current building stock in a way that would lead to a much more efficient use of energy (with conditions mature enough to transform these buildings into almost zero-energy buildings) and make the country independent of fossil fuels by 2050.

In all of this, the energy certification of buildings plays an important role, positioning itself as one of the most important tools of the energy policy for buildings in Lithuania. Over the period 2007–2021, 257,196 EPCs were issued and registered in Lithuania. For the purpose of this study, the data comprised the certification calculations of 5558 multi-apartment buildings registered between the period 2014–2020 [34]. Figure 7, to this effect, shows the distribution of energy performance certificates issued for these 5558 multi-apartment buildings.

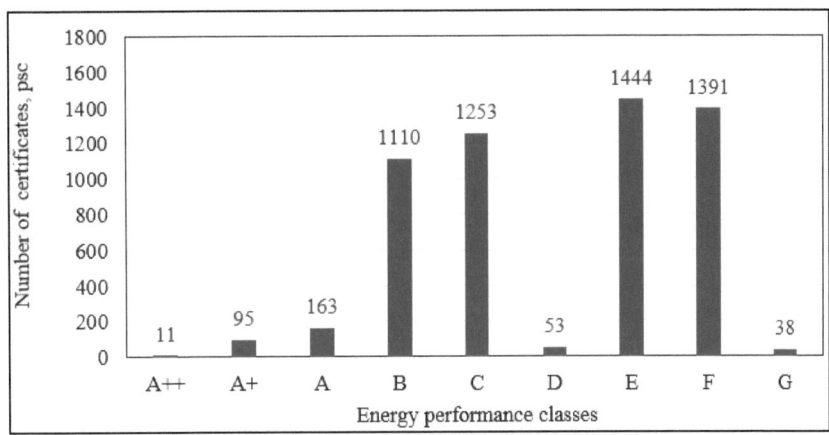

Figure 7. The number of multi-apartment building certificates analyzed from December 2014 to April 2020.

Most common energy renovation strategies taken in those buildings were the replacement of the heating system (or part of it), outer wall insulation, roof insulation/replacement, window replacement. The implementation of the Energy Performance of Buildings Directive (EPBD) in Lithuania started in 2007 and during a period of 10 years (from 2000 to 2010), the overall energy efficiency in Lithuania increased by about 20% (Norvaišienė, Karbauskaitė, and Bruzgevičius 2014), while during 2010 and 2016, it grew additionally by 6.3% (Statistics Lithuania, 2018). All buildings in Lithuania are classified into one of nine classes: A++, A+, A, B, C, D, E, F, or G, where class A++ represents the highest energy efficient building class or NZEB building, while class G refers to a building with poor energy efficiency. It should also be noted that for renovated (modernized) buildings in Lithuania, it is mandatory to achieve an energy efficiency class no lower than C from 2014 and D energy efficiency class till 2014. Over the period 2007–2021, 257,196 energy efficiency certificates (EPCs) were issued and registered in Lithuania. For the purpose of this study, the data comprised the certification calculations of 5558 multi-apartment buildings registered between the period 2014 and 2020, among which were 1253 of C class and 53 D energy efficiency class renovated multi-apartment buildings. Due to the increased level of thermal insulation of the building envelope, the average final energy consumption used for heating purposes decreased from 262 kWh/m^2.annum for buildings with an energy performance class G, compared to 78 kWh/m^2.annum for renovated multi-apartment buildings with an energy performance class D and 56 kWh/m^2 with an energy performance class C. Likewise, primary energy consumption decreased from 440 kWh/m^2.annum for buildings with an energy performance class G, compared to 102 kWh/m^2.annum for buildings with an energy performance class D and 74 kWh/m^2 with an energy performance class C.

And finally, taking, as an example, the primary energy used to heat renovated buildings in Lithuania with an energy performance class, class C utilize a share of around 35% of their total primary energy consumption for heating, 34% for domestic hot water preparation, 27% for lighting and electrical appliances, and 4% for cooling.

4. Discussion

Due to the limited number of studies of a similar nature carried out in Latvia, this study drew upon the framework of scientific studies on the same subject carried out in Sweden, Lithuania, and Poland.

A substantial amount of data on thermal energy consumption in partially or fully renovated multi-apartment residential buildings were collected over a 10-year period to objectively identify successful projects. Despite regional variations and a relatively slow pace of building renovations in both of the examined countries (Latvia and Lithuania), ongoing projects present opportunities to amend deficiencies and implement new technologies and materials. Renovated buildings were grouped based on structural features such as floor count, construction material, and living space area, facilitating comparative analysis.

The BETSI experiment (buildings, energy consumption, technical status, and indoor environment) conducted in Sweden, which involved numerous residential buildings, demonstrated that countries situated in a single climatic zone do not require dispersed national data acquisition [35,36]. Consequently, measurements can be organized in a selected inhabited area or city within Latvia, based on data availability. Daugavpils, the second most populous city in Latvia, experiences lower winter temperatures compared to the capital and other coastal cities, making indoor temperature readings more significant. The majority of housing in Daugavpils consists of series-type buildings constructed between 1945 and the late 1980s, further simplifying the grouping of renovated buildings. Subgroups were created to evaluate the effectiveness of renovation measures, such as facade insulation, heating system renovation, or full renovation.

Considering the volatile global energy market and economic calculations, prioritizing projects with shorter payback periods, such as heating system renovations, is recommended. Projects with longer repayment periods should be deferred until the State Treasury's discount rate stabilizes or until project co-financing support significantly reduces the repayment period. During the research, it was discovered that the air exchange systems in renovated residential buildings often lacked sufficient ventilation, resulting in compromised air quality in habitable rooms. This finding aligns with observations made by [37], emphasizing the overlooked aspect of ventilation during building renovations.

The determination of actual thermal energy consumption in buildings is influenced by factors such as the performed renovation package, calculation methodologies before and after the renovation, and individual heat energy metering. Installation of additional heat energy metering devices in buildings undergoing complete renovation or heating system renovation allows for the accurate measurement of thermal energy consumption. However, for buildings that do not have an integrated metering option, alternative approaches are required to determine actual thermal energy savings [38].

The findings of this study provide valuable insights into the effectiveness of the Energy Performance of Buildings Directive (EPBD) and building renovation packages in improving energy efficiency in Lithuania. The observed 20% increase in energy efficiency over a 10-year period, coupled with an additional 6.3% increase in subsequent years, suggests that the EPBD regulations have positively influenced the energy performance of buildings. The mandatory requirement for renovated buildings to achieve a minimum energy efficiency class has demonstrated significant reductions in energy consumption for heating purposes. The decrease in average final energy consumption from class G buildings to class C buildings highlights the impact of building renovation efforts on energy efficiency improvement. Furthermore, the distribution of primary energy consumption in renovated class C buildings, with a considerable share allocated to heating, domestic hot water preparation, lighting, electrical appliances, and cooling, indicates a balanced and efficient use of energy resources. The findings support the notion that building renovation packages, aligned with the EPBD regulations, play a crucial role in enhancing energy efficiency and promoting sustainability in the built environment. Future research could explore the long-term effects of these measures and assess the economic feasibility of building renovation packages in achieving even higher energy efficiency targets.

The study conducted by Yu et al. [39] serves as a valuable complement to our research, emphasizing the importance of considering multiple objectives and post-occupancy evaluation in building retrofits for more comprehensive and sustainable outcomes. By employing a post-occupancy evaluation approach and multi-objective optimization (MOO) techniques, the authors effectively address the challenges of achieving comfort and energy efficiency in retrofit projects. Their findings demonstrate the potential of integrating energy performance feedback, as demonstrated in our own research. This integration enables stakeholders to make more informed and efficient choices, resulting in sustainable and optimized retrofit solutions.

5. Conclusions

The present study examined the efficiency of building renovation measures across the building stock in the context of two Northern European countries—Latvia and Lithuania.

5.1. Latvia

The findings reveal that amendments to Latvian building regulations in the last 10 years, regarding the insulation thickness of the enclosing structure, did not significantly affect the average thermal energy consumption in fully renovated buildings. In buildings that were renovated in 2012 (featuring 100 mm thick mineral wool insulation), compared to buildings renovated in 2020 (featuring 150 mm thick rock wool insulation), the thermal energy consumption was very similar, despite the fact that the registered winter temperature averages have risen, highlighting the impact of regulatory building codes during full renovation. The majority of the examined buildings in the study were subject to facade renovations (11 of 13) and 10 out of 13 had their heating systems upgraded. The upgrades also involved transitioning from less efficient one-pipe distribution systems to more efficient two-pipe heating distribution systems. This change allows for better heat control and distribution throughout the buildings, contributing to improved energy efficiency and thermal comfort. The renovation measures resulted in significant reductions in thermal energy consumption. Average energy consumption for space heating decreased from 151.34 kWh/m^2 to 74.78 kWh/m^2, resulting in a remarkable 50.59% reduction in thermal energy consumption for space heating. These savings can be attributed to a combination of facade renovations, improved ventilation systems, and upgraded heating systems. The findings also emphasize the importance of heat metering. Installing a heat energy meter in the heating circuit prior to renovation greatly simplifies the assessment of actual thermal energy savings. Access to pre-renovation thermal energy consumption data during energy audits improves the accuracy of building models and helps mitigate the influence of regulatory changes on energy consumption calculations. In conjunction with the heat energy meter, measurements of indoor air temperature should be conducted. Residents' preferences for maintaining consistent temperature conditions indicate the need for regulatory standards on minimum room air temperatures to prevent excessive individual adjustments that affect neighboring apartments during prolonged absences. Insufficient attention has been addressed with respect to indoor air quality during renovation processes. Energy efficiency interventions in serial-type multi-apartment buildings primarily focus on reducing thermal energy consumption rather than enhancing the quality of living conditions. The installation of heat recovery mechanical ventilation systems is not feasible within simplified building renovation frameworks. Therefore, to improve indoor environmental quality, a shift from traditional radiator-based room-heating systems to air-heating/ventilation (hybrid) systems is suggested, albeit the integration of such systems is more complex and thus not commonly applied in the multi-apartment housing sector in Latvia.

5.2. Lithuania

The implementation of the Energy Performance of Buildings Directive (EPBD) in Lithuania has resulted in a notable increase in overall energy efficiency. Over a 10-year period (2000–2010), energy efficiency in Lithuania improved by 20%, with an additional

6.3% increase in subsequent years. The mandatory energy efficiency requirements for renovated buildings have been effective, as evidenced by the significant decrease in energy consumption for heating purposes. The distribution of primary energy usage in renovated buildings reflects a balanced and efficient utilization of resources. These findings highlight the positive impact of building renovation packages and the EPBD regulations in improving energy efficiency in Lithuania.

This study demonstrates the effectiveness of renovation measures in improving energy efficiency and reducing thermal energy consumption in buildings. Facade renovations, improved ventilation systems, and upgraded heating systems all contributed to significant energy savings. Future renovation projects should prioritize comprehensive measures that encompass all aspects of the heating and distribution systems to maximize energy efficiency and achieve substantial reductions in thermal energy consumption.

Our study contributes to highlighting building retrofit effectiveness by providing quantitative evaluations of thermal energy reduction achieved through retrofit measures. By analyzing data from real-world retrofit projects in Latvia and Lithuania, we offer empirical evidence of the energy-saving potential associated with specific strategies. Additionally, our research contextualizes retrofit practices within the Baltic countries, filling a gap in localized knowledge and guidelines. This tailored information enables stakeholders to make informed decisions, prioritize cost-effective measures, and achieve significant energy savings. Moreover, the findings contribute to sustainability efforts by quantifying the environmental impact reduction resulting from retrofit interventions. In summary, our study's insights support informed decision-making, advance sustainable practices, and address the unique retrofit challenges faced in Baltic countries.

Author Contributions: Conceptualization, A.P. and A.B.; methodology, S.Z. and A.B.; resources, S.Z., K.B., E.M. and R.N., validation—A.B.; writing—original draft preparation, A.P.; revision—A.B.; visualization, A.P. and S.Z.; supervision, A.B. All authors have read and agreed to the published version of the manuscript.

Funding: This research received no external funding.

Conflicts of Interest: The authors declare no conflict of interest.

References

1. Thonipara, A.; Runst, P.; Ochsner, C.; Bizer, K. Energy efficiency of residential buildings in the European Union—An exploratory analysis of cross-country consumption patterns. *Energy Policy* **2019**, *129*, 1156–1167. [CrossRef]
2. Shadram, F.; Mukkavaara, J. Exploring the effects of several energy efficiency measures on the embodied/operational energy trade-off: A case study of swedish residential buildings. *Energy Build.* **2019**, *183*, 283–296. [CrossRef]
3. Fernandes, D.V.; Silva, C.S. Open Energy Data—A regulatory framework proposal under the Portuguese electric system context. *Energy Policy* **2022**, *170*, 113240. [CrossRef]
4. Ekström, T.; Bernardo, R.; Blomsterberg, Å. Cost-effective passive house renovation packages for Swedish single-family houses from the 1960s and 1970s. *Energy Build.* **2018**, *161*, 89–102. [CrossRef]
5. Scorpio, M.; Ciampi, G.; Gentile, N.; Sibilio, S. Effectiveness of low-cost non-invasive solutions for daylight and electric lighting integration to improve energy efficiency in historical buildings. *Energy Build.* **2022**, *270*, 112281. [CrossRef]
6. Rademaekers, I.R.A.K.; Williams, R.; Yearwood, J. Boosting Building Renovation: What Potential and Value for Europe? 2016. Available online: https://www.europarl.europa.eu/RegData/etudes/STUD/2016/587326/IPOL_STU(2016)587326_EN.pdf (accessed on 25 July 2023).
7. D'Agostino, D.; Tzeiranaki, S.T.; Zangheri, P.; Bertoldi, P. Assessing Nearly Zero Energy Buildings (NZEBs) development in Europe. *Energy Strat. Rev.* **2021**, *36*, 100680. [CrossRef]
8. Deshko, V.; Bilous, I.; Sukhodub, I.; Yatsenko, O. Evaluation of energy use for heating in residential building under the influence of air exchange modes. *J. Build. Eng.* **2021**, *42*, 103020. [CrossRef]
9. Zemitis, J.; Terekh, M. Management of energy efficient measures by buildings' thermorenovation. *MATEC Web Conf.* **2018**, *245*, 06003. [CrossRef]
10. Krumins, A.; Lebedeva, K.; Tamane, A.; Millers, R. Possibilities of Balancing Buildings Energy Demand for Increasing Energy Efficiency in Latvia. *Environ. Clim. Technol.* **2022**, *26*, 98–114. [CrossRef]
11. Grazieschi, G.; Asdrubali, F.; Thomas, G. Embodied energy and carbon of building insulating materials: A critical review. *Clean. Environ. Syst.* **2021**, *2*, 100032. [CrossRef]

12. Zajacs, A.; Lebedeva, K.; Bogdanovičs, R. Evaluation of Heat Pump Operation in a Single-Family House. *Latv. J. Phys. Tech. Sci.* **2023**, *60*, 85–98. [CrossRef]
13. Geikins, A.; Borodinecs, A.; Jacnevs, V. Estimation of Energy Profile and Possible Energy Savings of Unclassified Buildings. *Buildings* **2022**, *12*, 974. [CrossRef]
14. Borodinecs, A.; Zemitis, J.; Dobelis, M.; Kalinka, M.; Prozuments, A.; Šteinerte, K. Modular retrofitting solution of buildings based on 3D scanning. *Procedia Eng.* **2017**, *205*, 160–166. [CrossRef]
15. Asdrubali, F.; Grazieschi, G. Life cycle assessment of energy efficient buildings. *Energy Rep.* **2020**, *6*, 270–285. [CrossRef]
16. Gulotta, T.; Cellura, M.; Guarino, F.; Longo, S. A bottom-up harmonized energy-environmental models for europe (BOHEEME): A case study on the thermal insulation of the EU-28 building stock. *Energy Build.* **2021**, *231*, 110584. [CrossRef]
17. EUBD. European Buildings Database. 2017. Available online: https://ec.europa.eu/energy/en/eu-buildings-database (accessed on 25 July 2023).
18. Tzeiranaki, S.T.; Bertoldi, P.; Diluiso, F.; Castellazzi, L.; Economidou, M.; Labanca, N.; Serrenho, T.R.; Zangheri, P. Analysis of the EU Residential Energy Consumption: Trends and Determinants. *Energies* **2019**, *12*, 1065. [CrossRef]
19. EM. Latvian Building Industry's Development Strategy 2017–2024. 2016. Available online: https://www.em.gov.lv/files/buvnieciba/BS_17.05.2017.pdf (accessed on 25 July 2023).
20. Deshko, V.; Bilous, I.; Sukhodub, I.; Yatsenko, O. Analysis of the Influence of Air Exchange Distribution between Rooms on the Apartment Energy Consumption. *Power Eng. Econ. Tech. Ecol.* **2021**, *49*, 39–50. [CrossRef]
21. Borodinecs, A.; Zemitis, J.; Sorokins, J.; Baranova, D.; Sovetnikov, D. Renovation need for apartment buildings in Latvia. *Mag. Civ. Eng.* **2017**, *68*, 58–64. [CrossRef]
22. Aslani, A.; Bakhtiar, A.; Akbarzadeh, M.H. Energy-efficiency technologies in the building envelope: Life cycle and adaptation assessment. *J. Build. Eng.* **2019**, *21*, 55–63. [CrossRef]
23. Eliopoulou, E.; Mantziou, E. Architectural Energy Retrofit (AER): An alternative building's deep energy retrofit strategy. *Energy Build.* **2017**, *150*, 239–252. [CrossRef]
24. Cho, K.; Yang, J.; Kim, T.; Jang, W. Influence of building characteristics and renovation techniques on the energy-saving performances of EU smart city projects. *Energy Build.* **2021**, *252*, 111477. [CrossRef]
25. Fan, Y.; Xia, X. Energy-efficiency building retrofit planning for green building compliance. *Build. Environ.* **2018**, *136*, 312–321. [CrossRef]
26. Copiello, S. Building energy efficiency: A research branch made of paradoxes. *Renew. Sustain. Energy Rev.* **2017**, *69*, 1064–1076. [CrossRef]
27. Röck, M.; Baldereschi, E.; Verellen, E.; Passer, A.; Sala, S.; Allacker, K. Environmental modelling of building stocks—An integrated review of life cycle-based assessment models to support EU policy making. *Renew. Sustain. Energy Rev.* **2021**, *151*, 111550. [CrossRef]
28. Toleikyte, A.; Kranzl, L.; Müller, A. Cost curves of energy efficiency investments in buildings—Methodologies and a case study of Lithuania. *Energy Policy* **2018**, *115*, 148–157. [CrossRef]
29. Attia, S.; Kurnitski, J.; Kosiński, P.; Borodinecs, A.; Belafi, Z.D.; István, K.; Krstić, H.; Moldovan, M.; Visa, I.; Mihailov, N.; et al. Overview and future challenges of nearly zero-energy building (nZEB) design in Eastern Europe. *Energy Build.* **2022**, *267*, 112165. [CrossRef]
30. Monstvilas, E.; Borg, S.P.; Norvaišienė, R.; Banionis, K.; Ramanauskas, J. Impact of the EPBD on Changes in the Energy Performance of Multi-Apartment Buildings in Lithuania. *Sustainability* **2023**, *15*, 2032. [CrossRef]
31. Šadauskienė, J.; Seduikyte, L.; Paukštys, V.; Banionis, K.; Gailius, A. The role of air tightness in assessment of building energy performance: Case study of Lithuania. *Energy Sustain. Dev.* **2016**, *32*, 31–39. [CrossRef]
32. LSTA. 2018 Overview of Lithuania's Centralized Heat Supply Sector. 2018. Available online: https://lsta.lt/wp-content/uploads/2019/10/LSTA_apzvalga_2018.pdf (accessed on 18 July 2023).
33. Janulis, M.; Bieksa, D. Lietuvos Ilgalaikė Renovacijos Strategija. Available online: https://energy.ec.europa.eu/system/files/2021-04/lt_2020_ltrs_0.pdf (accessed on 25 July 2023).
34. SPSC. Center of Certification of Construction Production. Available online: https://www.spsc.lt/cms/index.php?option=com_content&view=article&id=60&Itemid=267&lang=en (accessed on 25 July 2023).
35. Psomas, T.; Teli, D.; Langer, S.; Wahlgren, P.; Wargocki, P. Indoor humidity of dwellings and association with building characteristics, behaviors and health in a northern climate. *Build. Environ.* **2021**, *198*, 107885. [CrossRef]
36. Wang, J.; Norbäck, D. Subjective indoor air quality and thermal comfort among adults in relation to inspected and measured indoor environment factors in single-family houses in Sweden-the BETSI study. *Sci. Total. Environ.* **2022**, *802*, 149804. [CrossRef]
37. Cholewa, T.; Balaras, C.A.; Nižetić, S.; Siuta-Olcha, A. On calculated and actual energy savings from thermal building renovations—Long term field evaluation of multifamily buildings. *Energy Build.* **2020**, *223*, 110145. [CrossRef]

38. Cabovská, B.; Bekö, G.; Teli, D.; Ekberg, L.; Dalenbäck, J.-O.; Wargocki, P.; Psomas, T.; Langer, S. Ventilation strategies and indoor air quality in Swedish primary school classrooms. *Build. Environ.* **2022**, *226*, 109744. [CrossRef]
39. Yu, C.-R.; Liu, X.; Wang, Q.-C.; Yang, D. Solving the comfort-retrofit conundrum through post-occupancy evaluation and multi-objective optimisation. *Build. Serv. Eng. Res. Technol.* **2023**, *44*, 381–403. [CrossRef]

Disclaimer/Publisher's Note: The statements, opinions and data contained in all publications are solely those of the individual author(s) and contributor(s) and not of MDPI and/or the editor(s). MDPI and/or the editor(s) disclaim responsibility for any injury to people or property resulting from any ideas, methods, instructions or products referred to in the content.

Article

Application of Cluster Analysis to Examine the Performance of Low-Cost Volatile Organic Compound Sensors

Jakub Kolarik [1,*], Nadja Lynge Lyng [2], Rossana Bossi [3], Rongling Li [1], Thomas Witterseh [2], Kevin Michael Smith [1] and Pawel Wargocki [4]

1. Department of Civil and Mechanical Engineering, Technical University of Denmark, 2800 Kgs. Lyngby, Denmark; liron@dtu.dk (R.L.); kevs@dtu.dk (K.M.S.)
2. Building and Construction, Danish Technological Institute, 2630 Taastrup, Denmark; nal@teknologisk.dk (N.L.L.); twi@teknologisk.dk (T.W.)
3. Department of Environmental Science, Aarhus University, 4000 Roskilde, Denmark; rbo@envs.au.dk
4. Department of Environmental and Resource Engineering, Technical University of Denmark, 2800 Kgs. Lyngby, Denmark
* Correspondence: jakol@dtu.dk; Tel.: +45-45-25-19-27

Abstract: Airtight energy-efficient buildings of today need efficient ventilation to secure high indoor air quality. There is a need for affordable and reliable sensors to make demand control available in a broad range of ventilation systems. Low-cost metal oxide semiconductor (MOS) volatile organic compound (VOC) sensors offer such a possibility, but they are usually non-selective and react to broad range of compounds. The objective of the present paper was to use cluster analysis to assess the ability of five commercially available MOS VOC sensors to detect pollutants in a residential setting. We studied three scenarios: emissions from people (human bioeffluents), furnishing materials (linoleum), and human activity (surface cleaning with spray detergent). We monitored each scenario with five MOS VOC sensors and a proton-transfer-reaction–time-of-flight mass spectrometer (PTR-ToF-MS). We applied an agglomerative hierarchical clustering algorithm to evaluate the dissimilarity between clusters. Four of the five tested sensors produced signals in agreement with the concentration patterns measured with the PTR-ToF-MS; one sensor underperformed in all cases. Three sensors showed a very similar performance under all emission scenarios. The results showed that the clustering could help in understanding whether a particular sensor matched the intended emission scenario.

Keywords: indoor air quality; MOS VOC sensor; residential ventilation; cluster analysis

1. Introduction

Airtight energy-efficient buildings of today need efficient ventilation to secure high indoor air quality (IAQ). Current energy-efficient ventilation solutions frequently use the so-called demand-controlled ventilation (DCV) principle. This means that the system modulates airflows according to an immediate need expressed by different demand indicators. Those include, e.g., human presence, temperature and relative humidity, carbon dioxide (CO_2) concentration, or their combinations. It is common to use the CO_2 concentration as an indicator of IAQ. Yet it is often questioned whether it is sufficient, as CO_2 concentration mostly represents occupant-related pollution and is valid only in the presence of building occupants. The fact that measurements of CO_2 do not reveal the full picture regarding the indoor air pollution and call for other indicators has been discussed by, for example, Alonso et al. [1]. Recent advances in sensor technology have brought new types of sensors that can potentially replace or supplement CO_2 sensing to control ventilation. Metal oxide semiconductor (MOS) sensors for measuring volatile organic compounds (VOC) are an example [2]. When used in DCV systems, MOS VOC sensors do not only account for air pollution related to occupancy but also for diverse events that worsen IAQ. These events comprise cleaning by means of different detergents, cooking, use of personal cosmetics, or

Citation: Kolarik, J.; Lyng, N.L.; Bossi, R.; Li, R.; Witterseh, T.; Smith, K.M.; Wargocki, P. Application of Cluster Analysis to Examine the Performance of Low-Cost Volatile Organic Compound Sensors. *Buildings* **2023**, *13*, 2070. https://doi.org/10.3390/buildings13082070

Academic Editor: Lambros T. Doulos

Received: 19 June 2023
Revised: 8 August 2023
Accepted: 9 August 2023
Published: 15 August 2023

Copyright: © 2023 by the authors. Licensee MDPI, Basel, Switzerland. This article is an open access article distributed under the terms and conditions of the Creative Commons Attribution (CC BY) license (https://creativecommons.org/licenses/by/4.0/).

even a sudden release of unwanted chemicals. From the IAQ viewpoint, the advantage is clear—the ventilation system increases the outdoor air supply rate when the sensor detects pollutants other than CO_2. The MOS technology provides an opportunity to produce sensors that are more affordable than the state-of-the-art CO_2 sensors. That is why they have become ubiquitous among so-called low-cost sensors (LCS) [3]. Other advantages of MOS sensors include a low energy consumption, small size, and high durability. As a result, residential ventilation systems can utilize DCV control strategies at a lower cost. Large commercial systems can on the other hand utilize a larger number of sensors. Such possibilities fit very well with the concept of smart ventilation [4]. MOS sensors, like the one developed by Herberger et al. [5], can integrate the measurement of human-emitted VOCs and several other typical indoor pollutants so that there is no need for a CO_2 sensor. The interpretation of the signals from the low-cost sensors was extended to include the CO_2 concept. Using data from a study by Burdack-Freitag et al. [6], the measured VOC signal was correlated with anthropological emissions of CO_2. This resulted in the so-called CO_2 equivalent concentration. The reasoning behind this cross-correlation was that the term "CO_2 concentration" had become known to the public as an indicator of IAQ. As a consequence, VOC sensor signals could be more easily interpreted by building occupants.

The above-mentioned arguments speak in favor of MOS VOC sensor technology in comparison to the currently used CO_2 sensors. However, there are also several research studies [3,7] stating that MOS VOC sensors suffer from several drawbacks. Authors mention cross-sensitivity to relative humidity, low resolution, and an inability to measure the concentration of individual chemicals. MOS VOC sensors react to a broad variety of compounds, which can make their application for ventilation control challenging. As DCV control was almost exclusively based on CO_2 for decades [8], the amount of scientific literature related to DCV control based on VOCs or other pollutants is rather limited. Despite the limited research, a VOC-controlled DCV is being offered by an increasing number of ventilation producers. Moreover, MOS VOC sensors are frequently installed in internet-enabled indoor environmental quality monitors. These should still be considered as electronic gadgets rather than reliable monitoring instruments, but their popularity is increasing, driven by the general boom in "smart home technologies" [9]. Several studies attempted to characterize MOS VOC sensors' performance with respect to ventilation control. However, these studies focused mostly on the consequences of the MOS VOC-based control in terms of energy efficiency or IAQ. A study conducted by Kolarik [10] showed that signals from the VOC and CO_2 sensors installed in an office room agreed that ventilation was needed in the space for 49% of the occupied time, while for an additional 11% of the occupied time, only the VOC called for more ventilation. These results, together with results by Laverge et al. [11], indicate that MOS VOC sensors cannot be used directly as an alternative low-cost replacement for CO_2 sensors. Field tests conducted by Merzkirch et al. [12] showed that with selection of an appropriate ventilation control strategy, application of MOS VOC sensors decreased the overall ventilation flow (or operation time) and thus led to primary energy savings while maintaining acceptable IAQ. A study by Abdul-Hamid et al. [13] demonstrated that the positioning of MOS VOC sensors in ventilated spaces plays a significant role in the achieved IAQ. A study by De Sutter et al. [14] showed a notable increase in ventilation rates (and thus energy consumption) related to the sharp peaks in the MOS VOC signals when the system used the same set point for both CO_2 and MOS VOC control based on a CO_2 equivalent. The use of an MOS VOC sensor will consequently result in adjustments to the ventilation strategy.

The above-mentioned practical issues regarding MOS VOC sensor performance relate to the more fundamental aspect of their performance—they react to a wide range of VOCs [15,16]. Thus, they can be considered to produce an aggregated response to the VOCs present in the air. Several studies; for example, those by Kolarik et al. [15] or Demanega et al. [17], showed that the sensor response usually strongly correlated with measurements using laboratory-grade instruments, but for many sensors, there was poor quantitative agreement. These observations were made despite the fact that the majority of producers calibrate their sensors by exposing them to single compounds (e.g., ethanol or isobutylene) or pre-defined gas mixtures [18,19]).

An objective of the present paper was to study whether several commercially available MOS VOC sensors can detect VOCs during typical residential pollution emission scenarios. The aim was to combine detailed VOC measurements using laboratory-grade instruments with data mining techniques to overcome the fact that MOS VOC sensors react to a group of compounds rather than to an individual pollutant. Our hypothesis was that it is possible to use a cluster analysis on detailed VOC data obtained by a laboratory-grade instrument together with MOS VOC signals obtained under the same experimental conditions to identify compounds with a dominant influence on the MOS VOC signals. Such performance characteristics would determine the suitability of a particular MOS VOC sensor for a concrete application.

Application of a cluster analysis using data from five commercially available MOS VOC sensors together with pollutant concentration data measured with a PTR-ToF-MS showed agreement among four sensors. Their signals appeared in the same clusters as concentration patterns of VOCs characteristic of emission scenarios of human bioeffluents, linoleum, and cleaning. One of the sensors had significantly different response patterns, thus it was not suitable to detect pollutants representing the studied scenarios. The cluster analysis seemed to be useful to identify which compounds triggered the MOS VOC sensor response in different pollution situations. However, due to the nature of the cluster analysis, we recommend analyzing the absolute concentration levels for measured pollutants at the same time. This will ensure that the analysis considers pollutants that play a realistic role in the studied exposure.

2. Materials and Methods

We created different emission scenarios corresponding to typical polluting activities in residences. We conducted all measurements in a test room that allowed for controlled ventilation and thermal environment. We conducted measurements with five commercially available MOS VOC sensors as well as with a proton-transfer-reaction–time-of-flight mass spectrometer (PTR-ToF-MS). The PTR-ToF-MS was a laboratory-grade instrument capable of measuring the real-time concentration of VOC down to ppb levels. We performed a cluster analysis on the collected data.

2.1. Selected Sensors

During the preliminary market survey, we identified seventeen commercially available MOS VOC sensors from five different producers. We limited the final selection to sensors with a delivery time of less than three months and without minimum-order-quantity restriction. This led to a choice of five sensors that were tested. Table 1 summarizes their technical parameters. The sensors coded as A, B, D, and E were available as integrated modules that enabled pre-processing of the sensor signal. This pre-processing included the built-in algorithms for conversion of the measured sensor resistance change to a signal for the equivalent concentration of total volatile organic compounds (TVOCs). For some of the sensors, the pre-processing also included proprietary auto-calibration algorithms. Such algorithms have two purposes mainly. Firstly, they deal with the cross-sensitivity of the sensor to the water vapor content in the air. Secondly, they establish a sensor's baseline usually based on the lowest measured concentration without a change for a specific period. As the MOS VOC sensors produce relative measurements, they use the baseline to characterize "clean air". We knew such a determination of the baseline could be problematic from an IAQ standpoint. However, the present work did not deal with this issue, and our data analysis was not sensitive to the pre-processing algorithms. While sensors A and B represented integrated modules without casing, and their practical use would require further integration on a host circuit board providing a power supply and output connectors, sensors D and E represented the "ready to use" modules that could be directly connected to building automation system. Sensor C represented solely a sensor element and provided a raw voltage signal. Sensors A and B were integrated into a commercially available indoor climate measurement device, which at the same time measured the temperature (T), humidity (RH), and CO_2 concentration (CO_2); for details, see Section 2.3. The device was GSM-enabled and sent data into a cloud database every 5 min. Sensors C, D, and E connected to a laboratory data logger that was capable of continuously recording the voltage signal. We did not perform any additional calibration of the sensors. For sensors A, B, and C, we checked the intra-unit consistency, which was defined as the variability between signals from individual sensors < 20% [20].

Table 1. Technical parameters of investigated sensors based on manufacturer data sheets.

Abbreviation	A	B	C	D	E
Configuration	Sensor module	Sensor module	Sensor	Sensor module	Sensor module
Output (units)	TVOC eq. (ppb) [1] CO_2 eq. (ppm)	TVOC eq. (ppb) CO_2 eq. (ppm)	Voltage (V)	Voltage (V)	Voltage (V)
Sensing range	CO_2 eqv.: 400–2000 ppm TVOC: 0–1000 ppb	CO_2 eqv.: 450–2000 ppm TVOC: 125–600 ppb [2]	NH_3: 10–300 ppm [3] C_6H_6: 10–1000 ppm Alcohols: 10–300 ppm	0–100% VOC	0–100% VOC
Measuring accuracy	N/A	N/A	N/A	±20% of final value [5]	N/A
Measurement interval/response time	1 s/<5 s for TVOC	1 s/N/A	N/A	N/A/60 s	N/A/ <13 min, <3.5 min, <1 min [6]
Power supply	3.3 V DC ± 5%	3.3 V DC ± 0.1 V	5 V DC or AC ± 0.1 V	24 V ± 10% AC/DC	24 V ± 20% AC
Communication	I^2C bus	I^2C bus	analog	0–10 V or 4–20 mA	Analog: 0–10 V or 0–5 V DC
Warm up time	15 min	5 min	>24 h	1 h	N/A
Operation temperature range	0–50 °C	0–50 °C	−10–45 °C	0–50 °C	0–50 °C
Operation humidity range	5–95%, non-condensing	5–95%, non-condensing	<95%	N/A	0–95%, non-condensing
Automatic baseline correction	Yes [4]	Yes	N/A	Yes	Yes

[1] Isobutylene equivalent. [2] Relative measurement; values above the defined sensing range are provided as well. [3] Calibrated using 100 ppm NH_3 in clean air (T = 20 °C, RH = 65%); O_2 concentration 21%. [4] The manufacturer states that no calibration is needed. [5] The manufacturer specifies that the value refers to the calibration gas, but the type of calibration gas is not specified. [6] It is possible to define the response time during installation–setup.

2.2. Experimental Design

We exposed the MOS VOC sensors to emission scenarios representing typical activities in residences: (1) emission of human bioeffluents, (2) emissions from furnishing materials (linoleum), and (3) emissions from house cleaning with typical detergent. Table 2 summarizes the details regarding each scenario. We tested each scenario on a separate day. Mechanical ventilation was active both during and between the experiments to eliminate the accumulation of pollution from previous experiments.

We placed the MOS VOC sensors side by side in the middle of the test room (see Section 2.3) at a height of approximately 0.85 m above the floor. We powered the sensors 48 h before the first experiment, and they remained connected to electricity during the entire experimental campaign. We placed the PTR-ToF-MS in the test room with the sampling point just beside the investigated sensors. We initiated the PTR-ToF-MS measurement approximately 30 min before the actual measurement period for each scenario to sample the background in the test room. We continued the PTR-TOF-MS measurement for several hours after the measurement period to follow the decay of the released compounds.

Table 2. Summary of emission scenarios.

Scenario	Start of PTR-ToF-MS Measurement	Scenario Start	Scenario End	Description
Human bioeffluents	9:13 a.m.	9:47 a.m.	3:02 p.m.	Six adults were seated in the test room. They were instructed not to eat spicy food or use cosmetics before the experiment. Each person was equipped with a laptop and power supply. Persons performed sedentary work corresponding to a metabolic activity of 1.2 met. Persons could drink water but not consume any food in the test room. If one of the persons needed to leave, another adult was brought in the test room as a substitute.
Linoleum	9:58 a.m.	10:31 a.m.	1:47 p.m.	Linoleum flooring was used to represent emissions from typical furnishing materials. The surface area of the linoleum was 17 m^2, corresponding to half of the floor area of the test room. Linoleum strips were fixed against each other by the bottom surface so that only the upper surface of the material was exposed to air. Linoleum strips were hung on a steel rack.
Cleaning	10:03 a.m.	10:37 a.m.	10:52 a.m.	A solution consisting of 60 mL of universal citrus-scented detergent was mixed in 5 L of water as instructed by the manufacturer. Preparation of the solution took place outside the test room immediately before the activity. One adult washed all wall surfaces in the room with a cloth soaked with the solution; 240 mL of the solution was used. The cleaning took 15 min, and the remaining cleaning solution was then removed from the test room.

2.3. Experimental Facilities and Measuring Conditions

The test room was 7.0 m wide and 4.5 m deep, corresponding to a floor area of 31.5 m^2; the ceiling height was 2.6 m. The outer wall of the room consisted of the building façade (concrete elements with insulation) with windows, and the opposite wall facing a hallway area consisted of glass. Painted plasterboard formed the sidewalls of the room. The floor consisted of a wall-to-wall carpet on top of a vinyl flooring. The ceiling comprised a drywall suspended ceiling system with acoustic panels and built-in lighting fixtures.

The mechanical ventilation system supplied the fresh air through two chilled beams. The system worked with an outdoor airflow rate corresponding to 0.5 h^{-1} in the test room.

The controller adjusted the supply air temperature between 19 °C and 21 °C depending on the heat load in the test room. Five table fans ensured full mixing of the air in the test room throughout the experimental period. We used exterior perforated textile sunscreens to reduce the solar heat load on the south-facing façade. We used two thermostat-controlled electric radiators to ensure a minimum air temperature of 23 °C in the test room. We controlled the relative humidity using two ultrasonic steam humidifiers. The relative humidity set point was 50%.

We continuously measured the air temperature (accuracy: ±0.3 °C), relative humidity (accuracy: ±2%), and CO_2 concentration (accuracy: ±30 ppm ± 3% of reading) in the test room. We did not conduct any measurements outside the test room. Additionally, we measured an air change rate in the test room using the decay method according to ASTM standard E741-11 [21] using tracer gas (R134a). The method provides an average air change rate over a period of 1–3 h. We used a Brüel & Kjær Photoacoustic Gas Analyzer (model 1302) to monitor the concentration of the tracer gas.

2.4. PTR-ToF-MS Measurements

We used a proton-transfer-reaction–time-of-flight mass spectrometer (PTR-ToF-MS) to measure the VOC concentrations during the emission scenarios. A PTR-ToF-MS is an analytical measurement device that allows for on-line monitoring of VOC concentrations at low detection limits with a fast response time. A PTR-ToF-MS utilizes a proton transfer reaction from H_3O^+ to VOC with a proton affinity higher than that of water (166.55 kcal/mol). The charged VOC molecules are then detected by a ToF mass spectrometer [22]. The PTR-ToF-MS provides an on-line quantification and, at the same time, formula confirmation of VOCs. A PTR-ToF-MS in H_3O^+ ionization mode does not include the detection of alkanes. Moreover, a PTR-ToF-MS does not allow distinguishing between isomers. The device has been used with success in IAQ to characterize pollution sources and map chemical reactions occurring in indoor air [23,24].

The PTR-ToF-MS 8000 (Ionicon Analytik GmbH, Innsbruck, Austria) used in the experiments was operated with hydronium ions (H_3O^+) as a reagent, a drift tube temperature of 70 °C, a drift pressure of 2.80 mbar, and a drift tube voltage of 650 V leading to an E/N (electric field/density of the buffer gas in the drift tube) value of around 120 Townsend (Td). Mass spectra up to m/z = 430 Da were collected at a 5 s scan rate. The instrument inlet consisted of a PEEK capillary tube heated to 70 °C and a built-in permeation unit (PerMasCal; Ionicon Analytik), which emitted 1,3-diiodobenzene used for continuous mass scale calibration. Blank measurements were obtained by coupling a charcoal filter to the instrument's inlet tube. We processed the data generated by the PTR-ToF-MS with the software PTR-MS Viewer v. 3.2.12 (Ionicon Analytik). The PTR-MS Viewer automatically calculated the mass calibrations and the VOC mixing ratio. Compound names were assigned based on a comparison with the libraries from the PTR MS Viewer, Pagonis et al. [25] and the references therein, and a priori knowledge.

2.5. Cluster Analysis and Data Processing

We applied a clustering method to analyze the relationship between the VOC signals measured by the PTR-ToF-MS and the MOS sensors. Cluster analysis is a data mining method particularly suitable for the analysis of time series. It belongs to the data mining, pattern recognition, and statistical machine learning [26] methods. Clustering is an unsupervised data-mining method that is commonly used to discover patterns in data sets by dividing the data into several subgroups. The objective of clustering is to partition a data set into several groups with the observations in the same group as similar as possible while the observations in different groups are dissimilar to a maximum extent [27]. Clustering is commonly used to analyze time-series energy-consumption data to group similar profiles into the same subgroups and unveil the most typical load profiles. Clustering algorithms for time-series data are centroid-based methods such as k-means and k-medoids [28–30], hierarchical clustering with an agglomerative or divisive approach [28,30], and a self-organizing map (SOM) [31,32]. For data with a small sample size, hierarchical clustering algorithms are suitable [33].

The cloud database stored the data from sensors A and B in 5 min intervals. We downloaded the data and aggregated them with measurements of temperature, relative humidity, and CO_2. Data from sensors C, D, and E were stored in a laboratory data logger every 1 min. We processed these data to obtain 5 min mean values corresponding to the time step of the data from sensors A and B. We applied the same procedure to the data from the PTR-ToF-MS collected at approximately 1 s intervals.

We applied a clustering method to an aggregated data file comprising measurements by the MOS VOC sensors and PTR-ToF-MS, temperature, relative humidity, and CO_2 concentration. As the MOS VOC sensors provided a signal corresponding to a range of VOCs, their appearance in the same cluster together with specific VOCs measured by the PTR-ToF-MS would indicate their ability to detect these compounds. The statistical software R version 3.4.3 [34] was used for the analysis.

We normalized observations in the aggregated data set to avoid the influence of the absolute value of each observation. We normalized each observation against the difference of its maximum value and minimum value (the so-called min–max normalization) as shown in Equation (1):

$$y = (x - \min(x))/(\max(x) - \min(x)) \tag{1}$$

where x is the observation and y is its normalized value.

As the sample size was relatively small, we decided to apply an agglomerative hierarchical clustering algorithm using the R function NbClust [35] to identify the optimal number of clusters. We compared different cluster agglomeration methods, i.e., linkage methods to measure the dissimilarity between two clusters of observations. The comparison showed that the Ward linkage method [36] was the most suitable with respect to the analyzed data. The method minimizes the total within-cluster variance as it merges the pair of clusters with the minimum between-cluster distance at each computation step. As it was unclear whether the air temperature (T) and relative humidity (RH) could be influential in the clustering results, we conducted a preliminary analysis including T and RH, excluding T, excluding RH, and excluding T and RH. The presented results describe only the final analysis of the data.

3. Results

3.1. Environmental Conditions in the Test Room

Table 3 summarizes the measurements of the air temperature, relative humidity, and air change rate. The average room temperature was about 1.7 °C higher in the case of human bioeffluent activity. The intensive internal heat loads associated with the persons who emitted bioeffluents and their personal computers were likely the cause for the elevated air temperature. The relative humidity stayed in a relatively narrow range independent of the temperature in the test room. We measured the air change rate once during the scenario involving exposure to linoleum. During scenarios in which the test room door was opened several times (human bioeffluents and cleaning), we measured the air change repeatedly.

Table 3. Indoor environmental conditions. The values for air temperature and relative humidity represent the mean (min–max) corresponding to the period from the start of PTR-ToF-MS measurement until 10 p.m. each experimental day. The air change rate measurement was performed 1 to 3 times per activity.

Activity	Temperature (°C) Mean (Min–Max)	Relative Air Humidity (%) Mean (Min–Max)	Air Change Rate (h^{-1})
Human bioeffluents	24.4 (22.6–25.7)	45.5 (43.5–47.2)	0.7; 0.7; 0.6
Linoleum	22.8 (22.4–23.4)	45.1 (43.1–46.8)	0.7
Cleaning	22.7 (22.3–23.0)	45.1 (43.1–48.2)	0.7; 0.8

3.2. Compounds Identified by the PTR-TOF-MS

The following section summarizes the results of measurements made with the PTR-ToF-MS. We present additional details regarding the measured compounds in Appendix B. Table 4 presents compounds measured during the human bioeffluent emission scenario. It summarizes compounds whose concentration increased more than 50% compared with the background concentration prior to occupancy. The compound names are according to the libraries from the PTR-ToF-MS software v. 3.2.12 (Ionicon Analytik), and references are listed in Table 4. Acetone and methanol accounted for the majority of the total VOC concentration. Both compounds can be associated with metabolic processes and are mainly emitted during breathing by humans [37]. Additionally, the oxidation of squalene, a compound present on human skin [38], can also produce acetone. Alkyl fragment/propyne and propanol fragment/propene constituted about 23% of the total VOC concentration. The last two compounds may originate from in-source fragmentation of longer chain alkenes. Isoprene and the acids contributed slightly above 2% to the total VOC concentration during this emission scenario.

Table 4. Compounds detected and identified via PTR-ToF-MS measurements during the human bioeffluent emission scenario; the compounds whose concentration increased by ≥50% compared to the background concentration are presented. The VOCs are ranked according to their contribution to the total VOC concentration. References relate to association of the compound with the studied emission scenario.

Compound	Contribution to TVOCs (%)	Reference
Methanol	24.8	[39,40]
Acetone	23.1	[40]
Propanol fragment (-H_2O)/propene/cyclopropane	12.8	[40,41]
Alkyl fragment or propyne	9.8	[40,41]
Octanal	0.9	[38]
6-Methyl-5-hepten-2-one (6-MHO)	3.6	[38]
Formaldehyde [1]	2.6	-
Unsaturated carbonyl (e.g., methyl vinyl ketone)	0.6	[40,41]
Isoprene	2.3	[39,40]
Hydroxyacetone/propionic acid	2.2	[38,40,41]
1-Octen-3-ol fragment (-H_2O) + others	0.4	[40,41]
C6-carboxylic acid	0.4	[40]
C8 saturated carbonyl + 1-octen-3-ol	1.3	[40,41]
1,2-Propendiol [1]	0.2	-
Anisaldehyde + others	1.1	[40,41]
Acetylpropionyl + others	1.0	[40,41]
cis-3-Hexen-1-ol + others	1.0	[40,41]
Butyric acid	<0.1	[40,41]
C12-carboxylic acid	<0.1	[40]

[1] These compounds were detected during the human bioeffluent scenario, but there is no reference relating them to human presence.

Table 5 summarizes the compounds measured with the PTR-ToF-MS during the emission scenario with linoleum. Organic acids—from acetic and formic acid to hexanoic acid—characterized the profile of the linoleum emissions. The contribution of organic acids to the total VOCs was about 50%, and the relative contribution decreased with an increasing number of carbon atoms. The remaining part of the total concentration consisted mostly of isoprene, aldehydes, and ketones.

Table 5. Compounds detected and identified via the PTR-ToF-MS measurements during the linoleum emission scenario; the compounds whose concentration increased by ≥50% compared to the background concentration are presented. The VOCs are ranked according to their contribution to the total VOC concentration. References relate to the association of the compound with the studied emission scenario.

Compound	Contribution to TVOCs (%)	Reference
Acetic acid	28.0	[42–44]
Ketene [1]	15.3	-
Formic acid	13.7	[44]
Acetone [1]	13.1	-
Acetaldehyde [1]	12.8	-
Propionic acid	4.4	[44,45]
Propenal [1]	1.5	-
Isoprene [1]	1.4	-
Butyric acid	0.9	[43,44]
Pentanoic acid	0.6	[43,44]
C8-alkane [1]	0.3	-
Cyclohexane diones [1]	0.3	-
Cyclopentane carboxylic acid [1]	0.3	-
C7 aldehyde/ketone [1]	0.2	-
Cycloheptanone [1]	0.2	-
Heptanal [1]	0.2	-
Propanol fragment ($-H_2O$)/propene/cyclopropane [1]	<0.1	-
Hexanoic acid	<0.1	[44,45]

[1] These compounds were detected during the linoleum scenario, but there is no reference relating them to emission from linoleum.

Table 6 gives an overview of the compounds detected with the PTR-ToF-MS during the cleaning emission scenario. The compounds we could directly relate to the use of cleaning detergent were the monoterpenes and their fragments, which increased substantially (about 20× in comparison to the background concentration prior to this emission scenario) during the cleaning activity. As is clear in Table 6, we also detected compounds related to human presence, as there was a researcher performing the cleaning in the test room. The total concentration was dominated by the bioeffluents.

Table 6. Compounds detected and identified via PTR-ToF-MS measurements during the cleaning emission scenario; the compounds whose concentration increased by ≥50% compared to the background concentration are presented. The VOCs are ranked according to their contribution to the total VOC concentration. References relate to association of the compound with the studied emission scenario.

Compound	Contribution to TVOCs (%)	Reference
Acetone [1]	35.3	-
Methanol [1]	29.4	-
Formaldehyde [1]	7.7	-
Propanol_fragment_(-H$_2$O)/propene/cyclopropane [1]	5.6	-
Alkyl_fragment_or_propyne [1]	5.5	-
Monoterpene fragment	4.6	-
Monoterpene	3.1	[40,46]
Isoprene [1]	1.6	-
Cis-3-hexen-1-ol_+_others [1]	1.3	-
Toluene [2]	1.2	-
Phenol [2]	0.9	-
Acetonitrile [2]	0.9	-
Benzene [2]	0.9	-
C$_7$H$_{10}$H+ [2]	0.8	-
Nonanal [2]	0.5	-
Decanal [2]	0.4	-
1,2-propendiol	0.3	-

[1] These compounds had a high probability of being related to the presence of a human subject conducting the cleaning in the test room; see Table 4. [2] These compounds were detected in the cleaning scenario, but there is no reference relating them to the emissions from cleaning products.

3.3. MOS VOC Sensor Signals

Figure 1 shows the normalized MOS VOC signals for the different examined emission scenarios. There were rather consistent signals from all sensors, but sensor C characterized the exposure to human bioeffluents (Figure 1a). Most of the signals followed the build-up of the human bioeffluent concentration in the test room as well as the consequent decay when persons left the test room. The signal from sensor C had somewhat the same build-up pattern, but there seemed to be a certain delay in its second part. In the decay period, the C signal was noisier, and the decay was not as obvious as it was for the remaining sensors.

For the emission scenario with linoleum (Figure 1b), the sensor signals were much more dispersed than in the case of bioeffluents, suggesting that their response to the pollutants emitted from linoleum was different. Yet, all sensors but sensor C could be characterized by immediate build-up and decay. Sensor C again underperformed, and this time it was not responsive to the pollutants emitted from linoleum (a short visit from the experimenter in the test room probably caused the peak and consequent decay in signal of sensor C during the exposure). Sensor B presented the least noisy signal.

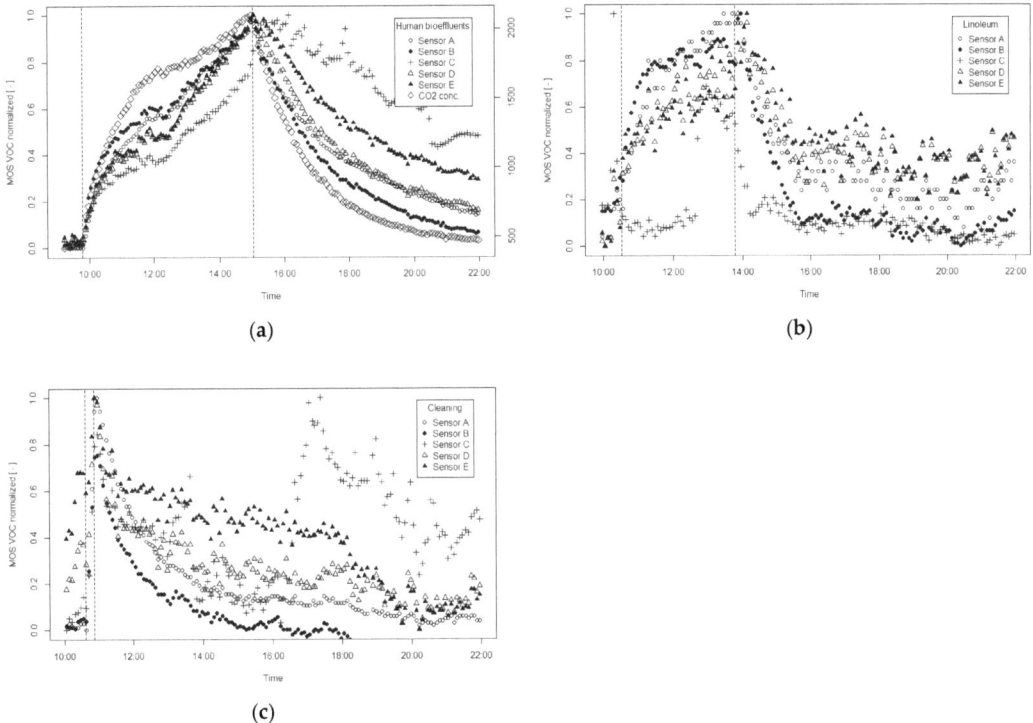

Figure 1. Normalized signals from five MOS VOC sensors for three emission scenarios: (**a**) human bioeffluents, (**b**) linoleum, (**c**) cleaning. Vertical dashed lines indicate the start (first line) and the end (second line) of the emission.

The emission scenario with air cleaning (Figure 1c) provided a sharp response and build-up immediately after starting the cleaning. All sensors reacted to the sudden release of pollutants. After the cleaning was finished, sensors A and B presented the most consistent decays. Decays of the signals from sensors D and E were noisier, with sensor E first presenting a very shallow decay followed by a steeper decay. The signal from sensor C was rather scattered with no clear decay pattern. Moreover, there was a sudden increase in the C signal during the later period of decay.

3.4. Cluster Analysis

First, we ran a cluster analysis including the T and RH signals. This preliminary analysis showed that neither the number of generated clusters nor the distribution of measured compounds and MOS VOC sensor signals changed when we added or removed the T and RH from the data set. We therefore concluded that the T and RH did not influence the clustering and excluded them from further analysis.

Figure 2 shows the results of the cluster analysis for the human bioeffluent emission scenario. The results are shown in the form of a dendrogram, which is a common practice to present clustering results displaying a close arrangement of observations with similar patterns; the colors indicate the different clusters. The dissimilarity or distance between clusters is shown as the height on the vertical axis. The data obtained during this scenario formed three clusters. The analysis placed sensors A, B, D, and E in the first cluster (marked with blue) together with compounds like acetone, isoprene, formaldehyde, and propyne (alkyl fragment; see Table 4). The second cluster (marked with a grey color) contained the signal from sensor C together with methanol, toluene, and benzaldehyde. The third and largest cluster (marked with an orange color) did not include any of the sensor signals, suggesting that the compounds identified in the orange cluster had concentration patterns that could not be associated with any signals obtained from the sensors. In other words, this result suggested that the sensors did not react to changes in concentrations for these compounds.

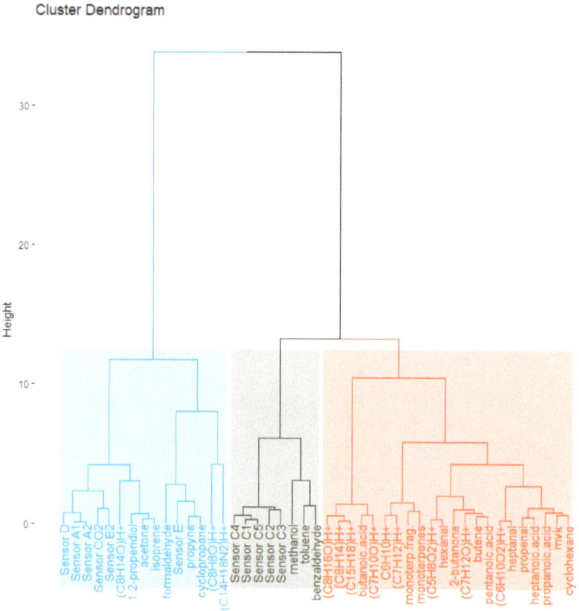

Figure 2. Cluster dendrogram for the human bioeffluent scenario. The clustering included all tested MOS VOC sensors; "mvk" stands for unsaturated carbonyl.

Dendrograms for the emission scenarios with linoleum and cleaning are presented in the Appendix A as Figures A1 and A2, respectively.

The dendrogram for the emission scenario with linoleum (Figure A1) showed that the clustering divided the data into two clusters, with the signal from sensor C being the only one in the first cluster (grey). The second cluster (blue) accommodated signals from sensors A, B, D, and E as well as all compounds measured by the PTR-ToF-MS. The main compounds appearing in this cluster were acetone, ketone, and acetic acid.

The dendrogram of the clustering result for the emission scenario with cleaning (Figure A2) revealed three clusters. The first cluster (blue) accommodated all sensor signals except sensors C and E. The most important compounds in this cluster were monoterpenes, formaldehyde, and methanol. The second cluster (grey) accommodated only the signal from sensor C, as no other measured signal had a pattern similar to the sensor C response. We found the signal from sensor E in the third cluster (orange), and the other major compounds in that cluster were acetone, propene, and propyne (akryl fragment).

4. Discussion

The studied MOS VOC sensors do not offer absolute and selective measurements of individual compounds but rather an indication of the relative change in concentration of a VOC mixture. As mentioned in the Introduction, several research studies have elaborated on this matter and presented it as a main limitation of using MOS VOC sensors. In the present paper, we did not focus on the fact that MOS VOC sensors provide relative measurements and not absolute concentrations. This certainly represents a challenge when the sensors are applied for ventilation control, but sensor manufacturers deal with this issue (with variable success) by post-processing the sensor signal to establish a "clean air" baseline. However, even if they manage to tune the baseline algorithm, the problem with non-selectivity of the MOS VOC sensor remains unsolved. All reputable producers present the calibration data indicating the change in sensor signal in relation to some reference gas, usually ethanol, isobutylene, or even gas mixtures according to ISO 16000-29:2014 [19]. They conduct these tests in laboratory conditions somewhat far from the exposures in real environments.

Our approach was to mimic the realistic exposures and make a connection between the detailed analytical measurements and MOS VOC signals. We selected three emission scenarios that could be typical for residential environments but also occur frequently in commercial buildings such as schools and offices. Cluster analysis was the method used to analyze the performance of the MOS VOC sensors. Table 7 provides a general summary of the cluster analysis. The PTR-ToF-MS identified many compounds during the investigated scenarios; however, concentrations of some of them were rather low, hence their impact on the IAQ was considered negligible. As the cluster analysis used normalized data, it did not account for the absolute values of concentrations. We do not regard this as a limitation of our approach because we aimed at studying comparable concentration patterns. Including compounds at very low concentrations makes a practical interpretation of results more difficult. This is because it leads to clusters "crowded" with compounds that may have the same concentration pattern as the investigated sensors, but their actual concentrations are negligible. Consequently, in Table 7, we present only those compounds that contributed more than 5% to the total volatile organic compound (TVOC) concentration as measured by the PTR-ToF-MS (see Tables 4–6).

Table 7. Relations among MOS VOC sensor signals and compounds contributing > 5% to the TVOC. The signal and compound concentration profile were related if they appeared in the same cluster under a particular scenario: h—human bioeffluents; l—linoleum; c—cleaning.

	Sensor A	Sensor B	Sensor C	Sensor D	Sensor E
Acetone	h/l	h/l	-[2]	h/l	h/l/c
Methanol	c	c	h	c	-
Acetic acid	l	l	-	l	l
Ketene	l	l	-	l	l
Formic acid	l	l	-	l	l
Propanol fragment [1]	h	h	-	h	h/c
Acetaldehyde	l	l	-	l	l
Alkyl fragment/propyne	h	h	-	h	h/c
Formaldehyde	c	c	-	c	-
CO_2	h/l/c	h/l/c	-	h/l/c	h/l

[1] (-H_2O)/propene/cyclopropane. [2] The dash indicates that the sensor and the compound never appeared in the same cluster.

Table 7 reveals that sensor C performed very differently from the rest of the tested sensors. The pattern of its signal was similar to methanol in the case of the emission scenario with the human bioeffluents. In the other scenarios, the signal of sensor C could not be linked to a change in the concentration of any compound listed in Table 7. Sensor E was shown through the cluster analysis as sensitive to acetone in all three scenarios. Sensors A, B, and D had a very comparable performance under different emission scenarios if the pollutants measured were considered. The signals from sensors A, B, and D also appeared in the same cluster as the CO_2 concentration in all tested scenarios. This suggests that sensors A, B, and D were responsive to pollutants whose concentration correlated with the concentration of CO_2 and could be used for control in case the human bioeffluent concentration was a determining factor. In the linoleum emission scenario, all sensors but sensor C could detect emissions of organic acids that dominated the emissions related to linoleum. More tests would be necessary to clarify whether this indicates suitability to track emissions from building materials in practice or that it is a salient feature for linoleum emissions. In our experiment, the linoleum was brought into the test room on a rack. Thus, the emission characteristics were a step-change rather than a slow, continual increase, which is typical for material emissions in real buildings.

Identification of the dominant compounds that influenced the response of MOS VOC sensors was also a focus of other studies. A recent study by Schultealbert et al. [47] showed that alcohols—especially ethanol—played a major role in TVOC measurements with low-cost MOS VOC sensors. The authors concluded that these sensors need to be scientifically validated because of their broad response; which methods should be used for such validation remains a question. The methodology used in our study as well as the one used by Schultealbert et al. offer possibilities. Both require advanced laboratory-grade instrumentation.

Using the method for examining the performance of low-cost sensors presented in our paper, one could perform an evaluation of many sensors without dedicating time to their calibration. The fact that the clustering worked with normalized signals eliminates the problem of the response shift among sensors from the same producer as well as the difference in output signals among sensors from different producers (voltage, ppb TVOC, "VOC index", etc.). Our results show that a majority of the tested sensors had a comparable performance. We could also clearly identify the sensor that did not detect the changes in IAQ at all (sensor C). As there are dozens of different MOS VOC sensors available on the market, a method allowing the screening of their basic detection capabilities seems to be necessary. Our method can be used for preliminary examination of the sensors, filtering out the sensors that underperform (in our case, it was sensor C), identifying sensors with a similar performance (in our case, these were sensors A, B, and D), and identifying sensors that specifically respond to a certain emission scenario (in our case, it was sensor E). Our method thus provides a simple yet very useful step in examining low-cost sensors before they are even examined for other features describing their performance.

Our work also had limitations. Firstly, the long-term performance of MOS VOC sensors is of high importance with respect to their suitability for ventilation control. Our experiments did not include any long-term exposures. At the same time, it is our judgement that long-term exposures would influence sensors' drift with respect to a "clean air" baseline or a factory calibration rather than the results of the cluster analysis. Secondly, the timeline and budget of the project did not allow the repetition of pollution scenarios or testing of a higher number of sensors. Nevertheless, we consider our study robust enough to demonstrate the application of a data-clustering approach to evaluate MOS VOC performance. Further experiments should cover more sensor types as well as more emission scenarios (e.g., cooking) in residences as well as other indoor environments.

5. Conclusions

- We used a cluster analysis to detect which of the five selected commercially available MOS VOC sensors produced signals in agreement with the concentration patterns of VOCs characteristic of three emission scenarios (human bioeffluents, cleaning, and linoleum) as measured by a laboratory-grade analytic instrument (PTR-ToF-MS).
- Four of the five tested sensors produced signals in agreement with the concentration patterns of characteristic VOCs. One sensor underperformed in all cases and was not able to detect the characteristic concentration patterns.
- Three sensors showed a similar performance, reacting in agreement to all emission scenarios.
- The compounds characteristic of human presence dominated the emission scenarios with human bioeffluents and cleaning. In the cleaning emission scenario, monoterpenes and their fragments characterized the emissions from the cleaning detergent. Organic acids dominated the emissions related to linoleum.
- We showed that a cluster analysis is a useful tool for examining the performance of low-cost MOS VOC sensors regarding their response to different emission scenarios. Consequently, even if the underlying pollutants responsible for the response are not

known, the sensors that are responsive to typical pollutant generating activities can be identified. Further studies supporting this observation and advancing the method would be useful.

Author Contributions: Conceptualization, J.K., P.W., T.W. and N.L.L.; methodology, J.K., P.W., N.L.L., R.B. and R.L.; software, R.L.; investigation, J.K., N.L.L., R.B. and K.M.S.; formal analysis, J.K., R.L. and R.B.; data curation, N.L.L., R.B., R.L. and J.K.; writing—original draft preparation, J.K., R.B., N.L.L. and R.L.; writing—review and editing, P.W., K.M.S. and T.W. All authors have read and agreed to the published version of the manuscript.

Funding: This research was funded by the Danish Energy Technology Development and Demonstration Programme (EUDP; https://www.eudp.dk/en), The Danish Energy Agency, Carsten Niebuhrs Gade 43, 1577 Copenhagen, Denmark under the grant number 64016-0042.

Data Availability Statement: Not applicable.

Conflicts of Interest: The authors declare no conflict of interest.

Appendix A

Clustering dendrograms for all tested pollution scenarios.

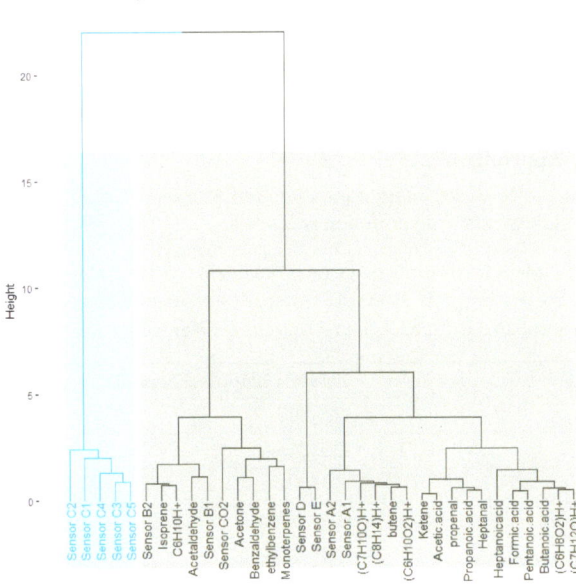

Figure A1. Cluster dendrogram for the linoleum emission scenario. The clustering included all tested MOS VOC sensors.

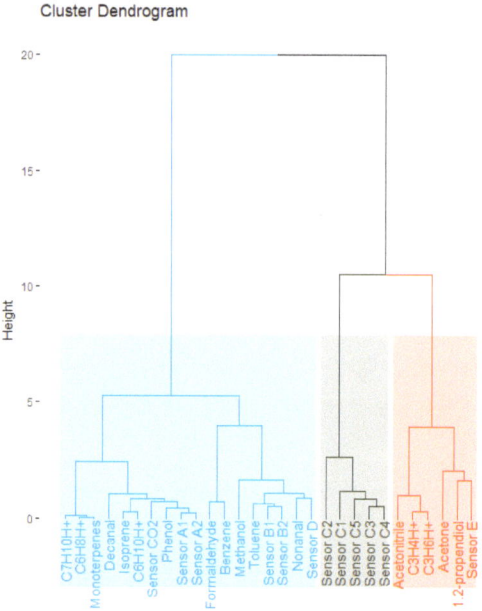

Figure A2. Cluster dendrogram for the cleaning emission scenario. The clustering included all tested MOS VOC sensors.

Appendix B

Mass-to-charge ratios for compounds detected and identified via PTR-ToF-MS measurements in all emission scenarios.

Table A1. Mass-to-charge (m/z) ratios for compounds detected and identified via PTR-ToF-MS measurements in all emission scenarios; the compounds whose concentration increased by $\geq 50\%$ compared to the background concentration are presented.

Compound	Possible Empirical Formula	Detected Ions (m/z)
Formaldehyde	CH_2OH^+	31.0178
Methanol	CH_4OH^+	33.0335
Alkyl fragment or propyne	$C_3H_4H^+$	41.0386
Acetonitrile	$C_2H_3NH^+$	42.0346
Ketene	C_2H_2O	43.01784
Propanol fragment (-H_2O)/propene/cyclopropane	$C_3H_6H^+$	43.0542
Acetaldehyde	$C_2H_4OH^+$	45.03349
Formic acid	$CH_2O_2H^+$	47.0127
Propenal	$C_3H_4OH^+$	57.0335
Acetone	$C_3H_6OH^+$	59.0491
Acetic acid	$C_2H_4O_2H^+$	61.0284
Isoprene	$C_5H_8H^+$	69.0699
Unsaturated carbonyl (e.g., methyl vinyl ketone)	$C_4H_6OH^+$	71.0491
Hydroxyacetone/propionic acid	$C_3H_6O_2H^+$	75.0440
1,2-Propendiol	$C_3H_8O_2H^+$	77.0597

Table A1. Cont.

Compound	Possible Empirical Formula	Detected Ions (m/z)
Benzene	$C_6H_6H^+$	79.05423
Toluene	$C_6H_5CH_3$	79.0548
Phenol	C_6H_6O	95.04914
Monoterpene fragment	$C_6H_8H^+$	81.0699
cis-3-Hexen-1-ol + others	$C_6H_{10}H^+$	83.0855
Butyric acid	$C_4H_8O_2H^+$	89.0597
Cyclopentylacetylene	$C_7H_{10}H^+$	95.08553
Acetylpropionyl + others	$C_5H_8O_2H^+$	101.0597
Pentanoic acid	$C_5H_8O_2H^+$	101.0597
Octanal	$C_7H_{10}OH^+$	111.0804
C7 aldehyde/ketone	$C_7H_{10}OH^+$	111.0855
1-Octen-3-ol fragment (-H_2O) + others/C8-alkane	$C_8H_{14}H^+$	111.1168
Cyclohexane diones	$C_6H_8O_2H^+$	113.0597
Cycloheptanone	$C_7H_{12}OH^+$	113.0961
C6-carboxylic acid/Cyclopentane carboxylic acid	$C_6H_{10}O_2H^+$	115.0753
Heptanal	$C_7H_{14}OH^+$	115.1117
Hexanoic acid	$C_6H_{12}O_2H^+$	117.0916
Anisaldehyde + others	$C_8H_8OH^+$	121.0670
6-Methyl-5-hepten-2-one (6-MHO)	$C8H14OH^+$	127.1150
C8 saturated carbonyl + 1-octen-3-ol	$C_8H_{16}OH^+$	129.1295
Monoterpene	$C_{10}H_{16}H^+$	137.1325
Nonanal	$C_9H_{18}O$	143.14360
Decanal	$C_{10}H_{20}O$	157.157
C12-carboxylic acid	$C_{12}H_{22}O_2H^+$	199.16953

References

1. Jousto Alonso, M.; Wolf, S.; Jørgensen, R.B.; Madsen, H.; Mathisen, H.M. A methodology for the selection of pollutants for ensuring good indoor air quality using the de-trended cross-correlation function. *Build Environ.* **2022**, *209*, 108668. [CrossRef]
2. Herberger, S.; Ulmer, H. Indoor Air Quality Monitoring Improving Air Quality Perception. *Clean-Soil Air Water* **2012**, *40*, 578–585. [CrossRef]
3. Kang, Y.; Aye, L.; Ngo, T.D.; Zhou, J. Performance evaluation of low-cost air quality sensors: A review. *Sci. Total Environ.* **2022**, *818*, 151769. [CrossRef]
4. Durier, F.; Carrié, R.; Sherman, M. *What Is Smart Ventilation?* Ventilation Information Paper n 38; INVIE EEIG: Brussels, Belgium, 2018.
5. Herberger, S.; Herold, M.; Ulmer, H.; Burdack-Freitag, A.; Mayer, F. Detection of human effluents by a MOS gas sensor in correlation to VOC quantification by GC/MS. *Build. Environ.* **2010**, *45*, 2430–2439. [CrossRef]
6. Burdack-Freitag, A.; Rampf, R.; Mayer, F.; Breuer, K. Identification of anthropogenic volatile organic compounds correlating with bad indoor air quality. In Proceedings of the 9th International Conference and Exhibition Healthy Buildings 2009, Syracuse, NY, USA, 13–17 September 2009.
7. Won, D.Y.; Schleibinger, H. *Commercial IAQ Sensors and their Performance Requirements for Demand-Controlled Ventilation*; Report no. IRC-RR-323; National Research Council Canada: Ottawa, ON, Canada, 2011.
8. Fisk, W.J.; Almeida, A.T.D. Sensor-based demand-controlled ventilation: A review. *Energy Build.* **1998**, *29*, 35–45. [CrossRef]
9. Mehmet, T. A low-cost air quality monitoring system based on Internet of Things for smart homes. *J. Ambient Intell. Smart Environ.* **2022**, *14*, 351–374.
10. Kolarik, J. CO_2 Sensor versus Volatile Organic Compounds (VOC) sensor—Analysis of field measurements and implications for Demand Controlled Ventilation. In Proceedings of the Indoor Air 2014, Hong Kong, China, 7–12 July 2014.

11. Laverge, J.; Pollet, I.; Spruytte, S.; Losfeld, F.; Vens, A. VOC or CO2: Are They Interchangeable As Sensors for Demand Control? In Proceedings of the Healthy Buildings Europe 2015, Eindhoven, The Netherlands, 18–20 May 2015.
12. Merzkirch, A.; Maas, S.; Scholzen, F.; Waldmann, D. A semi-centralized, valve less and demand controlled ventilation system in comparison to other concepts in field tests. *Build. Environ.* **2015**, *93*, 21–26. [CrossRef]
13. Abdul-Hamid, A.; El-Zoubi, S.; Omid, S. Evaluation of set points for moisture supply and volatile organic compounds as controlling parameters for demand controlled ventilation in multifamily houses. In Proceedings of the Indoor Air 2014, Hong Kong, China, 7–12 July 2014.
14. De Sutter, R.; Pollet, I.; Vens, A.; Losfeld, F.; Laverge, J. TVOC concentrations measured in Belgium dwellings and their potential for DCV control. In Proceedings of the 38th AIVC Conference, Nottingham, UK, 13–14 September 2017.
15. Kolarik, J.; Lyng, N.L.; Laverge, J. Metal Oxide Semiconductor Sensors to Measure Volatile Organic Compounds for Ventilation Control. Report from the AIVC Webinar: Using Metal Oxide Semiconductor (MOS) Sensors to Measure Volatile Organic Compounds (VOC) for Ventilation Control. Available online: https://www.aivc.org/resource/metal-oxide-semiconductor-sensors-measure-volatile-organic-compounds-ventilation-control (accessed on 13 June 2023).
16. Collier-Oxandale, A.M.; Thorson, J.; Halliday, H.; Milford, J.; Hannigan, M. Understanding the ability of low-cost MOx sensors to quantify ambient VOCs. *Atmos. Meas. Tech.* **2019**, *12*, 1441–1460. [CrossRef]
17. Demanega, I.; Mujan, I.; Singer, B.C.; Andelković, A.S.; Babich, F.; Licina, D. Performance assessment of low-cost environmental monitors and single sensors under variable indoor air quality and thermal conditions. *Build. Environ.* **2021**, *187*, 107415. [CrossRef]
18. Fahlen, P.; Andersson, H.; Ruud, S. *Sensor Tests, Demand Control Ventilation Systems*; SP Report; Swedish National Testing and Research Institute: Boras, Sweden, 1992; ISBN 91-7848-331-331-X.
19. *ISO 16000-29*; Indoor Air-Part 29: Test Methods for VOC Detectors. ISO: Geneva, Switzerland, 2014.
20. Justo Alonso, M.; Madsen, H.; Liu, P.; Jørgensen, R.B.; Jørgensen, T.B.; Christiansen, E.J.; Myrvang, O.A.; Bastien, D.; Mathisen, H.M. Evaluation of low-cost formaldehyde sensors calibration. *Build Environ.* **2022**, *222*, 109380. [CrossRef]
21. *ASTM E741-11*; Standard Test Method for Determining Air Change in a Single Zone by Means of a Tracer Gas Dilution. ASTM International: West Conshohocken, PA, USA, 2017.
22. Graus, M.; Müller, M.; Hansel, A. High resolution PTR-TOF: Quantification and Formula Confirmation of VOC in Real Time. *J. Am. Soc. Mass Spectrom.* **2010**, *21*, 1037–1044. [CrossRef]
23. Kolarik, B.; Wargocki, P.; Skorek-Osikowska, A.; Wisthaler, A. The effect of a photocatalytic air purifier on indoor air quality quantified using different measuring methods. *Build. Environ.* **2010**, *45*, 1434–1440. [CrossRef]
24. Schripp, T.; Etienne, S.; Fauck, C.; Fuhrmann, F.; Märk, L.; Salthammer, T. Application of proton-transfer-reaction-mass-spectrometry for Indoor air quality research. *Indoor Air* **2014**, *24*, 178–189. [CrossRef]
25. Jain, A.K.; Murty, M.N.; Flynn, P.J. Data clustering: A review. *ACM Comput. Surv.* **1996**, *31*, 264–323. [CrossRef]
26. Pagonis, D.; Sekimoto, K.; de Gouw, J. A Library of Proton-Transfer Reactions of H_3O^+ Ions Used for Trace Gas Detection. *J. Am. Soc. Mass. Spectrom.* **2019**, *30*, 1330–1335. [CrossRef]
27. Zhou, K.; Yang, S.; Shao, Z. Household monthly electricity consumption pattern mining: A fuzzy clustering-based model and a case study. *J. Clean. Prod.* **2017**, *141*, 900–908. [CrossRef]
28. Jin, L.; Lee, D.; Sim, A.; Borgeson, S.; Wu, K.; Spurlock, C.A.; Todd, A. Comparison of clustering techniques for residential energy behavior using smart meter data. In Proceedings of the 31st AAAI Conference on Artificial Intelligence, San Francisco, CA, USA, 4–9 February 2017.
29. Gianniou, P.; Liu, X.; Heller, A.; Nielsen, P.S.; Rode, C. Clustering-based analysis for residential district heating data. *Energy Convers. Manag.* **2018**, *165*, 840–850. [CrossRef]
30. Fernandes, M.P.; Viegas, J.L.; Vieira, S.M.; Sousa, J.M.C. Segmentation of residential gas consumers using clustering analysis. *Energies* **2017**, *10*, 2047. [CrossRef]
31. McLoughlin, F.; Duffy, A.; Conlon, M.A. Clustering approach to domestic electricity load profile characterisation using smart metering data. *Appl. Energy* **2015**, *141*, 190–199. [CrossRef]
32. Beckel, C.; Sadamori, L.; Santini, S. Towards automatic classification of private households using electricity consumption data. In Proceedings of the BuildSys 2012—4th ACM Workshop on Embedded Systems for Energy Efficiency in Buildings, Toronto, ON, Canada, 6 November 2012.
33. Dolnicar, S.A. Review of Unquestioned Standards in Using Cluster Analysis for Data-Driven Market Segmentation. In Proceedings of the Australian and New Zealand Marketing Academy Conference 2002, Melbourne, Australia, 2–4 December 2002.
34. R Core Team. R: A Language and Environment for Statistical Computing. R Foundation for Statistical Computing, Vienna, Austria. Available online: https://www.R-project.org/ (accessed on 13 June 2023).
35. Charrad, M.; Ghazzali, N.; Boiteau, V.; Niknafs, A. Nbclust: An R package for determining the relevant number of clusters in a data set. *J. Stat. Softw.* **2014**, *61*, 1–36. [CrossRef]
36. Ward, J.H. Hierarchical Grouping to Optimize an Objective Function. *J. Am. Stat. Assoc.* **1963**, *58*, 236–244. [CrossRef]
37. Fenske, J.D.; Paulson, S.E. Human breath emissions of VOCs. *J. Air Waste Manag. Assoc.* **1999**, *49*, 594–598. [CrossRef] [PubMed]
38. Wisthaler, A.; Weschler, C.J. Reactions of ozone with human skin lipids: Sources of carbonyls, dicarbonyls, and hydroxycarbonyls. *Proc. Natl. Acad. Sci. USA* **2010**, *107*, 6568–6575. [CrossRef]
39. Stönner, C.; Edtbauer, A.; Williams, J. Real-world volatile organic compound emission rate from seated adults and children for use in indoor air studies. *Indoor Air* **2018**, *28*, 164–172. [CrossRef] [PubMed]

40. Tang, X.; Misztal, P.K.; Nazaroff, W.W.; Goldstein, A.H. Volatile organic compounds emissions from human indoors. *Environ. Sci. Technol.* **2016**, *50*, 12686–12694. [CrossRef]
41. Liu, Y.; Misztal, P.K.; Xiong, J.; Tian, Y.; Arata, C.; Weber, R.J.; Nazaroff, W.W.; Goldstein, A.H. Characterizing sources and emissions of volatile organic compounds in a northern California residence using space- and time-resolved measurements. *Indoor Air* **2019**, *29*, 630–644. [CrossRef] [PubMed]
42. Wilke, O.; Jann, O.; Brödner, D. VOC- and SVOC-emissions from adhesives, floor coverings and complete floor structures. *Indoor Air* **2004**, *14*, 98–107. [CrossRef]
43. Han, K.H.; Zhang, S.; Wargocki, P.; Knudsen, H.N.; Guo, B. Determination of material emission signatures by PTR-MS and their correlation with odor assessment by human subjects. *Indoor Air* **2010**, *20*, 341–354. [CrossRef]
44. Krejcirikova, B.; Kolarik, J.; Wargocki, P. The effects of cement-based and cement-ash-based mortar slabs on indoor air quality. *Build. Environ.* **2018**, *135*, 213–223. [CrossRef]
45. Knudsen, H.N.; Clausen, P.A.; Wilkins, C.K.; Wolkoff, P. Sensory and chemical evaluation of odorous emissions from building products with and without linseed oil. *Build. Environ.* **2007**, *42*, 4059–4067. [CrossRef]
46. Höllbacher, E.; Ters, T.; Rieder-Gradinger, C.; Srebotnik, E. Emissions of indoor air pollutants from six user scenarios in a model room. *Atmos. Environ.* **2017**, *150*, 389–394. [CrossRef]
47. Schultealbert, C.; Baur, T.; Leidinger, M.; Conrad, T.; Amann, J.; Bur, C.; Schütze, A. Do alcohols dominate the VOC measurement of low-cost sensors? In Proceedings of the 18th Healthy Buildings Europe Conference, Aachen, Germany, 11–14 June 2023.

Disclaimer/Publisher's Note: The statements, opinions and data contained in all publications are solely those of the individual author(s) and contributor(s) and not of MDPI and/or the editor(s). MDPI and/or the editor(s) disclaim responsibility for any injury to people or property resulting from any ideas, methods, instructions or products referred to in the content.

Article

Preliminary Study on the Emission Dynamics of TVOC and Formaldehyde in Homes with Eco-Friendly Materials: Beyond Green Building

Chuloh Jung [1], Naglaa Sami Abdelaziz Mahmoud [2,*], Nahla Al Qassimi [3] and Gamal Elsamanoudy [2]

[1] Department of Architectural Engineering, College of Engineering, University of Sharjah, Sharjah P.O. Box 27272, United Arab Emirates; chuloh@sharjah.ac.ae
[2] Department of Interior Design, College of Architecture, Art and Design, Healthy and Sustainable Buildings Research Center, Ajman University, Ajman P.O. Box 346, United Arab Emirates; g.elsamanoudy@ajman.ac.ae
[3] Department of Architecture, College of Architecture, Art and Design, Ajman University, Ajman P.O. Box 346, United Arab Emirates; n.alqassimi@ajman.ac.ae
* Correspondence: n.abdelaziz@ajman.ac.ae

Abstract: This preliminary study investigates the emission characteristics of formaldehyde (HCHO) and total volatile organic compounds (TVOC) in indoor environments, comparing the effects of eco-friendly materials and general materials. The study analyzes the concentration changes over time in the living rooms of experimental units to assess the effectiveness of eco-friendly materials in reducing indoor air pollutants. The results show that eco-friendly materials exhibit lower initial emissions of TVOC than general materials, gradually decreasing over time. Compared to the eco-friendly material unit, the general material unit takes longer to reach acceptable TVOC concentrations. The emission pattern of HCHO differs from TVOC, with the highest peak occurring on the seventh day. Major individual VOCs, except for benzene, exhibit a similar decreasing trend for TVOC over time. Eco-friendly materials demonstrate significant reductions in emissions compared to general materials in various material applications, including parquet flooring, wallpaper, built-in furniture, and kitchen furniture. However, the difference in emissions for door and window frames using eco-friendly materials is minimal. These findings emphasize the effectiveness of eco-friendly materials in reducing indoor air pollutants and provide valuable insights for creating healthier living environments. Further research is needed to optimize the application of eco-friendly materials in specific components and investigate their long-term impact on indoor air quality and occupant health.

Keywords: indoor air quality (IAQ); eco-friendly materials; VOC emissions; HCHO emissions; Dubai

Citation: Jung, C.; Abdelaziz Mahmoud, N.S.; Al Qassimi, N.; Elsamanoudy, G. Preliminary Study on the Emission Dynamics of TVOC and Formaldehyde in Homes with Eco-Friendly Materials: Beyond Green Building. *Buildings* **2023**, *13*, 2847. https://doi.org/10.3390/buildings13112847

Academic Editors: Cinzia Buratti and Apple L.S. Chan

Received: 24 September 2023
Revised: 28 October 2023
Accepted: 12 November 2023
Published: 14 November 2023

Copyright: © 2023 by the authors. Licensee MDPI, Basel, Switzerland. This article is an open access article distributed under the terms and conditions of the Creative Commons Attribution (CC BY) license (https://creativecommons.org/licenses/by/4.0/).

1. Introduction

Recently, residential properties' insulation and airtightness requirements have become imperative for energy conservation [1]. Consequently, diverse construction techniques and the adoption of novel, high-efficiency, and multifunctional building interior materials have increased indoor air contaminants [2,3]. The building materials utilized in modern housing consist of intricate compounds and consequently release an array of perilous chemicals, including volatile organic compounds (VOCs) and formaldehyde (HCHO), which contribute to the degradation of indoor air quality (IAQ) [4,5]. Research is currently underway to investigate these emissions [6]. These harmful substances can potentially lead to various ailments among occupants, such as headaches, dizziness, nausea, drowsiness, and diminished concentration [7,8]. While not always directly or linearly correlated, these symptoms can collectively define sick building syndrome (SBS) [9]. Sick building syndrome (SBS) refers to non-specific health complaints often linked to exposure to indoor and outdoor air pollutants, including symptoms such as fatigue, headaches, irritations of the eyes, nose, and throat, dry cough, parched or itchy skin, dizziness, and difficulty maintaining concentration. The emission of hazardous chemicals indoors has given rise to numerous predicaments in the daily activities of inhabitants [10]. To address these issues, installing and operating appropriate ventilation systems is crucial, as is employing environmentally friendly materials while constructing new edifices and advancing the development of low-pollutant-emitting materials [11,12]. Desperate efforts are needed to tackle these challenges.

Various sources influence indoor air quality, each contributing to the presence of volatile organic compounds (VOCs) and formaldehyde (HCHO) in indoor environments. Construction materials, with their adhesives, sealants, and finishes, significantly release VOCs and HCHO indoors. However, they are not alone in this; furniture and furnishings made of composite wood products, household products such as cleaning agents, and even personal care items can emit these pollutants. Moreover, cooking and heating appliances, particularly gas stoves, can introduce combustion byproducts, including VOCs, into indoor air [13]. Outdoor sources such as traffic emissions and industrial activities can infiltrate indoor spaces, while human activities such as smoking and hobbies involving solvents add to the mix. Outdoor air quality also influences indoor air quality, and effective ventilation can help dilute pollutants and mitigate their impact. Managing indoor air quality effectively necessitates a holistic approach that addresses these diverse sources and employs strategies such as source control, ventilation, and air purification to ensure healthier indoor environments [14].

Extensive research conducted in Dubai has highlighted the critical nature of indoor air quality (IAQ) and its profound impact on residents [15,16]. Kim et al. (2022) conducted a study in Dubai, revealing that 15% of the city's population has reported experiencing symptoms associated with sick building syndrome (SBS). This syndrome is characterized by various non-specific health complaints often linked to exposure to a complex mix of indoor and outdoor air pollutants [17]. SBS is a multifaceted issue highlighting the importance of indoor air quality in urban settings, where individuals spend a significant portion of their lives in various environments. Understanding the factors contributing to SBS, including specific pollutant sources, ventilation systems, and building design, is crucial for promoting healthier indoor environments and the well-being of urban populations. The symptoms of SBS include fatigue, headaches, irritations of the eyes, nose, and throat, dry cough, parched or itchy skin, dizziness, and difficulty maintaining concentration. To recognize these concerns and ensure compliance with IAQ standards, the Dubai Municipality has outlined specific concentration thresholds [18]. These mandates stipulate that HCHO levels should not exceed 0.08 parts per million (ppm), total volatile organic compounds (TVOC) should be maintained below 300 $\mu g/m^3$, and particulate matter (PM_{10}) should be limited to 150 $\mu g/m^3$. These measurements are obtained through continuous monitoring over 8 h before newly constructed houses are occupied [19].

Arar et al. (2022) surveyed between December 2021 and January 2022, targeting residents of townhouses in Dubai, and revealed a notable level of awareness regarding SBS [20]. A significant 95% of respondents indicated having above-average knowledge of SBS [21]. However, despite this awareness, a majority demonstrated limited knowledge or indifference towards methods to improve IAQ [22]. It was observed that individuals who spent substantial amounts of time indoors, such as housewives and children, were the most adversely affected [23]. Furthermore, Carrer and Wolkoff (2018) identified a trend of increased vigilance among individuals in assessing IAQ before relocating [24]. However, once settled, there was a lack of guidance and systems for maintaining healthy living conditions [25].

Regarding construction practices, most developers in Dubai have adopted eco-friendly materials, established ventilation systems, and implemented pre-occupancy bake-outs to ensure compliance with recommended indoor air quality standards in new apartment buildings [26]. However, after moving in, residents are responsible for actively enhancing the indoor air environment [27]. This can be achieved through diligent utilization of ventilation facilities and restricting the use of household items that contribute to indoor pollutant generation, thereby reducing pollutant concentrations [28].

Jung et al. (2021) focus on evaluating the indoor environment within specific developments, such as The Springs, an iconic townhouse-type residential complex in Dubai [29]. This research aims to discern residents' preferences concerning various indoor environmental factors, including thermal comfort, indoor air quality, lighting, and acoustics [30]. Preliminary findings indicate that, during the summer, thermal comfort emerges as the foremost concern for living rooms and master bedrooms. In contrast, indoor air quality assumes greater significance during winter [31,32]. The outcomes of this research are expected to guide future renovation guidelines to enhance indoor environments, particularly in buildings nearing the twenty-year mark, thereby preventing complications associated with SBS [33].

Furthermore, to enhance the quality of final interior finishing materials, utilizing substances that possess diminished levels of hazardous chemicals is also necessary [34]. Additionally, there is a growing need to systematically evaluate the efficacy of employing these materials [35].

This research employs a meticulous selection process to identify eco-friendly and conventional materials as primary candidates for indoor finishing materials, known to be the primary culprits behind indoor air pollution [36]. Subsequently, experiments are conducted to examine the emission of harmful chemicals and assess the effectiveness of implementing eco-friendly building materials [37]. The specific objectives encompass two principal aims: firstly, to evaluate the performance of experimental houses constructed using both eco-friendly materials and conventional materials [38]; secondly, to ascertain the emission characteristics of HCHO and VOCs within the living spaces, accounting for the location of each construction material [39,40].

2. Materials and Methods

2.1. Research Methods and Procedures

As shown in Table 1, four mockup test units, labeled 4A, 6B, 8C, and 10D, were meticulously constructed at the Sobha Hartland One Park Avenue construction site before the completion of the apartment [41].

Table 1. Composition of the four experimental units.

Sobha Hartland One Park Avenue	Material	Experiment Contents Duration	Evaluation Criteria
Unit 4A	General material	81 days	Concentration changes over time
Unit 6B	Eco-friendly material		
Unit 8C	General material	14 days for each material	Changes in indoor pollutant concentration according to the construction location for each material
Unit 10D	Eco-friendly material		

Experimental unit 4A was erected using conventional materials, while experimental unit 6B employed eco-friendly materials, enabling the assessment of long-term reductions in HCHO and VOCs (Figure 1) [42,43]. Furthermore, experimental units 8C and 10D were dedicated to general and eco-friendly materials [44]. Sequentially, these units involved the installation of wallpaper (including adhesive), floor materials (including adhesive), general furniture, kitchen furniture, and wooden window and door materials at approximately two-week intervals [45]. The concentrations of formaldehyde and VOCs were measured on the first day and the fifth to eighth days after each construction phase, employing a repetitive construction cycle, measurement, and demolition [46].

To elaborate, the wallpaper experiment commenced with an initial measurement of background concentration, followed by the installation of wallpaper on the walls and ceiling of the living room, each room, and the kitchen in the experimental units [47]. The emissions were monitored and recorded before material removal [48]. Upon removal, thorough ventilation was conducted by fully opening the doors of all units for a specific duration [49]. Subsequently, the background concentration was remeasured before proceeding with the construction of subsequent materials [50].

Figure 1. Unit 6B in Sobha Hartland One Park Avenue. Measurement point: Grey Wolf device.

The order of construction materials was determined based on the ease of dismantling and minimal residual impact [51]. General furniture, kitchen furniture, wooden windows, wallpaper, and flooring materials were successively installed [52]. Moreover, when calculating the final concentration, the background concentration measured before construction was subtracted and considered [53].

In this study, eco-friendly materials encompass substances specifically developed to reduce hazardous chemicals compared to conventional materials [54]. Table 2 provides detailed information regarding their specific composition.

Table 2. Comparison of the composition of finishing materials.

Classification			General Material	Eco-Friendly Material
Wallpaper (adhesive)			PVC-based wallpaper (general adhesive)	PP-based wallpaper (HCHO low-emission adhesive)
Parquet flooring (adhesive)			General flooring (Oil-based epoxy adhesive)	Hazardous chemical substance reduction floor (Urethane adhesive)
Window frame	Core material		Laminated wood E2 grade	Laminated wood E1 grade
	Surface material		HDF E2 grade	HDF E1 grade
Built-in Furniture	Core material	Body frame	LVL E2 grade	LVL E1 grade
		Door	PB E2 grade	PB E1 grade
	Surface material		MDF E2 grade	MDF E1 grade
			PVC wrapping	LPM
Kitchen furniture	Core material	Body frame	PB E2 grade	PB E1 grade
		Door	MDF E2 grade	MDF E1 grade
	Surface material		Laquer paint	UV paint

Wallpaper was affixed to the walls and ceiling of the living room, as well as the walls and ceiling of each of the three individual rooms and the kitchen [55]. Conventional materials employed polyvinyl chloride (PVC)-based resin for the wallpaper, while eco-friendly materials employed polypropylene (PP)-based resin [56]. Flooring materials were installed in the living room and kitchen. Conventional floorboards utilized oil-based epoxy resin-based adhesives, whereas eco-friendly floorboards utilized urethane resin-based adhesives [52,57].

General furniture includes a shoe rack, a dressing table in the master bedroom, a closet in the dressing room, and a decorative cabinet in the living room [58]. Regarding the materials utilized for general furniture, the core body employed a particle board (PB), while the doors were constructed using medium density fiberboard (MDF) [59]. It is worth noting that these furniture materials adhered to the E2 grade for HCHO emissions, ensuring limited radiation of HCHO [60]. Conversely, eco-friendly furniture materials adhered to the E1 grade, signifying a higher level of environmental friendliness [61]. Regarding surface treatment, adhesives were circumvented using the PVC wrapping technique for general furniture materials [62]. On the other hand, the low-pressure laminate (LPM) processing method was employed for eco-friendly furniture materials [63].

The kitchen area was furnished with general furniture comprising a core body of PB E2-grade and MDF E2-grade materials to ensure compliance with the specified formaldehyde (HCHO) radiation standards [64]. Furthermore, a membrane finish was applied for a polished appearance [65]. On the other hand, eco-friendly kitchen furniture featured PB E1-grade and MDF E1-grade materials and a coating of ultraviolet curing (UV) paint, signifying a commitment to environmentally conscious practices [66].

Wood windows and doors encompass the materials utilized for each room's doors, doorframes, and window frames. The core component of general and eco-friendly materials comprises E2-grade laminated wood and a high-intensity fiberboard (HDF) surface layer. In the case of general materials, veneer lumber (LVL) adhered to the E2 grade, while HDF met the E1 grade requirements. Eco-friendly materials, on the other hand, adhered to E2 grade specifications. The experiment entailed a repetitive process of constructing and demolishing the aforementioned materials in each designated area, carried out sequentially.

2.2. Target Building Status

The focus of measurement encompassed the two-bedroom units (102.13 m^2) within the Sobha Hartland One Park Avenue apartment complex situated in Mohammad Bin Rashid Al Maktoum City (MBR), Dubai [67]. The experiment was conducted over the period spanning from November 2022 to December 2022. The room conditions were diligently maintained during the experiment at an air temperature of 25 °C. Figure 2 illustrates the experimental layout, showcasing the plan view and the positioning of the measuring points.

Figure 2. Measurement point.

The measurement locations were determined based on the living room, serving as the primary measurement point [68]. Each material was measured at its main construction location (e.g., bedroom, kitchen). To clarify further, the living rooms of units 4A and 6B were measured. In contrast, in units 8C and 10D, the living rooms were assessed for flooring, wallpaper, general furniture, and wooden window and door installations. In contrast, the living rooms and bedrooms were examined for kitchen furniture.

The measurement points were positioned at the center, with a minimum distance of 1 m from the walls, and the height was set between 1.2 and 1.5 m from the floor [69]. Indoor air collection was measured following a process adhering to the World Health Organization's (WHO) IAQ testing method [70]. This entailed 30 min of ventilation followed by 5 h of sealing. The concentration of VOCs emitted by each building material was calculated by determining the concentration of individual VOCs and compounds identified through analysis. For compounds that could not be specifically identified, they were converted to the concentration of toluene, and subsequently, the concentration of TVOC was calculated by summing the two concentrations.

2.3. Measurements

Specialized measurement sensors were employed to collect data on indoor air quality meticulously. These sensors were thoughtfully selected based on their exceptional accuracy and reliability in quantifying the concentrations of formaldehyde (HCHO) and volatile organic compounds (VOCs) in the indoor environment.

A highly sensitive approach was adopted for VOC measurements, utilizing a stainless tube filled with 200 mg of Tenex-TA (60/80 mesh, Supelco, Bellefonte, PA, USA) for solid adsorption. Similarly, for HCHO measurements, we employed a purified 2,4-DNPH Silica Cartridge (Supelco, S10, Bellefonte, PA, USA). To maintain the utmost precision in our measurements, a micro pump (Gilian, Pinellas County, FL, USA) was meticulously chosen for its minimal flow fluctuations before and after measurements, guaranteeing our data's accuracy and reliability. The flow rates for VOC and HCHO measurements were set at 50 mL/min and 250 mL/min, respectively. These flow rates were continually monitored using a digital flow meter (All-tech, Lexington, KY, USA), ensuring that fluctuations remained within the 5% range.

Our analytical arsenal was further bolstered by utilizing gas chromatography with mass spectrometry detection (GC/MSD) for VOC analysis. This state-of-the-art approach incorporated an HP-1 Capillary column (60 m × 0.32 mm × 5 μm) and adhered to rigorous analysis conditions. These conditions entailed maintaining the column temperature between 40 °C and 220 °C, maintaining a column flow rate of 1 mL/min, and sustaining a mass spectrometry detector (MSD) temperature of 230 °C.

HCHO analysis was performed using high-performance liquid chromatography (HPLC) with a C-18 column (3.9 × 300 mm) to ensure comprehensive analysis. The mobile phase consisted of acetonitrile and water in a precise ratio of 55:45, with detection conducted at a wavelength of 360 nm. The flow rate was methodically set at 1.0 mL/min, and each sample was injected using a consistent volume of 20 μL. The acetonitrile and water used in the analysis were procured from reputable suppliers to meet the highest analytical standards. These measurement sensors and analytical methods were thoughtfully selected for their exceptional precision and reliability, underscoring our unwavering commitment to ensuring the utmost accuracy in our data collection process.

2.4. Sample Collection and Analysis Method

The solid adsorption method used a stainless tube filled with 200 mg of Tenex-TA (60/80 mesh, Supelco, Bellefonte, PA, USA) to measure VOCs. For HCHO measurements, a purified 2,4-DNPH silica cartridge (Supelco, S10, Bellefonte, PA, USA) was utilized. A micro pump (Gilian, Pinellas County, FL, USA) with minimal flow fluctuations before and after measurements was utilized for VOC and HCHO measurements.

In the case of VOCs, a total volume of 1.5 L was measured over 30 min, with a flow rate of 50 mL/min. For HCHO measurements, a total volume of 7.5 L was measured over the same 30 min period at a flow rate of 250 mL/min. The flow rate before and after the measurement was assessed using a digital flow meter (Alltech, Lexington, KY, USA), ensuring that the variation in flow rate remained within 5%.

Before measurement, VOCs were thermally desorbed and conditioned using ATD-400 (Perkinelmer, Buckinghamshire, UK). After measurement, the VOC adsorption tube was securely sealed, protected from light, and stored in a cool and dark environment at temperatures below 4 °C until further analysis. The desorbed VOCs from the adsorption tube were separated using a BP-1 column as the stationary phase and detected using a mass spectrometry detector (MSD) (PerkinElmer, Buckinghamshire, UK). The following are the analysis conditions for GC/MSD (Table 3).

Table 3. Conditions for VOC analysis.

Equipment	Analysis Conditions
GC/MSD	HP 6890/HP-5973N Column: HP-1 Capillary column(60 m × 0.32 mm × 5 μm) Column temperature: 40 °C (5 min) >> 70 °C (5 min) >> 150 °C (5 min) >> 200 °C (5 min)->220 °C (5 min) Ramp rate: 5 °C/min to 200 °C, 10 °C/min to 220 °C Column flow: 1 mL/min MS ion source temp: 230 °C

During the measurement of HCHO, certain factors, such as ozone, sunlight, and moisture, can interfere with the derivatization reaction of aldehydes. To mitigate the impact of ozone, an ozone scrubber (Waters, Milford, MA, USA) was employed at the front end of the 2,4-DNPH cartridge. Additionally, the influence of sunlight was deemed negligible, as it did not directly affect the measurement point. Following the measurement, the sample was carefully sealed, shielded from light using aluminum foil, and stored in a cool, dark environment below 4 °C. The sample was then fixed within a sample extractor, namely, the Vacuum Elution Rack (Supelco, Bellefonte, PA, USA), and filtered using an oil-soluble filter (47 mm diameter, 0.45 μm pore size, PTFE), employing HPLC-grade acetonitrile (JTbaker, Phillipsburg, NJ, USA) solution. A volume of 5 mL was extracted for analysis. Sample analysis was conducted using high-performance liquid chromatography (HPLC). The following outlines the analysis conditions for HCHO (Table 4).

Table 4. Conditions for HCHO analysis.

Equipment	Analysis Conditions
HPLC	Column: C-18 column (3.9 × 300 mm) waters U.S.A. Mobile phase: acetonitrile/water = 55:45 UV detector: 360 nm Flow rate: 1.0 mL/min Sample injection amount: 20 μL

3. Results

3.1. Comparison of VOC and HCHO Emission Concentrations over Time between General and Eco-Friendly Material

The findings of this experiment elucidate the concentration changes over time in the living rooms of experimental units 4A and 6B. Figures 3 and 4 present the long-term variations in TVOC and HCHO concentrations for both the eco-friendly and general material units.

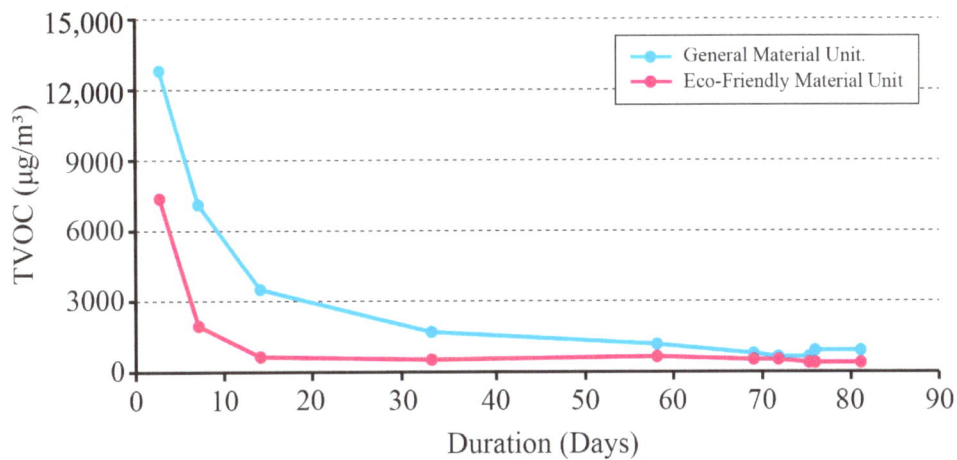

Figure 3. The concentration of TVOC in the general material unit and the eco-friendly unit.

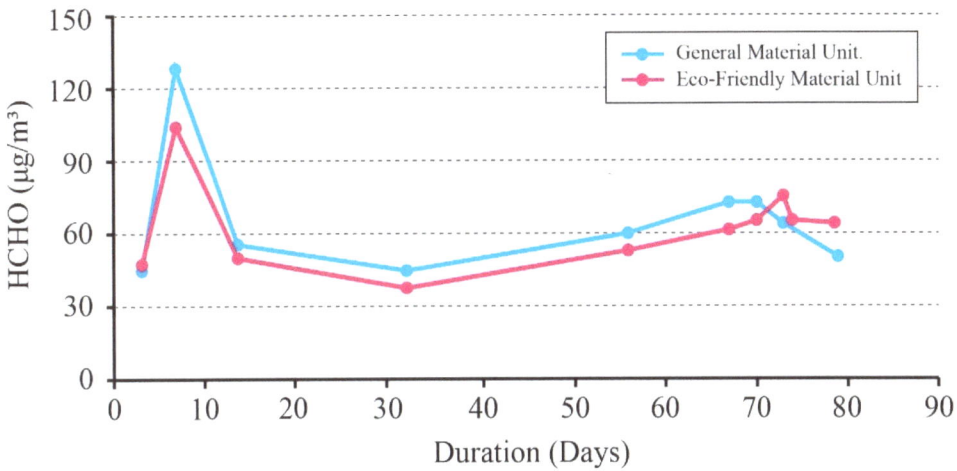

Figure 4. The concentration of HCHO in the general material unit and the eco-friendly unit.

The initial TVOC concentration in the general material unit was approximately 1.7 times higher than in the eco-friendly material unit. However, it exhibited a decreasing trend over time, indicating a high initial emission of pollutants that gradually diminished. Conversely, in the eco-friendly material unit, the TVOC concentration dropped below 1000 μg/m^3 after 14 days of construction and remained stable even after several tens of days.

In contrast, it took 69 days of construction for the concentration of the general material unit to fall below 1000 μg/m^3. The TVOC concentration in the general material unit was consistently lower than in the eco-friendly material unit, but it required more than 50 days to achieve such levels. Figure 3 illustrates a similar trend in TVOC emission concentration between the general and eco-friendly material units after several construction days have elapsed. Considering these observations, the application of eco-friendly materials proves effective in ensuring a more comfortable indoor air quality for residents when moving in, especially considering the typical occupancy timeline of 30 days after the completion of interior finishing materials.

Unlike TVOC, the maximum peak of HCHO emission concentration occurred on the seventh day, reaching 126 μg/m^3 in the general material unit. It is important to note that

HCHO concentrations may exhibit fluctuations depending on indoor temperature and humidity conditions. However, in this experimental setting, temperature fluctuations were minimal due to the consistent indoor temperature of 23 ± 1 °C maintained throughout the experiment (Figure 4). Nevertheless, it is worth mentioning that although the room temperature controller in the living room was set at 25 °C, the measured air temperature in the central breathing area of the living room was approximately 2 °C lower.

Upon analyzing the temporal variations in indoor concentrations of major individual volatile organic compounds (VOCs), it was observed that most substances exhibit a similar declining trend as that of the total volatile organic compounds (TVOC) (Figures 5–9). Notably, toluene has substantial emission levels and a decreasing pattern that is remarkably comparable to TVOC (Figure 6). However, benzene showcases a distinct behavior with low initial emission levels, displaying a cyclic pattern of fluctuations over time. In all cases, except for benzene, the initial emission levels (60 days before the start of the experiment) are considerably higher in general materials compared to eco-friendly materials, indicating a significant disparity. Therefore, eco-friendly materials are effective in mitigating initial emissions [53].

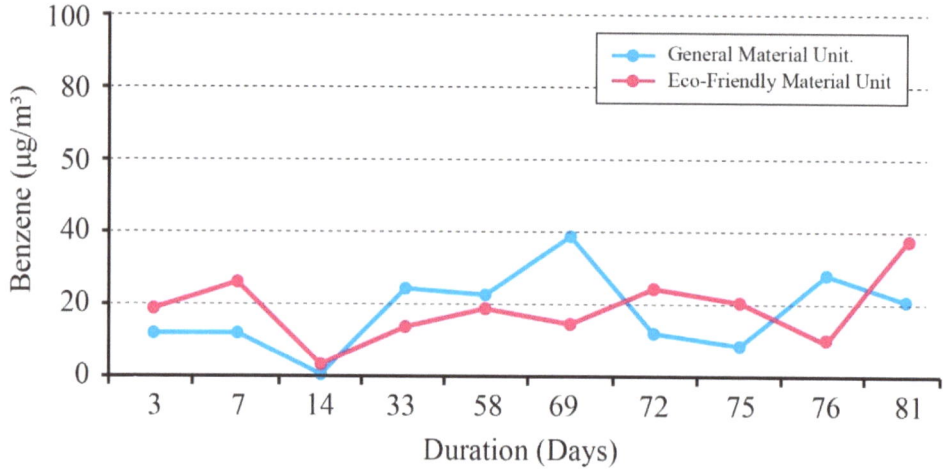

Figure 5. The concentration of benzene in the general material unit and the eco-friendly unit.

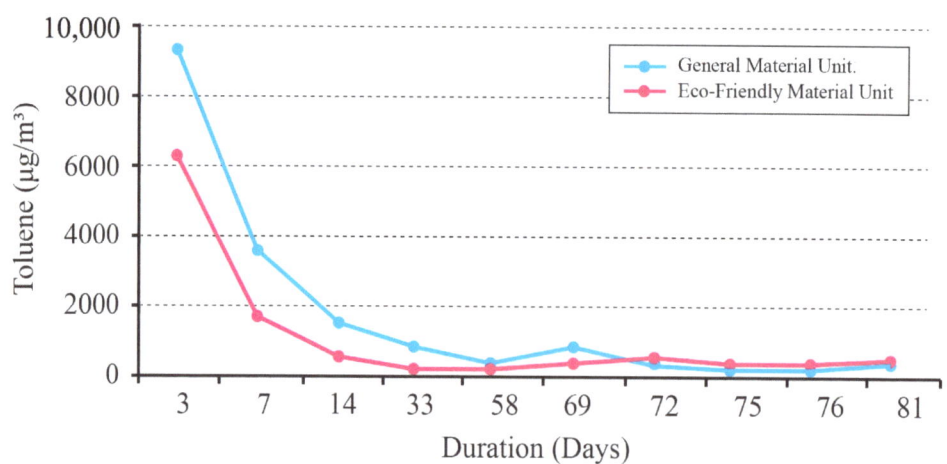

Figure 6. The concentration of toluene in the general material unit and the eco-friendly unit.

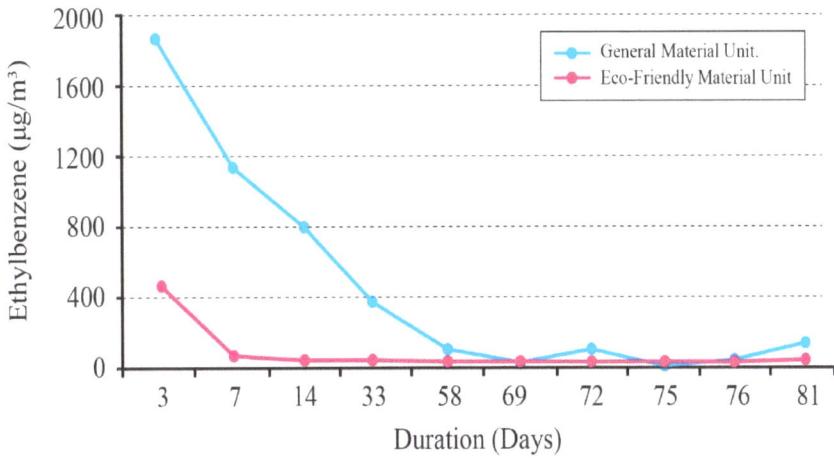

Figure 7. The concentration of ethylbenzene in the general material unit and the eco-friendly unit.

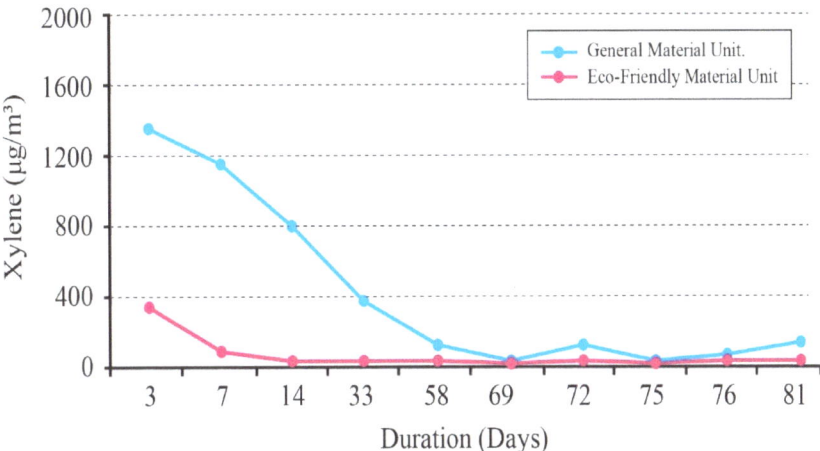

Figure 8. The concentration of xylene in the general material unit and the eco-friendly unit.

3.2. Comparison of TVOC and HCHO Emission Concentrations by Locations between General and Eco-Friendly Material

The outcomes of this study pertain to the experimental units 8C and 10D. The experiment aimed to ascertain the emission characteristics of HCHO and VOCs indoors concerning the construction location of each material. Furthermore, the efficacy of eco-friendly materials for each specific material was evaluated. Measurements were conducted at the central position within the living room area.

3.2.1. Parquet Flooring

Parquet flooring was meticulously installed in both the living room and kitchen areas. An adhesive containing oil-based epoxy resin was utilized for conventional flooring materials, while eco-friendly flooring materials employed a resin adhesive based on urethane. A comparative analysis of TVOC emissions reveals a substantial disparity. The construction of eco-friendly flooring materials results in significantly lower emission levels compared to the installation of conventional floor materials (Figure 10).

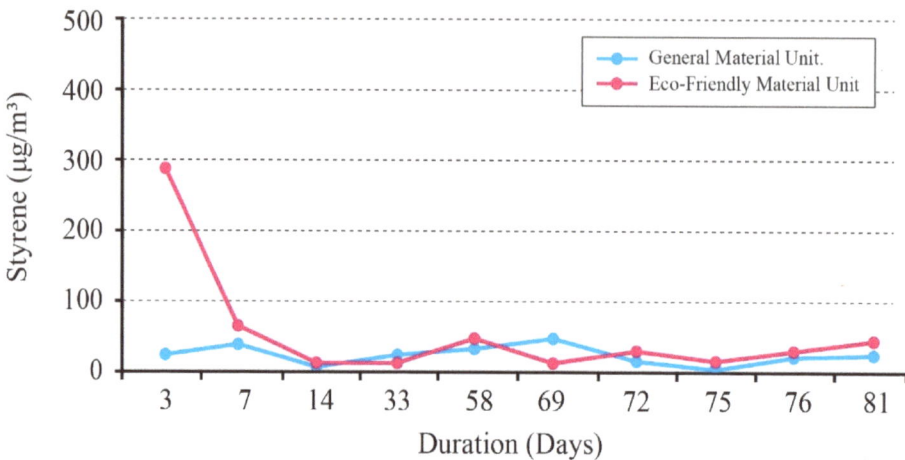

Figure 9. The concentration of styrene in the general material unit and the eco-friendly unit.

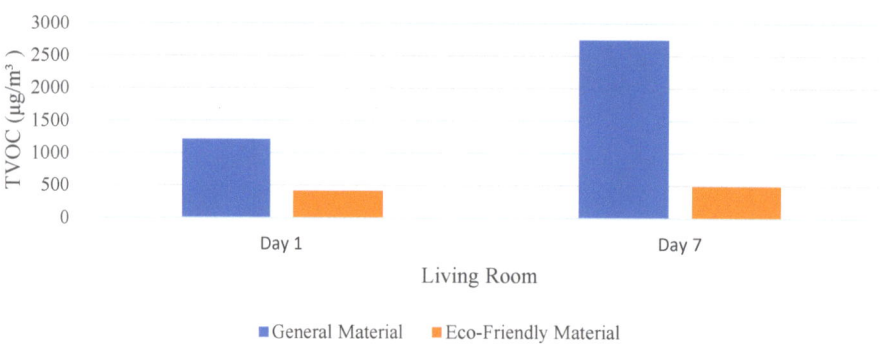

Figure 10. TVOC concentration according to the parquet flooring.

Regarding HCHO emissions, eco-friendly flooring materials exhibit lower overall levels than general flooring materials. Specifically, on the first day after construction, the difference in emissions was approximately twice as large. However, it should be noted that there was a slight increase in emission amounts over time, and the disparity with the emission levels of general flooring materials was not significantly large (Figure 11).

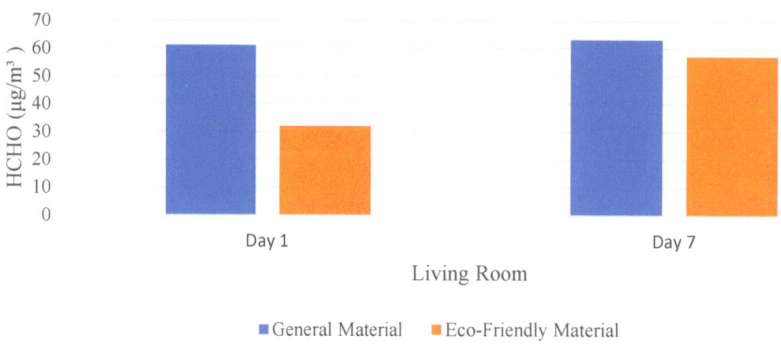

Figure 11. HCHO concentration according to the parquet flooring.

3.2.2. Wallpaper

The experiment focused on two types of wallpaper: one made with PVC (polyvinyl chloride) resin and the other made with PP (polypropylene) resin. The experiment results demonstrated a reduction in the emissions of TVOC and HCHO with the use of eco-friendly wallpaper (Figure 12). Specifically, on the first day after construction, a considerable concentration of emissions was observed, highlighting the notable difference between eco-friendly wallpaper and general wallpaper. By the sixth day, there was no significant emission level disparity between the general and eco-friendly wallpaper. However, it was evident that the emission amounts were significantly reduced compared to the first day, indicating the effectiveness of eco-friendly wallpaper in minimizing emissions over time (Figure 13).

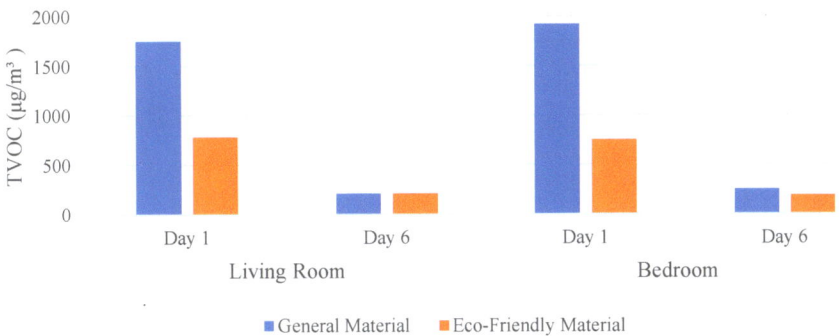

Figure 12. TVOC concentration according to the wallpaper.

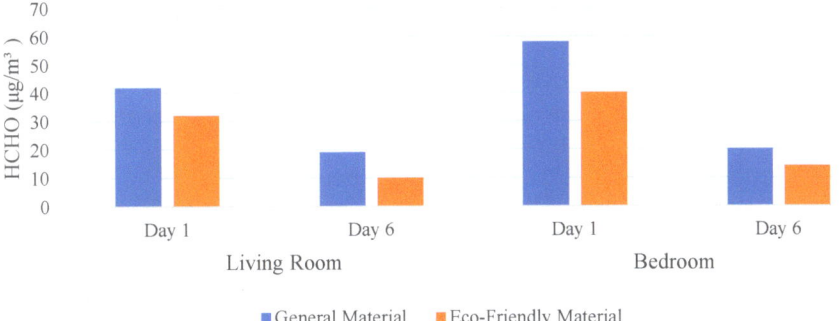

Figure 13. HCHO concentration according to the wallpaper.

3.2.3. Built-in Furniture

The emission levels were measured on the first and seventh days after installing general furniture in the experimental household, including shoe cabinets, dressing tables in the master bedroom, closets in the dressing room, and cabinets in the living room. Figures 14 and 15 illustrate the concentrations of TVOC in the living room and bedroom after general furniture and eco-friendly materials are installed. In the case of general furniture, it was observed that eco-friendly materials resulted in a significant reduction in TVOC concentrations. Regarding HCHO emissions, it was noted that on the first day in the bedroom, eco-friendly furniture exhibited slightly higher emissions than general furniture. However, after seven days of construction, it was confirmed that the emission levels were more than twice as low when using eco-friendly materials compared to general materials.

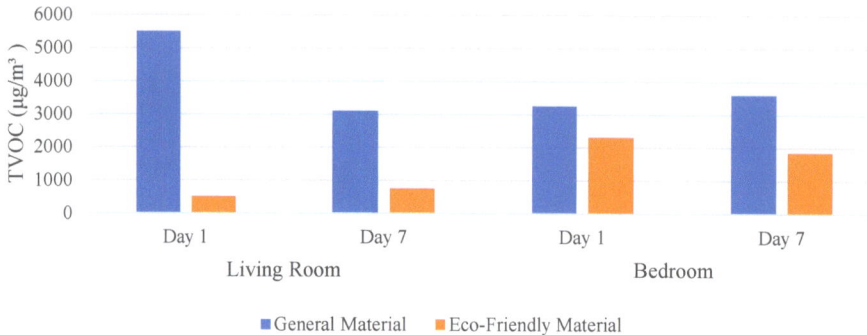

Figure 14. TVOC concentration according to the built-in furniture.

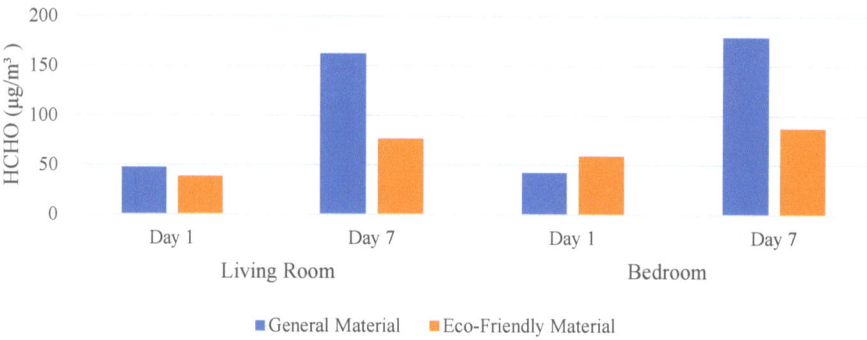

Figure 15. HCHO concentration according to the built-in furniture.

3.2.4. Kitchen Furniture

When considering kitchen furniture, eco-friendly subsidiary materials resulted in a more than two times reduction in TVOC emissions compared to general subsidiary materials (Figure 16). This reduction was observed at the beginning of construction and on the eighth day. In the case of eco-friendly kitchen furniture, the application of UV paint as the surface finish played a significant role in reducing VOC emissions. The paint-drying process during the molding stage contributed to lower VOC emissions after the furniture was installed in the experimental unit. Using eco-friendly subsidiary materials led to lower HCHO emissions than general subsidiary materials (Figure 17).

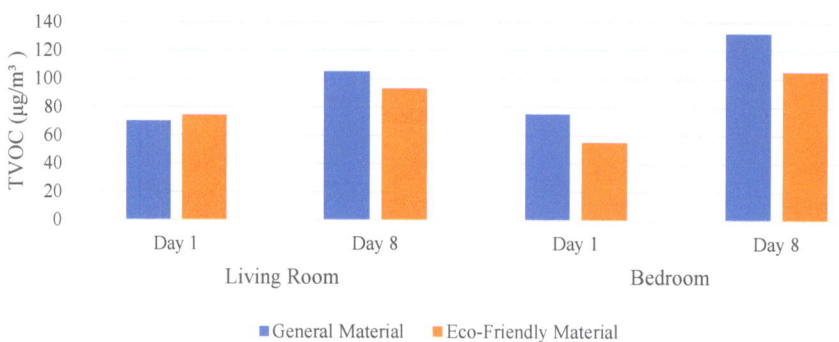

Figure 16. TVOC concentration according to the kitchen furniture.

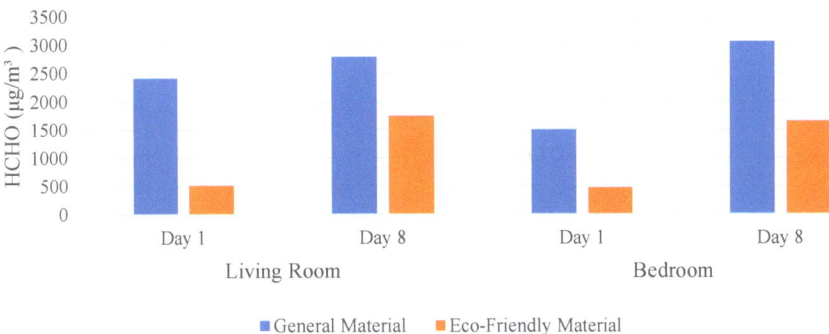

Figure 17. HCHO concentration according to the kitchen furniture.

3.2.5. Door/Window Frame

Regarding door and window frames, E2-grade laminated wood, HDF, and LVL were used with general auxiliary materials. In contrast, eco-friendly auxiliary materials were assigned an E1 grade to enhance performance. As a result, both TVOC and HCHO emissions were slightly lower when eco-friendly subsidiary materials were utilized. However, the difference in emissions was minimal, indicating the need for further emphasis on applying eco-friendly materials for wooden windows and doors (Figures 18 and 19).

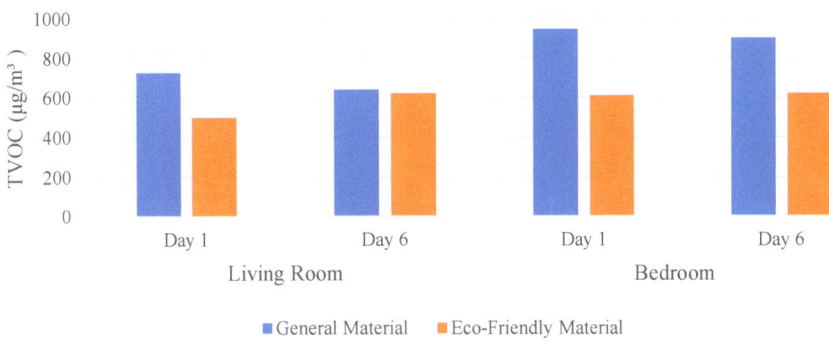

Figure 18. TVOC concentration according to the door/window frame.

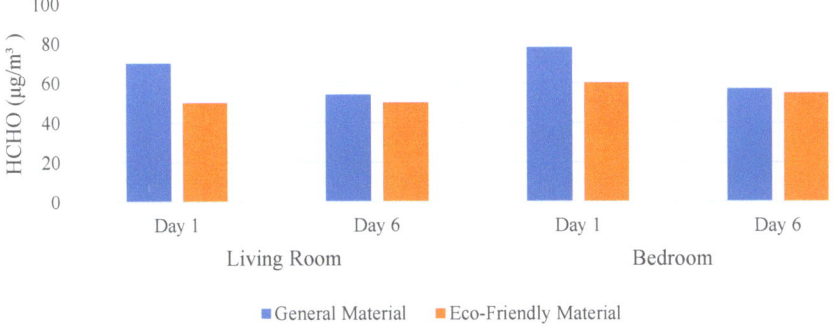

Figure 19. HCHO concentration according to the door/window frame.

4. Discussion

The findings of our experiment yield valuable insights into the emission characteristics of VOCs and HCHO in indoor environments, comparing the effects of general materials and eco-friendly materials. The temporal changes in concentrations within the living rooms of the experimental units serve as evidence of the efficacy of eco-friendly materials in reducing indoor air pollutants.

Our findings indicate that the initial concentration of TVOC in the general material unit was approximately 1.7 times higher than that in the eco-friendly material unit. However, the TVOC concentration in the general material unit gradually decreased over time. It's important to note that this reduction reflects the concentration levels, not necessarily the emission rate, which could also be influenced by factors such as increased ventilation or absorption by materials. In contrast, the eco-friendly material unit experienced a rapid reduction in TVOC concentration, reaching levels below 1000 $\mu g/m^3$ after 14 days of construction. The concentration remained stable even after an extended period. Conversely, the general material unit took significantly longer, approximately 69 days, to reach TVOC concentrations below the threshold. This stark difference underscores the effectiveness of eco-friendly materials in mitigating TVOC emissions, aligning with previous studies [61,63,66] that have reported lower TVOC emissions in buildings utilizing eco-friendly materials.

On the other hand, the emission pattern of formaldehyde (HCHO) differed from that of TVOC. The highest peak of HCHO concentration was observed on the seventh day, reaching 126 $\mu g/m^3$ in the general material unit. Fluctuations in HCHO concentrations were attributed to variations in indoor temperature and humidity conditions. Despite maintaining a controlled indoor temperature of 23 ± 1 °C throughout the experiment, it is noteworthy that the air temperature measured in the central breathing area of the living room was approximately 2 °C lower than the set temperature of 25 °C. This temperature difference may have contributed to the observed fluctuations in HCHO emissions.

The analysis of major individual VOCs, excluding benzene, exhibited a similar declining trend in concentration over time, paralleling the pattern observed for TVOC. Notably, toluene demonstrated a substantial emission concentration, closely following the decreasing trend of TVOC. In contrast, benzene displayed distinct behavior with low initial emission levels and cyclic fluctuations over time. Except for benzene, the initial emission levels were significantly higher in general materials compared to eco-friendly materials. This underscores the effectiveness of eco-friendly materials in mitigating initial VOC emissions, thereby suggesting their superiority in indoor air quality.

From a practical perspective, these findings have significant implications for indoor construction and renovation practices. As supported by this study, eco-friendly materials can drastically reduce indoor air pollution, ensuring a healthier living environment for occupants. This insight is particularly important for urban settings where residents often face air quality issues, both outdoors and indoors. Additionally, the reduced emission levels from eco-friendly materials can lead to decreased health-related expenditures in the long run.

Our study also examined the effects of eco-friendly alternatives in specific material applications. Parquet flooring constructed with eco-friendly materials exhibited significantly lower TVOC emissions than conventional flooring materials. Although there was a slight increase in HCHO emissions over time, the difference compared to general flooring materials was not substantial (Figures 10 and 11). Similarly, eco-friendly wallpaper significantly reduced TVOC and HCHO emissions compared to general wallpaper, particularly on the first day after construction. By the sixth day, emissions levels equalized between the two types of wallpaper, with both showing significant reductions compared to the initial values. This demonstrates the long-term effectiveness of eco-friendly wallpaper in reducing emissions.

Regarding built-in furniture, utilizing eco-friendly materials significantly reduced TVOC concentrations compared to general materials. Although slightly higher HCHO emissions were observed on the first day in the bedroom, after seven days of construction, the emission levels were more than twice as low when eco-friendly materials were employed (Figures 14 and 15). Eco-friendly kitchen furniture, constructed with subsidiary eco-friendly materials, exhibited over two times lower TVOC emissions compared to general subsidiary materials. Applying UV paint as a surface finish was crucial in reducing VOC emissions. The overall emissions of HCHO were also lower in eco-friendly kitchen furniture.

Regarding door and window frames, both TVOC and HCHO emissions were slightly lower when eco-friendly auxiliary materials were employed. However, the difference in emissions was minimal, suggesting the need for further emphasis on applying eco-friendly materials in these components.

Furthermore, these results also guide the selection of materials for construction professionals, architects, and interior designers, promoting eco-friendly choices that offer environmental and health advantages.

The current study provides compelling evidence for the effectiveness of eco-friendly materials in reducing indoor air pollutants, particularly TVOC and HCHO. The temporal analysis of concentrations in various material applications underscores the significant benefits of eco-friendly alternatives. Nevertheless, further research and development are necessary to optimize the application of eco-friendly materials in specific components, such as wooden windows and doors. The findings from this study contribute to the expanding body of knowledge on indoor air quality, offering valuable insights for architects, builders, and homeowners striving to create healthier living environments. Future studies should focus on refining eco-friendly materials and investigating their long-term impact on indoor air quality and occupant health.

Moreover, with the global shift towards sustainability and the increasing awareness of health implications due to indoor pollutants, eco-friendly materials can also present economic opportunities for manufacturers and suppliers. Therefore, this study serves as an impetus for healthier living conditions and a call for industries to invest and innovate in eco-friendly construction materials.

It is essential to acknowledge that this study is preliminary. The findings provide compelling evidence for the effectiveness of eco-friendly materials in reducing indoor air pollutants, particularly TVOC and HCHO. The temporal analysis of concentrations in various material applications underscores the significant benefits of eco-friendly alternatives. However, it is important to note that this study did not consider relative humidity, which is one of the limitations. Nevertheless, further research and development are necessary to optimize the application of eco-friendly materials in specific components, such as wooden windows and doors. The findings from this study contribute to the expanding body of knowledge on indoor air quality, offering valuable insights for architects, builders, and homeowners striving to create healthier living environments. Future studies should focus on refining eco-friendly materials and investigating their long-term impact on indoor air quality and occupant health.

5. Conclusions

In this study, we compared the emission concentrations of formaldehyde (HCHO) and volatile organic compounds (VOCs) over time between eco-friendly and general materials in an experimental house. We also assessed emission characteristics in different areas of the house. The key findings are as follows:

TVOC emissions initially peaked and gradually decreased in houses constructed with eco-friendly materials, with the general material unit taking over 50 days to reach stable pollutant levels below 1000 $\mu g/m^3$. This suggests that using eco-friendly materials can create a more comfortable indoor environment over time. Overall, eco-friendly materials effectively improve indoor air quality.

In contrast, HCHO emissions reached their highest level on the seventh day and then stabilized. Other major VOCs showed declining concentrations over time, with benzene exhibiting cyclic patterns and toluene being the dominant influencer. General materials had higher initial emissions than eco-friendly ones, highlighting the efficacy of eco-friendly materials in reducing initial emissions.

Overall emissions were lower when using eco-friendly materials throughout the house. Eco-friendly flooring, wallpaper, built-in furniture, and kitchen furniture consistently demonstrated reduced emissions, particularly in the initial phase, showcasing the long-term effectiveness of eco-friendly materials.

Eco-friendly subsidiary materials contributed to lower TVOC and HCHO emissions, with kitchen furniture exhibiting significant reductions compared to general materials. However, window and door frame materials showed insignificant differences.

In summary, eco-friendly materials reduce indoor air pollutants, particularly TVOC and initial emissions. Toluene plays a significant role in VOC emissions, and eco-friendly materials outperform general materials in improving indoor air quality.

Author Contributions: All authors contributed significantly to this study. C.J. and N.S.A.M. identified and secured the example buildings used in the study. The data acquisition system and installations of measurements were designed and installed by N.A.Q. and G.E. C.J. and N.S.A.M. were responsible for data collection. Data analysis was performed by G.E. The manuscript was compiled by C.J. and N.A.Q. and reviewed by G.E. and N.A.Q. All authors have read and agreed to the published version of the manuscript.

Funding: This research received no external funding.

Data Availability Statement: New data were created or analyzed in this study. Data will be shared upon request and consideration of the authors.

Acknowledgments: The authors would like to express their gratitude to University of Sharjah and Ajman University for APC support and providing great research environment.

Conflicts of Interest: The authors declare no conflict of interest.

References

1. Xu, J.; Sun, J.; Zhao, J.; Zhang, W.; Zhou, J.; Xu, L.; Guo, H.; Liu, Y.; Zhang, D. Eco-friendly wood plastic composites with biomass-activated carbon-based form-stable phase change material for building energy conversion. *Ind. Crops Prod.* **2023**, *197*, 116573. [CrossRef]
2. Qian, C.; Fan, W.; Yang, G.; Han, L.; Xing, B.; Lv, X. Influence of crumb rubber particle size and SBS structure on properties of CR/SBS composite modified asphalt. *Constr. Build. Mater.* **2020**, *235*, 117517. [CrossRef]
3. Jung, C.; Awad, J. The improvement of indoor air quality in residential buildings in Dubai, UAE. *Buildings* **2021**, *11*, 250. [CrossRef]
4. Surawattanasakul, V.; Sirikul, W.; Sapbamrer, R.; Wangsan, K.; Panumasvivat, J.; Assavanopakun, P.; Muangkaew, S. Respiratory symptoms and skin sick building syndrome among office workers at University Hospital, Chiang Mai, Thailand: Associations with indoor air quality, AIRMED Project. *Int. J. Environ. Res. Public Health* **2022**, *19*, 10850. [CrossRef] [PubMed]
5. Awad, J.; Jung, C. Evaluating the indoor air quality after renovation at the Greens in Dubai, United Arab Emirates. *Buildings* **2021**, *11*, 353. [CrossRef]
6. Alfuraty, A.B. Sustainable Environment in Interior Design: Design by Choosing Sustainable Materials. *IOP Conf. Ser. Mater. Sci. Eng.* **2020**, *881*, 012035. [CrossRef]
7. Samuel, A.K.; Mohanan, V.; Sempey, A.; Garcia, F.Y.; Lagiere, P.; Bruneau, D.; Mahanta, N. A Sustainable Approach for a Climate Responsive House in UAE: Case Study of SDME 2018 BAITYKOOL Project. In Proceedings of the 2019 International Conference on Computational Intelligence and Knowledge Economy, ICCIKE 2019, Dubai, United Arab Emirates, 11–12 December 2019; pp. 816–823. [CrossRef]
8. Taleb, H.M. Using passive cooling strategies to improve thermal performance and reduce energy consumption of residential buildings in U.A.E. buildings. *Front. Archit. Res.* **2014**, *3*, 154–165. [CrossRef]
9. Jiang, C.; Li, D.; Zhang, P.; Li, J.; Wang, J.; Yu, J. Formaldehyde and volatile organic compound (VOC) emissions from particleboard: Identification of odorous compounds and effects of heat treatment. *Build. Environ.* **2017**, *117*, 118–126. [CrossRef]
10. Maghrabie, H.M.; Abdelkareem, M.A.; Al-Alami, A.H.; Ramadan, M.; Mushtaha, E.; Wilberforce, T.; Olabi, A.G. State-of-the-art technologies for building-integrated photovoltaic systems. *Buildings* **2021**, *11*, 383. [CrossRef]
11. Arar, M.; Jung, C.; Qassimi, N.A. Investigating the influence of the building material on the indoor air quality in apartment in Dubai. *Front. Built Environ.* **2022**, *7*, 804216. [CrossRef]
12. Vu, H.T.; Liu, Y.; Tran, D.V. Nationalizing a global phenomenon: A study of how the press in 45 countries and territories portrays climate change. *Glob. Environ. Change* **2019**, *58*, 101942. [CrossRef] [PubMed]
13. Jung, C.; Al Qassimi, N.; Arar, M.; Awad, J. The Analysis of Indoor Air Pollutants From Finishing Material of New Apartments at Business Bay, Dubai. *Front. Built Environ.* **2021**, *7*, 765689. [CrossRef]
14. Al-Sallal, K.A.; Al-Rais, L.; Dalmouk, M.B. Designing a sustainable house in the desert of Abu Dhabi. *Renew. Energy* **2013**, *49*, 80–84. [CrossRef]
15. He, C.; Morawska, L.; Gilbert, D. Particle deposition rates in residential houses. *Atmos. Environ.* **2005**, *39*, 3891–3899. [CrossRef]

16. *ASHRAE Standard 62.1-2019*; Ventilation for Acceptable Indoor Air Quality. American Society of Heating. Refrigerating and Air-Conditioning Engineers: Atlanta, GA, USA, 2005.
17. Kakoulli, C.; Kyriacou, A.; Michaelides, M.P. A review of field measurement studies on thermal comfort, indoor air quality and virus risk. *Atmosphere* **2022**, *13*, 191. [CrossRef]
18. Han, Y.; Yang, M.S.; Lee, S.-M.; Investigation on the Awareness and Preference for Wood Culture to Promote the Values of Wood: III. Living Environment and Trend of Wood Utilization. *J. Korean Wood Sci. Technol.* **2022**, *50*, 375–391. [CrossRef]
19. Burnard, M.D.; Kutnar, A. Wood and human stress in the built indoor environment: A review. *Wood Sci. Technol.* **2015**, *49*, 969–986. [CrossRef]
20. Ikei, H.; Song, C.; Miyazaki, Y. Physiological effects of wood on humans: A review. *J. Wood Sci.* **2017**, *63*, 1–23. [CrossRef]
21. Jung, C.; Mahmoud, N.S.A. Extracting the Critical Points of Formaldehyde (HCHO) Emission Model in Hot Desert Climate. *Air Soil Water Res.* **2022**, *15*, 11786221221105082. [CrossRef]
22. Kelly, F.J.; Fussell, J.C. Improving indoor air quality, health and performance within environments where people live, travel, learn and work. *Atmos. Environ.* **2019**, *200*, 90–109. [CrossRef]
23. Arar, M.; Jung, C. Analyzing the Perception of Indoor Air Quality (IAQ) from a Survey of New Townhouse Residents in Dubai. *Sustainability* **2022**, *14*, 15042. [CrossRef]
24. Mushtaha, E.; Salameh, T.; Kharrufa, S.; Mori, T.; Aldawoud, A.; Hamad, R.; Nemer, T. The impact of passive design strategies on cooling loads of buildings in temperate climate. *Case Stud. Therm. Eng.* **2021**, *28*, 101588. [CrossRef]
25. Mushtaha, E.; Alsyouf, I.; Al Labadi, L.; Hamad, R.; Khatib, N.; Al Mutawa, M. Application of AHP and a mathematical index to estimate livability in tourist districts: The case of Al Qasba in Sharjah. *Front. Archit. Res.* **2020**, *9*, 872–889. [CrossRef]
26. Elnaklah, R.; Walker, I.; Natarajan, S. Moving to a green building: Indoor environment quality, thermal comfort and health. *Build. Environ.* **2021**, *191*, 107592. [CrossRef]
27. Ebrahimi, P.; Khajeheian, D.; Fekete-Farkas, M. A SEM-NCA approach towards social networks marketing: Evaluating consumers' sustainable purchase behavior with the moderating role of eco-friendly attitude. *Int. J. Environ. Res. Public Health* **2021**, *18*, 13276. [CrossRef]
28. Elsaid, A.M.; Ahmed, M.S. Indoor air quality strategies for air-conditioning and ventilation systems with the spread of the global coronavirus (COVID-19) epidemic: Improvements and recommendations. *Environ. Res.* **2021**, *199*, 111314. [CrossRef]
29. Leo Samuel, D.G.; Dharmasastha, K.; Shiva Nagendra, S.M.; Maiya, M.P. Thermal comfort in traditional buildings composed of local and modern construction materials. *Int. J. Sustain. Built Environ.* **2017**, *6*, 463–475. [CrossRef]
30. Hussien, A.; Saleem, A.A.; Mushtaha, E.; Jannat, N.; Al-Shammaa, A.; Ali, S.B.; Assi, S.; Al-Jumeily, D. A statistical analysis of life cycle assessment for buildings and buildings' refurbishment research. *Ain Shams Eng. J.* **2023**, *14*, 102143. [CrossRef]
31. Michael, A.; Demosthenous, D.; Philokyprou, M. Natural ventilation for cooling in mediterranean climate: A case study in vernacular architecture of Cyprus. *Energy Build.* **2017**, *144*, 333–345. [CrossRef]
32. Saljoughinejad, S.; Rashidi Sharifabad, S. Classification of climatic strategies, used in Iranian vernacular residences based on spatial constituent elements. *Build. Environ.* **2015**, *92*, 475–493. [CrossRef]
33. Hayles, C.S. Environmentally sustainable interior design: A snapshot of current supply of and demand for green, sustainable or Fair Trade products for interior design practice. *Int. J. Sustain. Built Environ.* **2015**, *4*, 100–108. [CrossRef]
34. Yang, S.; Wi, S.; Lee, J.; Lee, H.; Kim, S. Biochar-red clay composites for energy efficiency as eco-friendly building materials: Thermal and mechanical performance. *J. Hazard. Mater.* **2019**, *373*, 844–855. [CrossRef] [PubMed]
35. Krishna, B.V.; Reddy, E.R. Applications of green materials for the preparation of eco-friendly bricks and pavers. *Int. J. Eng. Technol.* **2018**, *7*, 75–79. [CrossRef]
36. Khoshnava, S.M.; Rostami, R.; Mohamad Zin, R.; Štreimikienė, D.; Mardani, A.; Ismail, M. The role of green building materials in reducing environmental and human health impacts. *Int. J. Environ. Res. Public Health* **2020**, *17*, 2589. [CrossRef]
37. Wi, S.; Kim, M.G.; Myung, S.W.; Baik, Y.K.; Lee, K.B.; Song, H.S.; Kwak, M.-J.; Kim, S. Evaluation and analysis of volatile organic compounds and formaldehyde emission of building products in accordance with legal standards: A statistical experimental study. *J. Hazard. Mater.* **2020**, *393*, 122381. [CrossRef]
38. Nasr, M.S.; Shubbar, A.A.; Abed, Z.A.A.R.; Ibrahim, M.S. Properties of eco-friendly cement mortar contained recycled materials from different sources. *J. Build. Eng.* **2020**, *31*, 101444. [CrossRef]
39. Liu, Y.; Misztal, P.K.; Xiong, J.; Tian, Y.; Arata, C.; Weber, R.J.; Nazaroff, W.W.; Goldstein, A.H. Characterizing sources and emissions of volatile organic compounds in a northern California residence using space-and time-resolved measurements. *Indoor Air* **2019**, *29*, 630–644. [CrossRef]
40. Hussien, A.; Jannat, N.; Mushtaha, E.; Al-Shammaa, A. A holistic plan of flat roof to green-roof conversion: Towards a sustainable built environment. *Ecol. Eng.* **2023**, *190*, 106925. [CrossRef]
41. Sobha Realty. Sobha Hartland One Park Avenue. 2023. Available online: https://www.sobharealty.com/apartments-for-sale-in-dubai/one-park-avenue/ (accessed on 12 April 2023).
42. Hong, Q.; Liu, C.; Hu, Q.; Zhang, Y.; Xing, C.; Su, W.; Ji, X.; Xiao, S. Evaluating the feasibility of formaldehyde derived from hyperspectral remote sensing as a proxy for volatile organic compounds. *Atmos. Res.* **2021**, *264*, 105777. [CrossRef]
43. Huang, K.; Sun, W.; Feng, G.; Wang, J.; Song, J. Indoor air quality analysis of 8 mechanically ventilated residential buildings in northeast China based on long-term monitoring. *Sustain. Cities Soc.* **2020**, *54*, 101947. [CrossRef]

44. Chen, Q.; Xiao, R.; Lei, X.; Yu, T.; Mo, J. Experimental and modeling investigations on the adsorption behaviors of indoor volatile organic compounds in an in-situ thermally regenerated adsorption-board module. *Build. Environ.* **2021**, *203*, 108065. [CrossRef]
45. Liu, C.Y.; Tseng, C.H.; Wang, H.C.; Dai, C.F.; Shih, Y.H. The study of an ultraviolet radiation technique for removal of the indoor air volatile organic compounds and bioaerosol. *Int. J. Environ. Res. Public Health* **2019**, *16*, 2557. [CrossRef] [PubMed]
46. Bahri, M.; Haghighat, F.; Kazemian, H.; Rohani, S. A comparative study on metal organic frameworks for indoor environment application: Adsorption evaluation. *Chem. Eng. J.* **2017**, *313*, 711–723. [CrossRef]
47. Chang, T.; Wang, J.; Lu, J.; Shen, Z.; Huang, Y.; Sun, J.; Xu, H.; Wang, X.; Ren, D.; Cao, J. Evaluation of indoor air pollution during decorating process and inhalation health risks in Xi'an, China: A case study. *Aerosol Air Qual. Res.* **2019**, *19*, 854–864. [CrossRef]
48. Caron, F.; Guichard, R.; Robert, L.; Verriele, M.; Thevenet, F. Behaviour of individual VOCs in indoor environments: How ventilation affects emission from materials. *Atmos. Environ.* **2020**, *243*, 117713. [CrossRef]
49. van den Broek, J.; Cerrejon, D.K.; Pratsinis, S.E.; Güntner, A.T. Selective formaldehyde detection at ppb in indoor air with a portable sensor. *J. Hazard. Mater.* **2020**, *399*, 123052. [CrossRef]
50. Harashima, H.; Sumiyoshi, E.; Ito, K. Numerical models for seamlessly predicting internal diffusion and re-emission of leaked liquid toluene from indoor mortar materials. *J. Build. Eng.* **2022**, *57*, 104976. [CrossRef]
51. Geldermans, B.; Tenpierik, M.; Luscuere, P. Circular and flexible indoor partitioning—A design conceptualization of innovative materials and value chains. *Buildings* **2019**, *9*, 194. [CrossRef]
52. Hoang, C.P.; Kinney, K.A.; Corsi, R.L. Ozone removal by green building materials. *Build. Environ.* **2009**, *44*, 1627–1633. [CrossRef]
53. Liu, S.; Song, R.; Zhang, T.T. Residential building ventilation in situations with outdoor PM2.5 pollution. *Build. Environ.* **2021**, *202*, 108040. [CrossRef]
54. Alonso, M.J.; Moazami, T.N.; Liu, P.; Jørgensen, R.B.; Mathisen, H.M. Assessing the indoor air quality and their predictor variable in 21 home offices during the COVID-19 pandemic in Norway. *Build. Environ.* **2022**, *225*, 109580. [CrossRef]
55. Marques, A.C.; Mocanu, A.; Tomić, N.Z.; Balos, S.; Stammen, E.; Lundevall, A.; Abrahami, S.T.; Günther, R.; de Kok, J.M.M.; de Freitas, S.T. Review on adhesives and surface treatments for structural applications: Recent developments on sustainability and implementation for metal and composite substrates. *Materials* **2020**, *13*, 5590. [CrossRef] [PubMed]
56. Lee, K.E. Environmental sustainability in the textile industry. In *Sustainability in the Textile Industry*; Springer: Singapore, 2017; pp. 17–55.
57. Blanchet, P.; Pepin, S. Trends in chemical wood surface improvements and modifications: A review of the last five years. *Coatings* **2021**, *11*, 1514. [CrossRef]
58. Rudawska, A.; Szabelski, J.; Miturska-Barańska, I.; Doluk, E. Biological Reinforcement of Epoxies as Structural Adhesives. In *Structural Adhesives: Properties, Characterization and Applications*; Wiley: Hoboken, NJ, USA, 2023; pp. 31–104.
59. Ahmad, J.; Arbili, M.M.; Deifalla, A.F.; Salmi, A.; Maglad, A.M.; Althoey, F. Sustainable concrete with partial substitution of paper pulp ash: A review. *Sci. Eng. Compos. Mater.* **2023**, *30*, 20220193. [CrossRef]
60. Krzyżaniak, Ł.; Kuşkun, T.; Kasal, A.; Smardzewski, J. Analysis of the internal mounting forces and strength of newly designed fastener to joints wood and wood-based panels. *Materials* **2021**, *14*, 7119. [CrossRef]
61. Foong, S.Y.; Liew, R.K.; Lee, C.L.; Tan, W.P.; Peng, W.; Sonne, C.; Tsang, Y.F.; Lam, S.S. Strategic hazard mitigation of waste furniture boards via pyrolysis: Pyrolysis behavior, mechanisms, and value-added products. *J. Hazard. Mater.* **2022**, *421*, 126774. [CrossRef] [PubMed]
62. Raydan, N.D.V.; Leroyer, L.; Charrier, B.; Robles, E. Recent advances on the development of protein-based adhesives for wood composite materials—A review. *Molecules* **2021**, *26*, 7617. [CrossRef]
63. Sharma, B.; Shekhar, S.; Sharma, S.; Jain, P. The paradigm in conversion of plastic waste into value added materials. *Clean. Eng. Technol.* **2021**, *4*, 100254. [CrossRef]
64. Gonzalez-Martin, J.; Kraakman, N.J.R.; Perez, C.; Lebrero, R.; Munoz, R. A state–of–the-art review on indoor air pollution and strategies for indoor air pollution control. *Chemosphere* **2021**, *262*, 128376. [CrossRef]
65. Hernández-Gordillo, A.; Ruiz-Correa, S.; Robledo-Valero, V.; Hernández-Rosales, C.; Arriaga, S. Recent advancements in low-cost portable sensors for urban and indoor air quality monitoring. *Air Qual. Atmos. Health* **2021**, *14*, 1931–1951. [CrossRef]
66. Akinyemi, B.A.; Kolajo, T.E.; Adedolu, O. Blended formaldehyde adhesive bonded particleboards made from groundnut shell and rice husk wastes. *Clean Technol. Environ. Policy* **2022**, *24*, 1653–1662. [CrossRef]
67. Sobha Realty. One Park Avenue at Sobha Hartland–MBR City. 2023. Available online: https://sobha-residences.ae/one-park-avenue.php (accessed on 24 April 2023).
68. Huang, K.; Song, J.; Feng, G.; Chang, Q.; Jiang, B.; Wang, J.; Sun, W.; Li, H.; Wang, J.; Fang, X. Indoor air quality analysis of residential buildings in northeast China based on field measurements and longtime monitoring. *Build. Environ.* **2018**, *144*, 171–183. [CrossRef]
69. Ismaeel, W.S.; Mohamed, A.G. Indoor air quality for sustainable building renovation: A decision-support assessment system using structural equation modelling. *Build. Environ.* **2022**, *214*, 108933. [CrossRef]
70. Alonso, M.J.; Wolf, S.; Jørgensen, R.B.; Madsen, H.; Mathisen, H.M. A methodology for the selection of pollutants for ensuring good indoor air quality using the de-trended cross-correlation function. *Build. Environ.* **2022**, *209*, 108668. [CrossRef]

Disclaimer/Publisher's Note: The statements, opinions and data contained in all publications are solely those of the individual author(s) and contributor(s) and not of MDPI and/or the editor(s). MDPI and/or the editor(s) disclaim responsibility for any injury to people or property resulting from any ideas, methods, instructions or products referred to in the content.

Article

Construction Solutions, Cost and Thermal Behavior of Efficiently Designed Above-Ground Wine-Aging Facilities

María Teresa Gómez-Villarino [1], María del Mar Barbero-Barrera [2], Ignacio Cañas [1], Alba Ramos-Sanz [3], Fátima Baptista [4] and Fernando R. Mazarrón [1,*]

[1] Department of Agroforestry Engineering, Escuela Técnica Superior de Ingeniería Agronómica, Alimentaria y de Biosistemas, Universidad Politécnica de Madrid, 28040 Madrid, Spain; teresa.gomez.villarino@upm.es (M.T.G.-V.); ignacio.canas@upm.es (I.C.)

[2] Department of Construction and Technology in Architecture, Escuela Técnica Superior de Arquitectura, Universidad Politécnica de Madrid, 28040 Madrid, Spain; mar.barbero@upm.es

[3] Consejo Nacional de Investigaciones Científicas y Técnicas (CONICET), San Juan 5400, Argentina; aramossanz@faud.unsj.edu.ar

[4] Departamento de Engenharia Rural, Escola de Ciências e Tecnologia, MED—Mediterranean Institute for Agriculture, Environment and Development & CHANGE—Global Change and Sustainability Institute, Universidade de Évora, 7006-554 Évora, Portugal; fb@uevora.pt

* Correspondence: f.ruiz@upm.es; Tel.: +34-910671001

Abstract: The wine industry requires a considerable amount of energy, with an important fraction corresponding to the cooling and ventilation of above-ground aging warehouses. The large investments made in aging facilities can compromise the viability and competitiveness of wineries if their design is not optimized. The objective of this study was to provide guidance for the efficient design of new above-ground warehouses. To this end, multiple construction solutions (structure, envelopes, levels of integration, etc.) were characterized, and their costs and the resulting interior environments were analyzed. The results offer a comprehensive view of potential construction solutions and benchmark price ranges for viable and profitable designs. With a total cost of 300 EUR/m^2, an average damping of 98% per day can be achieved. Increasing the costs does not imply better effectiveness. A double enclosure with internal insulation—with or without an air chamber—can achieve excellent results. Greater integration as a result of several enclosures being in contact with other rooms and/or the terrain allows for a high effectiveness to be achieved without air conditioning. Perimeter glazing and ventilation holes can reduce the effectiveness of the construction, resulting in greater instability and a lower damping capacity.

Keywords: wine; aging; above-ground warehouse; construction; cost; damping

1. Introduction

The wine industry is one of the industries that is the most affected by climate change [1], but on the other hand, it contributes significantly to global warming [2], and it can be considered an energy-intensive industry, as it produces approximately 0.3% of the annual global greenhouse gas emissions [3].

Some of these emissions are due to energy consumption in wineries, which is mainly associated with the process used in aging rooms to cool and ventilate warehouses [4]. This is because wine requires very strict environmental conditions for its aging and maturation, which must be maintained throughout many months of the year. Indeed, although an optimal interval has not been established, various authors have pointed out that if the temperature rises above 18–20 °C, the quality of the wine decreases [5–7] and evaporation losses occur [8]. It is also accepted that temperatures below 4–5 °C slow down the aging of wine [9]. Frequent temperature changes are also harmful and may compromise the wine's longevity [7,10]. In addition, ventilation must be promoted to avoid the appearance of harmful mold [11,12]. In wineries with natural ventilation systems, critical factors emerge, such as mold growth or wine evapotranspiration, and ventilation has been proven to be poorly designed, as it is either insufficient or excessive [13]. Therefore, the use of climate control equipment is common in aging rooms that do not passively meet the described requirements. HVAC is the most in-demand equipment for reducing temperatures [14].

The challenge faced by many wineries is the high cost of climate control to ensure optimal conditions for wine aging. This issue has been exacerbated in recent years due to the increase in energy prices. For instance, the energy bill of a Spanish winery doubled in just one year from 2021 to 2022 [15]. The prospects of energy price instability and high costs have the potential to impact the competitiveness and viability of numerous wineries.

Under these circumstances, various strategies for reducing the energy bills of wineries have been examined. Thus, for example, in a Greek winery, replacing the air-to-liquid cooling unit with PV panels demonstrated 54.7% energy savings and a payback period of 3.6 years [16]. The integration of a Portuguese winery into an energy community with collective photovoltaic self-consumption in a small city promoted a higher penetration of photovoltaic capacity (up to 23%) and achieved a modest reduction in the overall cost of electricity (up to 8%) [17]. The amount of biogas generated in a wine production plant was sufficient to supply all of the necessary energy used by a Chilean winery [18].

Long before the implementation of renewable energies, a sector of the wine industry embraced nearly zero-energy buildings, which prevented the need to invest in climate control during wine aging. Over the last two decades, numerous studies have analyzed the effectiveness of constructions with high thermal inertia, such as underground, sheltered, and basement designs. In Italy, a partially underground solution produced a reduction in the cooling energy demand of 75% in comparison with an above-ground solution, reaching 100% with an underground solution [19]. Ideal temperatures can be reached inside an underground cellar without air-conditioning expenses in several locations across the world by adjusting the depth according to the ground properties and exterior conditions [20]. Basement constructions have an acceptable capacity for reducing outdoor variations and can be an inexpensive solution for indoor climate control. On the contrary, above-ground constructions without air-conditioning systems or ground thermal mass present lower capacities for reducing outdoor variations, as they have less control over climate conditions [21].

Despite the energy advantages offered by constructions with high thermal inertia, the increased initial investment required, potential issues due to high relative humidity, and other operational aspects (the use of forklifts, presence of pillars, increased difficulty in handling barrels, etc.) continue to make above-ground wineries the more common solution. For unconditioned warehouses in Italy, several variations in architectural elements and retrofit interventions have been studied through simulations with the aim of enhancing buildings' responses and minimizing energy consumption. The results showed that, in general, roof and wall interventions were more effective than orientation and solar shading, and the combination of more strategies allowed for improved results to be achieved [22,23]. Mazarrón et al. [24] characterized the indoor environment of a typical non-climate-controlled winery in Spain and analyzed its thermal response in different locations of the world through simulations, concluding that the hygrothermal conditions could go well beyond acceptable limits, especially during the summer months. High temperatures can lead to the rapid aging of wine, reducing its quality and increasing wine losses. In these types of warehouses, natural lighting can play a key role due to its low thermal inertia and its influence on the variance in temperature and humidity [25].

Faced with the future scenario of climate change, scientific studies aimed at reducing energy requirements in aging rooms have gained significance. A recent study conducted on nearly zero-energy-consumption buildings used for the aging of sherry wine, known as cathedral wineries, demonstrated that climate change could render these above-ground buildings less effective, causing them to require climate control [26].

With regard to the impact of the design on winery costs, the precedents are scarce. Accorsi et al. [27] delved into the design of warehouse buildings aimed at reducing their cycle time, total expenses, and carbon footprint. The results highlighted that the total cost and the carbon footprint functions led to similar warehouse configurations that were distinguished by a compact vertical structure. Ramos-Sanz [28] conducted an analysis of the cost-effectiveness of passive strategies that were applied to the envelope of a winery situated in Argentina. It was concluded that the incorporation of thermal insulation in the walls and adiabatic reinforcement in the roof was effective to the extent that the minimum thickness of the insulation was greater than or equal to 0.12 m, achieving savings of 54% in the total cooling demand. Gómez-Villarino et al. [29] demonstrated that in constructions with high thermal inertia, there were notable differences in cost, damping effectiveness, and the resulting hygrothermal environment depending on the type of building. The correlation between performance and construction costs showed large differences in cost per degree of damping achieved. The average cost was 0.7 ± 0.2 EUR/m^2 for buried warehouses, 1.1 ± 0.5 EUR/m^2 for basement warehouses, and 2.5 ± 0.5 EUR/m^2 in the case of underground constructions. In the case of above-ground warehouses, Beni et al. [19] quantified the cost of one in Italy and compared it with other high-thermal-inertia designs by estimating the thermal behavior through simulations. Aside from reducing the energy demand, the underground building solutions that were analyzed involved significantly higher construction costs, with increases ranging from 12% to 27% in comparison with those of above-ground constructions. This is one of the reasons why cheaper and less efficient buildings have been preferred for wine production in recent decades.

There is no record of previous studies breaking down the costs of multiple construction solutions used in above-ground warehouses where quality wine is produced and correlating them with the actual thermal response. Given that this typology represents the most common design of wine-aging facilities, having a benchmark to aid in the efficient design of new aging rooms and their subsequent management is essential. Hence, this study examined multiple construction solutions (facades, roofs, and floors) used in above-ground warehouses where quality wine is produced and characterized their cost, thermal behavior, and effectiveness. The main objective was to provide guidance for the efficient design of new above-ground warehouses and the improvement of existing ones.

2. Materials and Methods

Multiple factors influence the design of above-ground wineries. In this study, the most significant ones were analyzed, which were the cost, insulation, and thermal inertia of construction solutions used in the envelopes (roof, enclosures, and floor), the structure, the level of integration (exposure to the exterior), and the presence of ventilation and lighting openings.

To achieve the proposed objective, it was necessary to identify a large number of wineries and select several in which quality wine was produced. Subsequently, a characterization of the construction of the aging rooms and general aspects of the wineries was carried out, and the current construction cost was quantified to enable comparison within a consistent framework. A monitoring system was installed in each aging room to characterize the indoor and outdoor environments throughout the year for a subsequent analysis of the internal stability and the effectiveness of mitigating external variations.

2.1. Analyzed Warehouses

For the selection of the analyzed warehouses, many of the wineries present in Spain were identified. After making contact with 3863 wineries via telephone and/or email, 389 wineries agreed to conduct a subsequent survey on their general characteristics. A final selection of 10 above-ground wineries was made by taking the construction solution, the winery's interest in collaboration, and the availability of a construction project on which to base the study into account.

The selected wineries produced high-quality wines. They were scattered among referential producing regions, such as La Ribera del Duero, Rioja, and Madrid (Figure 1). In particular, the monitored warehouses belonged to the Alvarez Alfaro, Castillejo de Robledo,

Emina, Hermanos Pascual Miguel, Legarís, Mauro, Moradas de San Martín, Murua, Resalte de Peñafiel, Ribera, and Valdelosfrailes wineries.

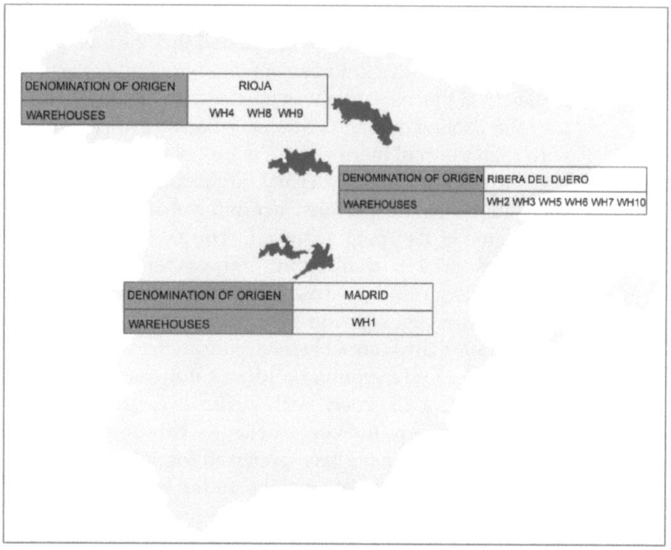

Figure 1. Locations of the analyzed warehouses.

2.2. Characterization of Construction

The characterization of the construction included both constructive aspects (materials, properties, thicknesses, construction systems, etc.) and their costs based on information provided by the wineries (projects, work certifications, etc.), as well as information collected through non-destructive methods during visits.

For each warehouse, a file with general data on the winery and specific data on the aging room was generated (Figure 2). The aging room was understood as a space that was partially or totally separated from the rest of the buildings in the winery in which barrels were permanently housed to age the wine. The file was completed with elevation plans, designs, and structures; together with the construction details, these data allowed for the construction costs of the wine warehouses to be updated.

The detailed information in the files included the following: identification of the winery, location, description, microclimate, photos of the aging room, exterior photos, construction characteristics, facilities, budget, structure, plant, elevation, section, construction details, plan with sensor locations, and observations. The following details of the construction characteristics were included as follows: foundation, floor, structure, pavement, roof, enclosure, carpentry, and construction details.

In order to compare the construction costs of all of the warehouses within the same frame of reference, all of the original budgets for the same year were calculated. Together with the original plans, the construction details allowed us to obtain the budget items and measurements necessary to calculate an updated budget of the construction costs of each aging room (Figure 3). For this purpose, the CYPE 2023 budget program and the Presto 17.03 software, which are reference software for budgets, measurements, and cost control for buildings, were used, as they allowed for the costs of the different work units and the budgets to be systematically updated in a standardized way.

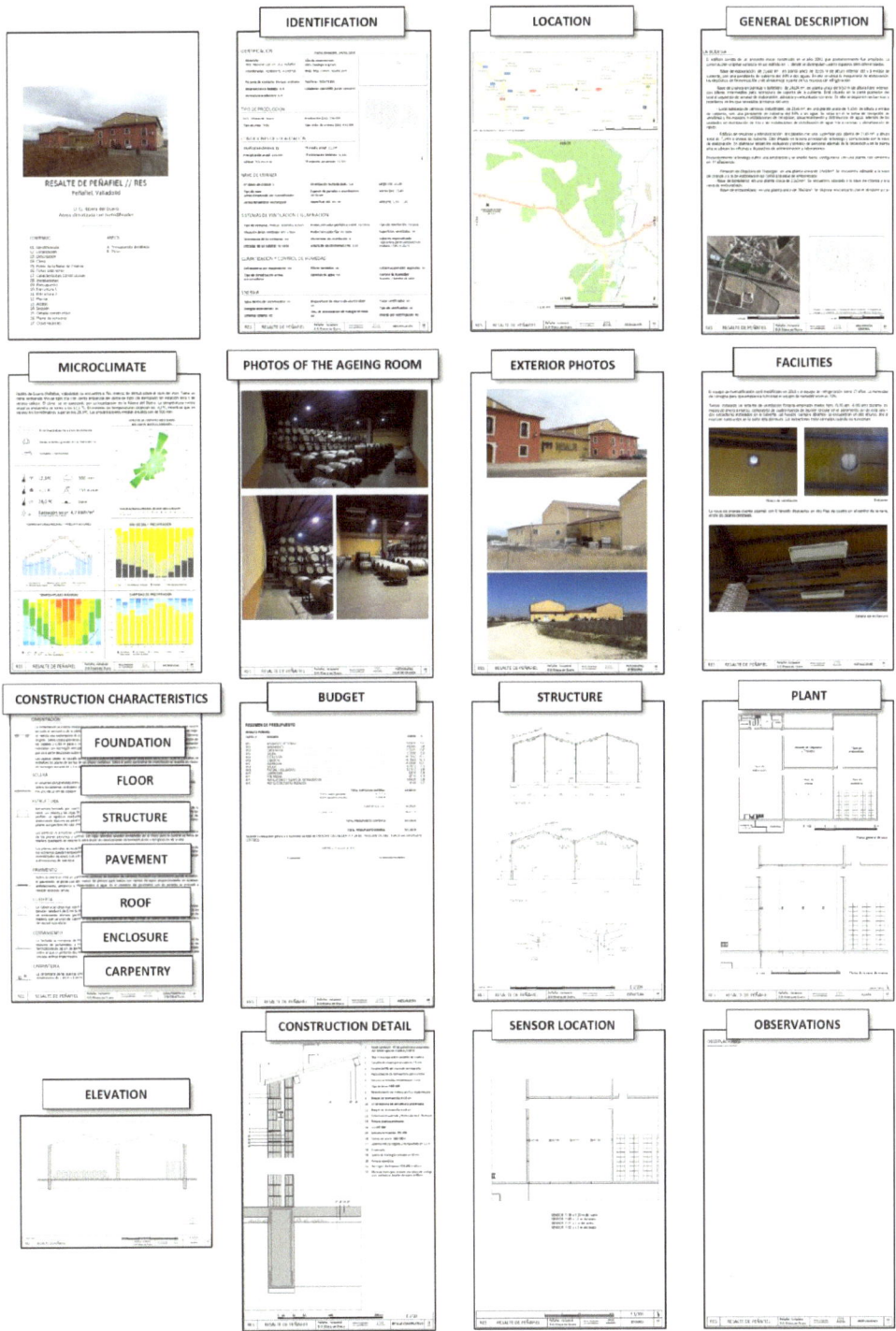

Figure 2. Example of the different pages that made up a construction characterization file.

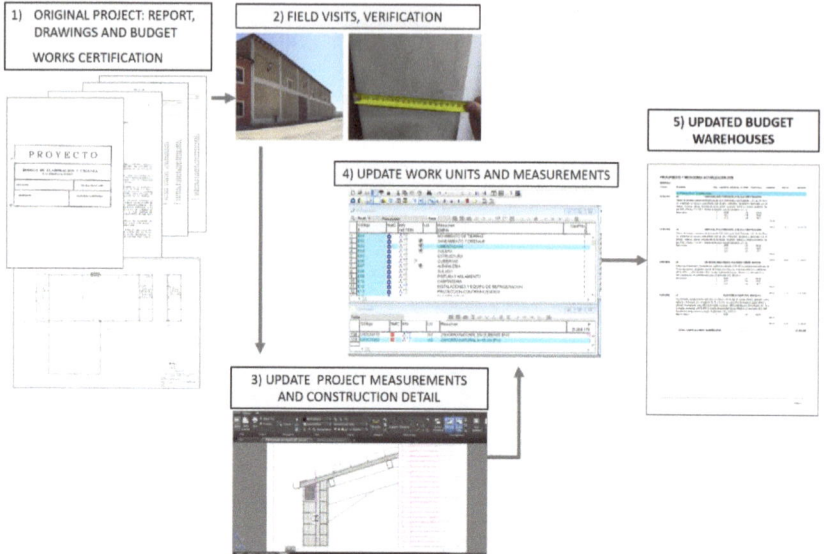

Figure 3. Example of the process of updating the budget of a warehouse.

2.3. Thermal Behavior and Effectiveness

The monitoring was carried out using Hobo Pro v2 (Hobo®) temperature recorders to monitor the outdoor environment (accuracy of ± 0.18 °C at 25 °C and resolution 0.02 °C) and OM-92 (Omega) recorders for indoor temperature monitoring (accuracy of ±0.3 °C from 5 to 60 °C and resolution of 0.01 °C). The long experience of the research team in the monitoring of warehouses (more than a decade) has shown that the interior environment of warehouses is very uniform on the horizontal plane, with marked differences in the vertical plane, which has a strong stratification in the summer months [30]. Therefore, in each warehouse, sensors were installed at various heights at a central point of the barrel room, and the average temperature to which the barrels were subjected was calculated. The monitoring period was 1 year, and a measurement interval of 15 min was used.

The analyzed wineries were spread throughout Spain and, thus, were subject to different external conditions. Therefore, the effectiveness of the warehouses was quantified according to their ability to dampen outside temperature variations, which was equivalent to previous work in warehouses [20,21,24,29]. Specifically, thermal damping was calculated as a percentage of the external variation:

$$\text{Damping}(\%) = \frac{\Delta T \text{ exterior} - \Delta T \text{ interior}}{\Delta T \text{ exterior}} \quad (1)$$

Considering that most above-ground warehouses where quality wines are made have powerful air-conditioning equipment that controls the increase in temperature in the summer months (and, in some cases, at other times of the year), the annual damping is not reliable for the determination of the effectiveness of a construction solution. Therefore, daily damping was used while considering periods without air conditioning.

The thermal stability of the warehouses was also analyzed, and it was calculated as the difference between the average temperatures of two consecutive weeks. This indicator provided an insight into the rate at which the temperature changed indoors.

3. Results and Discussion

Firstly, the construction solutions used in the facades, partitions, roofs, floors, and structural elements of the selected wineries (producers of quality wines) were characterized, and their impacts on the total costs of the aging facilities were determined. The level of integration of the warehouse within the winery and its exposure to the external environment were also assessed. The aim was to provide the sector with benchmark values for designing new wineries and improving existing ones.

Subsequently, a characterization of the resulting indoor environment was conducted by quantifying the damping and stability achieved to provide the optimal conditions required for wine aging. These data can serve as key performance indicators (KPIs) for wineries in this sector to gauge their situation and room for improvement.

Finally, through an analysis of periods without climate control, relevant conclusions were drawn regarding the effectiveness of the construction solutions, the level of integration, and other pertinent parameters. The insights extracted should be taken into consideration in the design of new efficient above-ground wineries.

3.1. Construction Solutions

3.1.1. Envelopes

The envelope in any construction is an intermediary between the outside climate and the environment within the structure. The analysis of the envelopes of the aging rooms made it possible to identify the various solutions used in the analyzed warehouses (Figure 4), and their theoretical construction costs (per m^2 of enclosure surface) were quantified. Aesthetic elements that were not relevant to the model and affected the final cost (stone finishes, special paintings, etc.) were eliminated.

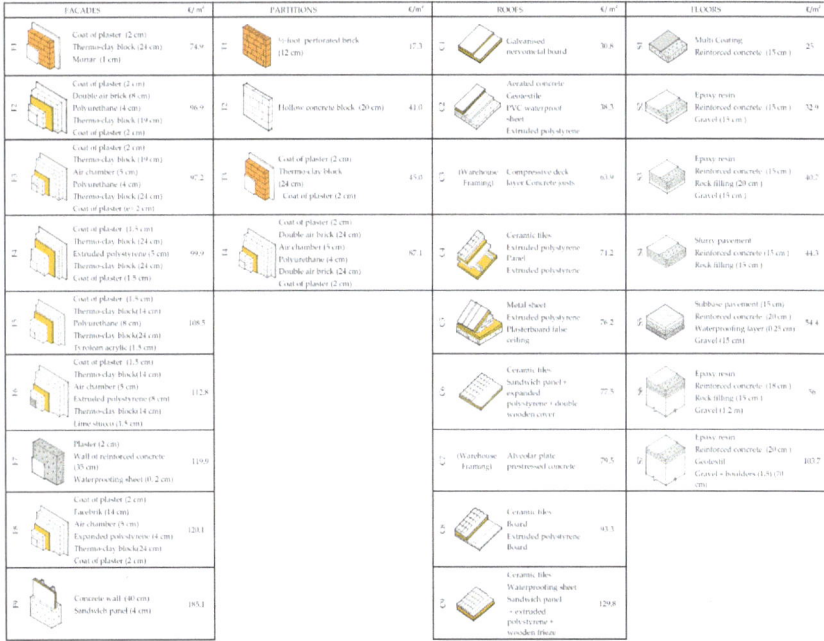

Figure 4. Standard construction solutions that made up the envelopes of the aging rooms.

These construction solutions were grouped according to common features in the facades, floors, and roofs. Thus, five models of facades were identified (Table 1), and their costs varied between 75 EUR/m^2 for the simplest enclosures composed of a rendered prefabricated element (27 cm) and 185 EUR/m^2 for solutions with a greater cost that included concrete walls and sandwich panels with a thickness of 44 cm. However, the most commonly used enclosure, which had a width that ranged between 35 and 50 cm, consisted of two layers (exterior and interior) of brick or thermo-clay and an insulating element inside that could also include a chamber of air, with an average cost of 106 EUR/m².

Table 1. Identified facades grouped according to their common characteristics.

Facade Model	Facades	Description	Thickness cm	Price EUR/m²
I	F_1	Prefabricated and rendered board	27	75
II	F_2 F_4 F_5	Sandwich wall composed of two thermo-clay or brick walls with a thermal insulator inside and exterior cladding	35–56	97–110
III	F_3 F_6 F_8	Sandwich wall composed of two thermo-clay or brick walls with a thermal insulator and air chamber inside and exterior cladding	44–56	97–120
IV	F_7	Reinforced concrete wall	37	120
V	F_9	Concrete wall and sandwich panel	44	185

The floors were arranged directly on the ground without insulation, and they were grouped into six models (Table 2). The minimum thickness was 15 cm in the simplest solution, which consisted of a massive concrete slab plus a basic cement pavement, which had a cost of 25 EUR/m². The maximum thickness reached 90 cm for a concrete floor on a level mounted on compacted stuffing and extended bolting, which had a cost of 104 EUR/m². The most common solution was gravel, reinforced concrete, and pavement, with a cost that varied depending on the pavement used—from basic cement pavement to an epoxy resin.

Table 2. Identified floors grouped according to their common characteristics.

Model Floor	Floors	Description	Thickness cm	Price €/m²
I	S_1	Concrete + pavement	15	25
II	S_2	Gravel + concrete + pavement	30	33
III	S_4	Rock filling + concrete + pavement	30	44
IV	S_5	Gravel + waterproofing layer + concrete + pavement	35	54
V	S_3 S_6	Gravel + filling + concrete + cement—chalky sand/pavement	50	40–56
VI	S_7	Gravel and boulders + geotextile + concrete + pavement	90	104

The roofs had a greater heterogeneity (Table 3). They mostly consisted of a sandwich panel or board that incorporated some insulation and were finished with ceramic tiles. Their price had a wide range, from 31 EUR/m² for a very simple solution of a deck roof with insulation boards to the most expensive and complex solution formed by a wooden frieze, double insulation layer with a sandwich panel, and ceramic tiles with a cost of 130 EUR/m². However, most of the solutions showed an average cost of 80 EUR/m² and consisted of a sandwich panel or a board with insulation and were finished off with tiles.

Table 3. Identified roofs grouped according to their common characteristics.

Roof Model	Roof	Description	Price EUR/m²
I	$C_1 \cdot C_2 \cdot C_4 \cdot C_6 \cdot C_8 \cdot C_9$	Insulated	31–129
II	C_5	With insulation and false ceiling	76
III	$C_3 \cdot C_7$	Warehouse framing	64–80

3.1.2. Structural Elements

In terms of structural elements, there was a huge diversity of solutions depending on the type of soil, the topography of the ground, and the type of structural solution. This was reflected in the dispersion of the costs, which ranged from 61 EUR/m^2 to 306 EUR/m^2, with an average cost of 119 EUR/m^2.

In general, the warehouses were adapted to the terrain, and the earthworks had an average cost of 16 EUR/m^2, except for cases in which important earth movements were carried out, where a maximum of 63 EUR/m^2 was reached. The impact of excavation on the building costs was defined in previous studies [29]. This impact is especially significant when an aging room is built on sloped land to take advantage of the ground to achieve better conditions with fewer influences from exterior fluctuations. In these cases, the structure requires the movement of a considerable amount of soil and the construction of a reinforced concrete retaining wall, which drives up the costs. Indeed, this is reflected in the foundations costs, which greatly vary depending on the type of soil, the location of the warehouse, and the most suitable depth. In this case, the costs varied from 10 EUR/m^2 to 168 EUR/m^2.

3.1.3. Level of Integration of the Aging Room in the Warehouse as a Whole

The range of options related to the integration of the aging room in the rest of the construction is very diverse. On one hand, independent warehouses, with all of their facades and roofs exposed to the outside, can be found; on the other hand, there are warehouses that are integrated into a set with a single exposed facade, with the rest of the facades and roofs bordering other warehouses or the land (Table 4).

Table 4. Level of integration and cost of the warehouses analyzed (E: outdoor environment, T: terrain, N: warehouse).

	Construction Cost EUR/m^2 Warehouse			Code of the Construction Solution							Envelope Contour		
	Total	Envelope	% Total Reduction	Facades		Roof		Floor		Lateral	Upper	Lower
WH1	558	249	1	II	F4+T1	I	C8	II	S2	3E/0T/1N	E	T
WH2	287	177	4	III	F6	II	C5	II	S2	3E/0T/1N	E	T
WH3	455	224	5	V	F9+T3	I	C1	II	S2	3E/0T/1N	E	T
WH4	471	268	2	I	F1	I	C9	V	S6	3E/0T/1N	E	T
WH5	440	194	3	II	F2	I	C8	V	S3	2E/0T/2N	E	T
WH6	363	212	8	II	F5	I	C6	IV	S5	1E/0T/3N	E	T
WH7	270	115	6	II	F2	I	C2	II	S2	1E/0T/3N	E	T
WH8	350	183	10	III	F8+T2	I	C4	I	S1	0.5E/0.5T/3N	E	T
WH9	771	254	6	III	F3	III	C7	VI	S7	1E/1T/2N	N	T
WH10	325	198	8	IV	F7+T4	III	C3	III	S4	1E/2T/1N	N	T

An increase in the level of integration resulted in a decrease in the cost of the aging room, since the cost of common envelopes was divided between the two rooms. Thus, the impact on the cost of the common envelopes between the rooms that shared them implied a reduction in the cost of the aging room by 1% to 10% depending on the shared area, the unit cost of the shared elements, and the percentage represented by the envelope with respect to the total budget of the warehouse (Table 4). Another aspect that reduced the cost of the aging room being integrated was the possibility of using partitions instead of walls, as they had a lower cost of construction (Figure 4).

The structure was the part of the construction system that, together with the facades, had the greatest influence on the total cost (Figure 5). On average, the structure represented 33% of the budget, the facade represented 20%, and the roof represented 18%.

Focusing on the cost associated with the envelope, the facade usually represented the most important cost, making up, on average, 40% (22–70%) of the total cost of the envelope, while the roof represented 34% (13–56%), and the floor represented 26% (14–47%). Therefore, a reduction in the facade surface implied a significant reduction in the cost of constructing the envelope.

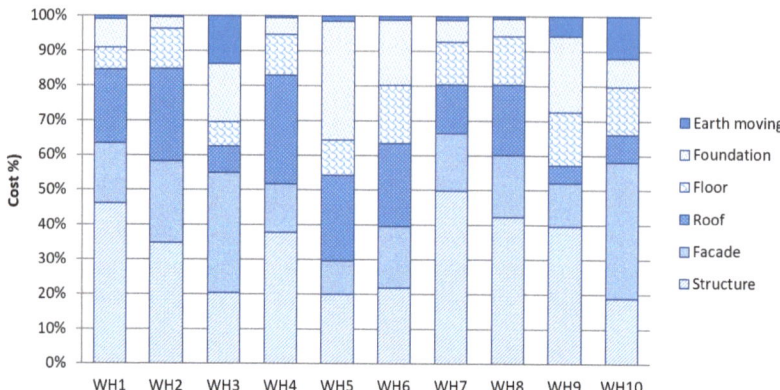

Figure 5. Percentages of costs associated with the main items in the budgets of wine cellars.

The average cost per square meter of the aging room when considering the envelope was 207.4 EUR/m^2 (115–268 EUR/m^2); the average cost for the facades was 120.6 EUR/m^2 (185–75 EUR/m^2), that for roofs was 84.5 EUR/m^2 (31–129 EUR/m^2), and that for the floor was 56.5 EUR/m^2 (25–103 EUR/m^2).

3.2. Thermal Behavior of the Warehouses

The thermal behavior of the warehouses was characterized through an analysis of the evolution of the indoor temperature throughout the year, its relationship with the external environment, the percentage of damping achieved, and the interior stability (the speed at which the temperature changed).

3.2.1. Temperature Inside the Warehouses

Every warehouse, excluding WH10, required climate control systems during the summer months to maintain the interior temperature within the limits established by oenologists. The operational period and the set target temperature varied according to each winery's specific requirements. For instance, in warehouses 2 and 6, the maximum temperature was set to 15 °C, which implied the need for climate control from July to October. On the contrary, warehouse 9 only activated its HVAC for a mere 2 months in an attempt to maintain the interior temperature at 19 °C (Figure 6), as could be observed in August and September.

In spite of that, the indoor temperatures of WH2 and WH6 remained 5 to 10 degrees lower than the outdoor temperature for most of the summertime (weeks 25 to 37). In both cases, the combination of the facades and roofs provided thermal inertia through the incorporation of an insulation with a high thickness. In the remaining warehouses, the most critical period was the summertime, as the temperature difference went over 5 °C during the three hottest summer weeks (Figure 7).

Moreover, certain warehouses, such as WH5 and WH7, also needed climate control during the winter months to maintain the minimum temperatures set by oenologists based on technical criteria or to conduct malolactic fermentation (Figure 6). Therefore, during certain weeks in winter, the indoor temperature exceeded the outdoor temperature by more than 10 °C, whereas in some warehouses, the difference was closer to 5 °C (Figure 7).

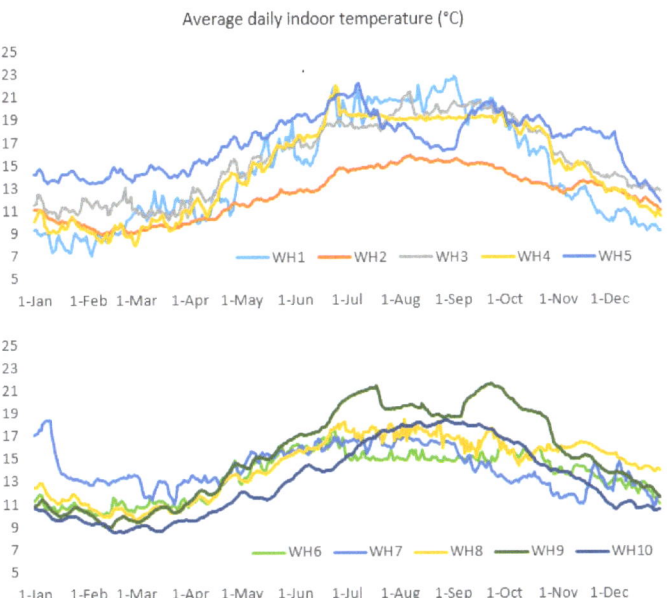

Figure 6. Average daily temperatures inside the monitored wineries.

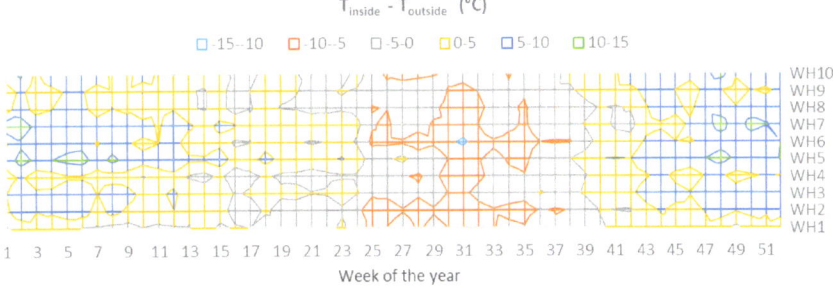

Figure 7. Average weekly temperature difference between inside and outside (°C).

3.2.2. Damping

Most of the wineries were effective in dampening the variations in the temperature of the outdoor environment, which meant that their wine was subjected to stable conditions that allowed proper aging. The annual damping values were 50% (WH1), 64% (WH4), 72% (WH3 and WH9), 76% (WH5), 77% (WH8), 79% (WH10), 80% (WH7), 81% (WH6), and 84% (WH2). The damping values that were obtained (except for that of WH1) were significantly higher than the 53% damping obtained in a previous study conducted in an above-ground warehouse without air conditioning [24]. The excessive ventilation of WH1 resulted in an effectiveness in dampening external variations that was equivalent to that of an unconditioned cellar. The range of values obtained in the remaining above-ground warehouses aligned closely with those of basement warehouses (between 67% and 85%) [21,29], which was attributed to the implementation of climate control. In optimal scenarios, parity was achieved with buried warehouses (83–88%) [21,29] and certain subterranean ones (78–92%) [20,21,29], even when they did not reach the maximum values.

When the annual damping was analyzed, only the case of WH10 led to unequivocal conclusions regarding the effectiveness of the construction solutions, as the values in the rest of the warehouses were greatly influenced by climate control in the summer period. In this case, a huge amount of thermal inertia in all of the construction systems justified the performance, with a clear delay in the daily temperature curve, which was characteristic of these types of buildings. Therefore, it was essential to analyze the daily damping achieved during periods without climate control.

Every warehouse except WH1 showed a mean daily damping that was greater than 90%, reaching 99% in some cases; that is, the difference between the maximum and minimum temperatures recorded each day represented between 1% and 10% of the measurements outside. Specifically, the values were 79 ± 12% (WH1), 98 ± 1% (WH2), 91 ± 7% (WH3), 92 ± 3% (WH4), 98 ± 2% (WH5), 93 ± 4% (WH6), 91 ± 5% (WH7), 96 ± 5% (WH8), 98 ± 3% (WH9), and 99 ± 1% (WH10). In most of the warehouses, the capacity for dampening fluctuations in the outdoor temperature remained relatively constant throughout the year both with and without air conditioning (Figure 8). Only warehouse 1 exhibited variations that were significantly above the average. However, this lack of efficiency was due to the fact that it was frequently opened, which seemed to cause excessive ventilation. Therefore, daily damping stands as a potential indicator of the efficiency of warehouse construction solutions, contingent upon it not being compromised by excessive ventilation.

Figure 8. Daily average damping in each of the analyzed warehouses.

3.2.3. Stability

Similarly to the damping case, most of the analyzed warehouses maintained good levels of thermal stability, providing constant thermal conditions for the wine. The rate at which the mean temperature changed was low, averaging below 1 °C per week for a significant portion of the year (87% of weeks), with 56% of weeks experiencing variations below 0.5 °C (Figure 9).

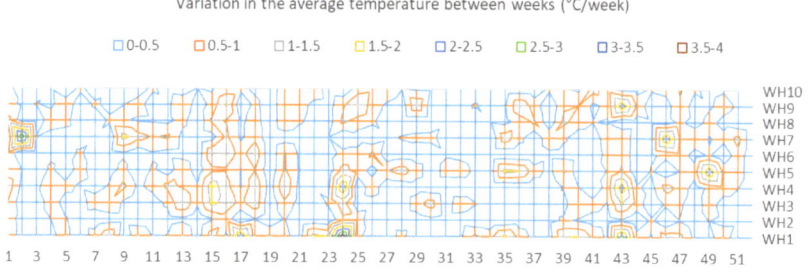

Figure 9. Weekly variations in average temperature when considering absolute values (°C/week).

The highest stability was observed in warehouses WH2 (annual average of 0.3 ± 0.2 °C/week), WH6 (0.4 ± 0.4 °C), WH08 (0.5 ± 0.3 °C), and WH10 (0.4 ± 0.3 °C). In those cases, the combination of the thermal inertia in the construction systems and the appropriate position and thickness of the thermal insulation was the key to this performance. Slightly higher values were observed in warehouses WH3 (0.6 ± 0.5 °C/week), WH4 (0.6 ± 0.6 °C), WH5 (0.6 ± 0.5 °C), WH7 (0.6 ± 0.6 °C), and WH9 (0.6 ± 0.5 °C). In most warehouses, there were no significant differences in stability between periods with and without climate control, yielding similar annual averages when the summer months were excluded. Therefore, stability stands as a potential indicator of the efficiency of warehouse construction solutions.

The least favorable performance was, once again, evident in WH1, where faster fluctuations in temperature (0.8 ± 0.7 °C/week) were caused by excessive ventilation. At certain times of the year, its weekly average temperature could experience variations of almost 4 °C (Figure 9). The flow rates of the climate control equipment and/or exterior ventilation might have compromised the potential stability of the warehouse.

3.3. Effectiveness of the Construction Solutions

The diverse comfort intervals set by oenologists and the consequent activation of climate control systems complicated the quantification of the effectiveness of the adopted construction solutions. Nevertheless, significant insights into the design and subsequent management can be derived by analyzing the data at daily and monthly intervals, with a particular emphasis on periods without climate control.

Within the envelope of a warehouse, the roof is one of the most important elements because of its exposure to the outdoor environment. In this case, the incorporation of thermal inertia is recommended to damp thermal fluctuations and stabilize indoor conditions, and it should be combined with thermal insulation to avoid excessive thermal fluxes of energy. All of the analyzed warehouses that had daily damping values greater than 90% had a slab floor (a massive indoor element) and included a thermal insulating layer with or without an air chamber. The incorporation of a space cushion could reduce the thermal increase and contribute to temperature damping. In addition, it was also clear when the space located in an aging warehouse was a living one due to the limitation of energy transfers towards the outside.

Regarding the facades, the warehouses that had a double layer of thermo-clay or hollow brick with a high thickness and a minimum of 4 cm of insulation (WH8, WH9, WH13, WH14, etc.) achieved high damping values. In these cases, as mentioned above, the combination of thermal inertia and thermal insulation was fundamental for the optimization of the performance by reducing gains and losses of energy while stabilizing the indoor conditions.

Equivalently, the warehouses with the highest levels of integration and with facades that were in contact with the ground (WH8, WH9, WH10) also achieved high levels of damping of oscillations in the external environment, with daily damping values that were greater than 95%, which was in line with previous research on high-thermal-inertia designs [29]. Warehouse WH10 was the only one that did not need air-conditioning equipment to achieve high damping in the summer months (99% daily damping).

In general, the combination of thermal inertia and thermal insulation achieved with thermo-clay walls, thermal insulation, and different specialized layers with insulation in the roof allowed for good levels of damping to be achieved. This was the case of WH4 (92% daily damping), which had the lowest facade cost of those analyzed (75 EUR/m^2), where the influence of the construction of the roof was clear.

There did not seem to be a clear relationship between the cost of the envelope of the aging room and the daily damping that it achieved (R^2 below 0.2 in a linear regression) or between the cost and the stability indicator (R^2 below 0.2 in a linear regression), since both depended on multiple parameters. Increasing the costs does not imply better damping; on the contrary, it was sometimes associated with relatively low damping values. From a total cost of 268 EUR/m^2 and 115 EUR/m^2 for the envelope, daily damping values of 98% could be achieved.

4. Conclusions

A comparison among a great diversity of construction solutions used in above-ground warehouses for the aging of quality wines was performed. It provided benchmark values for the cost of enclosures and structural elements to help in the design and expense control for new wineries.

- The average cost per square meter obtained for the envelope of an aging room was 207.4 EUR/m^2 (115–268 EUR/m^2); the average cost obtained for the facades was 120.6 EUR/m^2 (185–75 EUR/m^2), that of roofs was 84.5 EUR/m^2 (31–129 EUR/m^2), and that of the floor was 56.5 EUR/m^2 (25–103 EUR/m^2).
- The total cost increased unevenly depending on the solutions used in the structure and other construction elements, with the total price with respect to the area (square meters) of the warehouse ranging between 270 EUR/m^2 and 771 EUR/m^2.
- The level of integration of the warehouse in the entirety of the construction was associated with a lower cost of the envelope, with the added advantage that the surface exposed to the outside environment was reduced.

The characterization of the construction was complemented by an analysis of the thermal behavior and effectiveness of the solutions that were employed. This comprehensive approach allowed for the extraction of benchmark values and key factors to consider in the design of new wineries.

- The external temperature buffering achieved in most above-ground wineries was equivalent to that of basement, buried, and some underground wineries. However, this required climate control for 2 to 4 months per year. The lower initial investment was offset by the energy costs needed throughout its lifespan.
- Most of the analyzed constructions exhibited a good capacity for buffering against daily external fluctuations (average buffering between 91 and 99%), remaining relatively consistent throughout the year. Similarly to the damping case, most of the analyzed warehouses maintained good levels of thermal stability (average variations close to 0.5 °C per week), providing suitable aging conditions.
- Perimeter glazing and infiltration through ventilation can reduce the effectiveness of the construction. The above-ground wineries appeared to be highly sensitive to excessive ventilation, making this a crucial factor in both their design and subsequent management.
- A high level of integration achieved by having several enclosures in contact with other rooms (including the roof) and/or the ground minimized the exposure to outdoor conditions and allowed high damping to be achieved without air-conditioning equipment. This type of design can be a good alternative to basement or underground wineries while maintaining the advantages of above-ground facilities.
- A larger investment does not imply the better performance of a warehouse in terms of damping or the stability of the temperature.
- A lack of thermal insulation on the floor allows for the use of the thermal inertia of the ground and the stabilization of the temperature (as an energy sink). However, perimetral insulation can reduce the exchange of the envelope and should be taken into consideration.
- In all cases, a combination of thermal inertia and a suitable insulation thickness (which could be as low as 4 cm) was the most suitable solution for reducing the energy transfer through the envelope while preserving the indoor hygrothermal conditions. The roof had a notable effect on the thermal stability of the buildings. In facades, a double layer (thermo-clay or hollow brick blocks, not midfoot) with internal insulation—with or without an air chamber—achieved good damping results.

Author Contributions: Conceptualization, I.C. and F.R.M.; methodology, F.R.M., M.T.G.-V. and M.d.M.B.-B.; software, M.T.G.-V. and M.d.M.B.-B.; validation, M.T.G.-V. and M.d.M.B.-B.; formal analysis, M.T.G.-V., M.d.M.B.-B., F.B. and A.R.-S.; investigation, M.T.G.-V., M.d.M.B.-B., F.B. and

A.R.-S.; resources, I.C.; writing—original draft preparation, M.T.G.-V., M.d.M.B.-B. and F.R.M.; writing—review and editing, F.B. and A.R.-S.; visualization, F.B. and A.R.-S.; supervision, I.C. and F.R.M.; project administration, I.C. and F.R.M.; funding acquisition, I.C. All authors have read and agreed to the published version of the manuscript.

Funding: This research was carried out as part of the research project BIA2014-54291-R, entitled "Bioclimatic design strategies in wine cellars as nearly zero-energy buildings models", funded by the Ministry of Economy and Competitiveness of Spain.

Data Availability Statement: The raw data supporting the conclusions of this article will be made available by the authors on request.

Acknowledgments: We appreciate the invaluable help of all of the wine cellars that altruistically collaborated in this project. We would like to thank the Ibero-American Postgraduate University Association (AUIP) for financing the contact between researchers from Argentina and Spain.

Conflicts of Interest: The authors declare no conflicts of interest. The funders had no role in the design of the study, in the collection, analyses, or interpretation of data, in the writing of the manuscript, or in the decision to publish the results.

References

1. Hannah, L.; Roehrdanz, P.R.; Ikegami, M.; Shepard, A.V.; Shaw, M.R.; Tabor, G.; Zhi, L.; Marquet, P.A.; Hijmans, R.J. Climate change, wine, and conservation. *Proc. Natl. Acad. Sci. USA* **2013**, *110*, 6907–6912. [CrossRef]
2. Ponstein, H.J.; Meyer-Aurich, A.; Prochnow, A. Greenhouse gas emissions and mitigation options for german wine production. *J. Clean. Prod.* **2019**, *212*, 800–809. [CrossRef]
3. Rugani, B.; Vazquez-Rowe, I.; Benedetto, G.; Benetto, E. A comprehensive review of carbon footprint analysis as an extended environmental indicator in the wine sector. *J. Clean. Prod.* **2013**, *54*, 61–77. [CrossRef]
4. Arredondo-Ruiz, F.; Canas, I.; Mazarron, F.R.; Manjarrez-Dominguez, C.B. Designs for energy-efficient wine cellars (ageing rooms): A review. *Aust. J. Grape Wine Res.* **2020**, *26*, 9–28. [CrossRef]
5. Yravedra, M.J. *Arquitectura y Cultura del Vino. Andalucía, Cataluña, la Rioja y Otras Regiones*; Munilla-Lería: Madrid, Spain, 2003.
6. De Rosa, T. *Tecnología del Vino Tinto*; Mundi-Prensa: Madrid, Spain, 1988.
7. Troost, G. *Tecnología del Vino*; Omega: Barcelona, Spain, 1985.
8. Ruiz de Adana, M.; López, L.M.; Sala, J.M. A fickian model for calculating wine losses from oak casks depending on conditions in ageing facilities. *Appl. Therm. Eng.* **2005**, *25*, 709–718. [CrossRef]
9. Mazarron, F.R.; Lopez-Ocon, E.; Garcimartin, M.A.; Canas, I. Assessment of basement constructions in the winery industry. *Tunn. Undergr. Space Technol.* **2013**, *35*, 200–206. [CrossRef]
10. Plasencia, P.; Villalón, T. *Manual de los Vinos de España*; Editorial Everest: León, Spain, 1999.
11. Ocón, E.; Gutiérrez, A.R.; Garijo, P.; Santamaría, P.; López, R.; Olarte, C.; Sanz, S. Influence of winery age and design on the distribution of airborne molds in three wine cellars. *Am. J. Enol. Vitic.* **2014**, *65*, 479–485. [CrossRef]
12. Garijo, P.; Santamaria, P.; Lopez, R.; Sanz, S.; Olarte, C.; Gutierrez, A.R. The occurrence of fungi, yeasts and bacteria in the air of a spanish winery during vintage. *Int. J. Food Microbiol.* **2008**, *125*, 141–145. [CrossRef] [PubMed]
13. Santolini, E.; Barbaresi, A.; Torreggiani, D.; Tassinari, P. Numerical simulations for the optimisation of ventilation system designed for wine cellars. *J. Agric. Eng.* **2019**, *50*, 180–190. [CrossRef]
14. Malvoni, M.; Congedo, P.M.; Laforgia, D. Analysis of energy consumption: A case study of an italian winery. In Proceedings of the 72nd Conference of the Italian-Thermal-Machines-Engineering-Association (ATI), Lecce, Italy, 6–8 September 2017; pp. 227–233.
15. Garcia, J.L.; Perdigones, A.; Benavente, R.M.; Mazarron, F.R. Influence of the new energy context on the spanish agri-food industry. *Agronomy* **2022**, *12*, 977. [CrossRef]
16. Panaras, G.; Tzimas, P.; Tolis, E.I.; Papadopoulos, G.; Afentoulidis, A.; Souliotis, M. Combined investigation of indoor climate parameters and energy performance of a winery. *Appl. Sci.* **2021**, *11*, 593. [CrossRef]
17. Pontes Luz, G.; Amaro e Silva, R. Modeling energy communities with collective photovoltaic self-consumption: Synergies between a small city and a winery in portugal. *Energies* **2021**, *14*, 323. [CrossRef]
18. Montalvo, S.; Martinez, J.; Castillo, A.; Huilinir, C.; Borja, R.; Garcia, V.; Salazar, R. Sustainable energy for a winery through biogas production and its utilization: A chilean case study. *Sustain. Energy Technol. Assess.* **2020**, *37*, 100640. [CrossRef]
19. Benni, S.; Torreggiani, D.; Barbaresi, A.; Tassinari, P. Thermal performance assessment for energy-efficient design of farm wineries. *Trans. Asabe* **2013**, *56*, 1483–1491.
20. Mazarron, F.R.; Cid-Falceto, J.; Canas, I. An assessment of using ground thermal inertia as passive thermal technique in the wine industry around the world. *Appl. Therm. Eng.* **2012**, *33–34*, 54–61. [CrossRef]
21. Mazarrón, F.R.; Cid-Falceto, J.; Cañas, I. Ground thermal inertia for energy efficient building design: A case study on food industry. *Energies* **2012**, *5*, 227–242. [CrossRef]

22. Barbaresi, A.; Dallacasa, F.; Torreggiani, D.; Tassinari, P. Retrofit interventions in non-conditioned rooms: Calibration of an assessment method on a farm winery. *J. Build. Perform. Simul.* **2017**, *10*, 91–104. [CrossRef]
23. Torreggiani, D.; Barbaresi, A.; Dallacasa, F.; Tassinari, P. Effects of different architectural solutions on the thermal behaviour in an unconditioned rural building. The case of an italian winery. *J. Agric. Eng.* **2018**, *49*, 52–63. [CrossRef]
24. Mazarron, F.R.; Cid-Falceto, J.; Canas-Guerrero, I. Assessment of aboveground winery buildings for the aging and conservation of wine. *Appl. Eng. Agric.* **2012**, *28*, 903–910. [CrossRef]
25. Correia, J.; Mourao, A.; Cavique, M. Energy evaluation at a winery: A case study at a portuguese producer. In Proceedings of the 21st Innovative Manufacturing Engineering and Energy International Conference (IManE and E), Iasi, Romania, 24–27 May 2017.
26. Navia-Osorio, E.G.; Porras-Amores, C.; Mazarron, F.R.; Canas, I. Impact of climate change on sustainable production of sherry wine in nearly-zero energy buildings. *J. Clean. Prod.* **2023**, *382*, 135260. [CrossRef]
27. Accorsi, R.; Bortolini, M.; Gamberi, M.; Manzini, R.; Pilati, F. Multi-objective warehouse building design to optimize the cycle time, total cost, and carbon footprint. *Int. J. Adv. Manuf. Technol.* **2017**, *92*, 839–854. [CrossRef]
28. Ramos-Sanz, A.I. Rentabilidad de las estrategias pasivas de eficiencia energetica para la industria del vino. Análisis termo-energético y económico wine industry. A thermal-energy economic. *Rev. Habitat Sustentable* **2018**, *8*, 90–103. [CrossRef]
29. Gómez-Villarino, M.T.; Barbero-Barrera, H.D.M.I.; Mazarrón, F.R.; Cañas, I. Cost-Effectiveness Evaluation of Nearly Zero-Energy Buildings for the Aging of Red Wine. *Agronomy* **2021**, *11*, 687. [CrossRef]
30. Porras-Amores, C.; Mazarrón, F.R.; Cañas, I. Study of the vertical distribution of air temperature in warehouses. *Energies* **2014**, *7*, 1193–1206. [CrossRef]

Disclaimer/Publisher's Note: The statements, opinions and data contained in all publications are solely those of the individual author(s) and contributor(s) and not of MDPI and/or the editor(s). MDPI and/or the editor(s) disclaim responsibility for any injury to people or property resulting from any ideas, methods, instructions or products referred to in the content.

Article

Enhanced Indoor Air Quality Dashboard Framework and Index for Higher Educational Institutions

Farah Shoukry [1], Sherif Goubran [2,*] and Khaled Tarabieh [2]

[1] Environmental Engineering Program, School of Sciences and Engineering, The American University of Cairo, Cairo 11835, Egypt
[2] Department of Architecture, School of Sciences and Engineering, The American University of Cairo, Cairo 11835, Egypt
* Correspondence: sherifg@aucegypt.edu

Abstract: This research proposes a 10-step methodology for developing an enhanced IAQ dashboard and classroom index (CI) in higher educational facilities located in arid environments. The identified parameters of the enhanced IAQ dashboard–inspired by the pandemic experience, result from the literature review and the outcome of two electronic surveys of (52) respondents, including health professionals and facility management experts. On the other hand, the indicators included in the CI are based on (80) occupant survey responses, including parameters related to IAQ, Indoor Environmental Quality (IEQ), and thermal comfort, amongst other classroom operative considerations. The CI is further tested in four learning spaces at the American University in Cairo, Egypt. The main contribution of this research is to suggest a conceptual visualization of the dashboard and a practical classroom index that integrates a representative number of contextual indicators to recommend optimal IAQ scenarios for a given educational facility. This study concludes by highlighting several key findings: (1) both qualitative and quantitative metrics are necessary to capture indoor air quality-related parameters accurately; (2) tailoring the dashboard as well as the CI to specific contexts enhances its applicability across diverse locations; and finally, (3) the IAQ dashboard and CI offer flexibility for ad-hoc applications.

Keywords: dashboard; monitoring; arid environment; classroom index; facility management; indoor air quality; indoor environmental quality; comfort

Citation: Shoukry, F.; Goubran, S.; Tarabieh, K. Enhanced Indoor Air Quality Dashboard Framework and Index for Higher Educational Institutions. *Buildings* **2024**, *14*, 1640. https://doi.org/10.3390/buildings14061640

Academic Editors: Lina Šeduikytė and Jakub Kolarik

Received: 12 April 2024
Revised: 10 May 2024
Accepted: 22 May 2024
Published: 3 June 2024

Copyright: © 2024 by the authors. Licensee MDPI, Basel, Switzerland. This article is an open access article distributed under the terms and conditions of the Creative Commons Attribution (CC BY) license (https://creativecommons.org/licenses/by/4.0/).

1. Introduction

1.1. Background

Controlling buildings' ventilation systems improves the quality of breathable air and reduces associated health risks. Flu viruses, such as COVID-19—and its variants—spread more effectively in crowded, inadequately ventilated spaces, where people spend long periods in proximity. Viral transmission from one person to another most commonly happens via respiratory droplets or aerosols [1]. Indoor Air Quality (IAQ) has gained wide attention because of the extensive time people spend indoors, and the focus on IAQ research has been gaining significant momentum since the start of the pandemic.

In educational spaces, such as universities, vigilantly monitoring IAQ is essential for various reasons. Students and faculty spend significant time in enclosed learning spaces with a high occupancy rate. Poor IAQ considerably impacts learning productivity and concentration levels, and prolonged exposure time can lead to cognitive health hazards, declining concentration levels, and other respiratory-related health diseases [2–8]. By regularly monitoring the status of IAQ in academic institutions, early identification of contaminants is possible, enhancing the thermal comfort of occupants and understanding energy-related implications on Heating, Ventilation, and Air-Conditioning (HVAC) systems. Also, by having readily available data on IAQ, facility managers can make informed decisions regarding the welfare of students by improving the indoor conditions of learning spaces.

1.2. Research Problematic

The pandemic served as a wake-up call, highlighting the critical importance of managing indoor air quality in university classrooms. The pandemic caused the closure of classrooms worldwide, forcing 1.5 billion students and 63 million educators to suddenly modify their face-to-face academic practices wherever possible [9]. With the reopening of university campuses, precautionary measures were adopted at varying degrees to limit infection rates. The urgency to adapt stringent methods to stop the rapid viral transmission during the daily usage of classes and laboratories was thus a pressing matter. Contingent upon the new norm post-pandemic, the current rate of digital innovation in IAQ monitoring systems is unprecedented [10–18]. Ensuring the health of occupants has become an urgent priority for facility managers. One way to accomplish this is to control the building's ventilation systems and the resultant air quality.

1.3. IAQ Recommendations and Guidelines—Post Pandemic

A large body of literature was published following the pandemic, providing recommendations to building operators and facility managers on improving IAQ levels within enclosed spaces. By comparing the pre- and post-COVID IAQ recommendations within published guidelines, it becomes apparent that policymakers and academics are revising their recommendations to improve IAQ levels best. For example, pre-COVID, IAQ monitoring recommendations were mainly preventative measures to cater to occupants' health and comfort. An important IAQ standard in this regard is the ASHRAE 62.1 standard, which specifies minimum ventilation rates together with the ASHRAE EPIDEMIC TASK FORCE to provide acceptable IAQ to occupants and minimize adverse health effects [19,20]. Post-COVID, IAQ authorities—including the World Health Organization (WHO), the Centers for Disease Control and Prevention (CDC), the American Society of Heating, Refrigerating and Air-Conditioning Engineers (ASHRAE), and the Federation of European Heating, Ventilation and Air Conditioning associations (REHVA)—have all provided guidance documents and addenda to issued standards to combat the pandemic in terms of reducing the viral transmission within indoor spaces to prioritize the health of occupants. Comfort and environmental considerations of energy performance are not among the strategic objectives of such guidelines—refer to Table S1-I in Supplementary Materials.

The academic literature builds on such recommendations where popular research topics include occupants' health and thermal comfort, HVAC systems maintenance and operations, modes, and rates of viral transmission, and the role of technology post the pandemic in improving IAQ. Though the reviewed literature does not necessarily study the status of IAQ in academic institutions, the scientific findings apply to all non-medical institutions. The following subsections briefly summarize what we have learned about the IAQ-instigated recommendations post-pandemic and what they mean for the academic setting.

1.3.1. Occupant Health and Thermal Comfort

Researchers are assessing the effect of high ventilation rate requirements on occupants' health and thermal comfort [2,21–26]. Most published recommendations favor increasing fresh air intake by both allowing natural ventilation in spaces and operating HVAC systems under full capacities–given that naturally ventilated buildings can rarely meet thermal comfort requirements [23,27,28]. In turn, well-ventilated indoor spaces–following IAQ guidelines during the pandemic–undervalue the thermal comfort of occupants. This translates to thermally undermined learning environments due to the opening of windows and an increased percentage of fresh air in HVAC systems: cool indoor environments in the winter season and exceeding the thermal threshold in arid climates–regardless of the weather conditions.

Thermal comfort directly impacts physical health and occupant productivity [22,29]. Following the logic that clean air regulates the human body's metabolism, conversely, high exposure to indoor air pollutants (such as aerosols, particulate matter (PM), and nitrogen compounds) can cause acute and chronic health effects. Thus, measuring and monitoring IAQ in high-performance buildings is imperative to achieve good IAQ and optimize ventilation systems for energy reductions [22,24,30]. Many researchers have studied the effect of IAQ on occupants' performance in an educational setting, and thus, maintaining a good quality IAQ in classrooms is essential for successful learning dynamics.

1.3.2. Mechanical Ventilation

Recommendations published within the academic literature point toward measures to improve HVAC system efficiencies and maintenance schemes and simulate the effects of viral transmission through mechanical ventilation systems [14,22,25,31–33]. Increasing the ventilation rate–synonymous with fresh air intake and using high-efficiency air purifiers–is one of the most prominent recommendations to reduce infection rates in mechanically ventilated spaces. Such recommendations were applied by many academic institutions around the globe, which, in turn, had implications for higher energy consumption rates when campuses returned to full operations.

1.3.3. Viral Transmission

Best practices to reduce viral transmission include both pharmaceutical and non-pharmaceutical measures [25,34–40]. Non-medical measures other than HVAC systems and ventilation considerations include reducing occupant capacity within indoor spaces, limiting pollutants emissions, prohibiting air recirculation–especially in highly contaminated environments, and maintaining the facilities' hygiene and cleanliness requirements.

1.3.4. Role of Technology Post-Covid to Improve IAQ

The role of technology in improving IAQ has seen many applications post the pandemic [41–44]. Examples of technical innovations include sensors to track occupants' movement via carbon dioxide monitoring, and hence improve HVAC efficiencies; decentralized IAQ monitoring systems (e.g., mobile applications) that rely on artificial intelligence (AI) and machine learning algorithms for data collection and processing, as well as HVAC-related innovations to reduce energy consumption and increase operational efficiencies. During the pandemic, many institutions monitored the health of students, faculty, and staff via mobile applications, online surveys, and tracking devices, and the use of technology to monitor health has resulted in an exponential rise in technological solutions [10,45–47]. Many of the solutions available are yet to be re-directed for use after the pandemic, which has opened a dialogue on the issues of data privacy and security. After the pandemic, academic institutions are now more aware of their digital footprint and are taking more proactive measures for information security. Moreover, many universities have built a digital infrastructure to track the health status of occupants; this infrastructure has yet to be re-directed for other uses in the aftermath of the pandemic.

1.4. Selected Commercially Available IAQ Monitoring Devices

Various systems for monitoring different air parameters—particulate matter, volatile organic compounds, and detecting harmful gases—help evaluate the IAQ and provide real-time readings. When applied in an academic setting, this data-driven approach allows universities to promptly identify deviations from healthy air standards and proactively maintain a safe and comfortable environment for students, faculty, and staff. There are many commercially available models by which the IAQ monitoring systems are designed [48–53]—refer to Table S1-II in Supplementary Materials. Though they are not all specific to academic buildings, their functionalities can be applied in the same way.

1.5. IAQ Parameters and Probable Health Risks

Before designing or selecting a monitoring solution, it is crucial to have a clear understanding of the specific parameters that need to be monitored to address the unique needs and potential challenges of the university environment, ensuring that the chosen system provides accurate and relevant data for effective decision-making and proactive air quality management [13,54]. Several IAQ standards identify the measurable parameters of good quality air, including the World Health Organization [55] and the American Society of Heating, Refrigerating and Air-Conditioning Engineers [20]. Such parameters are identified based on whether they exceed a certain threshold, in which case there will be consequential health risks. Parameters include PM, CO_2, NOX, SOX, and Volatile Organic Compounds (VOCs). Refer to Table S1-III in Supplementary Materials: Literature Review for more detailed information on probable health risks.

1.6. Identifying Relevant IAQ Dashboard Parameters within Selected Literature Studies

To identify relevant IAQ dashboard parameters, an extensive examination of pertinent literature focused on monitoring systems within recently published journal articles is summarized in Table S1-IV in Supplementary Materials. One evident trend is using the Internet of Things (IoT) to build the IAQ monitoring system, a cost-effective air quality system for real-time monitoring. These systems can detect harmful gases in indoor environments, such as propane, ethanol, carbon monoxide, and nitrogen dioxide. There are other models available that have been tested in an educational setting, where IoT architecture is used to create a Wi-Fi module containing a multi-gas sensor and microcontroller [56] and can be designed to have low-cost sensors that rely on natural convection to move air to the sensor passively [15]. In the educational setting, sampling usually starts early in the morning and finishes when the pupils leave the classroom each day [18]. These systems generally monitor CO_2, PM2.5, total volatile organic compounds (tVOCs), aldehydes, temperature, and relative humidity [10]. Many commercially available IAQ monitoring solutions are available on the market. Refer to Table S1-II in Supplementary Materials. This shows that various IAQ monitoring systems are available to measure various parameters.

1.7. Air Quality Index

Taking one step backward to give a more universal context for air quality monitoring involves first introducing the most common index: an overall indicator used to measure ambient air quality. The Air Quality Index (AQI) is a numerical index used to report daily air quality. It aims to provide simple information about local air quality, how clean or unhealthy the air is, and what associated health effects might be a concern. Several types of IAQ indexes are available, depending on the specific type of pollutant being measured. Amongst the most well-known is the EPA AQI. The US Environmental Protection Agency (EPA) has a particular AQI. The AQI is calculated for five major air pollutants regulated by the clean air act: ground-level ozone, particle pollutants, carbon monoxide, Sulphur dioxide, and nitrogen dioxide. Refer to Table S1-V in Supplementary Materials for more detailed information on the US-EPA's Air Quality Index.

On the other hand, indoor air quality indices lack the standardization seen in ambient air quality measures. Efforts are frequently tailored to specific building types, occupancy rates, and local pathogens [57–60]. Parameter selection for these indices is often contingent upon pollutant concentrations relevant to the building's activities or specific local pollutants. Moreover, such IAQ indices are not restricted to IAQ parameters alone [59]. A recent example from the literature presents an Indoor Environmental Quality index, which includes thermal comfort, acoustic comfort, and illumination levels [58]. In another study, users' perception was an integral part of the index and did not rely on the measurable indicators alone [61]. In another study, the use of IoT was exercised by leveraging a blend of environmental parameters and an adaptive neuro-fuzzy interference system to improve computing accuracy [62]. These collective efforts highlight the growing recognition of the multifaceted nature of IAQ assessment and the evolving landscape of research aimed at improving indoor environmental conditions.

1.8. Identified Gap, Aims, and Contribution

The notable observation in commercially available IAQ monitoring systems is that their focus is solely on-air quality parameters, whereby they do not take into consideration the health status of occupants nor the quantification of energy consumption rates because of the operating HVAC systems, nor are the spatial parameters of the built environment considered. As a result, a literature gap, as well as a market gap, was identified.

This research aims to provide an expert-validated framework for an augmented IAQ dashboard and classroom index that would allow facility managers to monitor and track IAQ levels and make informed decisions regarding the safety and security of students in higher education spaces, specifically in arid environments. The specific concerns of air environments lie in the high energy consumption due to air conditioning, the thermal (dis)comfort due to the large temperature difference between outdoor and the indoors, and high PM2.5 concentrations due to environmental context. Therefore, the research objectives are to (1) identify, through the literature and expert responses, the IAQ and IEQ as well as thermal comfort-related parameters that are of most importance in higher education spaces in arid environments, (2) develop the framework and conceptual visualization of an enhanced IAQ dashboard and, and (3) enable facility managers within educational facilities to compare between learning spaces–via a generated classroom index–for improved IAQ conditions.

This research contributes indirectly to decreased rates of infection and an improvement in occupants' health through an IAQ dashboard that balances IAQ, energy, comfort, the health of occupants, and spatial parameters. This research introduces an expert-validated framework for an advanced IAQ dashboard and classroom index, with a focus on arid educational environments. By identifying crucial IAQ parameters, conceptualizing the dashboard, and designing the CI, this study offers practical tools for facility managers to improve the overall learning ability of students by optimizing operative variables in a classroom environment.

1.9. Research Structure

This paper reviews the current pertinent literature before presenting the research methodology. The research methodology (Section 2) is divided into two main sections: one dedicated to the enhanced IAQ dashboard (Section 2.1) and the second to the CI (Section 2.2). Section 3 presents the scope and limitations of the research study. Section 4 is focused on presenting the enhanced IAQ dashboard in terms of discussing the results obtained and presenting a summary of the key features of the proposed IAQ dashboard. Section 5 presents the main indicators constituting the classroom index and the lessons learned upon utilizing the CI when comparing learning spaces to one another. Finally, the conclusion summarizes the key findings of the research, presents some suggestions for future research, and makes recommendations for facility managers in higher education institutions.

2. Methodology

The research methodology utilizes qualitative tools and analytical methods. The research methodology consists of 10 sequential steps to arrive at the proposed classroom index. This methodology outlines a systematic approach to data-driven framework development for understanding and improving indoor air quality and related parameters affecting students' learning abilities. The methodological process was designed to ensure the research's reliability and practical relevance of the framework in learning spaces. The ten-step process includes desktop review, survey design and dissemination, survey validation, initial framework development, dashboard visualization, index conceptualization, index development, index visualization, index implementation, and iteration. The Delphi method inspired the two-step survey and expert panel process to offer validation for the findings. Finally, this paper provides a synthesis and consolidation of the results to propose the framework and conceptual visualization of the enhanced IAQ dashboard and classroom index. Figure 1 visualizes this methodological process. Further details on each methodology step are provided in the following subsections.

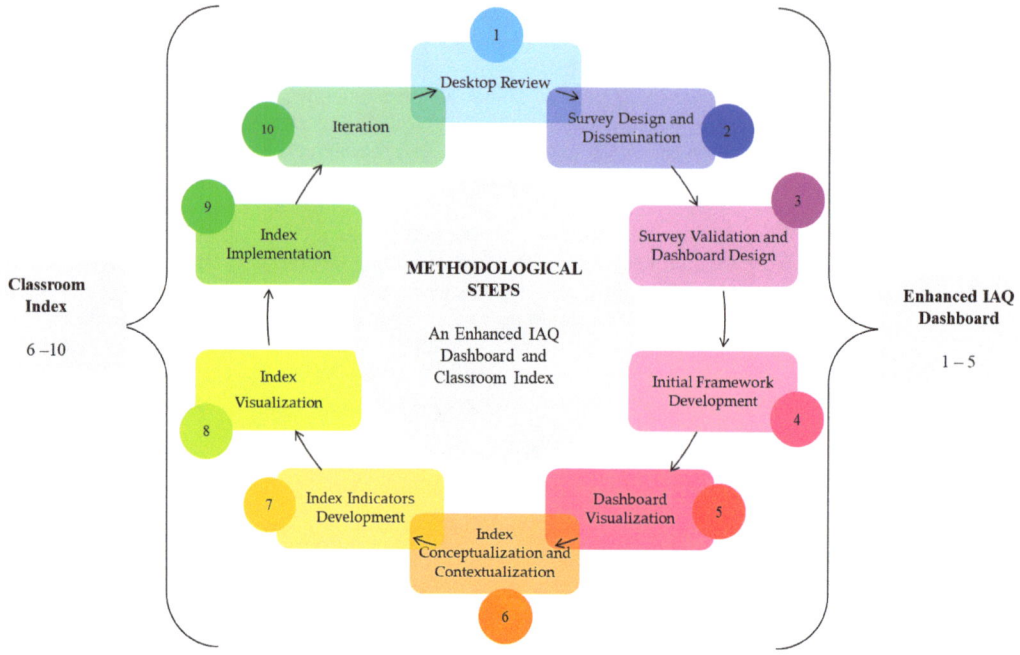

Figure 1. Research Methodology (Self-Produced).

2.1. Designing an Enhanced IAQ Dashboard

2.1.1. Step 01: Literature Review

The literature review contains an analytical review of selected academic journals addressing the issue of IAQ monitoring systems in the aftermath of the pandemic. Moreover, a desktop survey of available market technical innovations in IAQ monitoring systems of software and hardware applications was conducted. The academic and market desktop review aims to scan scholarly articles to identify IAQ parameters relevant to the dashboard design. Refer to the Literature Review Key Findings section in the Results section, which is further supported by S1—Literature Review in Supplementary Materials.

2.1.2. Step 02: Survey Design and Dissemination

After reviewing the literature, identifying the relevant research gaps, and elaborating on the research question, a pilot survey was designed and tested on (19) respondents. The pilot survey tested the survey design and explored the audience's general feedback on their understanding of the survey questions–refer to S2—Pilot Survey in Supplementary Materials.

Afterward, an expert survey was designed based on the pilot survey responses and testing. Based on the insights gained from the pilot survey, iterative refinements were made to the survey–this included adding a ranking of importance questions to the identified IAQ parameters in question, rephrasing selected questions, and removing secondary questions, resulting in a more detailed and robust survey instrument for subsequent data collection. Thirty-three (33) responses were collected from industry, health, and facility management experts. The leading target group is facility managers located in Egypt, whose information was found through the Egyptian Facility Management Association network. Emails were sent to (58) participants, with a low % response rate of 15% (9 responses). The other 24 responses are a result of a targeted survey to facility managers working in Egypt via LinkedIn—a professional social media networking platform—targeting the profiles of environmental consultants, health professionals, academic researchers, and experts within the authors' network–refer to Experts' Survey Results in the Results section which is further supported by S3—Experts Survey—Supplementary Materials. The survey respondents are highly accomplished professionals, including facility managers, architects, software consultants, health professionals, environmental consultants, academics, and top management in corporations of related industries—Figure 2.

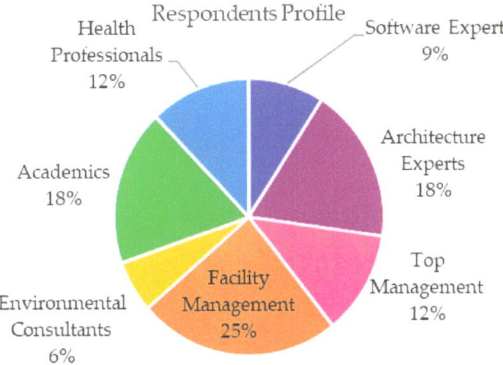

Figure 2. Respondents' Profile (Self-Produced).

In summary, the survey covered the following aspects:
- IAQ Parameters. The respondents gave weight to the importance of the relevance of the IAQ dashboard parameters and the degree to which they are willing to invest in monitoring equipment and data management systems to acquire such information.
- Air-Quality Related Concepts. It also discussed their awareness of the Air Quality Index (AQI) and their knowledge about carbon footprint and energy efficiency related to thermal comfort and HVAC ventilation. They offered new information and knowledge regarding IAQ perceptions and understanding among the respondents.
- Dashboard Design. The last set of questions of the survey was dedicated to gaining a deeper understanding of the relevance of the dashboard to facility management in terms of objectives of use, practical design considerations, and perceived operational risks and their associated mitigation measures.

2.1.3. Step 03: Experts' Panel for Survey Validation and Dashboard Design

Based on these survey findings and the insights collected, the experts' panel worked on developing a theoretical framework and conceptual visual design for the dashboard. The online experts' panel targeted top-tier professionals who responded to the survey. The experts' panel has two activities, labeled part 1 and part 2. Part 1 was to gain the experts' insights (7 participants) on the survey responses for validation. Part 2 was to carry out a brainstorming activity to design a conceptual landing page for the proposed IAQ dashboard while discussing its main features. Figure 3 shows a snapshot of the dashboard brainstorming exercise–using Google Slides. The PowerPoint slide shows the actual parameters that were selected to be included on the landing page (light yellow background). The icons on the margins were options for participants to select from. They are divided into the following categories: charts and metrics (top left), parameters (bottom left), information icons (top right), and action items (bottom right). The instructions for carrying out the brainstorming activity are written at the top of the page (top right).

Excerpts from the experts' panel are shown in the analysis section of this research paper. Refer to S4 in Supplementary Materials, which summarizes the comments and suggestions put forth by the participants, focusing on the IAQ parameters and the proposed features for the dashboard's user experience (UX) and user interface (UI). The panel discussion was conducted via Zoom software, allowing video conferencing, sharing pre-prepared presentations with all participants, and recording the session. The discussion was carried out mainly in Arabic to accommodate the seven participants, and the moderator was an Egyptian national for whom Arabic is a first language—refer to Table 1. Technical jargon remained in English as many of its words were not easily translated into Arabic but were used as loan words from English. The panel was transcribed using Sonix software, which detects multiple languages. Selected quotes were translated—refer to Experts' Panel Results in the Results section, which is further supported by S4 in Supplementary Materials.

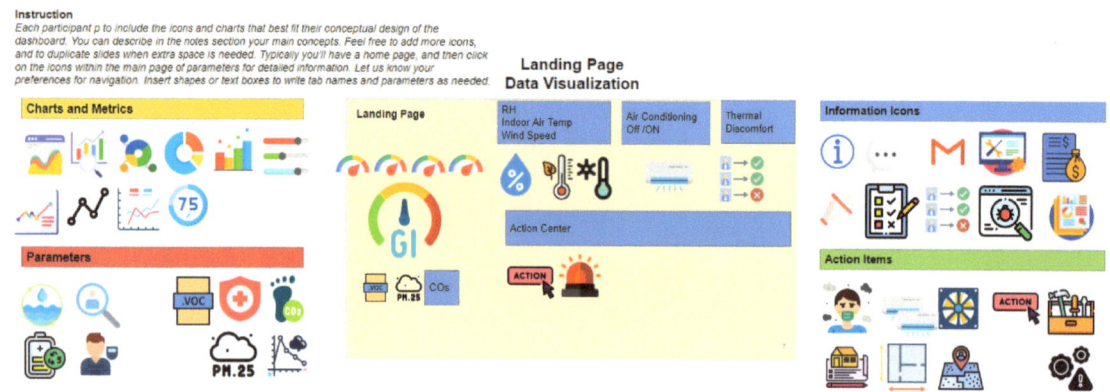

Figure 3. Experts' Panel Brainstorming Exercise (Screenshot of the Google Slides—Self-Produced).

Table 1. Experts Panel Profile.

No.	Title	Field	Yrs. of Exp.
1	Environmental Consultant—Managing Director	Environment	34+
2	Principle Architect—LEED AP—Managing Director	Architecture	25+
3	Professor Emeritus—Uppsala University	Sustainability	45+
4	Project Architect—Specialization Healthcare	Architecture and Healthcare	10+
5	Research Associate—Specialization Architecture	Architecture	10+
6	Mechanical Engineer	Facility Management	10+
7	Academic Advisor	Building Performance	10+

2.1.4. Step 04: Initial Framework Development

The fourth stage consolidates the research findings related to developing the IAQ dashboard framework. The proposed IAQ dashboard framework used the outcomes of the above three methodological steps to produce the initial framework. Refer to S5 in Supplementary Materials for Steps 1 to 4 of the methodology. The initial proposition includes indicators for each of the identified key features and proposed suggestions for the frequency of monitoring of each indicator. This is performed while considering the practicalities of data collection within a university setting, reflecting on the user experience requirements, and identifying the reference for each proposed indicator—refer to Section 4.3.

2.1.5. Step 05: Dashboard Visualization

The fifth step is concerned with the dashboard visualization. The process includes designing and developing a dashboard interface to visualize key metrics and insights derived from these data. In other words, we adopted visualization techniques such as charts, graphs, and UI/UX techniques to effectively communicate information to identified user groups.

2.2. Developing a Classroom Index

2.2.1. Step 06: Index Conceptualization and Contextualization

What we can consider as the core foundation of this work is a simplified Classroom Index (CI), which considers both qualitative (QL) and quantitative (QN) methods in assessing the indoor environmental quality inside learning spaces, i.e., classrooms. The index was conceptualized in terms of the broad categories that it takes into consideration or foresees as a prerequisite for a healthy university classroom. The index was also contextualized, meaning that it is specific to learning spaces, and the identified parameters are focused on the operative conditions. The focus of the selected indicators is on the operative mode of classrooms, meaning that it would be used to assess existing classrooms. The core purpose of this index is to assess the classroom operative performance post-COVID in terms of thermal performance, air quality, and environmental conditions. It is also used to understand the overall health conditions of occupants and the impact of indoor environmental conditions on their ability to learn.

2.2.2. Step 07: Index Indicators Development

The sixth step is defining the indicators and setting appropriate measurable values. The core purpose of the classroom is to assess classroom air quality, occupant thermal comfort, and indoor environmental quality with respect to operational effectiveness. It also relates the classroom conditions to the health of occupants and how this influences their learning ability.

2.2.3. Step 08: Index Visualization

The seventh step was to visualize the index and set the scoring scale. Also, the initial framework was refined and expanded based on feedback from stakeholders and additional insights gained through contextualization.

2.2.4. Step 09: Index Implementation

Sample spaces were selected as a case study to pilot the index and compare between selected spaces. The case study spaces–four in total–are located at the American University in Cairo, Egypt. Data were collected for selected learning spaces, a case study in the American University in Cairo, Cairo, Egypt: two medium-sized classrooms, which are mechanically ventilated, and two design studios, which are mechanically ventilated. Data for the index were obtained and collected via multiple sources.

(1) First, regarding an occupant survey, refer to [63] for detailed information about the survey.
(2) Second, IAQ monitors (Qingping Air Quality Monitor, Light, Model Number: CGS. Each portable IAQ monitor—Was installed in selected learning spaces and connected to the wireless network. The automatic logging of IAQ parameters occurred at a 15-min interval for three consecutive months (February to March 2022)), and logged IAQ-related parameters at 15-min intervals were installed in the selected classrooms for a duration of three months (February 2022–May 2022). The analysis conducted for the outcome of the IAQ monitoring and data logging employed statistical analysis techniques such as descriptive statistics and analysis pertaining to ventilation violation rates to identify patterns and trends within the logged data of Carbon Dioxide, Particulate Matter, Total Volatile Organic Compounds, Temperature, and Relative Humidity. A detailed analysis of the collected IAQ data samples can be referred to in [64].
(3) Third, an equation to estimate energy efficiency was utilized and measured for each selected space. This equation is based on a parametric model simulation of a university

classroom in Cairo, Egypt [65]. Finally, the outcome of these collected data was used to assess classroom quality and effectiveness.

2.2.5. Step 10: Iteration

The final step is an iteration of the overall dashboard design visuals and classroom indicators. This incorporated insights from implementation experiences and stakeholder input to refine the framework's utility and effectiveness.

3. Scope and Limitations

The proposed IAQ framework and CI are not without limitations. Acknowledging this study's constraints, particularly in its scope and generalizability is essential.

(1) The assumptions taken in hand of the proposed framework do not take into account the full potential of the product's commercialization process.
(2) The experts' panel is considering a first round of validation and consolidating the research findings. Further future validation is to take place after producing the first full revised version of the dashboard visualization. This would include both the research input of relevant metrics, indicators, type of infographics, and the user's experience (UX), as well as feedback regarding the user interface (UI)
(3) Moreover, the technical roadmap is left out of the scope of this research as many directions are not bound to the proposed framework metrics and indicators but rather to the technical infrastructure and available fiscal resources.
(4) Backend design is yet another element that is left out of the scope of the dashboard.
(5) The proposed Classroom Inex (CI) does not include score-weighting as it is left to the end-user–i.e., facility managers–to extract the relevant criteria to enable them to monitor IAQ-related parameters efficiently within their institutions.

Despite these limitations, the research represents a significant step forward in IAQ management strategies for higher education in arid climates.

4. What Main Features Did Experts Suggest for the Enhanced IAQ Dashboard?

The following section capitalizes on the validation workshop outcome by interpreting the comments and suggestions put forth by the participants, focusing on the IAQ parameters and the proposed features for the dashboard's user experience (UX) and user interface (UI) to come up with a framework for the enhanced IAQ dashboard. The subsections below discuss the findings and process, concluding with a concise proposal.

4.1. Air Quality Parameters

During the workshop, participants highlighted various IAQ parameters that should be included in the dashboard for effective monitoring and decision-making in a university setting, commenting on the survey findings (33 responses). While some parameters were deemed crucial for immediate monitoring, others required further integration or consideration.

Carbon Dioxide (CO_2) emerged as an important parameter to monitor in real-time, according to the experts' panel and the survey findings. Its importance is directly linked to the willingness to invest in equipment for measuring IAQ parameters. The experts' panel mentioned that high CO_2 levels can indicate poor ventilation, prompting the need for immediate action, such as opening windows or adjusting airflow systems. In line with this, participants suggested integrating sensors that could automate window operations when CO_2 levels exceed certain thresholds. Volatile Organic Compounds (VOCs) were identified as significant pollutants, but participants noted that instantaneous monitoring might be challenging. However, including VOC, specifically TVOCs, monitoring capabilities was recommended for a comprehensive IAQ assessment, albeit with the understanding that real-time measurements may not be feasible. Finally, Particulate Matter (PM2.5) was also recognized as an essential parameter to monitor, particularly in relation to dust accumulation resulting from open windows or construction activities. Participants emphasized the importance of incorporating housekeeping details into the dashboard, ensuring that cleanliness and maintenance practices are considered to maintain optimal IAQ.

4.2. Thermal Comfort Parameters

Thermal comfort emerged as a key consideration, but participants acknowledged the challenges of accurately assessing it using surveys, particularly with students. Standards such as LEED (Leadership in Energy and Environmental Design), in which software known as the ARC system, were mentioned to be used as a reference. The importance of incorporating thermal comfort surveys aligned with the LEED standard was emphasized, urging serious participation and prior notifications to ensure reliable data.

4.3. Carbon, Energy, and Ventilation

The status of air conditioning, whether it is on or off, was suggested as another parameter to include in the dashboard. Additionally, participants highlighted the relevance of including parameters that measure air conditioning performance in terms of energy efficiency. Monitoring the status of ventilation systems and their relation to energy efficiency was the inspiration for linking a more detailed reporting system on Carbon Footprint (CF) accounting. Further, the dashboard customization can link to previously developed CF accounting calculators, where facility managers can consider energy consumption not only for HVAC systems but also for the entire building infrastructure. The conceptual framework emphasizes that each institution will utilize additional indicators and readings from a carbon footprint calculator. This approach facilitates comprehensive data collection and calculations to estimate the carbon footprint, encompassing various factors contributing to emissions within the building's operations and activities.

4.4. Health Indicators

Health indicators such as the number of sick days were among the propositions detected in the survey findings and confirmed in importance during the experts' panel. Further suggestions were to link to the medical reports–for example, issued by the university clinic–to monitor the rate of infections during the onset of highly infectious airborne-related viral transmissions.

4.5. Spatial Considerations

Among the spatial parameters to be considered by the dashboard is optimizing for both occupant density and IAQ parameters. An evident case in university classrooms where prolonged use of spaces affects the quality of air. The other is the wall-to-window ratio, and this is among many other attributes that would be included in the backend calculations to enhance the energy efficiency performance of facilities.

4.6. Index

The proposed IAQ dashboard was envisioned to have an integrated index similar to the AQI of the US-EPA but for indoor spaces by the experts' panel. It considers multiple parameters, providing users with an overall assessment of indoor conditions. Furthermore, participants emphasized the significance of including an action center within the dashboard. The action center would facilitate demand control of ventilation and economizer systems, enabling users to make necessary adjustments or alert facility managers for appropriate actions. Interestingly, among the survey respondents, 40% are familiar with the concept of an Air Quality Index (AQI). However, a higher percentage, 70% of the survey respondents, agree that implementing an AQI could serve as a beneficial visualization tool for the IAQ dashboard indicators. Considering that the survey respondents are experts in the field, this signifies that there is limited awareness of the AQI and its interpretation, indicating the need for educational efforts to improve public understanding of air quality metrics.

Moreover, the proposed framework IAQ dashboard is suggested to include a cutting-edge interactive mapping feature designed to empower end users, i.e., facility managers.

This feature was suggested during the experts' panel as a way to allow navigation between different zones within the building, offering a visual representation of the spatial layout. This feature is suggested to enable real-time data access of installed monitors, and thus, would enable facility managers to keep-track of critical IAQ metrics and indicators specific to each zone.

4.7. Action Centre

Participants recognized the need for the IAQ dashboard to allow for a feedback system through the incorporation of an action center that serves as a centralized hub for notifications and messages. This feature streamlines communication and responsiveness by consolidating all relevant alerts, updates, and messages in one accessible location. The action center not only ensures that users promptly receive crucial notifications but also facilitates a seamless transition from receiving information to taking immediate action.

For example, the experts' panel participants recommended incorporating an alarm feature that triggers when the CO2 threshold reaches a predefined limit; an arbitrary threshold level was given as an example: 2000 ppm to address high carbon dioxide levels. This alarm would notify facility management and provide them with a designated timeframe, as suggested 15 min, to respond appropriately. Otherwise, the alarm would go off.

4.8. Other Design Considerations

The experts' panel was in agreement with the survey participants' feedback on what counts as important for their facility's dashboard design and customization. Linking to existing software systems was the number one motivation for adopting a new dashboard for IAQ monitoring. This integration would enable users to access detailed insights and take informed actions accordingly. They recommended implementing measures such as controlling the sources of pollution, ensuring adequate ventilation, and utilizing supplemental air cleaning and filtration systems. Another suggestion was to periodically sample occupants' annoyance levels to verify the effectiveness of the IAQ system. Integration of openings was also emphasized, along with addressing potential EMI interference. Setting alarms for emission levels and monitoring energy consumption were deemed important. Finally, respondents proposed linking the IAQ tool with an Action Plan Agenda for further enhancing indoor air quality.

The suggestions by the experts' panel complemented the feedback provided by the survey respondents and emphasized the importance of complying with industry standards to facilitate integration with current systems. Defining a standard API and protocols aligned with the existing IoT infrastructure would promote the hardware and software of IAQ devices, enabling cloud integration, online monitoring, alerting, and historical analysis in a separate database. The IAQ dashboard should have a detailed structure and integrate within a predefined industry standard framework, considering cost analysis and pricing structure, which would aid decision-makers in integrating the solution with their existing infrastructure. Collaboration with industry leaders and compliance with IoT and cloud platform standards are key factors in promoting the IAQ dashboard. Other risk mitigation strategies voiced by the survey respondents–including complying with industry standards for interfacing, IoT, and automation, particularly regarding Building Management Systems Integration for large areas or complexes.

On another note, the survey respondents highlighted the relevance of integrating the IAQ dashboard with quality of life (QoL), particularly in educational facilities, to raise awareness and encourage its application in other settings.

While the design of the dashboard framework pilots higher educational institutions (HEIs), which was given as a point of reference during the experts' panel, it is applicable to a wide range of building typologies. HEIs–as in an academic setting–were the starting point for conducting the initial research, as there is a documented need that inspired the embedded functionalities of the dashboard, such as reporting on carbon footprint and impact on performance. Carrying out the framework for other types of facilities would entail customization to suit the specific needs of the facility in question.

Additional suggestions included conducting a cost lifecycle analysis, ensuring user-friendliness, focusing on priority pollutants, examining the relationship between IAQ and organizational profitability, and emphasizing the importance of data analysis for guiding preventive actions against adverse health impacts. These suggestions aim to enhance the usability, effectiveness, and awareness of the IAQ dashboard.

While cost was acknowledged as a concern, the experts proposed a cost-effective solution by incorporating commercially available IAQ monitors within sample spaces in a large educational institution. This attempts to study critical spaces where specific measured parameters are a concern, and later, when the root cause issue is resolved, the sensors are transferred to another space to be monitored via the dashboard.

4.9. Reference, Context, and Scale

When the experts' panel was asked to comment on the survey results, they emphasized several key points. They suggested utilizing existing tools, such as the LEED ARC software as a reference, which focuses on Indoor Environmental Quality (IEQ) and indicators related to building performance, as a reference for developing the dashboard. Understanding the reasons behind respondents' relevance ratings for each parameter was deemed crucial, as it relies on the specific contextual conditions of the facility and the aspects that can be effectively monitored. Although the IAQ dashboard was initially examined in the context of the pandemic, it was advised that its design should also consider the post-pandemic situation, taking into account the transition of facilities from the extreme measures implemented during the peak time of COVID-19.

4.10. Enhanced IAQ Dashboard Framework Summary

Considering the insights of the pilot survey received, the proposed IAQ dashboard attempts to consider AQ, energy, comfort, health and well-being, and spatial parameters. The strategic aim of the dashboard is to enable building operators to monitor their facility, interact with occupants, and take informed decisions for ensuring a safe indoor environment within their facilities.

Moreover, the proposed IAQ dashboard design parameters attempt to include an IAQ index as well as an action center as key design features—see Section 5. The IAQ index is a benchmark of the identified parameters and weights given to the Key Performance Indicators (KPIs). Further, the IAQ dashboard is suggested to include an action center composed of interactive design features (e.g., in-app messaging, notification center) that are in line with the outcome of the decision-making tool managed by the building operator.

The proposed initial framework for an enhanced IAQ dashboard sets an example for many facilities in developing countries to adapt, which can inform both facility managers and occupants on the prevailing IAQ conditions as well as environmental and health indicators. The primary goal is to raise awareness among the population about the importance of indoor air quality and its direct impact on health. By promoting a better understanding of IAQ, the dashboard aims to empower individuals to take proactive measures to improve the air quality in educational spaces. The proposal can be customized and applied to homes, schools, workplaces, and other indoor spaces. Through the proposed methodology, the dashboard aims to become a valuable tool for decision-makers to implement effective policies and regulations to improve the status of air quality in developing countries. The IAQ dashboard must be aligned with the facility's emissions and air quality monitoring program and customized to its specific operations.

Table 2 aims to summarize the key features of the practical preposition of the dashboard for university settings, and it acknowledges the source of the suggestions provided either by survey respondents or experts of the focus group.

Table 2. Proposed Framework for Enhanced IAQ Dashboard.

No.	Key Feature	Frequency of Monitoring	UX/UI Comments	Insights
1.0			**Air Quality Parameters**	
1.1	Carbon Dioxide	Instantaneous	• Each of the selected parameters is to be provided in the form of a gauged meter	All sources
1.2	VOCs	Periodic		Experts' panel
1.3	PM 2.5	Instantaneous		Experts' panel
2.0			**Thermal Comfort Parameters**	
2.1	Temperature	Instantaneous	• Individual Readings • + to reflect the thermal comfort zone chart	Experts' Survey
2.2	Relative Humidity	Instantaneous		Experts' Survey
2.3	Wind Speed	Periodic		Experts' panel
	Thermal Comfort Survey	Periodic	• Connected to an electronic survey to integrate feedback from occupants	Experts' panel
3.0			**Carbon, Energy, and Ventilation**	
3.1	Windows open/closed	Instantaneous	• Individual Reading for selected windows via a motion sensor	Experts' panel
3.2	HVAC on/off	Instantaneous	-	Experts' panel
3.3	Energy Efficiency of HVAC System	Periodic	• Individual Reading for Selected HVAC Units	Experts' panel
3.4	Carbon Footprint Indicator	Instantaneous	• Individual Reading • Based on HVAC operations, energy losses (windows opening during HVAC operations)	Experts' Survey
4.0			**Health Indicators**	
4.1	Number of Sick Days of Occupants	Daily	• Individual Reading • Input to be provided by a clinic/or medical facility connected to the academic institution	Pilot Survey
5.0			**Spatial Considerations**	
5.1	Occupancy Rate	Instantaneous	• Individual Reading/Space to show on the Interactive Map of the Facilities • Input according to Classroom Schedule	Experts' panel
5.2	Wall to Opening Ration	Instantaneous		Experts' panel
6.0			**Index**	

6.1	Enhanced Indoor Air Quality Index	Instantaneous	• A sum of all dashboard parameters at a given point in time.
			 • The Enhanced IAQ Index takes the form of an integrated gauged meter–with three main subthemes:
			  ○ AQ
			  ○ Environmental
			  ○ Health
				All sources

6.2	Interactive Mapping of the Academic Facility	Updated Periodically	• Interactive map showing the location of monitored parameters	Experts' panel

Table 2. *Cont.*

No.	Key Feature	Frequency of Monitoring	UX/UI Comments	Insights
7.0			**Action Center**	
7.1	Notifications	Instantaneous	• In case of threshold increase in studied parameters	Experts' panel

7.2	Messages	Instantaneous	• To include options to remedy problematic parameters–examples of action messages:
			  ○ Turn on/off AC
			  ○ Turn on a heating device
			  ○ Turn off a cooling device
			  ○ Turn on Humidifier
			  ○ Open/Close Windows
				Commercially Available IAQ Monitors–Reference

With such a framework in mind, the conceptual visualization of the enhanced IAQ Dashboard shows a user-friendly digital interface of a given facility, where the user can navigate between the different spaces and can access real-time and historical information about the proposed parameters–including air quality, thermal comfort, carbon/energy/ventilation, health indicators; all through perceiving the spatial layout of the facility. The IAQ Dashboard visually presents these data through intuitive charts, graphs, and color-coded indicators, allowing the user/facility manager to quickly assess the current state of studied parameters—see Figure 4.

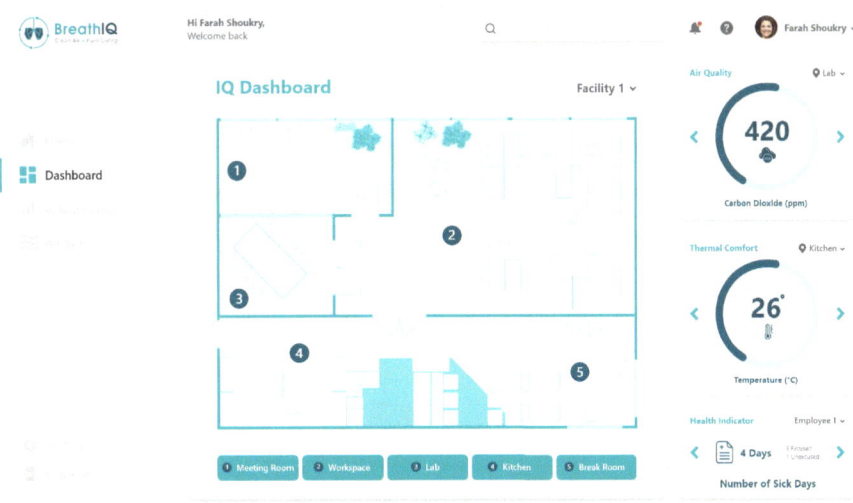

Figure 4. Dashboard Visualization (Self-Produced).

5. What Is the Main Concept behind the Classroom Index? How Does It Work?

This section presents the development process of the classroom index, shedding light on its conceptual framework, indicator selection process, and scoring system. The section ends by piloting the index via a comparative analysis across four distinct learning spaces.

5.1. Integrating the Healthy Design Framework of University Classrooms

While a comprehensive design framework to assess healthy university classroom adequateness is beyond the scope of this research, an operational framework is. More specifically, this scope considers that the evaluated classrooms are existing classrooms and are in 'good' standing in terms of architectural and educative overall conceptual criteria mentioned in the Healthy Design Framework of a university classroom. The healthy design framework refers to the existing classroom conditions, which advocates that the spatial and digital design of the classroom is optimized for occupants' comfort, well-being, and learning. It comprises five main pillars: architecture, education, technology, human aspects, and contextual factors. This framework is based on [66] and is under review.

5.2. Main Indicators Considered

The proposed CI is short for the following nomenclature: Classroom Indoor Environmental Quality Index (CIEQI). The CIEQI considers five main groups of indicators. Figure 5 shows the correlation between the healthy design framework of a university classroom and the CI. The three core indicators are Thermal Comfort Parameters, Indoor Air Quality, and Indoor Environmental Quality—Operational efficiency. The secondary layer is occupants' health. The fifth and final metric is the 'Ability to Learn' or 'Impact on Learning.' The group of indicators is beneficial in comparing a set of classrooms to one another to understand how to make the indoor environment more conducive to occupants' well-being and ability to learn—refer to Table 3. Nine out of the thirteen indicators are based on data acquired through occupant surveys. Three indicators are based on actual measurements (Classroom Presets, Thermal Comfort Parameters of Relative Humidity, Temperature, and Windspeed, and one indicator is based on an equation (Predictive Equation for Annual Mechanical Ventilation Energy Efficiency). Each indicator is defined, assigned to a research question, and normalized to a scale from 1 to 5. The breakdown of the indicators, definition, research question, and scale are included in S6—Supplementary Materials.

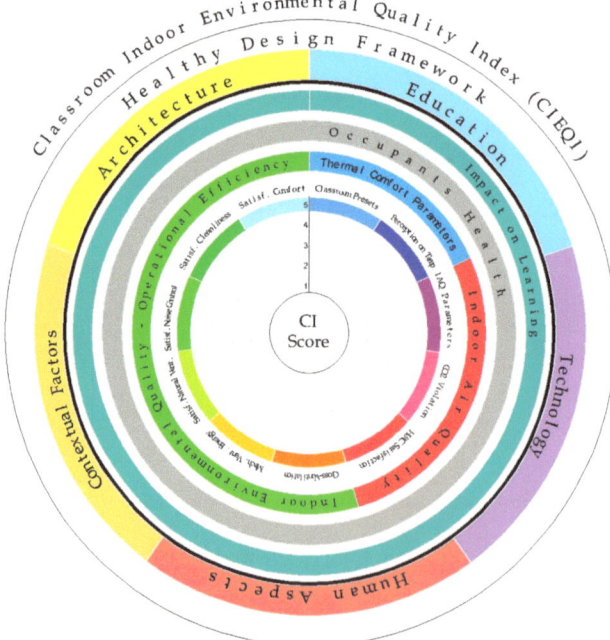

Figure 5. Conceptual Diagram for the Classroom Indoor Environmental Quality Index (CIEQI)—(Self-Produced).

Table 3. Classroom Index—Number of Indicators.

No.	Category	Number of Indicators
1.0	*Thermal Comfort Parameters*	
1.1	Measurement of Classroom Presets	2
1.2	Occupants' Perception of Temperature	
2.0	*Indoor Air Quality*	
2.1	Measuring Indoor Air Quality Parameters	
2.2	Measuring Carbon Dioxide Violation Rates in Selected Classroom	4
2.3	Occupant Satisfaction Regarding Mechanical Ventilation Levels	
2.4	Occupant Satisfaction Regarding Cross-Ventilation	
3.0	*Indoor Environmental Quality—Operational Efficiency*	
3.1	Annual Mechanical Ventilation Energy	
3.2	Occupant Satisfaction Regarding Natural Ventilation	
3.3	Occupant Satisfaction Regarding Noise Control	5
3.4	Occupant Satisfaction Regarding Classroom Cleanliness	
3.5	Occupant Satisfaction Regarding Overall Comfort in the Classroom	
4.0	*Health Indicators*	
4.1	Occupants' Perception of Heat Stress-Related Disorders	1
5.0	*Impact on Learning*	
5.1	Occupant Perception of the Ability to Learn	1
	Total Number of Indicators	13

5.3. Scoring and Assessment

When evaluating classrooms based on the CI, we propose a 5-point score for each indicator, resulting in a maximum of 65-point system. We proposed five evaluation categories, representing the final score for the CI. Depending on the rating, a classroom can, at best, attain an A and, at worst, attain an F score—refer to Table 4.

Table 4. Evaluation Tiers.

Category	Evaluation	Score
A	Classroom Operative Conditions are in good standing	65–52
B	Classroom Operative Conditions are acceptable	51–39
C	Classroom Operative Conditions need minor improvements	38–26
D	Classroom Operative Conditions need major improvements	25–13
F	Classroom Operative Conditions are not acceptable	12–0

5.4. Comparing Learning Spaces Using Classroom Index

We find the following if we conduct a trial run of the Classroom Index to compare selected classrooms. Only one classroom has attained an A score, and three classrooms have received a B score—refer to Figures 6–9 and Table 5. The breakdown of the scoring criteria for the selected learning spaces can be found in Annex F.

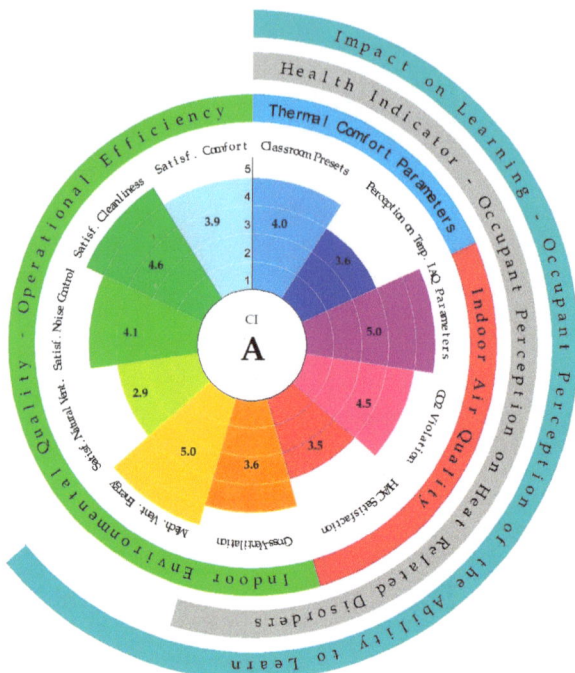

Figure 6. CIEQI—Classroom 01 (Self-Produced).

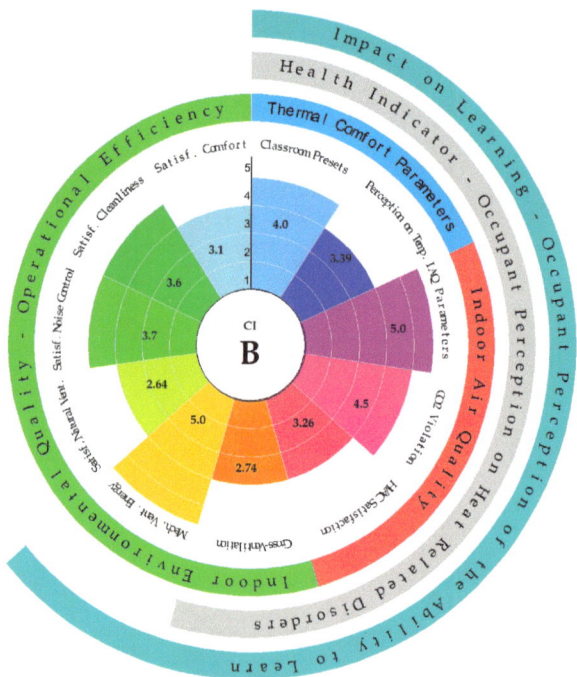

Figure 7. CIEQI—Classroom 02 (Self-Produced).

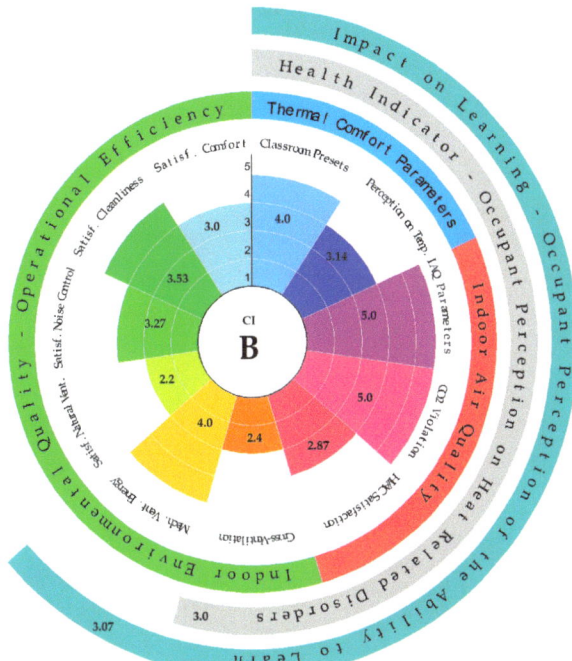

Figure 8. CIEQI—Studio 01(Self-Produced).

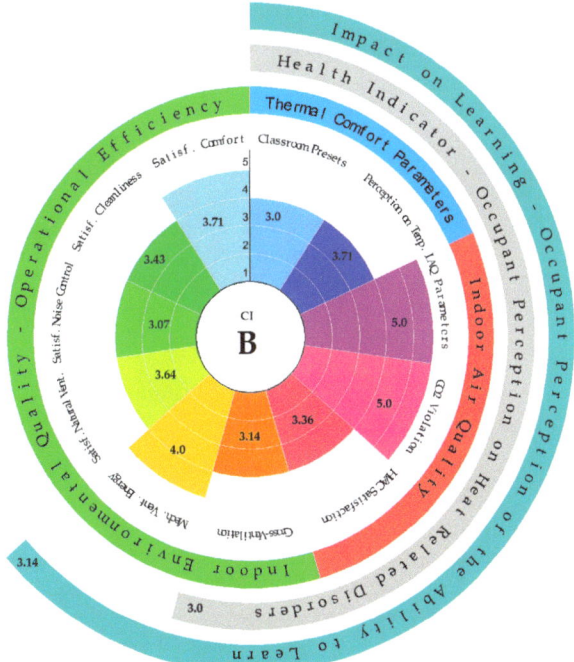

Figure 9. CIEQI—Studio 02 (Self-Produced).

Table 5. Pilot run of CI to compare selected learning spaces based on the 13 indicators.

No.	Category	Max Score	Class 01	Class 02	Studio 01	Studio 02
1.0	Thermal Comfort Parameters	10	7	7.39	7.14	6.71
1.1	Measurement of Classroom Presets	5	4	4	4	3
1.2	Occupants' Perception of Temperature	5	3.62	3.39	3.14	3.71
2.0	Indoor Air Quality	20	16.66	15.5	15.27	16.5
2.1	Measuring Indoor Air Quality Parameters	5	5	5	5	5
2.2	Measuring Carbon Dioxide Violation Rates in Selected Classroom	5	4.5	4.5	5	5
2.3	Occupant Satisfaction Regarding Mechanical Ventilation Levels	5	3.54	3.26	2.87	3.36
2.4	Occupant Satisfaction Regarding Cross-Ventilation	5	3.62	2.74	2.4	3.14
3.0	Indoor Environmental Quality—Operational Efficiency	25	20.5	18.05	16	16.85
3.1	Annual Mechanical Ventilation Energy	5	5	5	4	4
3.2	Occupant Satisfaction Regarding Natural Ventilation	5	2.9	2.65	2.2	2.64
3.3	Occupant Satisfaction Regarding Noise Control	5	4.1	3.7	3.27	3.07
3.4	Occupant Satisfaction Regarding Classroom Cleanliness	5	4.6	3.6	3.53	3.43
3.5	Occupant Satisfaction Regarding Overall Comfort in the Classroom	5	3.9	3.1	3	3.71
4.0	Health Indicators	5	3	3	3	3
4.1	Occupants' Perception of Heat Stress-Related Disorders	5	3	3	3	3
5.0	Impact on Learning	5	3.38	2.87	3.07	3.14
5.1	Occupant Perception of the Ability to Learn	5	3.38	2.87	3.07	3.14
	Total Number of Evaluated Indicators			13		
	Total Score	65	50.54	46.81	44.48	46.2
	CI Score		A	B	B	B

Several key findings emerge when comparing the four learning spaces using the CI. First, all spaces scored relatively high in terms of thermal comfort parameters, with Studio 01 achieving the highest score of 7.39. However, when examining indoor air quality (IAQ), Class 01 and Class 02 outperformed the design studios, indicating potential areas for improvement in IAQ management within the two studio environments. Class 01 and Class 02 exhibited higher satisfaction levels regarding mechanical ventilation compared to the design studios. Regarding operational efficiency and indoor environmental quality, Class 01 demonstrated the highest score, highlighting its effectiveness in providing a conducive learning environment.

Conversely, Studio 01 and Studio 02 received lower scores in these categories, suggesting that the volume of the space was a primary deciding factor in the consumption rate of mechanical ventilation energy. Regarding health indicators and impact on learning, all spaces showed similar scores, indicating comparable levels of occupant perception. Facility managers can leverage these findings to prioritize improvement efforts, focusing on areas with lower scores to enhance overall IAQ and optimize learning environments.

6. Conclusions

The enhanced IAQ dashboard framework aims to provide a tool for facility managers to both monitor and control the air quality, inform on energy performance, measure thermal comfort, and detect potential health hazards among occupants. Developing an IAQ dashboard customized to arid environments presents a critical opportunity to tackle the issues of IAQ, IEQ, and the thermal comfort of occupants. By raising awareness and providing real-time data on IAQ parameters, the proposed IAQ dashboard can empower individuals as end-users and facility managers as decision-makers to take proactive action to improve air quality. The comparative analysis of four learning spaces using the CI has yielded interesting insights, pinpointing areas for improvement and guiding targeted interventions. Facility managers are left to problem-solve for energy efficiency measures of HVAC systems, remedy IAQ parameters that surpass acceptable threshold levels, and read between the lines to understand occupants' concerns regarding thermal comfort indicators. The proposed methodology has provided a means to tailor the dashboard to specific contexts, applying it to diverse locations. This would enable future cross-sectional studies to be carried out for comparison purposes and validation of impacts. What we can deduce as a conclusion about the classroom index is that quantitative metrics are not enough to validate the adequacy of the space in terms of its indoor air quality conditions. User inputs are essential to capture a realistic snapshot of perception, satisfaction levels, and expectations of the CIEQI levels. Lessons learned from the pandemic emphasize the importance of simple, clear communication between end users and ensuring widespread adoption.

The proposed IAQ dashboard and the CI provide ample opportunities for ad-hoc applications that can further provide invaluable insights to building operators and AQ practitioners, such as:

- Informing on the inter-relatedness between occupant health and environmental qualities of the built environment.
- Providing dynamic IAQ monitoring data to optimize building performance, which can be further processed via machine learning technologies.
- Predicting the carbon footprint of a given indoor environment.
- Availing technical and financial benchmark performance indicators.
- Generating audit reports on IAQ and energy efficiency performance.

Supplementary Materials: The following supporting information can be downloaded at: https://www.mdpi.com/article/10.3390/buildings14061640/s1, The supplementary material is organized into six sections: S1 expands on the literature review with additional references and analyses; S2 includes the complete pilot survey questions and detailed results; S3 presents the experts' survey questions and results; S4 documents the Indoor Air Quality Workshop where an Experts' Panel validated survey results and discussed dashboard features; S5 covers steps 1 to 4 of the dashboard design process; and S6 details the development of the Classroom Index, including its conceptual framework, calculation methods, and data curation log.

Author Contributions: Conceptualization, F.S., S.G. and K.T.; Methodology, F.S. and S.G.; Software, F.S.; Formal analysis, F.S.; Investigation, F.S.; Resources, F.S. and S.G.; Data curation, F.S.; Writing—original draft, F.S.; Writing—review & editing, F.S. and S.G.; Supervision, S.G. and K.T.; Project administration, S.G. and K.T.; Funding acquisition, S.G. and K.T. All authors have read and agreed to the published version of the manuscript.

Funding: This research was funded by the American University in Cairo, PhD Graduate Research Grant (R45). Supplementary funding from the Office of the Associate Provost for Research, Innovation, and Creativity at the American University in Cairo funded the APC.

Institutional Review Board Statement: This study was approved by the Institutional Review Board (IRB) of the American University in Cairo (Case# 2022-2023-010).

Informed Consent Statement: All subjects gave their informed consent—written or verbal—for inclusion before they participated in the study.

Data Availability Statement: Data supporting this study is available from the first author pending justifiable reasoning for use.

Acknowledgments: The authors would like to acknowledge Noha Osama for her research assistance and support on this project. We want to thank the Egyptian Facility Management Association for their support and for providing feedback on the disseminated survey. Farah Shokry would like to thank the Dean of Graduate Studies for the funding received to support this work and partially the APC. Sherif Goubran would also like to acknowledge the funding from the American University in Cairo, which partially covered the APC.

Conflicts of Interest: The authors declare no conflicts of interest.

Abbreviations

AQI	Air Quality Index
ASHRAE	The American Society of Heating, Refrigerating and Air-Conditioning Engineers
CDC	Center of Disease and Control
CI	Classroom Index
CIEQI	Classroom Indoor Environmental Quality Index
EPA	Environmental Protection Agency
HEI	Higher Educational Institution
HVAC	Heating, Ventilation, and Air Conditioning
IAQ	Indoor Air Quality
LEED	Leadership in Energy and Environmental Design
REVAH	Federation of European Heating, Ventilation, and Air Conditioning
UI	User Interface
UX	User Experience
WHO	World Health Organization

References

1. WHO. Technical Guidance. 2021. Available online: https://www.who.int/emergencies/diseases/novel-coronavirus-2019/technical-guidance (accessed on 23 September 2021).
2. Abdul-Wahab, S.A.; Chin Fah En, S.; Elkamel, A.; Ahmadi, L.; Yetilmezsoy, K. A review of standards and guidelines set by international bodies for the parameters of indoor air quality. *Atmos. Pollut. Res.* **2015**, *6*, 751–767. [CrossRef]
3. Gola, M.; Settimo, G.; Capolongo, S. No Impacts on Users' Health: How Indoor Air Quality Assessments Can Promote Health and Prevent Disease. In *Integrating IoT AI Indoor Air Quality Assessment*; Saini, J., Dutta, M., Marques, G., Halgamuge, M.N., Eds.; Springer International Publishing: Cham, Switzerland, 2022; pp. 43–54. [CrossRef]
4. Kallawicha, K.; Chao, H.J. Bioaerosols: An Unavoidable Indoor Air Pollutant That Deteriorates Indoor Air Quality. In *Integrating IoT AI Indoor Air Quality Assessment*; Saini, J., Dutta, M., Marques, G., Halgamuge, M.N., Eds.; Springer International Publishing: Cham, Switzerland, 2022; pp. 27–41. [CrossRef]
5. Khan, T.; Lawrence, A.J. Health Risk Assessment Associated with Air Pollution Through Technological Interventions: A Futuristic Approach. In *Integrating IoT AI Indoor Air Quality Assessment*; Saini, J., Dutta, M., Marques, G., Halgamuge, M.N., Eds.; Springer International Publishing: Cham, Switzerland, 2022; pp. 149–167. [CrossRef]
6. Massawe, E.; Vasut, L. Promoting Healthy School Environments: A Step-by-Step Framework to Improve Indoor Air Quality in Tangipahoa Parish, Louisiana. *J. Environ. Health* **2013**, *76*, 22–31. [PubMed]
7. Wardhani, D.K.; Susan, S. Adaptation of Indoor Health and Comfort Criteria to Mitigate Covid-19 Transmission in the Workplace. *Humaniora* **2021**, *12*, 29–38. [CrossRef]
8. Zhang, S.; Mumovic, D.; Stamp, S.; Curran, K.; Cooper, E. What do we know about indoor air quality of nurseries? A review of the literature. *Build. Serv. Eng. Res. Technol.* **2021**, *42*, 603–632. [CrossRef]
9. Valverde-Berrocoso, J.; Garrido-Arroyo, M.d.C.; Burgos-Videla, C.; Morales-Cevallos, M.B. Trends in Educational Research about e-Learning: A Systematic Literature Review (2009–2018). *Sustainability* **2020**, *12*, 5153. [CrossRef]
10. Adzic, F.; Mustafa, M.; Wild, O.; Cheung, H.Y.W.; Malki-Epshtein, L. Post-Occupancy study of indoor air quality in university laboratories during the pandemic. In Proceedings of the 17th International Conference of the International Society of Indoor Air Quality and Climate, Kuopio, Finland, 12–16 June 2022.
11. Balta, D.; Yalçın, N.; Balta, M.; Özmen, A. Online Monitoring of Indoor Air Quality and Thermal Comfort Using a Distributed Sensor-Based Fuzzy Decision Tree Model. In *Integrating IoT AI Indoor Air Quality Assessment*; Saini, J., Dutta, M., Marques, G., Halgamuge, M.N., Eds.; Springer International Publishing: Cham, Switzerland, 2022; pp. 111–134. [CrossRef]
12. Hafizi, N.; Vural, S.M. An IoT-Based Framework of Indoor Air Quality Monitoring for Climate Adaptive Building Shells. In *Integrating IoT AI Indoor Air Quality Assessment*; Saini, J., Dutta, M., Marques, G., Halgamuge, M.N., Eds.; Springer International Publishing: Cham, Switzerland, 2022; pp. 89–109. [CrossRef]
13. Marzouk, M.; Atef, M. Assessment of Indoor Air Quality in Academic Buildings Using IoT and Deep Learning. *Sustainability* **2022**, *14*, 7015. [CrossRef]

14. Pastor-Fernández, A.; Cerezo-Narváez, A.; Montero-Gutiérrez, P.; Ballesteros-Pérez, P.; Otero-Mateo, M. Use of Low-Cost Devices for the Control and Monitoring of CO_2 Concentration in Existing Buildings after the COVID Era. *Appl. Sci.* **2022**, *12*, 3927. [CrossRef]
15. Pietrogrande, M.C.; Casari, L.; Demaria, G.; Russo, M. Indoor Air Quality in Domestic Environments during Periods Close to Italian COVID-19 Lockdown. *Int. J. Environ. Res. Public Health* **2021**, *18*, 4060. [CrossRef]
16. Saini, J.; Dutta, M.; Marques, G. Indoor Air Quality and Internet of Things: The State of the Art. In *Internet of Things for Indoor Air Quality Monitoring*; Saini, J., Dutta, M., Marques, G., Eds.; Springer International Publishing: Cham, Switzerland, 2021; pp. 33–50. [CrossRef]
17. Saini, J.; Dutta, M.; Marques, G. Indoor Air Quality: Impact on Public Health. In *Internet of Things for Indoor Air Quality Monitoring*; Saini, J., Dutta, M., Marques, G., Eds.; Springer International Publishing: Cham, Switzerland, 2021; pp. 1–14. [CrossRef]
18. Szabados, M.; Csákó, Z.; Kotlík, B.; Kazmarová, H.; Kozajda, A.; Jutraz, A.; Kukec, A.; Otorepec, P.; Dongiovanni, A.; Di Maggio, A.; et al. Indoor air quality and the associated health risk in primary school buildings in Central Europe—The InAirQ study. *Indoor Air* **2021**, *31*, 989–1003. [CrossRef]
19. ASHRAE Resources Available to Address COVID-19 Concerns. Available online: https://www.ashrae.org/about/news/2020/ashrae-resources-available-to-address-covid-19-concerns (accessed on 13 August 2021).
20. Standards 62.1 & 62.2. Available online: https://www.ashrae.org/technical-resources/bookstore/standards-62-1-62-2 (accessed on 17 December 2023).
21. Adeleke, J.; Moodley, D. An Ontology for Proactive Indoor Environmental Quality Monitoring and Control. In Proceedings of the 2015 Annual Research Conference on South African Institute of Computer Scientists and Information Technologists, Stellenbosch, South Africa, 28–30 September 2015; pp. 1–10. [CrossRef]
22. Agarwal, N.; Meena, C.S.; Raj, B.P.; Saini, L.; Kumar, A.; Gopalakrishnan, N.; Kumar, A.; Balam, N.B.; Alam, T.; Kapoor, N.R.; et al. Indoor air quality improvement in COVID-19 pandemic: Review. *Sustain. Cities Soc.* **2021**, *70*, 102942. [CrossRef]
23. Alonso, A.; Llanos, J.; Escandón, R.; Sendra, J.J. Effects of the COVID-19 Pandemic on Indoor Air Quality and Thermal Comfort of Primary Schools in Winter in a Mediterranean Climate. *Sustainability* **2021**, *13*, 2699. [CrossRef]
24. Anastasi, G.; Bartoli, C.; Conti, P.; Crisostomi, E.; Franco, A.; Saponara, S.; Testi, D.; Thomopulos, D.; Vallati, C. Optimized Energy and Air Quality Management of Shared Smart Buildings in the COVID-19 Scenario. *Energies* **2021**, *14*, 2124. [CrossRef]
25. Ding, J.; Yu, C.W.; Cao, S.-J. HVAC systems for environmental control to minimize the COVID-19 infection. *Indoor Built Environ.* **2020**, *29*, 1195–1201. [CrossRef]
26. Huang, M.; Liao, Y. Development of an indoor environment evaluation model for heating, ventilation and air-conditioning control system of office buildings in subtropical region considering indoor health and thermal comfort. *Indoor Built Environ.* **2022**, *31*, 807–819. [CrossRef]
27. Aviv, D.; Chen, K.W.; Teitelbaum, E.; Sheppard, D.; Pantelic, J.; Rysanek, A.; Meggers, F. A fresh (air) look at ventilation for COVID-19: Estimating the global energy savings potential of coupling natural ventilation with novel radiant cooling strategies. *Appl. Energy* **2021**, *292*, 116848. [CrossRef] [PubMed]
28. Švajlenka, J. Aspects of the Internal Environment Buildings in the Context of IoT. In *Integrating IoT AI Indoor Air Quality Assessment*; Saini, J., Dutta, M., Marques, G., Halgamuge, M.N., Eds.; Springer International Publishing: Cham, Switzerland, 2022; pp. 55–72. [CrossRef]
29. Chhikara, A.; Kumar, N. COVID-19 Lockdown: Impact on Air Quality of Three Metro Cities in India. *Asian J. Atmos. Environ. AJAE* **2020**, *14*, 378–393. [CrossRef]
30. Mouftah, H.T.; Erol-Kantarci, M.; Rehmani, M.H. Environmental Monitoring for Smart Buildings. In *Transportation and Power Grid in Smart Cities: Communication Networks and Services*; Wiley: Hoboken, NJ, USA, 2019; pp. 327–354. [CrossRef]
31. Cai, J.; Li, S.; Hu, D.; Xu, Y.; He, Q. Nationwide assessment of energy costs and policies to limit airborne infection risks in U.S. schools. *J. Build. Eng.* **2022**, *45*, 103533. [CrossRef]
32. Sodiq, A.; Khan, M.A.; Naas, M.; Amhamed, A. Addressing COVID-19 contagion through the HVAC systems by reviewing indoor airborne nature of infectious microbes: Will an innovative air recirculation concept provide a practical solution? *Environ. Res.* **2021**, *199*, 111329. [CrossRef]
33. Wu, J.; Chen, J.; Olfert, J.S.; Zhong, L. Filter evaluation and selection for heating, ventilation, and air conditioning systems during and beyond the COVID-19 pandemic. *Indoor Air* **2022**, *32*, e13099. [CrossRef] [PubMed]
34. Anchordoqui, L.A.; Chudnovsky, E.M. A Physicist View of COVID-19 Airborne Infection through Convective Airflow in Indoor Spaces. *SciMed. J.* **2020**, *2*, 68–72. [CrossRef]
35. Azuma, K.; Yanagi, U.; Kagi, N.; Kim, H.; Ogata, M.; Hayashi, M. Environmental factors involved in SARS-CoV-2 transmission: Effect and role of indoor environmental quality in the strategy for COVID-19 infection control. *Environ. Health Prev. Med.* **2020**, *25*, 66. [CrossRef] [PubMed]
36. Bhagat, R.K.; Davies Wykes, M.S.; Dalziel, S.B.; Linden, P.F. Effects of ventilation on the indoor spread of COVID-19. *J. Fluid Mech.* **2020**, *903*, F1. [CrossRef] [PubMed]
37. Bueno de Mesquita, P.J.; Delp, W.W.; Chan, W.R.; Bahnfleth, W.P.; Singer, B.C. Control of airborne infectious disease in buildings: Evidence and research priorities. *Indoor Air* **2022**, *32*, e12965. [CrossRef] [PubMed]
38. Nembhard, M.D.; Burton, D.J.; Cohen, J.M. Ventilation use in non-medical settings during COVID-19: Cleaning protocol, maintenance, and recommendations. *Toxicol. Ind. Health* **2020**, *36*, 644–653. [CrossRef] [PubMed]

39. Noorimotlagh, Z.; Jaafarzadeh, N.; Martínez, S.S.; Mirzaee, S.A. A systematic review of possible airborne transmission of the COVID-19 virus (SARS-CoV-2) in the indoor air environment. *Environ. Res.* **2021**, *193*, 110612. [CrossRef] [PubMed]
40. Tzoutzas, I.; Maltezou, H.C.; Barmparesos, N.; Tasios, P.; Efthymiou, C.; Assimakopoulos, M.N.; Tseroni, M.; Vorou, R.; Tzermpos, F.; Antoniadou, M.; et al. Indoor Air Quality Evaluation Using Mechanical Ventilation and Portable Air Purifiers in an Academic Dentistry Clinic during the COVID-19 Pandemic in Greece. *Int. J. Environ. Res. Public Health* **2021**, *18*, 8886. [CrossRef] [PubMed]
41. Saini, J.; Dutta, M.; Marques, G. Future Directions on IoT and Indoor Air Quality Management. In *Internet of Things for Indoor Air Quality Monitoring*; Saini, J., Dutta, M., Marques, G., Eds.; Springer International Publishing: Cham, Switzerland, 2021; pp. 69–82. [CrossRef]
42. Saini, J.; Dutta, M.; Marques, G. *Internet of Things for Indoor Air Quality Monitoring*; Springer International Publishing: Cham, Switzerland, 2021. [CrossRef]
43. Son, J.; Son, Y.-S. A Correlation Analysis of Indoor Environmental Quality and Indoor Air Quality using IoT. In Proceedings of the 2019 International Conference on Information and Communication Technology Convergence (ICTC), Jeju, Republic of Korea, 16–18 October 2019; pp. 977–979. [CrossRef]
44. Tahmasebi, F.; Wang, Y.; Cooper, E.; Godoy Shimizu, D.; Stamp, S.; Mumovic, D. Window operation behaviour and indoor air quality during lockdown: A monitoring-based simulation-assisted study in London. *Build. Serv. Eng. Res. Technol.* **2022**, *43*, 5–21. [CrossRef]
45. Afshari, R. Indoor Air Quality and Severity of COVID-19: Where Communicable and Non-communicable Preventive Measures Meet. *Asia Pac. J. Med. Toxicol.* **2020**, *9*, 1–2. [CrossRef]
46. García-Morales, V.J.; Garrido-Moreno, A.; Martín-Rojas, R. The Transformation of Higher Education After the COVID Disruption: Emerging Challenges in an Online Learning Scenario. *Front. Psychol.* **2021**, *12*, 616059. [CrossRef]
47. Gravett, K.; Baughan, P.; Rao, N.; Kinchin, I. Spaces and Places for Connection in the Postdigital University. *Postdigit. Sci. Educ.* **2022**, *5*, 694–715. [CrossRef]
48. Onset HOBO and InTemp Data Loggers. Available online: https://www.onsetcomp.com/ (accessed on 20 December 2022).
49. GrayWolf Sensing Solutions. IAQ Meter—Indoor Air Quality Meters. Available online: https://graywolfsensing.com/iaq/ (accessed on 20 December 2022).
50. Prana Air. Air Quality Sensor for PM & Gases, Highly Accurate. Available online: https://www.pranaair.com/air-quality-sensor/ (accessed on 20 December 2022).
51. Amazon.com: Air Quality Monitor Accurate Tester for CO2 Formaldehyde(HCHO) TVOC/AQI Multifunctional Air Gas Detector Real Time Data & Mean Value Recording for Home Office and Various Occasion—Silver: Industrial & Scientific. Available online: https://www.amazon.com/Accurate-Formaldehyde-Multifunctional-Detector-Recording/dp/B09MQ2N6MB/ref=zg_bs_5006564011_sccl_8/135-4182504-5243159?pd_rd_i=B0B5644GGT&th=1 (accessed on 20 December 2022).
52. TechHive. uHoo Smart Air Monitor Review: Great Sensor, but a Bad App. Available online: https://www.techhive.com/article/815402/uhoo-smart-air-monitor-review.html (accessed on 20 December 2022).
53. HibouAir. Indoor Air Quality Monitoring Device. Available online: https://www.hibouair.com/ (accessed on 20 December 2022).
54. Mumtaz, R.; Zaidi, S.M.H.; Shakir, M.Z.; Shafi, U.; Malik, M.M.; Haque, A.; Mumtaz, S.; Zaidi, S.A.R. Internet of Things (IoT) Based Indoor Air Quality Sensing and Predictive Analytic—A COVID-19 Perspective. *Electronics* **2021**, *10*, 184. [CrossRef]
55. World Health Organization. *WHO Global Air Quality Guidelines: Particulate Matter (PM2.5 and PM10), Ozone, Nitrogen Dioxide, Sulfur Dioxide and Carbon Monoxide*; World Health Organization: Geneva, Switzerland, 2021.
56. Saini, J.; Dutta, M.; Marques, G. Indoor Air Quality Monitoring Systems and COVID-19. In *Emerging Technologies during the Era of COVID-19 Pandemic*; Arpaci, I., Al-Emran, M., Al-Sharafi, M.A., Marques, G., Eds.; Springer International Publishing: Cham, Switzerland, 2021; Volume 348, pp. 133–147. [CrossRef]
57. Mohammed, A.M.; Saleh, I.A.; MAbdel-Latif, N. Review article: Air quality and characteristics of sources. *Int. J. Biosens. Bioelectron.* **2020**, *6*, 85–91. [CrossRef]
58. Mujan, I.; Licina, D.; Kljajić, M.; Čulić, A.; Anđelković, A.S. Development of indoor environmental quality index using a low-cost monitoring platform. *J. Clean. Prod.* **2021**, *312*, 127846. [CrossRef]
59. Hobeika, N.; García-Sánchez, C.; Bluyssen, P.M. Assessing Indoor Air Quality and Ventilation to Limit Aerosol Dispersion—Literature Review. *Buildings* **2023**, *13*, 742. [CrossRef]
60. De Capua, C.; Fulco, G.; Lugarà, M.; Ruffa, F. An Improvement Strategy for Indoor Air Quality Monitoring Systems. *Sensors* **2023**, *23*, 3999. [CrossRef]
61. Langer, S.; Psomas, T.; Teli, D. I-CUB: 'Indoor Climate-Users-Buildings': Relationship between measured and perceived indoor air quality in dwellings. *J. Phys. Conf. Ser.* **2021**, *2069*, 012245. [CrossRef]
62. Rastogi, K.; Lohani, D. An IoT-based Framework to Forecast Indoor Air Quality using ANFIS-DTMC model. *Int. J. Next-Gener. Comput.* **2020**, *11*, 76.
63. Shoukry, F.; Goubran, S.; Tarabieh, K. Perceived Comfort in University Classrooms Post the Pandemic: Interpretations Considering the Carbon Footprint of Learning Spaces. In Proceedings of the 8th International ICARB Conference, Edinburgh, Scotland, 25–26 September 2023; pp. 508–532.
64. Shoukry, F.; Marey, A.; Goubran, S.; Tarabieh, K. *Indoor Air Quality Monitoring in Learning Spaces: A Case Study in Cairo, Egypt*; Track Future Cities; KSA: Riyadh, Saudi Arabia, 2024.

65. Shoukry, F.; Raafat, R.; Tarabieh, K.; Goubran, S. Indoor Air Quality and Ventilation Energy in University Classrooms: Simplified Model to Predict Trade-Offs and Synergies. *Sustainability* **2024**, *16*, 2719. [CrossRef]
66. Shoukry, F.; Goubran, S. Learning Spaces in the Post-Pandemic Era: A Framework for Healthy University Classroom Design. *Discov. Educ. J.* 2024, *under review*.

Disclaimer/Publisher's Note: The statements, opinions and data contained in all publications are solely those of the individual author(s) and contributor(s) and not of MDPI and/or the editor(s). MDPI and/or the editor(s) disclaim responsibility for any injury to people or property resulting from any ideas, methods, instructions or products referred to in the content.

Article

Proposal of a Sensorization Methodology for Obtaining a Digital Model: A Case Study on the Dome of the Church of the Pious Schools of Valencia

Luis Cortés-Meseguer [1,*] and Jorge García-Valldecabres [2]

1 Centro de Investigación de Tecnología de la Edificación (CITE), Universitat Politècnica de València, 46022 Valencia, Spain
2 Centro de Investigación en Arquitectura, Patrimonio y Gestión para el Desarrollo Sostenible (PEGASO), Universitat Politècnica de València, 46022 Valencia, Spain
* Correspondence: luicorme@upv.es

Abstract: The Church of the Pious Schools of Valencia (18th century) has the largest Valencian dome ever constructed, with its 24.5 m span, and it is included among the prestigious great European domes, inspired by the Pantheon and belonging to neoclassicism. Currently, this monument is undergoing a thorough study and restoration process to improve its management, especially to halt its deterioration due to moisture and cracks. An initial study included in the Master Plan (1995) determined that these cracks were caused by thermal effects, but recently, other studies have suggested that these failures originated from the walls. Additionally, environmental impacts and thermal behavior are among the causes, as excessive humidity due to high interior occupancy can cause damage to the dome, which has historic coatings. As a result of this study process, we propose sensorizing the dome of the church in order to enable comprehensive control of the temperature, humidity, and CO_2, as well as installing accelerometers to monitor the movements of the structure. With this, after the restoration of the dome, the potential effects of temperature, humidity, and CO_2 on the dome's surfaces will be controlled, in addition to verifying if there is any correlation between the cracks and the temperature.

Keywords: heritage; study; crack

Citation: Cortés-Meseguer, L.; García-Valldecabres, J. Proposal of a Sensorization Methodology for Obtaining a Digital Model: A Case Study on the Dome of the Church of the Pious Schools of Valencia. *Buildings* **2024**, *14*, 2057. https://doi.org/10.3390/buildings14072057

Academic Editors: Lina Šeduikytė and Jakub Kolarik

Received: 31 May 2024
Revised: 25 June 2024
Accepted: 2 July 2024
Published: 5 July 2024

Copyright: © 2024 by the authors. Licensee MDPI, Basel, Switzerland. This article is an open access article distributed under the terms and conditions of the Creative Commons Attribution (CC BY) license (https://creativecommons.org/licenses/by/4.0/).

1. Introduction

Within architectural heritage, there are elements that, due to their scale and aesthetic beauty, form a unique entity; there is no person who does not marvel when entering through the portico of the Pantheon and observing the largest masonry dome in history with its nearly two millennia of existence and its 43.3 m span [1]. In this context, the spaces created and their structures form heritage that must be preserved, and thanks to current technology, we can access resources that ensure their maintenance and intervention.

It could be stated that pathology in historical constructions is defined by two main aspects: the presence of cracks and the presence of moisture. Both pathological lesions directly interfere with the useful life of the materials and the building itself, with preventive conservation being one of the ways to minimize future costly interventions, in addition to being an example of sustainability. Currently, the problem of CO_2 generation is the focus of the famous 2030 Agenda of the United Nations (UN) [2], with architectural restoration being a way to preserve buildings in such a way that environmental impacts are minimized and resource efficiency is maximized [3]. A sustainable approach to restoration includes preventive maintenance plans that ensure the longevity of structures [4].

The Church of the Escuelas Pías in Valencia (Spain) (Figure 1), built in 1771, is undergoing a comprehensive study with critical perspectives as its dome is currently being restored, funded by the Ministry of Transport and Sustainable Mobility (Government of Spain). This dome is the largest in the Valencian territory, with a span of 24.50 m [5], and is included among the great European domes, as well as being notable for its exemplarity within neoclassical architecture [6]. Its pathology is primarily due to the presence of four large cracks, each approximately 6 m long, in four of the ten sectors that compose the dome, in addition to moisture problems inside due to damage to its covering, which was made using the traditional masonry technique with blue glazed ceramic tiles.

Figure 1. Aerial view of the dome of Escuelas Pías.

Some authors, such as Soler and Benlloch [7], define the pathology of the dome as being due to thermal and rheological actions in a monument whose construction systems are made of brick and lime masonry. These authors base their findings on current research in other similar monuments, omitting the different construction materials, such as the major domes of Santa Maria del Fiore in Florence or the Pantheon of Agrippa in Rome. In the case of Santa Maria, Fanelli [8] conducted an exhaustive study of its cracks after having installed a monitoring system in 1987, concluding that the cracks extend from the base of the pillar to the top of the dome, dividing the structure into four substructures subjected to the action of their own weight and to eventual mutual reactions, i.e., reactions of compression. On the other hand, in the case of the Pantheon of Agrippa, Masi [9] concludes that meridional cracks may have been produced in the early stage of the dome's life by the action of concrete shrinkage and gravity.

Recently, with the increase in HBIM technology, we are seeing a rise in the monitoring of monuments for their management [10], especially for the control of wall surfaces and to prevent damage in areas where there are elements of interest, particularly paintings [11]. This has led to the creation of virtual twins to manage their maintenance and conservation [12]. However, when it comes to heritage, it is essential to consider that the overall stability of a structure is perhaps the most important aspect in the maintenance of the monument, as it can be the source of many issues and potential heritage loss [13]

For the study of cracks and dampness, as well as their causes, there currently exists a wide range of methodologies and tests, which can be either destructive or non-destructive. In the application of destructive tests, it is necessary to sacrifice part of the wall surface in order to test its resistance or physical and mechanical characteristics such as its composition and strength [14]. This category includes methods that use core samples, although some may be reversible. Conversely, there is a broad spectrum of non-destructive tests that also serve to determine certain aspects without the need to remove a piece from the structure for analysis [15]. These non-destructive methods include the application of ground-penetrating radar, ultrasound, thermography, tomography, etc. All of these testing methods are compatible with the parameterization and introduction of the obtained data into a repository following the HBIM methodology, representing the virtual DNA of the monument.

The main objective of this article is to establish, through case studies, a methodology for monitoring monuments, focusing on the most important heritage element of the Escuelas Pías Church in Valencia: its dome (Figure 2), which is within a small and prestigious group worldwide, as noted by Guastavino in 1893 [16]. Additionally, the proposed method aims to control the indoor air quality and improve the energy efficiency of the church, given its continuous use as the largest space in the school, filled with students almost every day. The first level houses a museum space that receives daily visits, and religious events are also held after school hours. This case serves as an example of knowledge transfer to society. Through innovative techniques applied experimentally in monument conservation, made possible by greater economic resources and the necessity of technology in monuments, these methods can later be exported to other existing or future buildings, thus serving as an example of sustainability.

Figure 2. Inside view of the church.

2. Study Object

The school and Church of the Pious Schools are located in the historic Velluters neighborhood in Valencia (Spain), occupying almost the entire rectangular block, excluding the eastern corner on Santa Teresa Street. The school building has a rectangular layout, with the church positioned at its eastern end. The church is bordered by the main facade with the bell tower to the south, two wings of the school cloister to the west, an alley to the north, and houses with a courtyard to the east. Architecturally, the church features a notable dome with a 24.5 m interior span and a total area of 1000 square meters, divided into ten sectors. This design is reminiscent of the Temple of Minerva Medica, though its architectural inspiration comes from the Pantheon of Agrippa [6].

Saint Joseph Calasanz founded the first public school in 1597 in Rome (Italy), naming it "Escuela Pía"; only the wealthy social classes had access to education [17]. The religious order, known as the Piarists or Escolapios, arrived in Valencia in 1737, where they initiated the construction of the school in 1739, finishing it in 1747 [18]. In 1767, construction of the church began according to the plans outlined by the master builder Josef Puchol. Archbishop Andrés Mayoral, the driving force behind the project, desired a church that would stand out from the Valencian architectural tradition, sending the architect to visit the church of the Bernardine Nuns in Alcalá (Madrid) [19].

The church's initial design and supervision were undertaken by José Puchol until he was replaced by Antonio Gilabert in 1768, who made further adjustments to the project. Construction was halted in 1769 due to Archbishop Mayoral's death and financial constraints. Despite these challenges, the church was eventually completed in January 1771 and consecrated in 1773. Puchol's original plan included a third section of greater height, but this was reduced in size for economic reasons [20].

The compositional scheme of the floor plan is similar to that of the Temple of Minerva Medica and derives from a circular plan. The first level of the temple's interior consists of ten radial spaces situated between the piers. Eight of these spaces are configured as exedras with concave back walls, while the remaining two have straight walls, designated for the atrium and the main chapel of the temple. The wall surface is structured with intercolumns flanked by fluted Corinthian pilasters. The intercolumns forming the fronts of the chapels are organized according to the well-known Palladian motif, which was highly innovative in Valencian architecture at the time, particularly for its serial application in the interior elevation of a temple. Seven of these spaces are intended for chapels and feature altars formed by two Corinthian columns on pedestals with alternating circular or triangular pediments. The remaining three spaces configured between the piers of the church are designated for entrances: one from the school cloister, one for access from the street, and one for access to the main altar. The main altar, located opposite the entrance, is larger than the others and connects the first level with the second.

The second level, lower in height than the first and encircled by an iron railing resting on a narrow walkway, accommodates galleries between the spaces left by the piers, all interconnected by narrow passages open within them. The vertical rhythm imposed in the first level by paired Corinthian pilasters is maintained in the second level, but there is a radical change in the composition of its intercolumns. The previous articulated elements give way to rigid lintel compositions here. This curved in antis composition employed in the Escuelas Pías serves as a columnar screen corresponding with the almost cubic framework of the galleries, but not with their lowered vaulting, as the straight entablature conceals the curvature of their covering.

The third level, formed above the upper cornice at 21 m, continues to maintain proportions and a distribution similar to that of the Pantheon. Enclosed by a metal railing and a narrow walkway, its surface is organized with ten rectangular windows above the galleries of the lower floor, niches in the piers, and decorative panels in the intermediate spaces. Statues of ten apostles are placed in the niches. From the cornice of the uppermost level rises the hemispherical dome, marked on its intrados by twenty paired ribs. Above the upper ring of the dome stands the lantern, divided into ten segments with windows.

3. Background: Previously Conducted Studies

More than a century has passed since Guastavino [16] studied the church of the Escuelas Pías, without alerting to any pathology in the dome. The Escuelas Pías was declared a National Artistic Historic Monument in 1982 after two hundred years of history, and due to the pathological damage of the dome, Rafael Soler's firm was commissioned to develop the Master Plan in 1993 [21] by analyzing the complex and focusing on the cracks of the dome (Figure 3) and their possible causes.

Figure 3. View of the dome and the cracks in four different sectors.

Since then, various reports and studies have been conducted to determine the cause of the cracks in the dome, while establishing a construction hypothesis for this masonry dome of approximately 48 cm in thickness. Initially, these studies focused only on the dome and attributed the origin to tension issues caused by thermal inertia, indicating that the thermal fluctuation stresses were not supported by the structure due to its material properties [21].

Various studies mention the behavior of the iron chains in the dome, especially Rodríguez and Gil [22], who suggested the possibility of spalling in a pattern parallel to the dome because of the oxidation of metal rings, initially mentioned by Zacarés [20]. They also conducted a planimetric survey with laser scanning (TLS), as well as preparing images and adjusting their color balance in order to observe the degradation and damages.

Marín [23] establishes an architectural comparison of the Escuelas Pías rotunda with other models, proposing a hypothesis of a trace analogous to that established by Carlo Fontana for the dome, while also highlighting that the cracks could have been caused by the settling of the masonry during the construction process due to the use of lime mortar. Similarly, he discusses the metal rings mentioned by Zacarés, and concludes, like the study by Alonso and Martínez [13], by suggesting that metal reinforcements are unnecessary once the dome is finished.

Other studies, both historical and structural, provide us with extensive knowledge about structural behavior throughout history. López [24] examines technical reports and studies conducted on masonry domes in the 18th century, highlighting how building masters applied principles of mechanics and material behavior. This connection links the study of domes (three-dimensional structures) with the theory of arches and vaults (two-dimensional structures). Huerta [25] studies and analyzes cases from different historical periods to conclude that the stability of arches, vaults, and domes relies on traditional calculations supported by geometry. He also correlates supported structures with load-bearing elements such as walls and buttresses.

Many historical and contemporary studies have predominantly focused on analyzing the behavior of cracks in domes, often overlooking investigations into the walls and foundations. However, when the ground or foundation settles, resulting in subsidence, cracks may manifest at the base of the walls, subsequently affecting the integrity of both the masonry and the dome. Instances such as that observed in the dome of St. Peter's Basilica in the Vatican, where cracks were evident at the level of the drum and dome [26], underscore the necessity of thoroughly examining the wall structure and foundation to comprehensively address structural concerns.

3.1. Sensorization between 1996 and 2003

Between 1996 and 2003, architect Rafael Soler's study carried out the sensorization of the dome in order to obtain the relationship between temperature and the movement (opening/closing) of the cracks, similar to what was carried out in the dome of Santa Maria del Fiore [27]. An analysis of the extensive documentation collected, along with the interpretation of the results that recorded the movements of the probes over the analyzed time period, confirmed the inverse correlation between temperature variation and crack opening: as the temperature increases, the crack distance decreases, and vice versa (Figure 4). It is worth noting that in the analysis of these data, a drift was observed, meaning that the crack openings increased over time due to thermal variations.

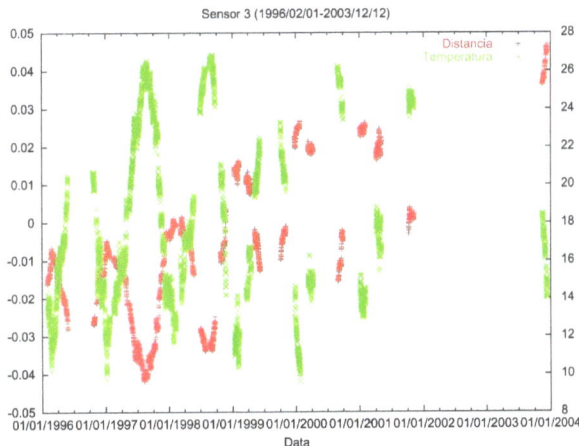

Figure 4. Scheme of the relationship between temperature and crack movement, with red representing movement and green representing temperature; 1996–2003.

3.2. Virtual Theoretical Study of the Dome

Alonso and Martínez [13] conducted a virtual theoretical study of thermal behavior due to temperature variations on the dome, concluding that the combination of thermal and gravitational loads leads to tensile stresses that greatly surpass the structural capacity of the material. However, they established that the damages observed in the model did not correspond to the actual state of fissures documented in the Master Plan [21]. In this study, they only analyzed the dome, without taking into account the supporting walls. However, Soler and Benlloch [7] stated that gravitational actions were not the cause of the injuries, given that neither their location in elements of very different sections nor the various studies conducted using graphic statics or finite elements indicated as such. They also established that seismic action did not seem to be the main cause of the damages, but perhaps it could have been, at some point in the past, a contributing factor to the fissures in two of the sectors, which are not the ones with the most damage.

Clearly, the cracks that exist in four out of the ten sectors of the drum, with lintels and jambs of the openings being split and the cracks extending to the upper ring and prior to the start of the dome (Figure 5), as well as the displacement of two ribs in two of the ten sectors, indicate some movement of the materials, and it can be affirmed that the cracks in the dome were not solely caused by thermal inertia. Following these findings, Cortés and Alonso [28] carried out a complete parameterization of the dome and walls to analyze the behavior due to settlements, using the finite-element method (Angle software 2021), and obtained results similar to those existing in the monument, providing an explanation for the cracks (Figure 6).

Figure 5. Cracks in the drum and the ring before the dome: (**a**) cracks in one lintel and vault of one of the four damaged sectors; (**b**) crack in the ring before the dome in sector G.

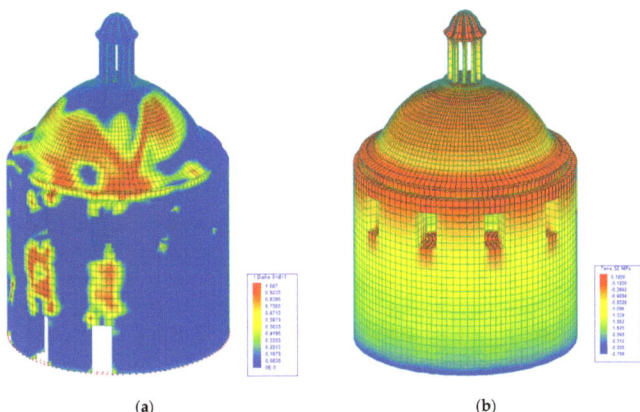

Figure 6. Cracks in the drum and the ring before the dome: (**a**) fissure pattern with the hypothesis of a settlement of the vertical wall; (**b**) stress state of the dome showing tension at the base of the dome.

3.3. Study of the Pathology

Both the interior and exterior pathology have been analyzed and documented in graphic and written reports, with special attention being given to the cracks as they represent the main problem of the building. The structural component is one of the fundamental aspects of architecture established by Vitruvius: venustas (beauty), firmitas (stability), and utilitas (utility).

Prior to the restoration, the dome exhibited four large cracks in the calotte that extended from the interior to the exterior, as well as in the drum. Additionally, there was a displacement of two roof tiles, allowing water to leak into the interior. The tile covering was in a very poor state of conservation, with water infiltration causing dampness, salt deposits, and loss of plaster, among other issues. The exterior showed numerous broken tiles, vegetation, missing tiles, and tiles with enamel loss. Furthermore, the lantern had had its openings blocked with a honeycomb brick infill and red waterproof paint.

Currently, the interior of the dome is undergoing restoration. As an initial step, the cracks in both the interior and exterior of the dome and its drum have been repaired. These cracks followed the meridian direction and were centrally located, extending from the base of the dome to approximately 3/5 of the length of the calotte, about 10.74 m, in four of the ten sectors of the dome (Figure 7). Associated with these cracks, there are numerous other fissures that exclusively pertain to the cracking of the coating.

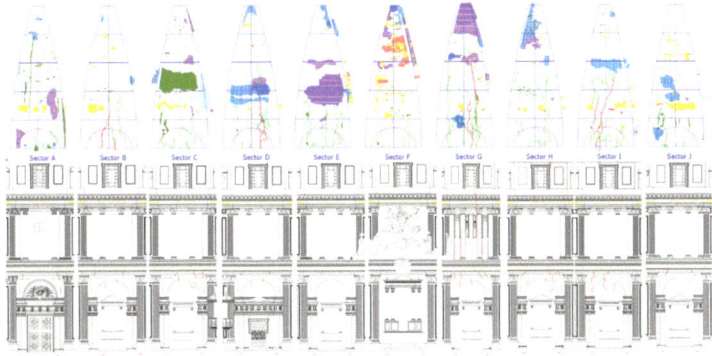

Figure 7. Study of the pathology of the interior, with the observation of damage in the dome derived from moisture and cracks.

3.4. Study of the Tilt

In light of the previously discussed condition of the dome and the drum, a more recent study was conducted to analyze tilts and settlements using orthoimages to establish a quantitative evaluation. This study took into account everything from the foundation to the highest part of the lantern.

The study began with a methodology that involved active sensor surveys, performing an exhaustive survey with laser scanning (TLS). For this 3D survey, the phase shift measurement system, which is faster and more precise, was used. A Leica scanner was utilized, conducting a total of 60 scans: 12 for the exterior and 48 for the interior space. In the subsequent office processing, the point clouds were downloaded and processed using Leica Cyclone Register360 software to generate a unified model, achieving a minimum precision of 1 mm for the exterior and 2 mm for the interior [22].

This study determined that the origin of the issues was the tilt of the drum, which consequently affected four sectors of the dome as it readjusted to its new state [28]. Additionally, the poor condition of the roof covering did not prevent water infiltration during rain, which affected the interior coating of the dome. This resulted in the appearance of dampness, salt, mold, and a partial detachment of the coating, impacting both the aesthetics and the healthiness of the building.

3.5. Geophysical Study

Zacarés [20] mentioned the existence of stone blocks in the dome where iron rings were fitted during the construction of the Escuelas Pías dome, specifying a weight of 613 @ (arroba), approximately 7790 kg. After two campaigns with ground-penetrating radar (GPR) technology, the existence of some metal rings was confirmed [29], similarly to what Guastavino [16] described as a "brick dome with iron rings", understanding that the depicted figure represents the dome of the Escuelas Pías de Valencia.

The objective of this geophysical study (Figure 8) was to search for metallic elements and locate rings and bars in the masonry dome (Figure 9) using a GSSI SIR3000 (Nashua, NH, USA) ground-penetrating radar system. Due to the range of thicknesses in the different sections that were to be studied in the dome, two antennas with central frequencies of 900 MHz and 400 MHz were selected for the location and detection of metallic elements up to approximately 2 m in depth:

- The 900 MHz antenna was used for location and detection up to 1 m in depth;
- The 400 MHz antenna was used for location and detection up to approximately 2 m in depth.

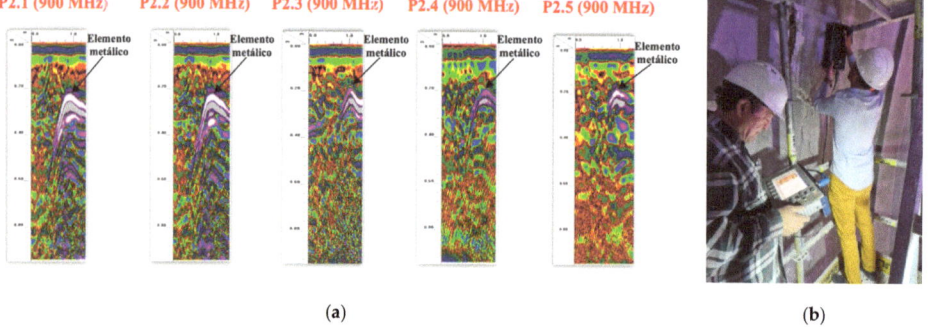

Figure 8. Results of the geophysical study using ground-penetrating radar to detect "anomalies" in the dome and an image depicting the progress of the study within the dome: (**a**) radargrams where anomalies in the waves can be observed; (**b**) image showing the execution of the geophysical study.

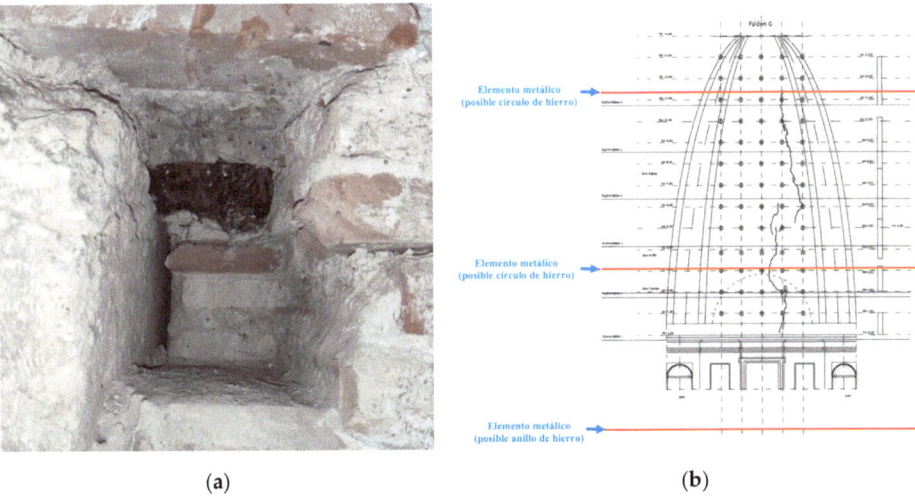

(a) (b)

Figure 9. With the geophysical study, it became possible to find the iron rings: (**a**) archaeological window with an iron ring; (**b**) representation of the grid with the red lines representing iron rings.

Two campaigns were carried out, conducting profiles both on the extrados and the intrados. On the extrados, a complete profile was conducted, finding metal elements at 11.20 m and 16.50 m, measured from the inner start of the dome. Internally, based on the platforms of the available scaffolding system and the accessibility to the dome elements, geophysical radar profiles were projected to search for and locate metallic elements in three sectors:

1. In sector G of the dome, 28 geophysical radar profiles were projected in 6 levels, corresponding to the 6 platforms of the available scaffolding. The 900 MHz antenna was used in all profiles, and the 400 MHz antenna was also used in the profiles between platform 1 and platform 2. These geophysical radar profiles were arranged longitudinally and parallel to each other, spaced 1 m apart.
2. Likewise, on platform 2 of sector I of the dome, two geophysical radar profiles were projected with the 900 MHz antenna to corroborate the results of the profiles on platform 2 of sector G.
3. In an area of the lintel above the columns of the second body, one geophysical radar profile was projected to detect a possible iron ring using the 900 MHz antenna (Figure 10).

In these three sectors of the dome, a total of 31 2D geophysical radar profiles were conducted, and an archaeological window was opened to observe the mentioned ring. This consisted of a 5 cm × 5 cm iron bar covered with braided esparto, embedded in stone blocks approximately every 5 m.

Study of Thermal Behavior

Given the importance of the thermal behavior of the dome due to its mass, with a perimeter of 77 m and approximate volume of 4500 m^3, a study of the thermal behavior of the dome has been conducted using Therm 8 software (Figure 11b) and sample collection with a thermal camera (Figure 11a) to enable its constructive parameterization. Various data captures have been made with the thermal camera, a Flir T530 camera, with the most satisfactory results occurring in winter due to the greater thermal contrast between the interior and exterior.

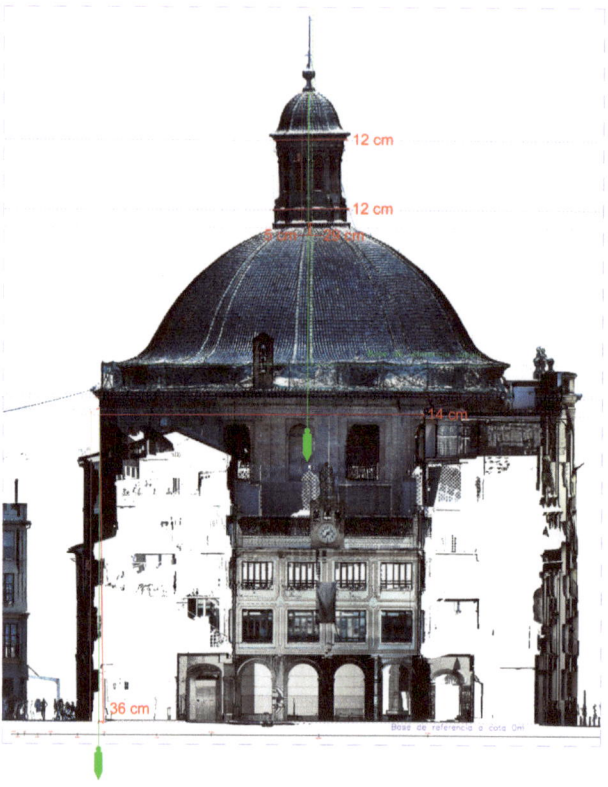

Figure 10. Study of the tilts using orthoimages obtained from 3D laser scanning (TLS).

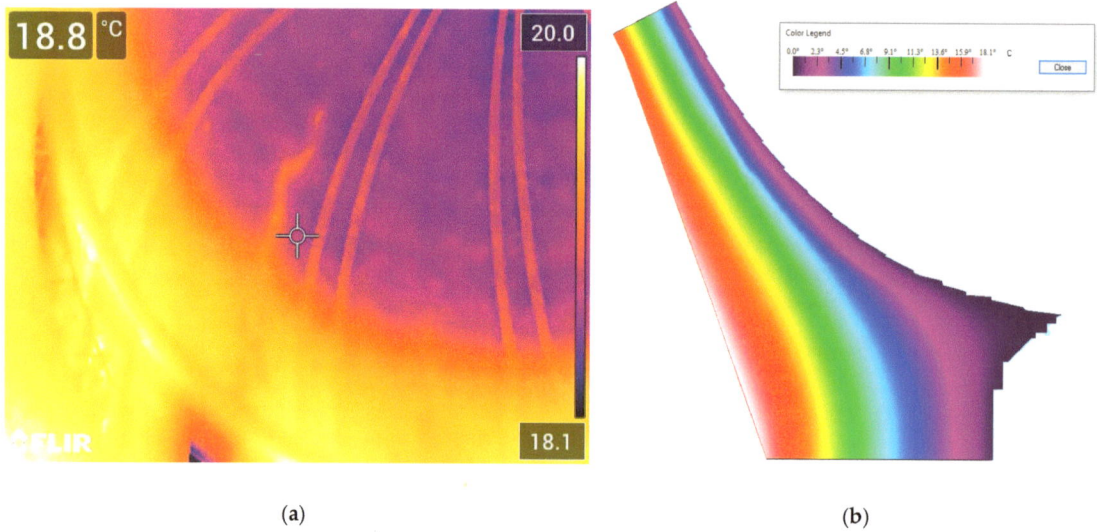

(a)　　　　　　　　　　　　　　　　　　(b)

Figure 11. Thermal behavior: (**a**) image from the thermal camera, where the crack shows a thermal bridge; (**b**) parameterization of thermal behavior in the Therm 8 software, with an external temperature of 0 °C and an internal temperature of 20 °C.

From the observations with the thermal camera, it was noted that across the dome, from the intrados starting from the drum, the ribs increase their thermal inertia, highlighting thermal bridges caused by the cracks, which have since been repaired and sealed.

The construction materials exhibit high conductivity, with no thermal insulation present, leading to rapid heat transmission. The thermal insulation of the church is determined by the thickness of the construction solution rather than the properties of the materials.

A verification of the limitation of interstitial and surface condensations of the dome (main thermal envelope and most delicate construction and structural elements) has been carried out using the eCondensa2 program according to the Technical Building Code (Spanish national regulations), document DA DB-HE/2 [30]. This verification established an indoor temperature of 20 °C and a relative humidity of 55%, and for the outdoor conditions, a temperature of 10.4 °C and a relative humidity of 63%, resulting in a thermal resistance of 0.532 m^2·K/W and a thermal transmittance of 1.881 W/m^2·K.

The verification of surface condensation limitation was based on a comparison of the surface temperature factor for the interior surface (fRsi) and the minimum surface temperature factor for interior surfaces (fRsi,min) for the corresponding indoor and outdoor conditions in January for the city of Valencia.

Thanks to its construction arrangement, including a 2 cm thick inner layer of plaster, a 45 cm thick solid brick dome, a layer of lime mortar, and glazed ceramic tiles, a more than satisfactory result was obtained. This is because the saturation pressure is much higher than the vapor pressure, and no condensation occurs or accumulates. The graph shows two parallel lines (Figure 12), indicating the absence of condensation accumulation.

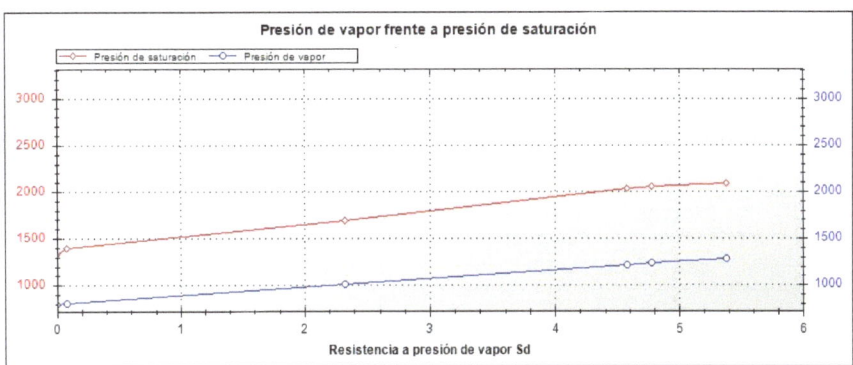

Figure 12. Study of the verification of interstitial and surface condensations of the dome.

4. Results and Discussion

These studies conducted to obtain a comprehensive understanding of the dome's construction are fundamental for addressing any other type of study, whether it be energy-related or related to structural behavior. Within this research, a new methodology must be established, which could include a thorough planimetric survey, a structural analysis, a pathological study, a study of collapses, and/or a geophysical study.

From the study of collapses, we know that the drum has a deviation of 36 cm from its base, and the horizontal planes in the dome bodies have rotated 14 cm. This study helps us address the disparity in opinions from all previous structural studies and explains the dome's cracks: as the drum deforms and there is a deviation, the ring cracks, and the dome readapts to the new stresses, resulting in cracks in the dome.

The pathological study provides all of the data needed to shape a project and make an architectural proposal, whether it involves repair, restoration, or consolidation. However, with the construction study and the complementary geophysical study, we can determine anomalies—in our case, the chains mentioned in the 19th century—without having to carry out costly destructive testing.

The thermal study serves to confirm what experience has granted over time, which is that the dome is a simple enclosure without thermal insulation, built with highly conductive materials, and with insulation provided by the thickness of the dome, which is 50 cm.

One of the results of this research is that it is unknown whether the walls follow a pattern of movement or remain fixed over time. Witnesses to the work conducted on the dome have confirmed that the cracks have not opened, but it is unknown if there are any possible movements of the building. Therefore, the following methodological proposal is made for the sensorization of the dome to obtain real-time data and obtain a digital twin, following the HBIM methodology [31].

Measuring CO_2 levels is a key indicator to assess indoor air quality [32]. Thanks to energy efficiency standards, improvements in insulation, and materials, energy efficiency conditions have substantially improved. However, this can lead to a dramatic worsening of indoor air quality conditions. On the other hand, in a heritage architectural context, an excessive concentration of CO_2 can affect artistic elements such as fresco paintings or coatings like plaster or stucco, detracting from the monument's appearance and leading to its heritage degradation. Therefore, active ventilation is necessary to maintain a healthy environment and ensure the proper preservation of artistic assets. Additionally, with the use of CO_2 sensors, it is possible to generate alerts if levels are excessive to ensure a healthy and safe environment for users. It is through these sensors that the concentration of CO_2 in the air can be measured.

It is important to know the temperature and humidity levels of materials because they reveal the technical behavior of the materials in normal conditions, under extreme temperatures, or during sudden changes in temperature or humidity. With these sensors, ambient humidity and temperature are also obtained.

The church facade that faces the alley is leaning because there is no building to buttress it against horizontal forces. The value of this slant is 36 cm over the height of slightly more than twenty-four meters. Simultaneously, the entablature of the first ring of the drum has a vertical tilt towards the aforementioned alley of 14 cm, 2 cm more than that of the base and the top of the lantern. Additionally, the lantern is displaced from its ideal axis. By using an inclinometer, it will be possible to determine whether the building is stable or if movements of the building and/or the terrain persist. In the case of this being affirmative, the appropriate stabilization or foundation reinforcement tasks will be carried out.

5. Conclusions

In light of all of the studies and tests conducted on this dome (Figure 13), and with the aim of adapting to the 21st century and the new requirements that must be imposed for the conservation of architectural heritage, it is necessary to sensorize the dome to obtain a virtual model and to understand the behavior of the dome in real time by obtaining its virtual twin.

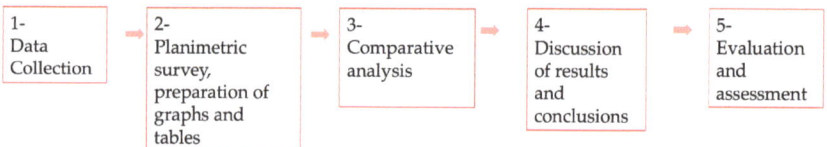

Figure 13. Phase chart.

From the records of the sensors in different orientations, it will be possible to evaluate the hygrothermal conditions of the enclosure. Different parameters of interest inside and outside the building, such as ambient temperature and relative humidity, will be obtained remotely, in real time, and continuously. Additionally, surface temperature and humidity sensors will also be installed to obtain a thermal profile. On the other hand, the analysis of data during one or several annual cycles will allow us to evaluate the moments when there is a higher risk of condensation, if the existing ventilation is adequate, or the thermal inertia of the dome.

Through the monitoring of the dome of the Church of the Pious Schools of Valencia, the continuous evaluation of the dome's condition and indoor health is intended to improve its heritage management. This church is in continuous use, as it is the meeting space for the school, houses a museum space, and also maintains daily liturgical use. With digital tools and computer applications, minute-by-minute information on its temperature, humidity, and CO_2 levels will be obtained, facilitating user-access management of the interior, improving the building's energy efficiency, and, above all, minimizing the environmental impact on the monument and its coatings.

Author Contributions: Conceptualization, L.C.-M. and J.G.-V.; methodology, L.C.-M.; software, L.C.-M.; validation, L.C.-M. and J.G.-V.; formal analysis, L.C.-M.; investigation, L.C.-M.; resources, L.C.-M. and J.G; data curation, L.C.-M. and J.G.-V.; writing—original draft preparation, L.C.-M. and J.G.-V.; writing—review and editing, L.C.-M. and J.G.-V.; visualization, L.C.-M. and J.G.-V.; supervision, L.C.-M. and J.G.-V.; project administration, L.C.-M. and J.G.-V.; funding acquisition, L.C.-M. and J.G.-V. All authors have read and agreed to the published version of the manuscript.

Funding: This research is part of the project "Restoration of the dome of the Church of the Pious Schools in Valencia", funded by the Ministry of Housing and Urban Agenda (Government of Spain), grant number 2022-17-10-0116.

Data Availability Statement: The data presented in this study are available on request from the corresponding author.

Conflicts of Interest: The authors declare no conflicts of interests.

References

1. Vighi, R. *Il Pantheon*; Tipografia Artistica: Rome, Italy, 1959; p. 28.
2. UN Sustainable Development Goals. Available online: https://www.un.org/sustainabledevelopment/ (accessed on 4 May 2024).
3. Harrison, R.; DeSilvey, C.; Holtorf, C.; Macdonald, S.; Bartolini, N.; Breithoff, E.; Fredheim, H.; Lyons, A.; May, S.; Morgan, J.; et al. *Heritage Futures: Comparative Approaches to Natural and Cultural Heritage Practices*; UCL Press: London, UK, 2020. [CrossRef]
4. Jokilehto, J. *A History of Architectural Conservation*; ICCROM: Rome, Italy, 2017.
5. Soler Verdú, R. *Cúpulas Históricas Valencianas in Soler Verdú, R. (director), Las Cúpulas azules de la Comunidad Valenciana*; Generalitat Valenciana: Valencia, Spain, 2006.
6. Bérchez, J. *Los Comienzos de la Arquitectura Académica en Valencia: Antonio Gilabert*; Editorial Federico Domenech S.A.: Valencia, Spain, 1987.
7. Soler Verdú, R.; Benlloch Marco, J. La cúpula de las Escuelas Pías: Causas de los daños e interpretación de las lesiones. In Proceedings of the Abierto por Obras, Con Motivo del Inicio de la Restauración de la Cúpula de la Rotonda, Real Academia de Bellas Artes de San Carlos, Valencia, Spain, 20 January 2023.
8. Fanelli, G.; Fanelli, M. *La cúpula de Brunelleschi. Historia y Futuro de una Grande Estructura*; Mandragora: Firenze, Italy, 2004.
9. Masi, F.; Stefanou, I.; Vannucci, P. On the Origin of the Cracks in the Dome of the Pantheon in Rome. hal-01719997v2. 2018. Available online: https://hal.archives-ouvertes.fr/hal-01719997v2 (accessed on 24 April 2024).
10. García-Valldecabres, J.; López-González, M.C.; Cortés-Meseguer, L. La conservación preventiva del patrimonio cultural. El estado de la cuestión en la adaptación a la metodología BIM. In Proceedings of the EUBIM 2021—BIM International Conference, Valencia, Spain, 27–29 October 2021.
11. García Diego, F.J. Sistema de monitorización de parámetros medioambientales en la cúpula renacentista de la catedral de Valencia. *Mètode* **2008**, *56*. Available online: https://metode.es/revistas-metode/monograficos/sistema-de-monitorizacion-de-parametros-medioambientales-en-la-cupula-renacentista-de-la-catedral-de-valencia.html (accessed on 30 May 2024).
12. Cortés Meseguer, L.; García Valldecabres, J. Digital Twins. HBIM information repositories to centralize knowledge and interdisciplinary management of architectural heritage. *VITRUVIO—Int. J. Archit. Technol. Sustain.* **2023**. [CrossRef]
13. Alonso Durá, A.; Martínez Boquera, A. Diagnóstico sobre el comportamiento estructural de la cúpula de las Escuelas Pías de Valencia. *Restauración Rehabil.* **2003**, *74*, 54–57.
14. Franco Gimeno, J.M. *Ensayos Destructivos Para Industria y Construcción*; Universidad de Zaragoza: Zaragoza, Spain, 1999.
15. De las Casas Gómez, A. Ensayos no-destructivos. In *Degradación y Conservación del Patrimonio Arquitectónico (Cursos de Verano de El Escorial de la Universidad Complutense de Madrid*; Editorial Complutense: Madrid, Spain, 1994; pp. 433–446.
16. Guastavino, R. *Essay on the Theory and History of Cohesive Construction*; Ticknor and Company: Boston, MA, USA, 1893.

17. Soler Blázquez, V.J. Origen y Establecimiento de las Escuelas Pías en Valencia (1735–1742). Ph.D. Thesis, Universidad CEU Cardenal Herrera, Valencia, Spain, 2017. Available online: https://repositorioinstitucional.ceu.es/jspui/bitstream/10637/8539/4/Origen_Soler_UCHCEU_Tesis_2017.pdf (accessed on 4 May 2024).
18. Bérchez, J. Iglesia de las Escuelas Pías. In *Catálogo de Monumentos y Conjuntos de la Comunidad Valenciana*; Generalitat Valenciana, Conselleria de Cultura, Educació i Ciència: Valencia, Spain, 1983; Volume 2, pp. 492–504.
19. Ponz, A. *Viage de España, ó cartas en que se dá noticia de las cosas mas apreciables, y dignas de saberse, que hay en ella*; Tomo Quarto; Joachin Ibarra: Madrid, Spain, 1774.
20. Zacarés, J.M. Antigüedades y bellezas de Valencia. Colegio Andresiano e Iglesia de las Escuelas Pías. *Rev. Edetana* **1849**, *11*, 497–498.
21. Soler Verdú, R. Plan Director de les Escoles Pies. Available online: https://dialnet.unirioja.es/servlet/autor?codigo=553390 (accessed on 4 May 2024).
22. Rodriguez-Navarro, P.; Gil-Piqueras, T. New Contributions on the Escuelas Pías Dome in Valencia. *Nexus Netw. J.* **2020**, *22*, 1081–1098. [CrossRef]
23. Marín Sánchez, R. El proyecto para la Rotonda de las Escuelas Pías de Valencia (1767–1773). Una mirada técnica. In *Geografías de la Movilidad Artística: Valencia en la Época Moderna*; Universidad de Valencia: Valencia, Spain, 2021; pp. 183–202.
24. López-Manzanares, G. Technical reports and theoretical studies about the structural behaviour of masonry domes in the 18th century. *Front. Archit. Res.* **2023**, *12*, 42–66. [CrossRef]
25. Huerta, S. *Arcos, Bóvedas y Cúpulas*; Instituto Juan de Herrera: Madrid, Spain, 2004.
26. López-Manzanares, G. La estabilidad de la cúpula de S. Pedro: El informe de los tres matemáticos. In *Actas del Segundo Congreso Nacional de Historia de la Construcción*; Instituto Juan de Herrera: A Coruña, Spain, 1998; pp. 285–294.
27. Fanelli, M. *Un Pòssibile Stato-Limite della cupole del Brunelleschi, Prima Volutazione Approxximata della Stabilita della Structura e dell'a apertura max. della Fissure*; Centro di Ricerca Idraulica e Strutturale, Enel Direzione Studi e Ricerche: Florence, Italy, 1985; pp. 56–78.
28. Cortés Meseguer, L.; Alonso Durá, A. Judgment of collapses and settlements from orthoimages: The method of pathological analysis in the Escuelas Pías church (Valencia). *DisegnareCON* **2023**, *16*, 8. [CrossRef]
29. García García, F. Georradar: Aplicación de sistema de georradar para la búsqueda de elementos metálicos y localización de anillos y barras en la calota de fábrica de la Iglesia de las Escuelas Pías de Valencia. Unpublished. 2023.
30. Código Técnico de la Edificación. Documento de Apoyo DA DB-HE/2. Available online: https://www.codigotecnico.org/pdf/Documentos/HE/DA-DB-HE-2_-_Condensaciones.pdf (accessed on 4 May 2024).
31. Liu, J.; Willkens, D.; Cortés-Meseguer, L.; García-Valldecabres, J.; Escudero, P.; Alathamneh, S. Comparative analysis of point clouds acquired from a TLS survey and a 3d virtual tour for HBIM development. *Int. Arch. Photogramm. Remote Sens. Spat. Inf. Sci.* **2023**, *XLVIII-M-2-2023*, 959–968. [CrossRef]
32. Galiano-Garrigós, A.; López-González, C.; García-Valldecabres, J.; Pérez-Carramiñana, C.; Emmitt, S. The Influence of Visitors on Heritage Conservation: The Case of the Church of San Juan del Hospital, Valencia, Spain. *Appl. Sci.* **2024**, *14*, 2065. [CrossRef]

Disclaimer/Publisher's Note: The statements, opinions and data contained in all publications are solely those of the individual author(s) and contributor(s) and not of MDPI and/or the editor(s). MDPI and/or the editor(s) disclaim responsibility for any injury to people or property resulting from any ideas, methods, instructions or products referred to in the content.

Article

Construction 4.0 in Refugee Camps: Facilitating Socio-Spatial Adaptation Patterns in Jordan's Zaatari Camp

Dima Abu-Aridah * and Rebecca L. Henn

Department of Architecture, The Pennsylvania State University, University Park, PA 16802, USA; rhenn@psu.edu
* Correspondence: dfa5278@psu.edu or dabuaridah@outlook.com

Abstract: Though refugee camps are by definition "temporary", many camps endure for decades, where individuals live full lives through childhood, marriage, children, grandchildren, and death. These settlements function no differently than cities in their social life, density, zoning, and operation, yet are "planned" through UNHCR (United Nations High Commissioner for Refugees) templates for camps. The Zaatari camp in Jordan for Syrian asylum seekers, for example, holds a population of 80,000. Rather than viewing refugee camps as temporary human warehouses, this article demonstrates that camps are spaces where individuals build social networks and economic activities flourish. As such, the camp planning templates should include adaptive Construction 4.0 technologies for more socially flexible settlements, even if the camps are considered "temporary". This case study research on the Zaatari camp illustrates how refugees adapt their built environment, identifying adaptation patterns that enhance both livability and sustainability. The work illustrates social and environmental changes that require adaptive housing configurations. The conclusion suggests linking modern tools in the construction industry to empirically derived planning objectives to be efficiently executed in moments of crisis.

Keywords: refugee camps; urban settlements; Construction 4.0; shape grammar; BIM; self-made communities; socio-spatial adaptation; shelter design; modular design

Citation: Abu-Aridah, D.; Henn, R.L. Construction 4.0 in Refugee Camps: Facilitating Socio-Spatial Adaptation Patterns in Jordan's Zaatari Camp. *Buildings* **2024**, *14*, 2927. https://doi.org/10.3390/buildings14092927

Academic Editors: Lina Šeduikytė and Jakub Kolarik

Received: 5 August 2024
Revised: 6 September 2024
Accepted: 10 September 2024
Published: 16 September 2024

Copyright: © 2024 by the authors. Licensee MDPI, Basel, Switzerland. This article is an open access article distributed under the terms and conditions of the Creative Commons Attribution (CC BY) license (https://creativecommons.org/licenses/by/4.0/).

1. Introduction

By mid-2023, the global population of forcibly displaced individuals had surged to 110 million, a staggering figure that includes refugees, asylum seekers, internally displaced persons, and those seeking international protection. The UNHCR defines refugee as "someone who has been forced to flee his or her country because of persecution, war or violence. A refugee has a well-founded fear of persecution for reasons of race, religion, nationality, political opinion or membership in a particular social group. Most likely, they cannot return home or are afraid to do so. War and ethnic, tribal and religious violence are leading causes of refugees fleeing their countries." People who are admitted to a country as refugees can claim permanent residency in the host country. An individual is called an asylum seeker until their request for refuge is processed and approved by the host country. If their application is denied, they lose the legal right to stay and may be deported [1]. Therefore, most "refugee camps" are actually camps for asylum seekers, as host countries want to neither grant the individuals permanent residency nor deport the individuals to their country of origin. In this article, we use the term "refugee" to include asylum seekers. Of the 110 million forcibly displaced people, refugees accounted for 36.4 million [2]. Remarkably, over half (52%) of the world's refugees came from just three countries: Syria, Afghanistan, and Ukraine [2]. The refugee numbers from these countries witnessed a rapid and dramatic escalation. For instance, the Ukrainian refugee population increased from 27,500 in 2021 to 5.90 million in 2023, a direct consequence of the Russian military buildup around Ukraine in 2021. Similarly, Syria's status shifted from being a major host of refugees, with 755,000 Iraqi refugees in 2011, to a major source of refugees, reaching 2.50 million by the end of 2013 and 6.50 million by 2023.

The future of many refugees remains uncertain. Out of the 110 million displaced individuals worldwide in 2023, only 59,500 resettled in a third country and 3.10 million returned to their home country. As of 2020, approximately 4.5 million refugees lived in planned and managed camps [3], which are just one of several sheltering solutions in humanitarian contexts.

While refugee camps are often viewed as temporary shelters for displaced populations who are typically isolated from broader society and treated primarily as humanitarian subjects and aid recipients, the camps frequently evolve into complex socio-spatial environments shaped by the needs and practices of their residents. Over time, often spanning a decade or multiple generations, refugees rebuild their social and spatial lives, transforming aid spaces into lived spaces and actively reshaping their identities despite the protracted political challenges governing their displacement. This socio-spatial transformation challenges the traditional perception of refugee camps as transient shelters and underscores the need for a deeper understanding of their long-term socio-spatial dynamics.

Refugee camps are designed following guidelines provided in emergency response handbooks such as the UNHCR Handbook of Emergencies, which provides insights into physical design features like topography, infrastructure, accessibility, proximity to services, climate, vegetation, environmental and gender considerations, and sanitation. For shelter design, the handbook offers guidance on area limitations, construction materials, and shelter options, including tents, shelter kits, plastic sheeting, temporary shelters, local materials, and refugee housing units. At the social level, the handbook recommends considering the social structures and relations within displaced communities and their hosting communities. This includes factors such as ethnic groups and family arrangements (UNHCR, 2023). However, many refugee camps are still designed with a grid-like layout that assumes that human needs are similar all around the world [4,5], without considering the social aspects mentioned in the guidelines. These guidelines are not comprehensive enough to support the design of socially resilient settlements. This research shows that refugees often work informally to compensate for shortcomings in the long-term planning and maintenance of their shelters and settlements. However, it takes them years of work, and they lack the expertise that this research aims to provide by incorporating Construction 4.0 technologies such as predictive analytics, BIM, and digital twins.

Some computational-based analysis and design methods such as shape-grammar formalisms, BIM, and predictive analytics can enhance flexibility in the design process, assist designers in tackling complex design problems, and aid in identifying spatial patterns in existing spatial configurations [6,7]. Construction 4.0 technologies would facilitate combining these tools with adaptive housing configurations and the associated social and environmental requirements prompting such changes in material and configuration. These Construction 4.0 technologies include Predictive Analytics, Building Information Modeling (BIM), and Digital Twin (DT) for the use of current data to predict future trends for better and efficient planning and design, and for real-time management of refugee shelters, making them more adaptable and resilient. Incorporating these technologies into the planning and development of refugee camps can enhance their adaptability and sustainability. Recognizing refugee camps as long-term communities rather than temporary shelters makes it possible to create more socially and spatially resilient environments for displaced populations.

Looking at the current research on integrating newer technology to aid in planning shelters and communities for refugees, recent studies explore various aspects of refugee camps, focusing on themes of materiality, housing quality assessment, participatory design, air quality, thermal performance, politics and symbolism, and access to land [8–13]. It is important to emphasize the architectural need for flexible, sustainable designs that use modular prefabricated materials while also promoting social interaction and climate protection [11]. For the Construction 4.0. technological aspects, Saad Alotaibi et al. [14], and Bazli et al. [15] explored the transformative potential of 3D printing (3DP), Artificial Intelligence (AI), and the Internet of Things (IoT) in post-disaster shelter construction, focusing on sustainability and efficiency. Alotaibi et al. emphasize how integrating these technologies enables rapid, cost-effective construction and smart energy management in homes, addressing the growing demand for eco-friendly solutions. However, they identify regulatory challenges and the need for skilled labor as barriers to adoption. Bazli et al. [15] similarly highlight the benefits of 3DP for remote and post-disaster housing, particularly in Australia's Northern Territory, where local material use and logistical efficiency could drive sustainability. Yet, they caution that technical limitations hinder widespread application, such as the lack of design guidelines for extreme environments.

More research is needed on the optimized layout and functional organization of shelter spaces that address the growing spatial and social needs of residents within refugee camps. While multiple studies address the materiality and architectural aspects of shelters, such as flexibility, sustainability, and construction materials, there's relatively limited explicit discussion regarding the spatial arrangement and detailed programming of these shelters to best accommodate the multifaceted needs of refugees and how Construction 4.0 technologies can be employed to enhance the process of creating better shelters for refugees even in protracted situations.

Understanding how shelters' spatial layout and zoning impact the daily lives, social interactions, and functionality of the refugee community (and vice-versa) is essential for designing adaptive and adaptable settlements that are more effective and efficient living environments within refugee camps. A more thorough investigation into the spatial configurations of shelters that are based on self-made adaptation patterns would contribute to the discourse on shelter design and enhancing livability and sustainability, potentially offering innovative solutions that enhance the space and community dynamics and improve the overall quality of life for refugees.

Based on the above, this study answers the following research questions:

- How do self-made adaptation patterns influence the spatial layout and functional organization of shelters within refugee camps?
- How can Construction 4.0 technologies, such as predictive analytics, Building Information Modeling (BIM), and digital twins, be integrated into the socio-spatial planning and development of refugee camps?

This study concludes that refugee camps, despite being intended as temporary solutions, often develop into permanent communities where individuals build social networks and engage in economic activities. The findings identify social attributes that can be quantified and used to perform predictive analysis to determine how a shelter and micro-community can provide the best spatial configuration to meet the residents' needs. The findings also suggest that integrating modern predictive analytics, BIM, DT, and construction tools and technologies into camp planning can lead to more adaptive and adaptable shelters and settlements. This research highlights the importance of linking construction industry advancements with humanitarian planning objectives to address the needs of displaced populations during crises efficiently.

The following sections provide background and related studies, including the literature on the architecture of refugee camps, the socio-spatial practices of refugees in camps, and the integration of Construction 4.0 technologies in these settings. The materials and methods section explains the selection of the case study, data collection methods, and the analytical framework. The results section presents both the spatial analysis findings and qualitative analysis outcomes. In the discussion section, socially and culturally informed spatial practices, environmentally conscious practices, the spatial grammar of Zaatari camp, and the potential for integrating Construction 4.0 in refugee camp design are explored. This paper concludes with a summary of findings, study limitations, and suggestions for future research.

2. Background and Literature Review
2.1. Refugee Camps

Refugee camps, a common spatial response to displacement crises, often present inadequate solutions. While seemingly addressing humanitarian emergencies, they perpetuate displacement due to complex management, competition for space, and a symbolic distance that normalizes suffering. This distance is further emphasized by researchers conforming to pre-defined categories. Filtering for relocation happens only when suffering reaches a critical point, raising questions about how displaced people are forced to conform and the perception of "deserving" refugees [16]. Further highlighting the shortcomings of camps, Paidakaki et al. [17] detail the harsh conditions in Greek camps—overcrowding, limited amenities, and restricted access to basic services. These factors worsen displacement and isolation, both physically and culturally. The authors stress the importance of empowering residents through adaptable, inclusive spaces designed with community involvement, highlighting the value of each individual's potential contribution. They call for holistic, flexible solutions, documentation of successful projects, and further research to refine the concept of resilient camps and understand the migration–resilience connection. Furthermore, Brankamp [18] argues that while refugee camps provide temporary protection from economic and political hardships, they also function as carceral spaces that cause long-term harm by perpetuating isolation, debilitation, and a form of "slow death." Camps inadequately address the root causes of displacement and instead, reinforce harmful structures of exclusion and control. Brankamp [18] emphasizes that, with the right strategies, camps can be transformed into spaces that truly support and empower their residents.

The study of space in refugee camps is based on the following key theoretical concepts, shown in Figure 1, which investigate the social and physical space in camps.

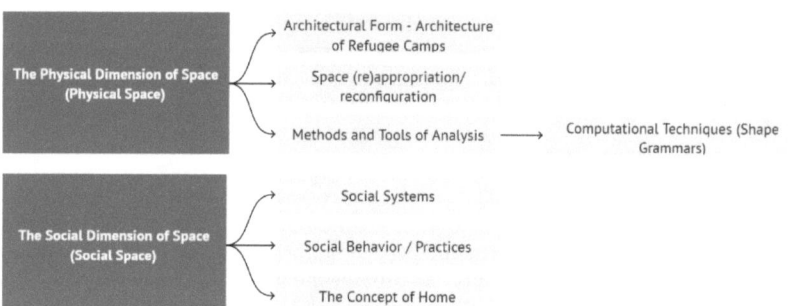

Figure 1. Key theoretical framework used to analyze the socio-spatial dynamics within refugee camps.

2.1.1. Architecture of Refugee Camps

Refugee architecture, the design and construction of shelters and settlements for displaced populations [19], is a field that should not only address immediate needs but also respect and incorporate cultural context, long-term functionality, and potential for adaptation [8,9,13]. By focusing on user needs, it ensures the effectiveness of creating spaces for displaced populations [10], which should evolve through the everyday practices of refugees, reflecting their agency and needs [20,21]. Refugee spaces should be designed to include areas for social interaction, livelihood activities, and cultural expression [21–23].

Standardized shelter solutions often fail to address user needs over time [8]. Emergency architecture needs to move beyond immediate response. It should focus on comprehensive planning for the long-term social inclusion of refugees [23,24]. This comprehensive planning is not just a luxury but a necessity, as it ensures that the needs of refugees are met not just in the short term but also in the long term. Dantas et al. [19] emphasize the necessity of tailored approaches for different types of emergency settlements. Their study reveals that many refugee settlements become quasi-permanent and require nuanced medium- and long-term policies and infrastructure planning. This complexity underscores the inadequacy of a one-size-fits-all strategy for planning and managing these settlements, advocating for urban planning strategies to enhance functionality and sustainability [19].

2.1.2. Spatial and Social Practices in Refugee Camps

Physical space is where people move, meet, live, and interact. It can be created spontaneously or through a process of design and planning [25–27]. According to Knoblauch and Löw (2020) [28], physical space can be understood through spatial logics, which include territory, trajectory, place, and network—dimensions they identify as the constitutive and rational aspects of socio-spatial relations. Spatial practices, in refugee camps, revolve around creating and recreating physical space. Nilsen [26] defines spatial reproduction as the design of the built environment and the establishment of physical settings for human activity, with an emphasis on the linkages between physical spaces and the planning and construction processes.

At the social level, social space is the product of group dynamics, with varying access to economic, social, and cultural capital shaping the spatial order. People are the key players in organizing and influencing space through their own social arrangements and behaviors. Factors that influence social behavior in space include creating boundaries, defining territories which refer to controlling what is inside and limiting access or excluding others, practices of place attachment, resistance, solidarity, and engagement in social and cultural activities [12,17,21,26,29,30]. Tayob [23] examines migrant-run markets in Bellville, Cape Town, as complex infrastructures crucial for new migrants. This study emphasizes the need for a nuanced understanding of these spaces, which includes understanding the social dynamics, economic activities, and cultural significance, and advocates for re-evaluating urban infrastructures and policies to better accommodate migrant communities

As an example of space appropriation, Ting [31] examines how Shan people displaced by conflict in Myanmar have built a sense of home despite their uncertain legal status and impermanent housing. The author argues that homemaking is an ongoing process for refugees, even in difficult circumstances. Shan people use everyday practices to create a sense of familiarity and community in their temporary dwellings. These dwellings become more than just shelters; they transform into homes over time through shared experiences and cultural practices. Their movement across borders allows them to maintain social connections and build a wider sense of "home territory" that transcends physical location. This challenges the idea that displacement is simply a state of being "out of place." Ting argues that homemaking for the displaced Shan is a complex process shaped by their mobility, their ongoing connection to their homeland, and the political forces they encounter.

Foundational texts by Bourdieu [32] and Lefebvre [33] argue that space is socially produced and is the result of human operations of synthesis (production). Sociological studies introduce social space as a realm of relations and interactions created by groups of people with varying access to economic, social, and cultural capital [34]. Similarly, Robinson [27] defines social space as a physical space imbued with meanings generated by the groups of people who use it. Kreuzer et al. [35] also note that social space describes relationships between individuals within physical spaces. Bourdieu conceptualized social space as structured so that groups located close to one another share proximity in this space, connecting it to physical (geographic) space through the concept of location or distance [32,36]. According to Bourdieu [32], people who are close in social space tend to be physically close in geographic space as well.

The concept of social capital is frequently discussed in the literature as a space-modifying factor. Social capital is crucial in influencing how social capital is a key concept that influences how the social space is produced or created [34,35,37–40]. It refers to the social relations between individuals and the dynamics that occur within these interactions [40].

Furthermore, the space we inhabit can be perceived as a system of social goods or as a structure composed of individuals who form the social order. It is shaped by the interaction between social activities and social structures [34,41,42]. Space is a product of human activities, as well as individual and social behavior [34,41]. According to Kühtreiber [41], people play a vital role in shaping space by organizing themselves and objects, which in turn impacts the space through these arrangements.

Exploring the relationship between social space and physical space, as well as refugees' social and spatial practices, is pivotal in evaluating the livability of camps. Recent studies have homed in on two key aspects of spatial practices: spatial appropriation practices and spatial reproduction patterns. Spatial appropriation practices involve human activities and daily routines that mold the use and meaning of existing or created spaces. Scholars underscore that spatial appropriation imbues physical and symbolic significance to spaces, potentially transforming the physical environment into meaningful places for inhabitants [30,43].

2.2. Technological Integration in Refugee Camp Design

The Fourth Industrial Revolution represents an evolution in science and technology, transitioning from traditional industrial operations to a new era of digitized and interconnected systems. This shift involves integrating advanced computational power, smart technologies, and intelligent machine learning algorithms, leading to fully intelligent and interconnected production environments. Technologies such as augmented reality (AR), system integration, cloud computing, big data mining, IoT, additive manufacturing, cyber security, predictive analytics, robotics, and simulation are central to this transformation. The integration of Industry 4.0 technologies in the Architecture, Engineering, and Construction (AEC) sector, often referred to as Construction 4.0, aims to address key challenges related to productivity and sustainability. The adoption of these technologies facilitates significant transformations across various phases of construction, including architectural and design planning, execution, and support process management [44,45].

2.2.1. Building Information Modeling (BIM) and Digital Twin (DT)

Building Information Modeling (BIM) tools and methods are increasingly utilized to manage data and information for the retrofit process, particularly with advancements in Digital Twin (DT) technology [46,47]. BIM extends beyond traditional 2D drawings by incorporating an integrated database capable of processing dynamic data and facilitating automated data exchange between sensors and the BIM model. This integration enables real-time visualizations and coordination of information by combining various types of models, including design, geometry, and behavioral models [48–50]. Digital twins, on the other hand, represent a digital counterpart of physical assets, including buildings and infrastructure. Digital twins provide real-time data on the condition and performance of assets, enabling predictive maintenance and operational optimization. The real-time nature of digital twins supports better decision-making and facility management, offering significant benefits throughout the asset's lifecycle [51,52].

2.2.2. Shape Grammar

Shape grammar, as discussed in the works of Haakonsen et al. [53], Paio et al. [54], Verniz and Duarte [55,56], E Costa et al. [57], and Barros et al. [58], refers to a formalism used in architectural and urban design to explore, analyze, and generate design alternatives based on a set of rules and shapes. It is a concept first introduced by Stiny and Gips in 1972. Employing shape-grammars-based design approaches can enhance flexibility in the design process, assist designers in tackling complex design problems, and aid in identifying spatial patterns in existing spatial configurations [6,7]. Shape grammars are rule-based formalisms that generate a design language through step-by-step processes, employing visual computations of shape production and transformation [59,60], making them a valuable tool for testing and evaluating different design alternatives [61] for predictive analysis.

As for refugee camps, such camps and settlements sometimes become similar to urban environments in terms of social life, congestion, planning, and service provision. The

Zaatari camp in Jordan is an example of how refugees have informally and independently re-created their shelter environment based on their needs and participated in making the camp an urban-like environment. Using shape grammar formalisms, this phenomenon was studied by Abu-Aridah et al. [29] in 2024, to analyze the self-made shelter layout, informal spatial patterns, typology, and adaptability.

The study by Abu-Aridah et al. [29] revealed the potential of informal flexible modularity in shelter design. This process, captured by shape grammar, allows for organic growth that aligns with refugees' preferences and circumstances. The grammar demonstrated the informal and adaptive nature of refugee-led design processes within the camp, showcasing its potential to create new shelter layouts that meet households' social and spatial needs.

3. Materials and Methods

This study employs an embedded mixed-methods research design, simultaneously collecting quantitative (spatial) and qualitative data and understanding the potential of integrating the outcomes in Construction 4.0 technologies, as illustrated in Figure 2. The research was conducted in three steps:

1. Simultaneous Analysis of Qualitative and Spatial Data: the first step involved studying both social (qualitative) and spatial (quantitative) data to identify common patterns in the camp.
2. Causal Analysis: next, the potential causality between social patterns and spatial configurations was analyzed.
3. Integration with Construction 4.0 Technologies: finally, this study explored how to integrate the findings from the social and spatial analyses into Construction 4.0 technologies, including predictive analytics for scenario planning for refugees considering their social structures and spatial needs by optimum use of land, Building Information Modeling, and Digital Twins.

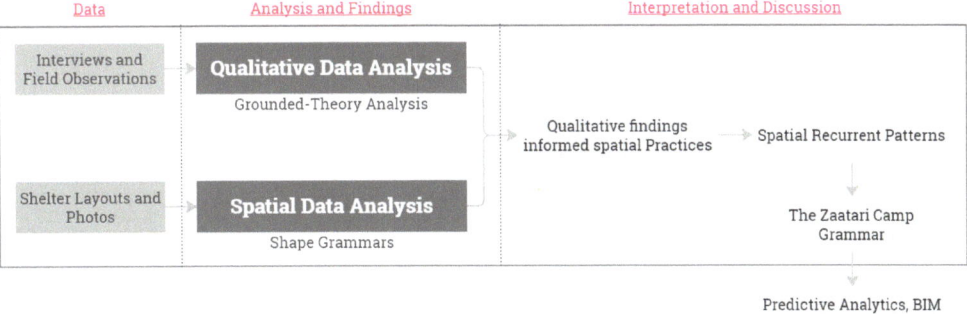

Figure 2. Analytical framework for the study outlining the key components and processes involved in the analytical approach used for this research.

3.1. Case Study Selection: The Zaatari Camp

This study uses the Zaatari camp in Jordan, which is one of the largest refugee camps in the world, as a case study. It is home to more than 80,000 Syrian refugees and it has more than 26,000 occupied shelters. It was established in 2012 close to Jordan's northern border with Syria [3]. Data for this paper are part of a larger research project collected through field observation, on-site documentation of shelter layouts, and face-to-face in-depth interviews with 64 households from the Zaatari camp. The sample size was determined based on previous studies of a similar qualitative nature by Steigemann and Misselwitz [21], Bilecen [62], Garcia and Haddock [63], Sanyal [64], Zíla [65] which demonstrated that 30 to 50 interviews are conducted to make the study outcomes generalizable.

The Zaatari camp, sprawling over 530 hectares of land owned by the Jordanian armed forces [3], was designed following a grid system based on the guidelines outlined in the UNHCR Handbook for Emergencies [66] and the Sphere Handbook [67]. The camp is divided into 12 districts as shown in Figure 3. Those districts are subdivided into blocks and sub-blocks using streets and circulation paths. The sub-blocks represent the smallest units within the layout of the camp, with each sub-block containing a certain number of shelters or housing units. At its inception, the camp used tents to shelter the refugees. Later, prefabricated units, locally referred to as 'Caravans', were arranged in rows within the grid system. However, what truly stands out is the significant changes made by the users to the original arrangement of shelters, a testament to their adaptability and resilience.

Figure 3. The Zaatari camp layout and districts.

3.2. Data Collection and Analysis

3.2.1. Data Collection

Data for this study were collected through field observation, on-site documentation of shelter layouts, and face-to-face in-depth semi-structured interviews with three households from the Zaatari camp. A stratified random selection technique was used to recruit participants. Interviews were conducted in July and August 2022 in the interviewees' shelters at a place of their choice. Three districts from the camp were selected based on the date of establishment of the district, the place of origin of hosted refugees, and the residential density of those districts. Districts No. 1, 4, and 12 were identified for data collection. The selection of the locations (districts) within the camp where data were collected was conducted with the help of the UNHCR's technical team to ensure that the selected districts are representative of the population of the camp. Multiple blocks were randomly selected from each district, and refugees were randomly selected from the selected blocks.

Interviews: Semi-structured interviews were conducted in Arabic to gather in-depth insights and perspectives from participants regarding their spatial and social experiences in the camp and their hometowns before displacement. The interviews were designed to be open-ended, allowing participants to express their thoughts, experiences, and viewpoints freely. Topics covered included participants' pre-displacement social and spatial experiences in their hometowns and their post-displacement experiences in the camp.

Field and Shelter Documentation: Field documentation involves recording observations, interactions, and events relevant to the research focus. Photography, sketches, video recordings, and audio recordings were utilized to complement observational notes and capture visual data for analysis. Moreover, shelter layouts were documented during the fieldwork using real-time sketching—as shown in Figure 4- and measurements were documented using a laser measurement tool. The shelter layouts presented in this article represent the conditions of the shelters as of August 2022.

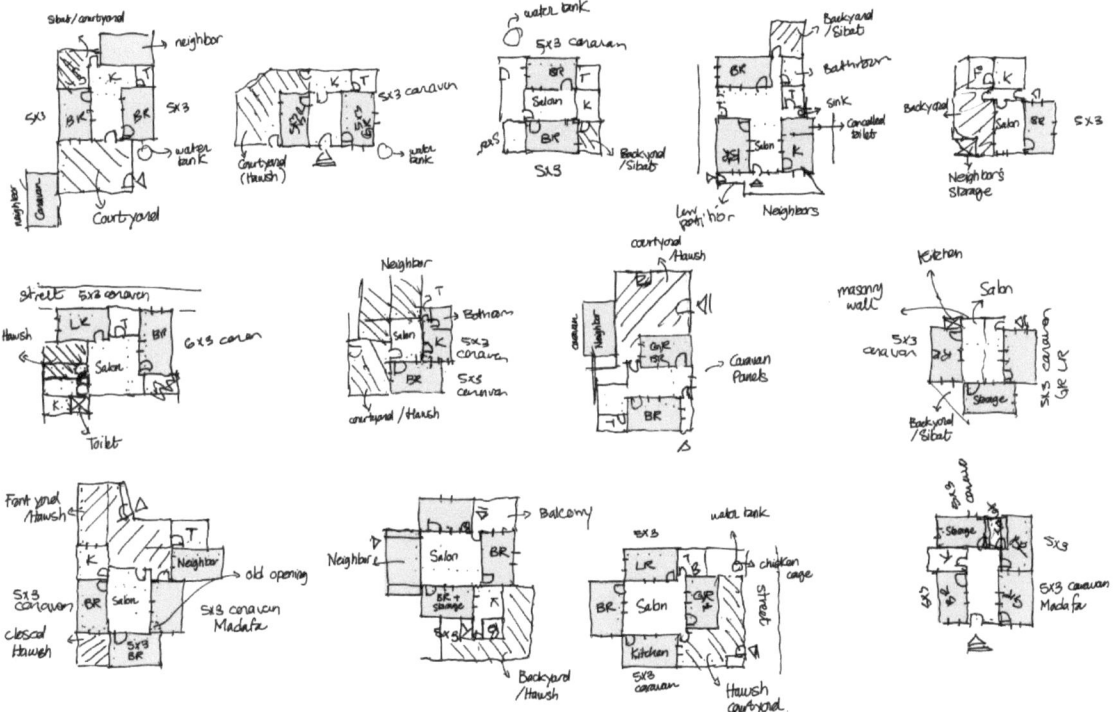

Figure 4. Examples of shelter layouts that were manually documented during the fieldwork.

3.2.2. Analytical Framework

Qualitative data analysis (Figure 5): The researcher manually transcribed the audio recordings of the interviews verbatim in Arabic. For data analysis, Atlas.ti 23 qualitative data analysis software was used to manage and analyze the data. The analysis process involved reading, re-reading, and coding the transcripts to identify significant categories and patterns. These categories were then classified into broader themes that aligned with this study's aims.

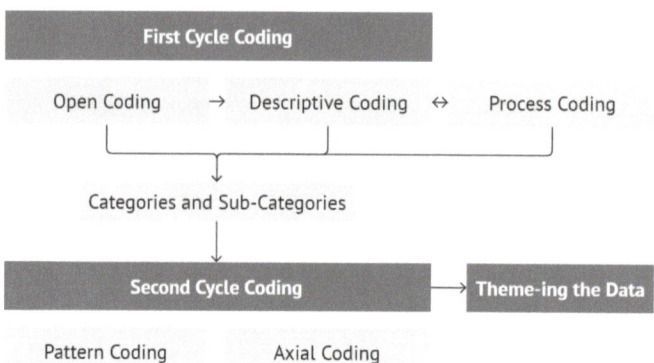

Figure 5. Qualitative data analytical framework.

Data were coded and analyzed using the coding methods described by Saldaña [68] following two cycles of coding. In the first cycle, open and descriptive coding was employed to break down the interview data into discrete parts, closely examining and comparing them for similarities and differences. Emotions and values coding were also incorporated in this cycle. In the second cycle, axial and pattern coding were used to group codes into categories and identify themes. This step involved identifying relationships among the open and descriptive codes to form categories and subcategories, aiding in understanding the broader context and connections between different codes.

Analysis of Shelter Layouts (Figure 6): Shelter layouts were studied to identify their main spatial relationships and structural components. First, all layout sketches were digitized in AutoCAD to prepare them for shape/layout analysis, which revealed recurrent patterns and configurations. Second, functional analysis was performed to understand the functional zones within shelters and understand how these zones are arranged within different shelter types. Third, key spatial recurring patterns were identified through the analysis. The following step involved contextual analysis to relate spatial patterns to social, environmental, and cultural factors influencing the shelter layouts.

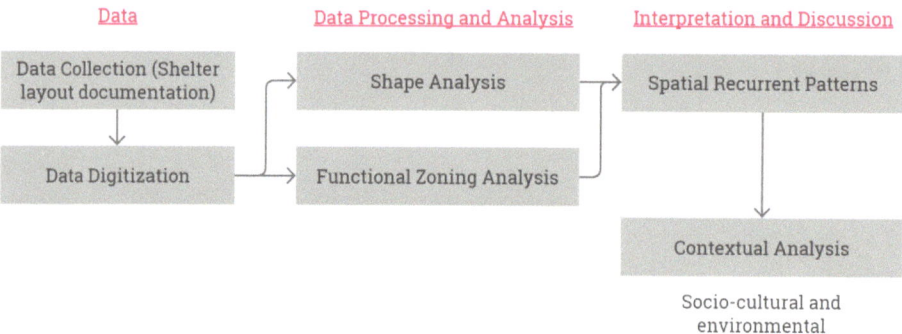

Figure 6. Spatial data analytical framework.

4. Results

4.1. Spatial Analysis Findings and the Zaatari Camp Grammar

4.1.1. Geometric/Shape Analysis

The shelter layouts in the Zaatari camp are dominated by rectangular shapes, with prefabricated structures (*Caravans*) consistently sized between 5 to 7.5 m in length and 3 m in width. These *Caravans* serve as the core units around which each shelter is organized. The layouts exhibit varied patterns, as shown in Figure 7, forming functional clusters that include living rooms, kitchens, bathrooms, sleeping areas, guest rooms, and outdoor spaces. These arrangements reflect different household sizes and needs while maintaining a consistent proportion and modularity in layouts.

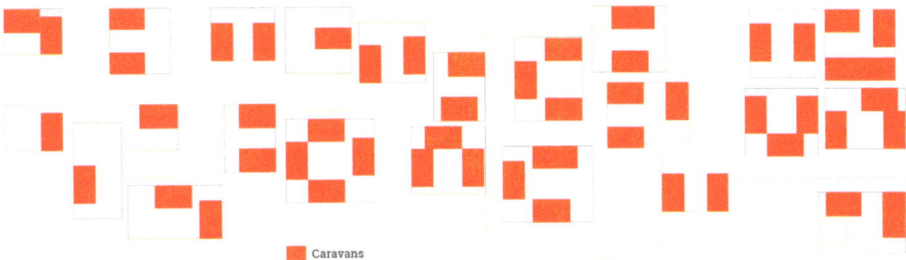

Figure 7. The positioning of Caravans in the Zaatari camp shelters.

Caravans are positioned to create clusters that facilitate the integration of indoor and outdoor spaces. Outdoor areas such as the *Sibat* (paved patios) and *Hawsh/Hakoura* (yards) are integral to the spatial configuration, providing additional areas for socializing and reflecting cultural preferences for outdoor living. Figure 8 illustrates the geometric relationships between indoor and outdoor spaces, highlighting the emphasis on ventilation, natural lighting, and social interaction.

Figure 8. Geometric/shape analysis of indoor and outdoor space relationships.

Outdoor spaces are often located adjacent to or between indoor areas, creating connections that enhance circulation and accessibility. Some layouts feature centralized outdoor spaces surrounded by indoor rooms, forming internal courtyards that offer privacy and access to open air. The integration of outdoor spaces within the layout underscores a refugees' practice and spatial configuration that values outdoor living, recreation, and cultural traditions. Functional zoning is evident, with specific types of indoor spaces linked to particular outdoor areas, creating organized and purposeful living environments.

4.1.2. Functional Analysis

The functional analysis of the Zaatari camp shelters reveals that the *Caravans* serve as central living spaces, accommodating multiple-purpose spaces such as sleeping, living, and guest areas. These multipurpose spaces are created to support daily household activities and provide rest at night. Temporary construction materials expand the core units, creating additional spaces for kitchens, toilets, and bathrooms [29], which are essential for cooking, hygiene, and food preparation.

Designated areas like the *Madafa* are used for hosting guests, while the *Sibat*, a front patio, serves as a space for outdoor seating and socializing, especially during cooler times of the day. The front and backyards, known as *Hawsh* or *Hakoura*, are created by constructing high fences, offering private outdoor spaces that reflect the pre-displacement rural lifestyle of many refugees, offering a sense of privacy and cultural continuity [29].

Access to shelters typically begins from the main street, leading through private outdoor spaces before reaching the indoor areas. The circulation layouts rely on the essential connection between indoor spaces and the private outdoor space (the *Hawash*). This connection is not just physical but also functional and cultural, as the private outdoor space is integral to the cultural background of the residents. It allows for easy movement between those indoor/outdoor spaces, facilitating social interactions and providing a sense of openness, yet is private. The *Sibat* and *Hawsh/Hakoura* facilitate smooth transitions and social interactions with close community members. Figure 9 illustrates the functional layout analysis of the shelters included in this study. In the Zaatari camp context, '*Sibat*' is used to describe a paved front patio or terrace. It serves as a transitional area between the indoors and the outdoors. It can be shaded or unshaded and is typically paved with a concrete screed. *Hawsh/Hakoura* is a term commonly used in rural regions of Syria, referring to farmland or a garden/farm attached to a house. In the context of the Zaatari camp, it is more than just a private outdoor space. It is a reflection of the pre-displacement rural lifestyle of many refugees, providing a sense of familiarity and comfort in their new environment. This concept is significant in rural areas, where many dwellings integrate agricultural spaces into their living environment [29].

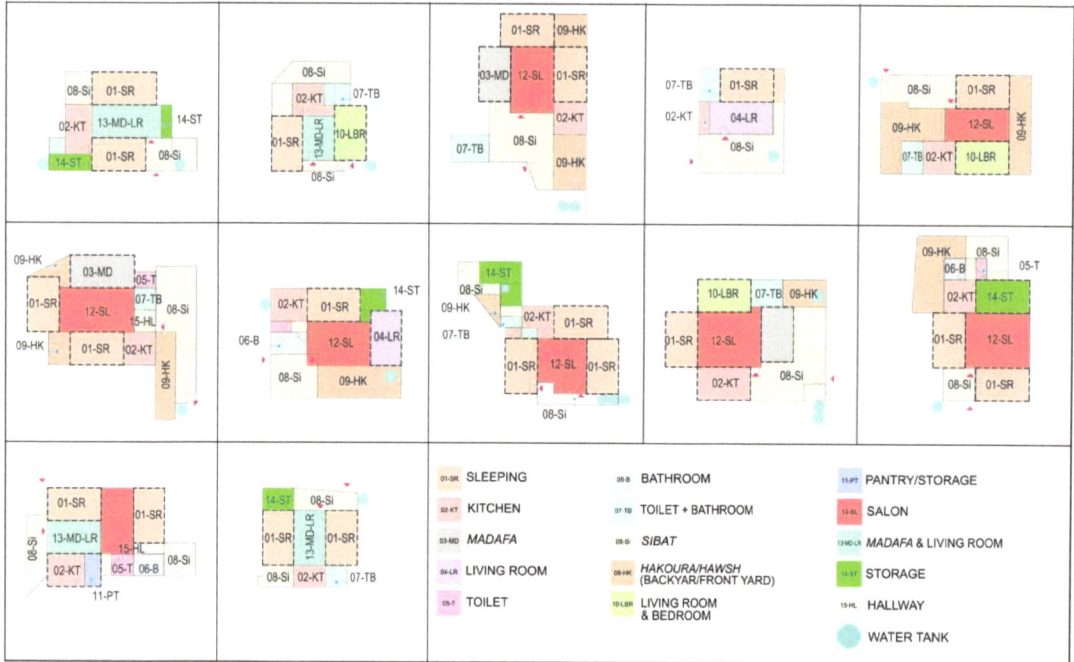

Figure 9. Example of the functional analysis performed for some of the Zaatari camp shelters based on the work of Abu-Aridah et al. [29].

4.1.3. Spatial Recurrent Patterns and the Zaatari Camp Grammar

The common spatial patterns observed in the Zaatari camp include clustered housing arrangements, fenced-in clusters, centralized space configurations, and modular layouts that integrate indoor and outdoor areas. These patterns reflect both functional requirements and cultural influences on spatial design. Figure 10 presents these spatial patterns, which are discussed further in Section 5.

Figure 10. Common spatial patterns observed in the Zaatari camp.

4.2. Qualitative Analysis Results

The analysis of the interviews highlights the complex and multifaceted experiences of refugees in camps. Key themes such as housing and infrastructure, community and social networks, and environment provide an understanding of their adaptation and social and spatial integration processes. These findings reveal that refugees undergo progressive adaptation, including adjusting to new living conditions, maintaining and re-establishing social ties, creating and recreating physical spaces, and finding new economic activities. Social and physical challenges such as social isolation, mental distress, and spatial constraints are also prominent. Figure 11 summarizes the qualitative research codes, categories, sub-categories, and themes extracted from the analysis process.

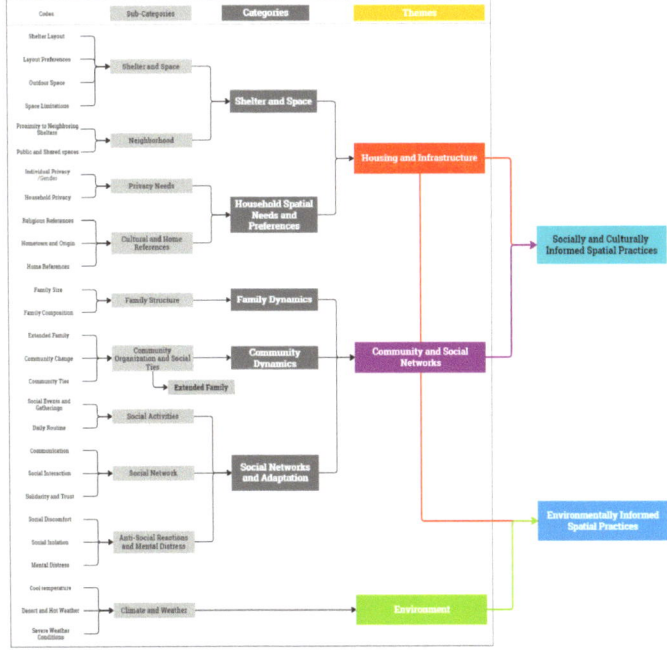

Figure 11. Summary of qualitative analysis codes, categories, sub-categories, and themes.

4.2.1. Housing and Infrastructure

The findings indicate that the layout of shelters and the organization of space within neighborhoods are critical for meeting privacy needs and maintaining cultural references, which are important for the inhabitants' sense of identity and well-being. Figure 12 shows the most common spatial features of the shelters in the Zaatari camp.

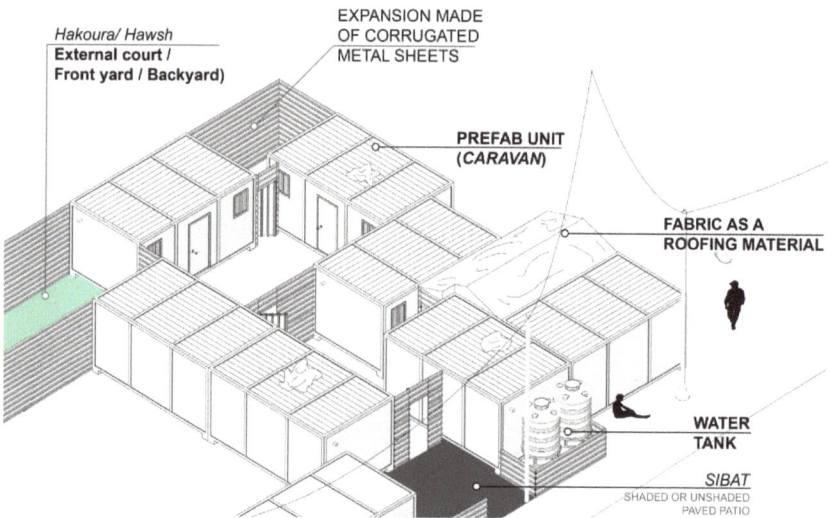

Figure 12. The most common spatial features of the shelters in the Zaatari camp.

1. Shelter and Space

Participants demonstrated resourcefulness in transforming their shelters into more livable spaces despite limited resources. Initially living in tents, households often replaced these with prefabricated units known as *"Caravans"*. Refugees used available materials such as corrugated sheets and leftover tent fabrics to adapt their shelters creatively. For example, some participants moved their prefabricated units by rolling them on empty cooking gas cylinders or lifting them onto locally made structures with wheels. These solutions illustrate resilience and adaptability in the face of adversity.

Challenges such as strict regulations against permanent construction materials and limited access to resources led participants to employ creative coping mechanisms. One participant utilized the boundaries of existing public structures, like communal kitchens, to establish initial boundaries for their own shelter. Another participant worked for an NGO and used the communal kitchen as a temporary shelter during severe weather, highlighting adaptability and resourcefulness.

Participants expressed preferences regarding the layout of their shelters, emphasizing the importance of spatial organization, layout preferences, and indoor–outdoor relationships to enhance usability. They often created multi-purpose areas due to space limitations, combining living, sleeping, and guest spaces. Participants also valued having their own outdoor spaces, preferring to stay indoors to avoid shared public spaces.

At the neighborhood level, familial relationships influenced neighborhood organization, with some districts being more crowded than others. Many residents chose to live close to relatives, maintaining strong ties within the community. However, overcrowding and inadequate spacing between units led to frustrations, impacting social interactions and overall comfort.

2. Household Spatial Needs and Preferences

Privacy emerged as a major concern, encompassing both social and spatial levels. Participants highlighted the importance of privacy in relation to gender- and age-related considerations when recreating their shelter spaces. Households with young children prioritized private sleeping areas, while those with older members emphasized gender separation and accommodating extended family.

Privacy needs extended to the neighborhood level, where participants created private outdoor spaces, such as the *Sibat*, an outdoor veranda surrounded by plants, for family gatherings, as explained by interviewees. The desire for privacy was driven by discomfort with the surrounding community and a need for spatial distancing.

Some participants expressed discomfort with the camp's diversity, indicating a need to recognize varied experiences within the camp setting. Efforts to recreate aspects of their homes were evident, with residents maintaining cultural practices from their hometowns in Syria such as frequent gatherings with close family members and acquaintances. These practices helped preserve some sort of cultural identity and provide a sense of continuity and normalcy at the smaller community level.

Religious considerations influenced shelter spaces, particularly regarding gender considerations. Participants emphasized the need for separate living spaces due to religious beliefs that command families to separate the sleeping areas of their kids after the age of seven.

4.2.2. Community and Social Networks

1. Family/Household Dynamics:

The analysis revealed themes related to family structure, including family gatherings, dynamics, and social connectivity. Family interactions and power dynamics highlighted how members navigate roles and responsibilities. Social connectivity underscored the importance of familial interaction with the broader social network. Camp life affected family dynamics, with an emphasis on self-sufficiency and privacy contrasting with the communal lifestyle in Syria. Isolation and change in the social structure in the camp led to fewer family gatherings and social events, but residents living among relatives maintained a sense of community and familiarity on a smaller community scale.

2. Community Dynamics

Whenever refugees had family or extended family members, they maintained social networks and support systems despite displacement. Community organization was influenced by extended family dynamics, emphasizing the importance of these relationships. Residents selectively engaged in social activities, prioritizing interactions with people from their own cultural backgrounds and maintaining traditional practices like family meals.

Community changes included decreased social interaction and increased safety concerns, particularly regarding children's freedom of movement. Cultural practices from Syria, such as family gatherings and community support, were upheld to recreate a sense of home and belonging among relatives and extended community members.

The analysis revealed how refugees maintain social networks and support systems despite being displaced from their homes. The emphasis on visiting and supporting each other during important life events underscores the cultural importance of community solidarity. Despite changes in their social environment, they continue to adhere to cultural norms and values, such as protecting their children and ensuring their socialization within trusted circles

3. Social Network and Adaptation

Social practices in the camp reflected a desire for interaction with peers, yet challenges related to building and maintaining social relationships persisted. Participants adapted social activities to camp constraints, holding weddings and events within familiar cultural settings.

While participants maintained close relationships with family, they exhibited a lack of trust in the broader community. Micro-community levels showed social solidarity and trust, absent at larger levels. Communication patterns changed, leading to more isolated and introverted social lives.

Social isolation and discomfort with diversity were prevalent, with participants avoiding conflicts and prioritizing family well-being. The lack of shared backgrounds contributed to social isolation and unfamiliarity among neighbors.

4. Mental Distress and Socio-cultural Change

The analysis revealed social, communal, and cultural changes due to the camp environment. Participants expressed frustration about shifts in gender roles, communal values, and social norms. Mental distress was prevalent, with emotions such as helplessness, anger, anxiety, and fear linked to safety concerns and an uncertain future.

4.2.3. Environment

The impact of climate and weather conditions is a significant factor affecting daily routines and overall comfort, influencing how people adapt their living spaces to cope with environmental challenges. Climate emerged as a significant factor in participants' narratives. Adaptation to harsh weather conditions was also a challenge for residents as they experienced difficulties with rain and cold, especially when living in tents. This code pertains to how climate conditions impact various aspects of creating more shaded areas, outdoor private spaces, and gathering spaces. One participant stated:

Sometimes, I mean, I place this chair outside; I sit there, and my neighbor comes. We sit and talk, and sometimes, my other neighbors join us! We arrange 3 or 4 chairs, and we sit together! We can either sit in the yard or outside the yard! It is usually cool outside— I mean, if there's no dust, it's cool and pleasant! We sit outdoors in the shade! At night, it gets very cold! Yesterday, I wore a jacket to fetch water at night! The water came last night.

One way to mitigate the effect of harsh environments is the households' effort to make green spaces within the boundaries of their shelters using their previous experience in Syria with self-sustenance through agriculture and planting.

Another environmental challenge that emerged from the analysis is water management, such as the need to secure more storage for water, as it is supplied only at limited times during the week. According to the participants, who came from the fertile agricultural regions of Syria, adjusting to the desert climate of the Zaatari camp proved to be a big challenge. The visible contrast between the agricultural green nature of their hometown and the dry, arid surroundings of the camp proved to be a difficult transition for them.

5. Discussion

Despite limited resources, the residents of Zaatari camp engaged in spatial practices that were socially, culturally, and environmentally driven, enhancing the livability of their physical spaces to the fullest extent possible with the resources available to them. Those spatial patterns were observed during the geometric and functional analysis of the shelter layouts included in this study. This aligns with the broader issues identified in the literature on refugee camps, which often fail to adequately support their residents' needs due to complex management and limited resources [12,13,22]. Figure 13 illustrates the main research outcomes that are discussed in this section, including the socially and culturally informed spatial practices and adaptation patterns, the environmentally informed spatial practices and adaptation patterns, the Zaatari Camp grammar, and the potential of Construction 4.0 technologies integration.

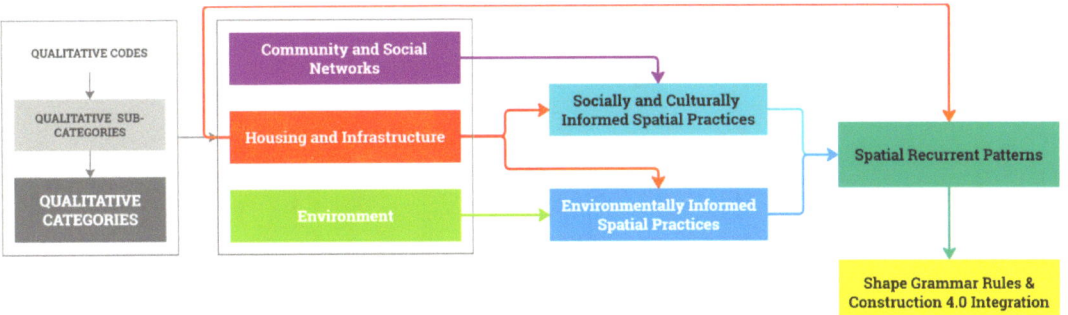

Figure 13. Summary of study's main themes and outcomes.

5.1. Socially and Culturally Informed Spatial Practices and Adaptation Patterns

The social and cultural aspects that significantly influenced the spatial patterns in the Zaatari camp include family size (as large households of five or more members are common), extended family structures and relationships, a preference for independent living, familial networks that assist with housing adjustments, and strong community involvement whenever possible. These elements are consistent with the findings of Tayob [23], who emphasized the importance of social dynamics and cultural significance in shaping spatial arrangements in migrant communities. Residents also strive to create a sense of family through traditional cultural practices and celebrations adapted to the camp environment. A strong attachment to their homeland leads them to replicate aspects of their spatial lives from home. This mirrors Ting's [31] observations on the ongoing process of homemaking among displaced populations, where familiar practices help recreate a sense of home despite displacement. The most common spatial configurations observed in the camp include clustered housing arrangements, fenced-in housing clusters, centralized space configurations, rectilinear and modular patterns, and integrated indoor–outdoor layouts.

Clustered Arrangements and Fenced-in Housing Clusters: Privacy, both personal and familial, was a primary catalyst for the development of clustered housing arrangements, as shown in Figure 14. This emphasis on privacy is indicative of the need for personal space in refugee settings, as discussed in the literature on refugee architecture [8,9,11–13]. Participants emphasized privacy as crucial in shaping their daily lives and interactions, requiring physical separation from neighbors while maintaining proximity to family. Concerns about privacy within and outside the home, coupled with a desire for a strong sense of community primarily among kin, led to the formation of these clustered living patterns. This is consistent with the findings of Paidakaki et al. [17], who noted that harsh conditions in camps often lead to social and cultural isolation.

Based on the social and physical circumstances faced by participants in the camp, there was a strong desire to stay indoors and socially isolate from the broader community. This behavior was likely due to discomfort with the surrounding environment and an attempt to maintain familiarity and security in their living spaces. Additionally, the participants' pre-displacement rural lifestyle, which was highly family-focused, significantly influenced their experiences and perceptions of the camp's environment, contributing to their discomfort and desire for isolation. This reflects the broader issues of social equity and livability in refugee camps, as highlighted in the literature [17]. As a result, they consciously attempted to establish micro-communities within the camp that mirrored their pre-displacement rural lifestyle. The spatial configuration of "fenced-in housing clusters," as depicted in Figure 15, is a clear manifestation of this phenomenon.

Furthermore, the refugees' tendency to prioritize their families and avoid interaction with the broader community resulted in specific behavioral patterns, including territoriality and a desire for more privacy. This was reflected in their spatial arrangements, with refugees constructing high fences around their shelters and creating large front and backyards enclosed by fences. These yards served as private spaces for families to spend time together without leaving the safety and familiarity of their territory.

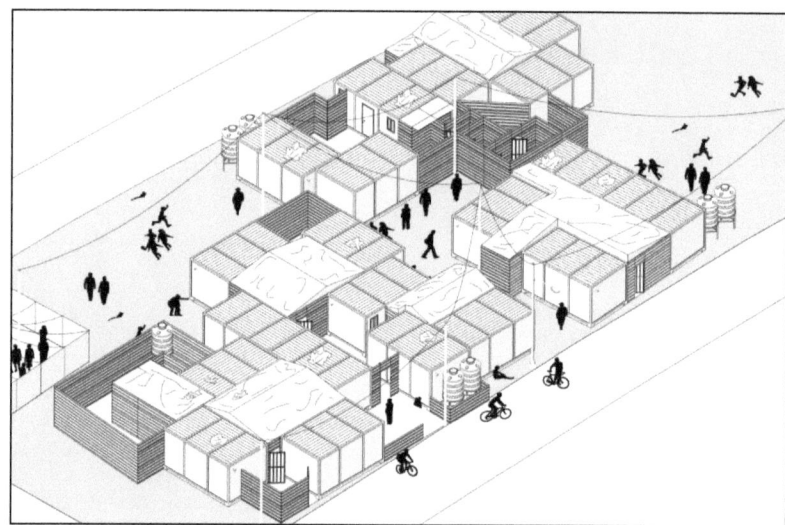

Figure 14. A common pattern of the clustered layout of shelters in one neighborhood/block in the Zaatari camp, as documented by the researcher.

Figure 15. Fenced-in housing clusters that are found in the camp.

Open-to-Inside Centralized Configurations and Rectilinear and Modular Layouts: The shelter units have a rectilinear shape with clear, modular divisions, as seen in Figure 16. This spatial configuration emerged as a response to using the *Caravans* as core structures, resulting in arrangements that maximize the efficiency of the available space while maintaining a sense of community and personal privacy for inhabitants. This adaptability and expandability are key characteristics of these units, reflecting the need for flexible and culturally respectful designs as emphasized in the literature on refugee architecture [10,16–19].

Figure 16. Clustering the Caravans about a central multi-purpose space.

Centralized indoor space is primarily utilized as a multi-purpose living area, combining functions such as sleeping, hosting guests, and daily living. The orientation of the housing units takes privacy into account, with entrances often facing away from the direct line of sight of other units or toward the central courtyards. Adaptability and expandability are key characteristics of these units.

Indoor-outdoor-integrated spatial layouts: In the Zaatari camp, the family structure typically includes multiple generations living together, usually within a housing cluster or a micro-community arrangement, reflecting the familial support systems common in many refugee communities. This reflects the importance of extended family structures in maintaining social ties and support networks, as observed in the literature [26,31].The presence of family gatherings and visits underscores the continued effort to preserve social ties and support networks, exemplified by the use of spaces like the *Sibat* and private indoor and outdoor gathering areas which resulted in the emergence of indoor–outdoor-integrated spatial layouts that can be seen in Figure 17. Interviewees often reflect on their homes in Syria, comparing them to life in the camp, which reveals cultural perceptions of what constitutes a home. There is a clear longing for familiar spaces and community dynamics from their homeland. The mention of extended family members, including children, siblings, and in-laws living nearby or in different districts of the camp, highlights the importance of extended family structures in their culture. These structures provide essential support and a sense of community continuity despite the displacement they have experienced.

Figure 17. Self-made outdoor space is an integral part of the housing patterns in the camp.

5.2. Environmentally Informed Spatial Practices and Adaptation Patterns

Climate and weather conditions significantly impact the daily lives of refugees in the Zaatari camp. The harsh desert climate necessitates creative adaptations to living spaces, such as the creation of shaded and outdoor gathering areas. These adaptations align with the need for environmentally sensitive design solutions highlighted in the literature [17]. Refugees draw on their agricultural experience from Syria to cultivate green spaces within their shelters, providing a sense of familiarity and resilience in the face of environmental challenges.

Water management emerges as a critical issue, with limited supply times necessitating innovative storage solutions. These environmental challenges highlight the need for sustainable design interventions that improve living conditions while respecting the cultural and social preferences of refugee populations. The need for such sustainable and efficient resource management solutions is consistent with the literature's emphasis on the integration of advanced technologies and sustainable practices in refugee camp design [44,45].

5.3. The Zaatari Camp Grammar

Abu-Aridah et al. [29] developed the Zaatari camp grammar, as shown in Figure 18. The figure illustrates some of the design and shape rules created for this grammar. This grammar, based on shape and contextual analysis, visualizes recurring spatial configurations and identifies the rules governing shelter layouts. It comprises five sets of rules detailing the step-by-step geometric and functional creation of shelters, including the placement of core units, additions of structures, assignment of functions, and installation of windows, doors, and other essential openings. Figure 19 demonstrates some design iterations of the application of these shape rules, highlighting how they guide the design and organization of shelter layouts in the Zaatari camp.

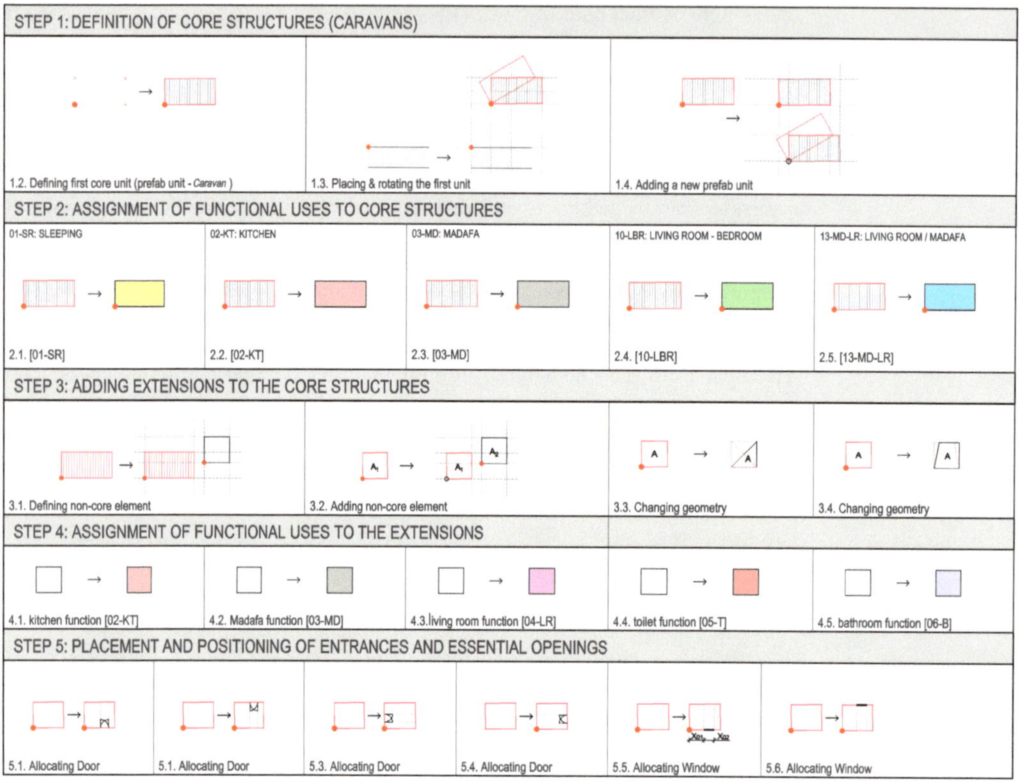

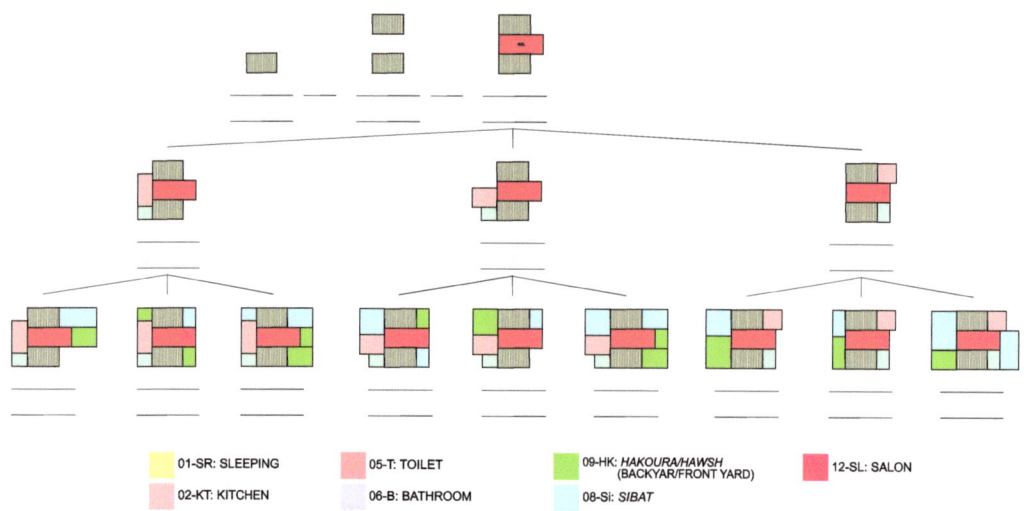

Figure 18. Some of the Shape rules for the Zaatari camp grammar [29].

Figure 19. Design iterations for shelters in the Zaatari camp using the Zaatari camp grammar.

The use of shape grammar to understand and generate spatial configurations aligns with the literature's discussion on the applicability of shape grammar in refugee camp design [33–43]. The grammar helps create new design iterations based on the studied spatial practices of the camp residents, providing a systematic approach to improving camp layouts and enhancing livability.

5.4. The Potential of Construction 4.0 Technologies Integration

The final step of this study involves integrating the outcomes of the social and spatial analysis with the Zaatari shape grammar into predictive analytics, BIM, and DT technologies. This study developed a four-step framework (shown in Figure 20) that will be transformed into a computational model to integrate social, environmental, and spatial data, as well as shape rules, with BIM and DT technologies. The first step is the data input using the social data that were extracted from the analysis of this research focusing on family size, and cultural preferences, the second step performing the predictive analytics, the third step is data processing and optimization, the fourth step is data integration in BIM, and the last step is data integration in DT. The four steps are as follows:

1. Data Input: this step includes identifying input data from social data sources, such as family size, cultural preferences, and environmental conditions.
2. Predictive Analysis: This step involves analyzing these data to understand privacy needs, spatial requirements, and community adaptation requirements. The integration of predictive analytics would benefit better allocation of the available limited resources, including construction materials, and access to infrastructure. It can also potentially improve shelter planning by determining the most suitable locations for shelters based on social dynamics, including community cohesion and social networks. Understanding social dynamics would also help camp managers create flexible, adaptive layouts that accommodate changes in population size and the growing needs of residents.
3. Data Processing and Optimization: this involves developing three algorithms:
   - Spatial Configurations Algorithm: identifies shelter and neighborhood spatial configurations based on the data input and analysis.
   - Environmental Adaptation Algorithm: identifies environmental needs for shelter creation.
   - Zaatari Camp Grammar Algorithm: identifies shape development requirements and applies shape rules.
4. BIM Integration: this comprises four steps:
   - Parametric Design with Revit Families: Involves creating Revit families with parametric shapes that respond to shape grammar rules. Adaptive components in Revit can create complex geometries that adapt based on points, allowing for dynamic shapes.
   - Using Dynamo for Computational Design: Starts with implementing shape grammar rules in Dynamo, a visual programming tool for Revit. Designers can define and manipulate geometry based on rules and parameters. The next step involves automating repetitive tasks, such as generating spatial patterns of shelters or urban layouts, using Dynamo scripts.
   - Design Automation and Iteration: Automates the generation of design alternatives using shape rules. This includes optimization and analysis, combining shape rules with optimization algorithms to find the best design solutions based on specific criteria (e.g., family size, maximizing space usage).
   - Visual Feedback: visualizes and adjusts shape rules in real time, allowing designers to see the impact of different rules on the overall design.
5. DT Integration: this comprises four steps:
   - Data Integration and Real-Time Updates: Includes dynamically adjusting shape grammar rules in the Revit model based on data from the physical twin.

- Predictive Analysis and Simulation: Involves scenario testing, where the digital twin simulates various scenarios using shape grammar rules to predict optimal outcomes and layouts, allowing for testing and refinement before making physical changes on-site.
- User Feedback: integrates occupant feedback into the digital twin, influencing adjustments to shape grammar rules, such as altering space layouts or environmental controls.
- Improvement and Optimization: Includes performance monitoring, where the digital twin continuously monitors shelter performance metrics and optimizes shape grammar rules for better performance. It also involves data-driven design iteration, with ongoing data collection and analysis informing future design improvements.

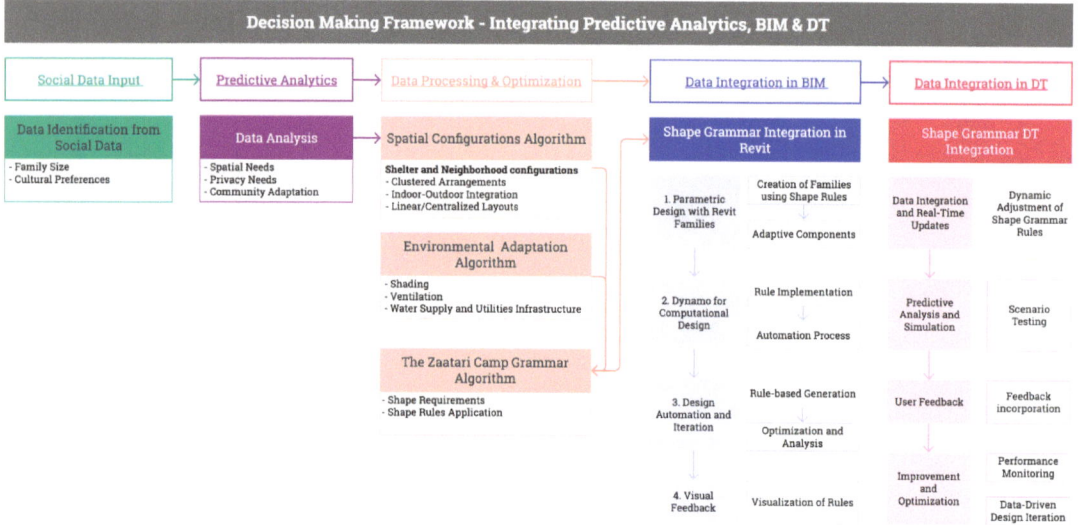

Figure 20. Decision-making framework for predictive analytics, BIM, and DT integration.

The integration of technologies such as Building Information Modeling (BIM) and Digital Twins (DTs) into refugee camp design offers significant opportunities for optimizing the design and management of these spaces. As highlighted in the literature [46,47,49,50,52], these technologies enhance the planning, coordination, and real-time management of refugee shelters, making them more adaptable and resilient. The discussion on the application of the Zaatari camp grammar emphasizes the importance of using such tools to create effective and sustainable living environments for displaced populations.

The step of DT integration comprises four steps:

- Data integration and real-time update, including the dynamic adjustment of shape grammar rules: based on the data received from the physical twin, shape grammar rules in the Revit model can be dynamically adjusted.
- Predictive analysis and simulation, which includes scenario testing, as the digital twin can simulate various scenarios using shape grammar rules, such as different configurations of shelter layouts, to predict optimum outcomes and layouts. This allows for testing and refinement before physical changes are made on-site.
- User feedback that incorporates the integration of occupant feedback that can be integrated into the digital twin. This feedback can influence the adjustment of shape grammar rules, such as altering space layouts or environmental controls.
- Improvement and optimization include performance monitoring, in which the digital twin can continuously monitor shelter performance metrics and optimize shape

grammar rules for better performance. They also include data-driven design iteration, where the integration of a digital twin allows for ongoing data collection and analysis, informing future design iterations and improvements.

The potential for integrating technologies such as Building Information Modeling (BIM) and Digital Twins (DTs) into refugee camp design offers opportunities for optimizing the design and management of these spaces. As highlighted in the literature [46,47,49,50,52], these technologies can enhance the planning, coordination, and real-time management of refugee shelters, making them more adaptable and resilient. The discussion on the Zaatari camp grammar's application underscores the importance of using such tools to create effective and sustainable living environments for displaced populations.

6. Conclusions

This study of the Zaatari camp's spatial configurations and refugee adaptation strategies highlights the complex interplay between social, cultural, and environmental factors. The findings reveal how residents, despite challenging conditions, employ spatial practices shaped by necessity and cultural continuity, demonstrating their resilience and adaptability. However, the lack of access to advanced expertise and resources could be addressed by integrating Construction 4.0 technologies, which offer the potential to enhance and optimize these practices.

Social and culturally informed spatial practices show that family size, privacy concerns, and community involvement significantly influence the camp's layout. Residents tend to create clustered housing arrangements and enclosed spaces to maintain family bonds, privacy, and security. These patterns underscore the importance of social and cultural factors in shaping refugee living spaces.

Environmental adaptation also plays a crucial role, with the camp's harsh desert climate driving the creation of shaded areas, green spaces, and innovative water management practices. These environmentally informed solutions highlight the residents' resilience and align with sustainable design practices that can improve living conditions.

The introduction of a Zaatari camp grammar provides a framework for understanding recurring spatial configurations and offers design rules for shelter layouts. This grammar enables systematic design generation, contributing to the development of culturally sensitive, adaptable, and livable spaces within the camp.

This study also emphasizes the potential of technological integration, particularly through Construction 4.0 technologies such as Predictive Analytics, Building Information Modeling (BIM), and Digital Twins (DTs). These tools offer significant opportunities for optimizing camp design and management, allowing for real-time data analysis, simulation of shelter configurations, and dynamic adaptation to changing needs. The proposed framework, which includes steps for data analysis, environmental adaptation, and the application of shape grammar rules, can enhance the flexibility and sustainability of refugee shelters.

In conclusion, this study underscores the need for a holistic approach to refugee camp design, one that incorporates both cultural and environmental factors alongside advanced technologies. Such an approach can create more humane, adaptable, and livable spaces for displaced populations, improving their overall well-being in challenging environments.

7. Study Limitations and Future Research

For future research, it is recommended that this research be expanded to other contexts and explored, based on refugees' socio-cultural preferences and experiences, how they recreate their built environment. Investigating sustainable design interventions to address environmental challenges, such as renewable energy sources and efficient water management systems, is also essential. Moreover, while this study outlines a framework for integrating predictive analytics, BIM, and DT technologies, future research should focus on practical implementation. This involves developing and testing computational models that integrate social, environmental, and spatial data with shape rules, providing real-time feedback and optimization.

Author Contributions: Conceptualization, D.A.-A. and R.L.H.; methodology, D.A.-A. and R.L.H.; validation, D.A.-A. and R.L.H.; formal analysis, D.A.-A.; investigation, D.A.-A.; data curation, D.A.-A.; writing—original draft preparation, D.A.-A.; writing—review and editing, D.A.-A. and R.L.H.; visualization, D.A.-A.; supervision, R.L.H. All authors have read and agreed to the published version of the manuscript.

Funding: This research received no external funding.

Data Availability Statement: The participants of this study did not give written consent for their data to be shared publicly, so due to the sensitive nature of the research, supporting data are not available.

Conflicts of Interest: The authors declare no conflicts of interest.

References

1. Chin, A.; Cortes, K.E. Chapter 12-The Refugee/Asylum Seeker. In *Handbook of the Economics of International Migration*; Chiswick, B.R., Miller, P.W., Eds.; North-Holland: Amsterdam, The Netherlands, 2015; Volume 1, pp. 585–658.
2. UNHCR. *Mid-year Trends 2023*; United Nations High Commissioner for Refugees: Copenhagen, Denmark, 2023.
3. UNHCR. Refugee Statistics. Available online: https://www.unrefugees.org/refugee-facts/statistics/ (accessed on 5 January 2022).
4. Alnsour, J.; Meaton, J. Housing Conditions in Palestinian Refugee Camps, Jordan. *Cities* **2014**, *36*, 65–73. [CrossRef]
5. Turner, S. What Is a Refugee Camp? Explorations of the Limits and Effects of the Camp. *J. Refug. Stud.* **2015**, *29*, 139–148. [CrossRef]
6. Duarte, J.P.; Beirão, J. Towards a Methodology for Flexible Urban Design: Designing with Urban Patterns and Shape Grammars. *Environ. Plan. B Plan. Des.* **2011**, *38*, 879–902. [CrossRef]
7. Lambe, N.R.; Dongre, A.R. A Shape Grammar Approach to Contextual Design: A Case Study of the Pol Houses of Ahmedabad, India. *Environ. Plan. B Urban Anal. City Sci.* **2019**, *46*, 845–861. [CrossRef]
8. Aburamadan, R.; Trillo, C.; Makore, B.C.N. Designing Refugees' Camps: Temporary Emergency Solutions, or Contemporary Paradigms of Incomplete Urban Citizenship? Insights from Al Za'atari. *City Territ. Archit.* **2020**, *7*, 12. [CrossRef]
9. Albadra, D.; Coley, D.; Hart, J. Toward Healthy Housing for the Displaced. *J. Archit.* **2018**, *23*, 115–136. [CrossRef]
10. Albadra, D.; Elamin, Z.; Adeyeye, K.; Polychronaki, E.; Coley, D.A.; Holley, J.; Copping, A. Participatory Design in Refugee Camps: Comparison of Different Methods and Visualization Tools. *Build. Res. Inf.* **2021**, *49*, 248–264. [CrossRef]
11. Ammoun, M.; Uzunoğlu, K. A Study on Flexible Cluster Units for Refugees Camps. *Eur. J. Sustain. Dev.* **2020**, *9*, 641–663. [CrossRef]
12. Dalal, A. The Refugee Camp as Site of Multiple Encounters and Realizations. *Rev. Middle East Stud.* **2020**, *54*, 215–233. [CrossRef]
13. Fosas, D.; Albadra, D.; Natarajan, S.; Coley, D.A. Refugee Housing Through Cyclic Design. *Arch. Sci. Rev.* **2018**, *61*, 327–337. [CrossRef]
14. Saad Alotaibi, B.; Ibrahim Shema, A.; Umar Ibrahim, A.; Awad Abuhussain, M.; Abdulmalik, H.; Aminu Dodo, Y.; Atakara, C. Assimilation of 3D printing, Artificial Intelligence (AI) and Internet of Things (IoT) for the Construction of Eco-Friendly Intelligent Homes: An Explorative Review. *Heliyon* **2024**, *10*. [CrossRef]
15. Bazli, M.; Ashrafi, H.; Rajabipour, A.; Kutay, C. 3D Printing for Remote Housing: Benefits and Challenges. *Autom. Constr.* **2023**, *148*, 104772. [CrossRef]
16. Kandylis, G. Accommodation as Displacement: Notes from Refugee Camps in Greece in 2016. *J. Refug. Stud.* **2019**, *32*, I12–I21. [CrossRef]
17. Paidakaki, A.; De Becker, R.; De Reu, Y.; Viaene, F.; Elnaschie, S.; Van den Broeck, P. How Can Community Architects Build Socially Resilient Refugee Camps? Lessons from the Office of Displaced Designers in Lesvos, Greece. *Archnet-IJAR Int. J. Archit. Res.* **2021**, *15*, 800–822. [CrossRef]
18. Brankamp, H. Camp Abolition: Ending Carceral Humanitarianism in Kenya (and Beyond). *Antipode* **2022**, *54*, 106–129. [CrossRef]
19. Dantas, A.; Banh, D.; Heywood, P.; Amado, M. Decoding Emergency Settlement through Quantitative Analysis. *Sustainability* **2021**, *13*, 13586. [CrossRef]
20. Singh, A.L. Arendt in the refugee camp: The political agency of world-building. *Political Geogr.* **2020**, *77*, 102149. [CrossRef]
21. Steigemann, A.M.; Misselwitz, P. Architectures of Asylum: Making Home in a State of Permanent Temporariness. *Curr. Sociol.* **2020**, *68*, 628–650. [CrossRef]
22. Avery, H.; Halimeh, N. Crafting Futures in Lebanese Refugee Camps: The Case of Burj El Barajneh Palestinian camp. *FormAkademisk* **2019**, *12*, 2. [CrossRef]
23. Tayob, H. Architecture-by-Migrants: The Porous Infrastructures of Bellville. *Anthropol. South. Afr.* **2019**, *42*, 46–58. [CrossRef]
24. Haggag, A.G.; Zaki, S.H.; Selim, A.M. Emergency camps design using analytical hierarchy process to promote the response plan for the natural disasters. In *Architectural Engineering and Design Management*; Taylor & Francis: London, UK, 2022; Volume 19, pp. 305–322. [CrossRef]
25. Gieseking, J.J.; Mangold, W.; Katz, C.; Low, S.; Saegert, S. *The People, Place, and Space Reader*; Routledge: London, UK, 2014.

26. Nilsen, A. The Scale of the City: The Social Dimension of Space in Theory and Method. In *Vernacular Buildings and Urban Social Practice: Wood and People in Early Modern Swedish Society*; Archaeopress Archaeology: Oxford, UK, 2021; pp. 6–18.
27. Robinson, E. Creating "People's Park": Toward a Redefinition of Urban Space. *Hum. Ecol. Rev.* **2019**, *25*, 87–110. [CrossRef]
28. Knoblauch, H.; Löw, M. The Re-Figuration of Spaces and Refigured Modernity—Concept and Diagnosis. *Hist. Soc. Res./Hist. Sozialforschung* **2020**, *45*, 263–292.
29. Abu-Aridah, D.; Henn, R.; Duarte, J.P. Using Shape Grammar as an Analytical Tool for Shelters in Protracted Refugee Camps: The Zaatari Camp Grammar. 2024; in press.
30. Dalal, A.; Fraikin, A.; Noll, A. Appropriating Berlin's Tempohomes. In *Spatial Transformations*, 1st ed.; Million, A., Haid, C., Ulloa, I.C., Baur, N., Eds.; Routledge: London, UK, 2021; pp. 285–293.
31. Ting, W.-C. Charting Interfaces of Power: Actors, Constellations of Mobility and Weaving Displaced Shan's Translocal 'Home' Territory along the Thai-Burma Border. *J. Refug. Stud.* **2018**, *31*, 390–406. [CrossRef]
32. Bourdieu, P. Social Space and Symbolic Power. *Sociol. Theory* **1989**, *7*, 14–25. [CrossRef]
33. Lefebvre, H. *The Production of Space*; Blackwell: Oxford, UK, 1991.
34. Löw, M. Foundations of a Sociology of Space—Summary. In *the Sociology of Space: Materiality, Social Structures, and Action*; Palgrave Macmillan US: New York, NY, USA, 2016; pp. 225–233.
35. Kreuzer, M.; Mühlbacher, H.; von Wallpach, S. Home in the Re-making: Immigrants' Transcultural Experiencing of Home. *J. Bus. Res.* **2018**, *91*, 334–341. [CrossRef]
36. Bourdieu, P. Physical Space, Social Space and Habitus. *Vilhelm Aubert Meml. Lect. Rep.* **1996**, *10*, 87–101.
37. Arvanitis, E.; Yelland, N. 'Home Means Everything to Me . . .': A Study of Young Syrian Refugees' Narratives Constructing Home in Greece. *J. Refug. Stud.* **2021**, *34*, 535–554. [CrossRef]
38. Mereine-Berki, B.; Malovics, G.; Cretan, R. "You Become One with the Place": Social Mixing, Social Capital, and the Lived Experience of Urban Desegregation in the Roma Community. *Cities* **2021**, *117*, 103302. [CrossRef]
39. Ruiu, M.L. The Social Capital of Cohousing Communities. *Sociology* **2016**, *50*, 400–415. [CrossRef]
40. Rutten, R.; Westlund, H.; Boekema, F. The Spatial Dimension of Social Capital. *Eur. Plan. Stud.* **2010**, *18*, 863–871. [CrossRef]
41. Kühtreiber, T. The Investigation of Domesticated Space in Archaeology-Architecture and Human Beings. In *Archaeology of Domestic Architecture and the Human Use of Space*, 1st ed.; Routledge: London, UK, 2014.
42. Million, A.; Haid, C.; Castillo Ulloa, I.; Baur, N. *Spatial Transformations: Kaleidoscopic Perspectives on the Refiguration of Spaces*, 1st ed.; Routledge: London, UK, 2022.
43. Huq, E.; Miraftab, F. "We are All Refugees": Camps and Informal Settlements as Converging Spaces of Global Displacements. *Plan. Theory Pract.* **2020**, *21*, 351–370. [CrossRef]
44. David, L.O.; Nwulu, N.I.; Aigbavboa, C.O.; Adepoju, O.O. Integrating Fourth Industrial Revolution (4IR) Technologies into the Water, Energy & Food Nexus for Sustainable Security: A Bibliometric Analysis. *J. Clean. Prod.* **2022**, *363*, 132522. [CrossRef]
45. Koh, L.; Orzes, G.; Jia, F.J. The Fourth Industrial Revolution (Industry 4.0): Technologies Disruption on Operations and Supply Chain Management. *Int. J. Oper. Prod. Manag.* **2019**, *39*, 817–828. [CrossRef]
46. Desogus, G.; Quaquero, E.; Rubiu, G.; Gatto, G.; Perra, C. BIM and IoT Sensors Integration: A Framework for Consumption and Indoor Conditions Data Monitoring of Existing Buildings. *Sustainability* **2021**, *13*, 4496. [CrossRef]
47. Palco, V.; Fulco, G.; De Capua, C.; Ruffa, F.; Lugarà, M. IoT and IAQ Monitoring Systems for Healthiness of Dwelling. In Proceedings of the 2022 IEEE International Workshop on Metrology for Living Environment (MetroLivEn), Cosenza, Italy, 25–27 May 2022; pp. 105–109.
48. Choi, E.J.; Park, B.R.; Kim, N.H.; Moon, J.W. Effects of Thermal Comfort-Driven Control Based on Real-Time Clothing Insulation Estimated Using an Image-Processing Model. *Build. Environ.* **2022**, *223*, 109438. [CrossRef]
49. Choi, H.; Crump, C.; Duriez, C.; Elmquist, A.; Hager, G.; Han, D.; Hearl, F.; Hodgins, J.; Jain, A.; Leve, F. On the Use of Simulation in Robotics: Opportunities, Challenges, and Suggestions for Moving Forward. *Proc. Natl. Acad. Sci. USA* **2021**, *118*, e1907856118. [CrossRef]
50. Edirisinghe, R.; Woo, J. BIM-based Performance Monitoring for Smart Building Management. *Facilities* **2021**, *39*, 19–35. [CrossRef]
51. Mancuso, I.; Petruzzelli, A.M.; Panniello, U. Industry 4.0 for AEC Sector: Impacts on Productivity and Sustainability. In *Architecture and Design for Industry 4.0: Theory and Practice*; Springer: Berlin/Heidelberg, Germany, 2023; pp. 33–50.
52. Santi, M. Digital Twins: Accelerating Digital Transformation in the Real Estate Industry. In *Architecture and Design for Industry 4.0: Theory and Practice*; Barberio, M., Colella, M., Figliola, A., Battisti, A., Eds.; Springer International Publishing: Cham, Germany, 2023; pp. 673–697.
53. Haakonsen, S.M.; Rønnquist, A.; Labonnote, N. Fifty years of Shape Grammars: A Systematic Mapping of its Application in Engineering and Architecture. *Int. J. Archit. Comput.* **2023**, *21*, 5–22. [CrossRef]
54. Paio, A.; Reis, J.; Santos, F.; Lopes, P.; Eloy, S.; Rato, V. Emerg.cities4all: Towards a Shape Grammar Based Computational System Tool for Generating a Sustainable and Integrated Urban Design. In Proceedings of the 29th Conference on Education in Computer Aided Architectural Design in Europe, eCAADe 2011, Ljubljana, Slovenia, 21–24 September 2011; pp. 152–158.
55. Verniz, D.; Duarte, J.P. Santa Marta Urban Grammar Towards an Understanding of the Genesis of Form. In Proceedings of the 35th International Conference on Education and Research in Computer Aided Architectural Design in Europe, eCAADe 2017, Rome, Italy, 20–22 September 2017; pp. 477–484.

56. Verniz, D.; Duarte, J.P. Santa Marta Urban Grammar: Unraveling the spontaneous occupation of Brazilian informal settlements. *Environ. Plann.* **2021**, *48*, 810–827. [CrossRef]
57. E Costa, E.C.; Verniz, D.; Varasteh, S.; Miller, M.; Duarte, J. Implementing the Santa Marta Urban Grammar a Pedagogical Tool for Design Computing in Architecture. In Proceedings of the 37th Conference on Education and Research in Computer Aided Architectural Design in Europe and 23rd Conference of the Iberoamerican Society Digital Graphics, Porto, Portugal, 11–13 September 2019; pp. 349–358.
58. Barros, P.S.; Beirão, J.N.; Duarte, J.P. The Language of Mozambican Slums Urban Integration Tool for Maputo's Informal Settlements. In Proceedings of the 31st International Conference on Education and research in Computer Aided Architectural Design in Europe, eCAADe 2013, Delft, The Netherlands, 18–20 September 2013; pp. 715–724.
59. Gips, J. Computer Implementation of Shape Grammars. In Proceedings of the NSF/MIT Workshop on Shape Computation, Chestnut Hill, MA, USA, January 1999; p. 56.
60. Stiny, G.; Gips, J. Shape Grammars and the Generative Specification of Painting and sculpture. In *Proceedings of the Information Processing 71*; North-Holland: Amsterdam, The Netherlands, 1972; pp. 1460–1465.
61. Schirmer, P.; Kawagishi, N. Using Shape Grammars as a Rule Based Approach in Urban Planning-a Report on Practice. In Proceedings of the 29th eCAADe Conference: Respecting Fragile Places, Ljubljana, Slovenia, 21–24 September 2011.
62. Bilecen, B. Home-making Practices and Social Protection Across Borders: An Example of Turkish Migrants Living in Germany. *J. Hous. Built Environ.* **2017**, *32*, 77–90. [CrossRef]
63. Garcia, M.; Haddock, S.V. Special Issue: Housing and Community Needs and Social Innovation Responses in Times of Crisis. *J. Hous. Built Environ.* **2016**, *31*, 393–407. [CrossRef]
64. Sanyal, R. Urbanizing Refuge: Interrogating Spaces of Displacement. *Int. J. Urban Reg. Res.* **2014**, *38*, 558–572. [CrossRef]
65. Zíla, O. The Myth of Return: Bosnian Refugees and the Perception of 'Home'. *Geogr. Pannonnica* **2015**, *19*, 130–145. [CrossRef]
66. UNHCR. *Handbook for Emergencies*, 3rd ed.; United Nations High Commissioner for Refugees: Geneva, Switzerland, 2007.
67. SphereAssociation. *The Sphere Hanbook: Humanitarian Charter and Minimum Standards in Humanitarian Response*; Sphere Association: Geneva, Switzerland, 2018.
68. Saldaña, J. *The Coding Manual for Qualitative Researchers*, 3rd. ed.; SAGE: Los Angeles, CA, USA, 2016.

Disclaimer/Publisher's Note: The statements, opinions and data contained in all publications are solely those of the individual author(s) and contributor(s) and not of MDPI and/or the editor(s). MDPI and/or the editor(s) disclaim responsibility for any injury to people or property resulting from any ideas, methods, instructions or products referred to in the content.

Systematic Review

Systematic Review of Factors Influencing Students' Performance in Educational Buildings: Focus on LCA, IoT, and BIM

Paulius Vestfal * and Lina Seduikyte

Faculty of Civil Engineering and Architecture, Kaunas University of Technology, LT-44249 Kaunas, Lithuania; lina.seduikyte@ktu.lt
* Correspondence: paulius.vestfal@ktu.edu

Abstract: In the evolving field of civil engineering studies, a significant transition is evident from fundamental to new-generation research approaches. This paper presents a systematic literature review aimed at analyzing these shifts, focusing specifically on the performance of students in educational buildings thought the integration of modern technologies such as the Internet of Things, life cycle assessments, and building information modeling. Covering the literature from the late twentieth century to the early twenty-first century, the review emphasizes advancements in sustainable infrastructure, eco-friendly designs, digitalization, and advanced modeling. A comparative analysis reveals that while the fundamental articles are primarily focused on indoor air quality parameters, the new-generation articles prioritize technological integration to address broader environmental concerns and for improved building performance. Challenges in the education sector, such as insufficient energy use, high maintenance costs, and poor working conditions, are also discussed, showcasing their impact on student learning outcomes. The methodology employed for this review included a comprehensive search in databases such as Scopus and Web of Science, using keywords such as "school buildings", "IoT", "BIM", and "LCA", ensuring a robust and diverse collection of academic articles. The findings show that new trends supplement existing topics, suggesting an integration rather than a replacement of traditional practices. Consequently, future research efforts will need to include a broader range of information to fully account for the evolving landscape in this field.

Keywords: LCA; BIM; IoT; CiteSpace; schools

1. Introduction

People spend 87% of their lifetime inside buildings [1], with approximately 30% of that time being spent by primary and secondary school students in a space dedicated to their learning, which is called the classroom [2]. Extensive research has demonstrated the negative impact of air pollution on health [3–9] and student academic accomplishments [10–12]. The classroom indoor air quality (IAQ) can influence the time students spend in classrooms and their academic performance due to illness-related attendance [13–17]. Multiple research studies have conducted assessments of classroom IAQ [18], revealing problems in poorly ventilated classrooms [19]. Many classrooms discussed in these studies did not have enough fresh air circulation. The circulation was below the levels recommended by the American Society of Heating, Refrigerating, and Air Conditioning Engineers (ASHRAE) for maintaining good indoor air quality [18,20].

A group of studies [21–26] looked into how the environment and features of school buildings affect student learning and success. In these previous studies, it was found that if indoor conditions such as ventilation [15,27–33] are inadequate, this could lead to health issues for students and staff, which can also result in reduced concentration, attendance, and academic performance. Air pollution has been shown to harm cognitive abilities, affecting memory, attention, visual processing, and problem-solving. In a recent study, it was discovered that if students are taught in premises with bad air quality, they are more likely to get lower grades, and this exposure to air pollution is linked to decreases in academic performance [32]. These findings indicated that environmental aspects such as indoor air quality and air pollution play a crucial role in student learning and success. It is suggested that making improvements in the quality of the school environment could enhance student outcomes and test scores [23,24,31]. Studies conducted in Europe and the USA have also shown a connection between students' academic performance and the ventilation rates in classrooms [34,35]. Wargocki et al. (2013) discovered that indoor air conditions, using CO_2 levels as an indicator of ventilation rates, affected student performance by causing more errors and slower task completion [35]. Additionally, the findings of Crosby et al.'s study emphasized the advantages of prioritizing energy efficiency in school infrastructure. They highlighted how this approach not only enhances the educational quality but also fosters a healthier learning environment, yielding numerous benefits [36]. Studies that highlighted the importance of various factors for student performance are shown in the Table 1.

The research by Economidou et al. (2011) revealed that within the European Union, educational structures encompass approximately 17% of the non-residential building inventory in terms of square meters. According to this statistic, educational buildings represent the third most substantial sector, trailing behind wholesale and retail buildings (at 28%) and offices (at 23%) [37]. However, not as many studies have looked into how the indoor environment affects well-being and work productivity in educational buildings compared to other types of structures. Mendell et al. (2005) [15] and Wargocki et al. (2013) [35] reviewed the existing research, combining the results from practical tests and experimental studies. They connected negative health effects and lower student performance to inadequate indoor temperature or air quality conditions [15,35]. This article extends their work by integrating modern technological trends such as IoT, BIM, and LCA, which were not the focus of earlier studies. While both this article and the earlier studies are focused on educational buildings, the key differences lie in the timelines and technological advancements, as the previous articles are over 10 years older. In contrast to office buildings, which are designed for the purpose of profit, educational buildings, created for non-profit purposes, have not been as thoroughly examined from the perspective of a life cycle cost/life cycle assessment (LCC/LCA) [38].

The fundamental articles primarily focused on parameters related to indoor air quality, particularly heating and ventilation. The earliest discussions about school ventilation dated back to the late 1800s, originating mainly within the medical field. While the authors of these articles primarily addressed heating and ventilation factors [39–41], they laid the background knowledge that would later influence discussions on indoor air quality in educational settings. However, only in the mid-1900s did articles specifically devoted to school ventilation begin to appear in engineering and environmental science journals. Since then, there has been a significant growth in the number of articles on this topic, and there have been two notable increases (Figure 1). The first increase occurred between 2007 and 2011; that is, around the time the era of Industry 4.0 began. Schwab (2017), in his book *The Fourth Industrial Revolution*, defined Industry 4.0 as the fourth industrial revolution, which involves the integration of smart, interconnected machines and systems (such as automated manufacturing robots and smart grids), the Internet of Things (IoT), and network-based operations (such as real-time data analytics and cloud computing) in various industries, including construction and education [42]. Next, there was a decrease, possibly because many researchers moved on to the Industry 4.0 field and started writing about that instead. The second increase occurred in 2019, during the time when the COVID-19 pandemic was affecting the world.

Sustainability aspects are important in all fields, including research. They are important because they help us balance the economy, people's well-being, and taking care of the environment. This represents an ability to meet today's needs without compromising the ability of future generations to meet their needs [43].

Figure 1. Number of publications by year in the Scopus database (keyword "school ventilation").

The usage of sustainable methods is crucial for schools. These methods help create schools that care about the environment and use resources wisely. There are different building certification systems such as Leadership in Energy and Environmental Design (LEED) [44], the Collaboration for High Performance Schools (CHPS) [45], the Building Research Establishment Environmental Assessment Method (BREEAM) [46], Green Star—Education V1 [47], and WELL [48] that give guidelines to make sure that the school environment is more sustainable. When certifying buildings using the mentioned systems, factors such as saving energy, using water wisely, using environmentally friendly materials, and making sure the IAQ is clean are taken into consideration.

The authors of this article conducted a systematic literature review, discussed in the Materials and Methods section, and looked into 25 published studies (Table 1), analyzing the various factors that affect students' academic performance. These 25 articles were selected based on the criteria listed in Table 1, which were chosen by the authors of this article. The table is part of an introduction and review of already existing articles to highlight the importance of investigating more parameters under one field of study in the future, as one parameter alone cannot impact students' performance as significantly as the combined effect of multiple parameters. The table could have included more articles if additional investigated parameters had been considered. The chosen articles specifically addressed parameters that influence students' performance, while other articles that were not included discussed topics such as the school surroundings, structures, BIM, and IoT in schools but did not focus on their impact on students' performance, addressing schools in general instead. However, no author has covered all of the factors mentioned in the table together in one study. The analyzed parameters were the reverberation time (RT), lighting (LI), indoor air quality (IAQ), temperature (T), CO_2 level, student testing (ST), short-term academic performance (S-TAP), cognitive performance or learning efficiency (CP), acoustics (A), air velocity (AV), and relative humidity (RH). Of the 25 published studies, two studies included a total of eight parameters, which was the highest number of parameters included. Additionally, three studies each covered seven, six, and five parameters, while two studies only covered four things. Three studies covered three parameters and one study covered two parameters, while eight studies covered one parameter.

After looking more closely, the authors found that different studies focused on different parameters (Table 1). The reverberation time was discussed in two publications, while lighting received attention in seven publications. Among the parameters studied, short-term academic performance and acoustics were addressed in four publications each, indicating a lower level of focus. The air velocity received slightly more attention and was discussed in five publications. CO_2 levels and student testing were equally covered, which were mentioned in 10 publications each. Moreover, the relative humidity was covered in 10 publications. Indoor air quality ranked third in popularity among the parameters studied and was discussed in 14 publications. The most popular parameter, however, was cognitive performance or learning efficiency, receiving extensive coverage across 17 publications; it was of critical importance in the academic research. These findings underscore the diverse exploration of parameters influencing student performance within the academic literature.

Table 1. Published studies about parameters that affects students' performance.

Reference	Investigated Parameter										
	RT	LI	IAQ	T	CO_2	ST	S-TAP	CP	A	AV	RH
Brink et al. (2023) [49]	+	+	+	−	+	+	+	+	−	−	−
Choi et al. (2014) [50]	−	+	+	+	−	−	+	+	−	−	−
Brink et al. (2021) [51]	−	+	+	+	+	+	+	+	+	−	−
Kim et al. (2012) [52]	−	+	+	+	−	−	−	−	+	−	−
Xiong et al. (2018) [53]	−	+	−	+	−	+	−	+	+	−	−
Calderón-Garcidueñas et al. (2008) [54]	−	−	−	−	−	−	−	+	−	−	−
Gardin et al. (2023) [55]	−	−	−	−	−	−	−	+	−	−	−
Duque et al. (2022) [56]	−	−	−	−	−	−	−	+	−	−	−
Kabirikopaei et al. (2021) [57]	−	−	+	+	+	−	−	+	−	+	+
Gaihre et al. (2014) [13]	−	−	−	+	+	−	−	+	−	−	+
Kielb et al. (2015) [14]	−	−	+	−	−	−	−	−	−	−	−
Mendell et al. (2005) [15]	−	−	+	+	+	+	−	+	−	−	+
Shendell et al. (2004) [16]	−	−	+	+	+	−	−	−	−	−	−
Wargocki et al. (2017) [17]	−	−	+	+	+	+	−	+	−	+	−
Requia et al. (2022) [58]	−	−	−	−	−	−	−	+	−	−	−
Guo et al. (2010) [59]	−	−	+	+	−	−	−	−	−	−	+
Richmond-Bryant et al. (2009) [60]	−	−	-	+	−	−	−	−	−	+	+
Rivas et al. (2014) [61]	−	−	+	−	−	−	−	−	−	−	−
Martínez-Lazcano et al. (2013) [62]	−	−	−	−	−	−	−	+	−	−	−
Forns et al. (2017) [63]	−	−	−	−	+	−	−	+	−	−	−
Benka-Coker et al. (2021) [64]	+	+	+	+	+	+	−	+	−	−	+
Choi et al. (2022) [65]	−	-	+	-	+	+	−	+	−	−	+
Wang et al. (2020) [2]	−	+	−	+	−	+	−	+	+	+	+
Shan et al. (2018) [38]	−	−	+	+	+	+	−	−	+	+	+
Ryan et al. (2022) [66]	−	−	−	+	−	−	−	−	−	−	+

The aim of this study was (i) to analyze patterns and trends in the scientific research related to school buildings, with a particular focus on thermal comfort and IAQ, as well as the integration of LCA, IoT, and BIM; (ii) to highlight the differences between the fundamental generation and new-generation articles; and (iii) to point out new trends in the research on school buildings in the civil engineering field, emphasizing the roles of LCA, IoT, and BIM.

2. Materials and Methods

We adopted a systematic literature review (SLR) approach [67,68] to establish a reliable evidence base for future research in schools. The systematic literature review approach is characterized as "a scientific process governed by a set of explicit and demanding rules oriented towards demonstrating comprehensiveness, immunity from bias, and transparency and accountability of technique and execution" [69]. This method has been criticized because it oversimplifies the research evidence, which might have reduce the results we obtained [70]. However, recently, there has been a tendency to include strong qualitative studies along with quantitative ones [71], which helps deal with this criticism to some extent.

The research question was formulated using the PICO method as outlined in Table 2. The complete research query was delineated as follows: "What differentiates the fundamental and new-generation approaches in the research of school buildings?" Additionally, in conducting the search, consideration was given to various filters pertinent to the investigation. These filters encompassed the research field, including engineering, construction, and technology, with a focus on publications in the English language, without restriction on the year of publication. In our methodology, we selected key themes such as LCA, IoT, and BIM to guide our systematic literature review. These themes were chosen due to their significance in the current research on school buildings and their potential to influence factors such as thermal comfort and indoor air quality (IAQ). The inclusion criteria were based on the relevance of these themes to the civil engineering field and their impact on educational environments. These themes were chosen to capture a broad spectrum of research studies related to technological advancements and methodologies in the construction and operation of educational facilities, ensuring comprehensive coverage of the relevant literature. The types of studies and publications included in the search were review papers, field studies, research papers, and technical reports to encompass both theoretical and practical insights.

Table 2. PICO table.

P	Population; problem; source of information	What population? What is the database? What is the source of the information?	Population: School buildings Database: Scopus and WOS Sources: Review papers, field studies, research papers, and technical reports
I	Intervention; factors	What interventions or factors are you interested in?	Differences between fundamental and new-generation topics in school buildings
C	Comparison; circumstances; situation	What circumstances are you interested in? What will you compare it to?	Comparison of fundamental and new-generation topics in school buildings.
O	Outcome; main point of interest	What do you expect to learn about? Dependent variable? Main focus?	To find out the differences and similarities between fundamental research topics and new-generation research topics. The main focus is parameters tested in schools and performance of students.

The search for relevant texts was conducted across databases including Scopus and Web of Science (WOS). Scopus was chosen for its extensive coverage of peer-reviewed literature in the fields of science, technology, and engineering, while Web of Science was selected for its multidisciplinary indexing and high-quality sources, providing a robust and diverse collection of academic articles [72]. These databases are widely recognized and respected in the academic community, ensuring that the search results are comprehensive and reliable.

The filters and key words were carefully selected to include relevant research studies from diverse but related fields, focusing on studies that explore the integration of advanced technologies in educational settings and their potential impact on various parameters affecting students' performance. By not restricting the year of publication, we ensured that both historical and contemporary perspectives were considered, allowing for a thorough understanding of the evolution and current state of research in this area.

The eligibility and inclusion and exclusion criteria were based on the relevance and quality of the selected studies. For the research area, only studies related to civil engineering (engineering, environmental sciences, and energy) were included. This focus ensured that the selected studies were related to the technical and environmental aspects of educational buildings, encompassing critical factors such as building design, construction materials, sustainability practices, and energy efficiency, which directly impact students' performance and well-being. Studies not related to civil engineering were excluded to maintain this article's connection to the field of civil engineering. Regarding the topic, the inclusion criteria encompassed studies on LCA, BIM, IoT, and educational buildings, as these areas are central to understanding how advanced technologies and methodologies can improve the design, operation, and impact of educational buildings and the performance of students. Studies focused on non-educational buildings, such as those related to industry, commercial, and residential buildings, were excluded to ensure this article was targeted and pertinent to its primary objective—educational buildings. The years of publication ranged from 1800 to 2023, allowing for a comprehensive historical perspective on the development and evolution of the research in this field, thereby identifying trends, advancements, and shifts in focus over time. The language criterion included only English-language publications, as English is the predominant language of the scientific literature in engineering and technology, ensuring accessibility and consistency in comprehension and interpretation. Non-English studies were excluded to avoid potential challenges related to translation and interpretation that could have affected the accuracy and consistency of this article. The inclusion criterion for the publication source specified peer-reviewed academic journals and technical reports to ensure the inclusion of high-quality, strictly checked research. While other sources such as conference papers can provide valuable insights, they often lack the same level of peer-reviewed scrutiny essential for maintaining the integrity and reliability of a literature review. Peer-reviewed journals and technical reports are recognized for their methodological rigor and scientific validity, which is crucial for a comprehensive and accurate analysis of the factors affecting students' academic performance. The inclusion and exclusion criteria are shown in Table 3.

Table 3. Eligibility and inclusion and exclusion criteria.

Criterion Type	Inclusion Criteria	Exclusion Criteria
Research area	Related to the civil engineering	Not Related to the civil engineering (e.g., the arts or humanities).
Topic	LCA, BIM, IoT, educational buildings	Not educational buildings (e.g., industrial, commercial, and residential buildings)
Year of publication	1800–2023	Outside the set range
Publication source	Peer-reviewed academic journals, technical reports	Other type of sources
Language	English	Other languages
Type of publication	Review papers, field studies, and research papers	Other types of publication

The following search string was used in the Scopus database to search for the articles: (TITLE-ABS-KEY (educational AND building AND bim) OR TITLE-ABS-KEY (school AND building AND bim) OR TITLE-ABS-KEY (school AND building AND iot) OR TITLE-ABS-KEY (educational AND building AND iot) OR TITLE-ABS-KEY (educational AND building AND lca) OR TITLE-ABS-KEY (school AND building AND lca). In the Scopus database, 895 documents were found before using the filters. In order to decrease the number of documents, several filters were applied. These included the subject areas of engineering, environmental science, and energy. Only articles and reviews written in English were considered. After the filters were applied, 304 documents were found. The search of the Scopus data base was performed on 17 March 2024.

When searching for documents in the WoS database, a slightly different query string was required compared to the one used in the Scopus database due to variations in their field tags. In WoS, the following string was used: "educational building bim" (All Fields) OR "school building bim" (All Fields) OR "school building iot" (All Fields) OR "educational building iot" (All Fields) OR "educational building lca" (All Fields) OR "school building lca" (All Fields). The search was carried out on 20 March 2024. In the WoS database, 8993 results were found. After the initial search, additional filters were applied. Specifically, document types such as articles, review articles, and data papers were included. Moreover, for the WoS categories, fields including civil engineering, construction building technology, environmental sciences, green sustainable science technology, engineering environmental, engineering multidisciplinary, and environmental studies were selected. After applying the filters, 3524 documents were found. Following the second search, additional filters were applied to further refine the results. Specifically, filters were applied to publication titles such as *Sustainability*, *Buildings*, *Journal of Cleaner Production*, *Journal of Building Engineering*, *Energy and Buildings*, *Journal of Construction Engineering and Management*, *Engineering Construction and Architectural Management*, and *Building and Environment* to reduce the number of documents to 1311.

Both the Scopus and WoS databases contained a total of 2206 publications, with only 59 duplicates identified. These duplicates were detected and excluded from the future SLR processes with the assistance of the Mendeley reference manager tool.

The second step of the database search was the title and abstract screening. Only the first 300 of the most relevant articles from the Scopus and WoS databases, with 150 from each database, related to IAQ and academic performance, which were included for further review. In Scopus and WoS, relevance is determined by the databases' built-in algorithms, which prioritize articles based on factors such as keyword matching, citation counts, publication recency, and overall impact in the field [42]. By selecting the most relevant articles, we ensured that the included studies were highly related to the research topic and likely to contribute valuable insights. This relevance sorting helped in identifying the most significant and influential studies, thereby enhancing the quality and focus of the review. The title and abstract selection process in the Scopus and WoS databases resulted in 150 publications from each database. Selecting 150 articles from each database, for a total of 300, was a strategic choice that balanced the need for comprehensiveness with practical limitations, ensuring a manageable yet representative and high-quality review process.

Further, the complete texts were read to evaluate whether the studies mentioned factors influencing students' performance. Following a thorough review, 58 publications were excluded. Ninety-two publications were chosen for the SLR.

Later, the data collection process involved using the bread-crumbing method, which is a technique where the references of a publication are checked to find additional relevant publications. Twenty-eight more publications were included.

After completing all data collection steps, a total of 120 publications were included in the SLR and analyzed (Figure 2).

The CiteSpace program was used to create visualizations. These visualizations were developed using data from the Web of Science database, focusing on exploring the dynamic landscape of life cycle assessments (LCAs), the Internet of Things (IoT), and building information modeling (BIM). A broader perspective was gained through a general visualization that encapsulated key aspects such as authorship, references, and cited authors. The study covered the period from January 2020 to December 2023, allowing for a comprehensive understanding of trends and contributions over time. Importantly, the scholarly impact assessment used a g-index with a scale factor of $k = 25$, adding a nuanced layer to the evaluation of significance in the fields of LCAs, IoT, and BIM. This method, combined with the visualization capabilities of CiteSpace, contributed to a deeper comprehension of the interconnected aspects in these fields, providing valuable insights for future research and strategic advancements.

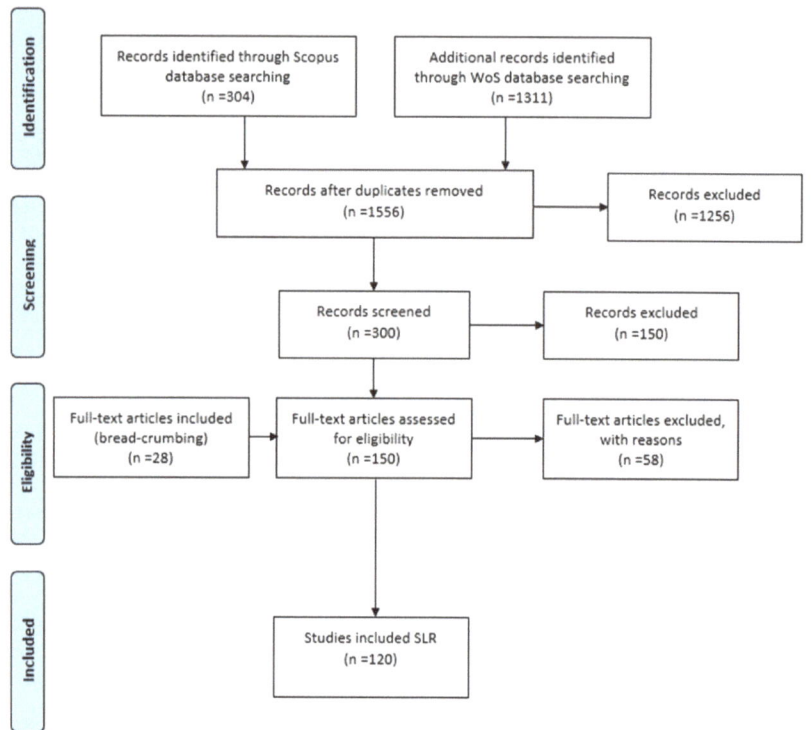

Figure 2. PRISMA 2009 flow diagram.

In the CiteSpace program, there are a few important concepts that help us understand scholarly networks and how they change. Clusters are groups of closely connected nodes, showing themes or research communities [73–75]. A burst, according to CiteSpace, is a sudden increase in the frequency of a particular type of events. It helps us see when there is a lot of activity or significance in a dataset. Centrality points out nodes that are important in the overall network, indicating their central position among many connected nodes. Lastly, CiteSpace uses the Sigma metric to measure the importance of a node in a network of cited references [76,77]. Sigma helps highlight structurally important nodes, showing rapid growth in citations, which is known as citation abundance [47]. This temporal aspect helps researchers identify nodes with growing influence, giving a detailed understanding of how specific scholarly works become more impactful and prominent over time.

Mongeon et al. (2016) screened the WoS and Scopus databases to see whether there were any unfair preferences. They found that both of these databases favored topics related to natural sciences, engineering, and biomedical research rather than social sciences and the arts and humanities [49]. Despite both of these databases having other advantages, such as WoS having good coverage going back to 1990, with most of its journals being in English and granting broader access to readers, Scopus covered more journals in total, which mostly included recent articles.

The literature review was carried out using both the WoS and Scopus databases. However, the visualizations were created only using the WoS database because it has more English publications. We chose to focus on visualizing IoT, BIM, and LCAs because these are significant trends influenced by Industry 4.0, reflecting current developments. The visualizations aimed to provide insights into these key areas, recognizing their importance in today's world.

3. Results

In the literature review, we distinguished two types of scientific papers for the analyzed topic—fundamental and new-generation articles (Figure 3). The term "new-generation articles" has come about as technology has advanced, particularly with the influence of the era of Industry 4.0. In contrast to the fundamental research focusing on basics such as CO_2, ventilation, heating, and IAQ, the new-generation articles highlight current trends such as IoT, digital twins (DTs), LCAs, BIM, and more. This categorization highlights the shift in research focus and emerging trends in the study of school buildings.

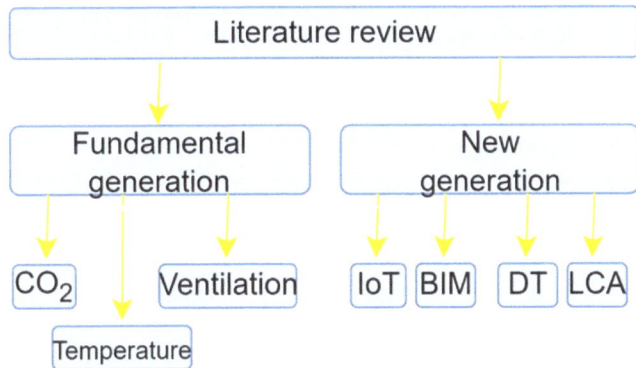

Figure 3. Differences between fundamental and new-generation articles.

3.1. Life Cycle Assessment (LCA)

Evaluating schools' sustainability and its impact on human well-being often involves an LCA, which is a method that helps achieving certifications such as the BREEAM and LEED. Vigovskaya et al. (2018) underlined the significance of LCAs in evaluating environmental impacts, emphasizing the crucial role of assessment results in adjusting the design and construction methods for improved energy efficiency and reduced environmental footprints [78]. Alshamrani et al. (2014) integrated LCAs with LEED to assess sustainability in 109 LEED-certified schools, revealing insights into energy, materials, and design choices. Their study recommended revisions in LEED integration for better functionality and proposed enhancements such as indoor air quality analyses and broader applicability across diverse climates and building types [79]. Meanwhile, Changyoon Ji et al. (2016) studied 23 buildings across South Korea, utilizing an LCA to evaluate impacts such as the global warming potential (GWP) and acidification potential (AP), finding that factors such as the gross floor area (GFA) and geographical location significantly influence these impacts [80].

Brás et al. (2015) researched mortars used in a 1980s school in Portugal, targeting thermal bridge issues. Their study compared various mortar types (cement–cork, cement–EPS, and hydraulic lime–cork), assessing the building's energy performance using original and new mortars, examining factors beyond the operational energy. According to their study, traditional mortars such as cement-based and hydraulic lime mortars significantly contributed to global warming, while cork-infused mortars, especially those with a 70% cork content, notably reduced CO_2 emissions by 30%. Cork-based mortars possessed lower embodied energy values, contrasting with EPS, which caused escalated energy consumption. Their research indicated that using mortars with less embodied energy, such as cement cork, helped reduce the operational energy requirements. Additionally, it emphasized a rapid decrease in heating needs and CO_2 emissions over time with cork-based options. In the initial eight years, embodied energy accounted for 30% of the school's operational energy [81].

The comparison by Pachta et al. (2015) of modern and historic school buildings revealed that the environmental performances vary, despite the similar location and operational demands. A modern school exhibited a significantly higher environmental impact, while a historic school consumed more operational energy due to its extended lifespan and absence of insulating materials [82]. Additionally, Gamarra et al. (2018), in their LCA research, examined a high school student's environmental impact over a school year, highlighting transport and mobility as the most significant contributors to climate change impacts (69%). Their study linked different impacts to material and energy consumption, emphasizing the role of educational activities in influencing various environmental aspects. Another study by Gamarra et al. (2018) focused on two pilot schools in Madrid, Spain, assessing the cumulated energy demand (CED), water resource depletion (WRD), and carbon footprint (CF) per student and per built gross area. The mentioned study stressed the importance of enhancing the conditioning and lighting efficiency to mitigate global warming effects and lower the overall energy consumption [83]. Furthermore, the analysis by Munoz et al. (2017) of an educational building's construction and operational aspects identified challenges in meeting the nZEB standards due to contributions from embodied materials and limitations in power generation. Their recommendations were intended to improve sustainability through a life cycle energy analysis (LCEA), advocating for materials with lower embodied energy levels and establishing clear classifications for non-residential buildings (NRBs) encompassing various types and energy services [84].

Figure 4 illustrates the principal clusters identified in the LCA visualization. The network discerns a total of six clusters, each encapsulating distinct facets of LCA research. This visualization provides a valuable overview, allowing for an in-depth exploration of the multidimensional landscape of LCA research, covering areas such as environmental sciences, materials science, construction and building technologies, and energy and fuels.

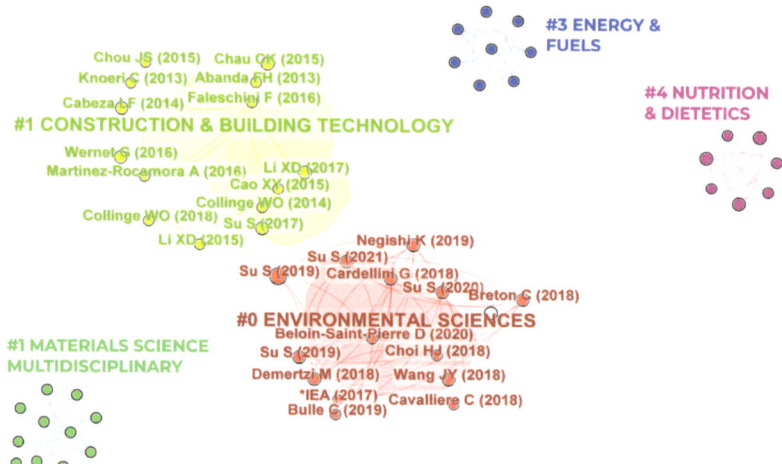

Figure 4. CiteSpace visualization of the multidimensional fields of LCAs and cluster analyses for SLR research [85–88].

The examination of the provided data revealed significant patterns in categories such as citation counts, bursts, degrees, centralities, and sigma values across various clusters. In clusters #0 and #2, as shown in Figure 4, the most cited author was Su et al. (2019) [85], Li et al. (2017) [86], Su et al. (2020) [87], and Su et al. (2021) [88], accumulating a total count of 16 citations. The citation for Su emphasized their considerable influence in environmental, construction, and building technology sciences. Notably, Su was consistently mentioned in each visualization category, indicating that Su held the utmost relevance in the field of LCA research.

3.2. Building Information Modeling (BIM)

In the architecture, engineering, and construction (AEC) sector, BIM serves various purposes, such as for 3D visualization; clash identification; feasibility assessments; cost estimations; scheduling; environmental, BREEAM, LEED, and other analyses; shop drawing generation; and facilities management [89–92]. BIM plays a role in simulating emergency scenarios such as fire evacuations, helping optimize school layouts for occupant safety and swift evacuations [93,94]. Additionally, BIM facilitates early-stage activities such as code compliance assessments, cost estimations, and sustainability analyses. It fosters collaborative efforts and empowers designers to evaluate building element performance and the environmental implications of sustainable design methods [92,95–97].

The study conducted by Zhuang et al. (2021) [98] introduced a framework called performance-integrated building information modeling (P-BIM), which focuses on designing for energy efficiency and environmental optimization. This framework organizes environmental data into various dimensions throughout the life cycle of a green building, ensuring standardized storage and interactions of this information. P-BIM is notable for freeing BIM platforms from data type constraints, allowing the integration of diverse performance data. It enhances the capacity of BIM for handling localized, customized, and big data, ensuring adaptability to evolving project requirements. The prototype demonstrated significant improvements in indoor environmental quality and cost reductions, highlighting the importance of detailed digital performance evaluations. By bridging the gap between the initial and late design stages, P-BIM empowers architects to better control the project life cycle. However, the study identified limitations, calling for further refinements, including refining the IEQ indicators, expanding the occupant satisfaction research, integrating more variables into the optimization model, broadening the application of P-BIM in the early design stages, and exploring energy efficiency during the construction and operation phases.

The challenge in this study was in managing fluctuating engagement levels and input across various stages of user participation in school design. While AEC professionals maintained a consistent and vested interest throughout (for instance, structural engineers overseeing the school building's structure at all phases), the involvement of school management professionals, teachers, and students proved significant in offering insights during development. For example, in an article written by Liu et al. (2018), during the design phase, students actively engaged in shaping the building's design through 3D walkthroughs, contributing ideas that influenced the final structure [92,99]. However, their participation tended to decrease during the realization phase, leading them to primarily receive what AEC professionals had crafted during the operational stage [100,101]. Ensuring the continual presence of management professionals, teachers, and students remained crucial in enhancing the efficacy and dependability of user engagement in the design process.

Figure 5 illustrates the main clusters identified through the BIM visualization. The complex network consists of a total of 11 clusters, each representing a different aspect of BIM application and research. In order to focus on the most significant contributors, the subsequent analysis zeroes in on the seven largest clusters, identified as #0 (FEMA p-58), #1 (building technology), #2 (industry foundation classes), #3 (risk assessments), #4 (educational training), #5 (team-based learning), and #6 (geometric quality inspections). These clusters cover various dimensions of BIM, ranging from its applications in risk assessments and building technology to educational training and team-based learning.

When looking into the details of each cluster, cluster #0, centered around FEMA p-58, addresses the complexities of disaster response and management within the BIM framework. Cluster #1 explores the realm of building technology, delving into innovative applications and advancements within the construction domain. The industry foundation classes in the cluster #2 highlight standardization and interoperability, playing a crucial role in enhancing collaborative efforts across BIM platforms.

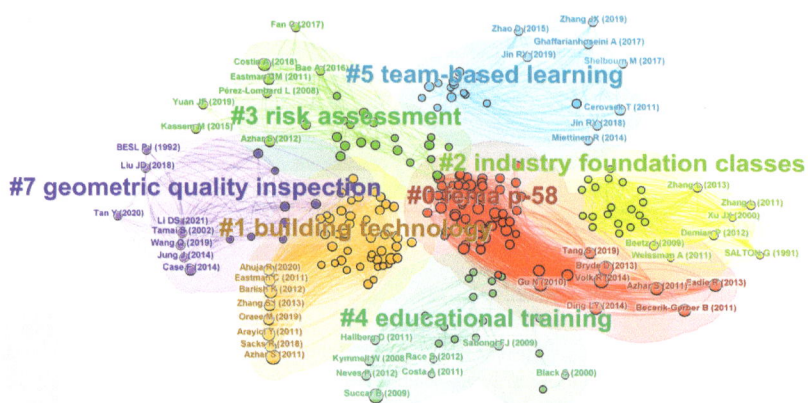

Figure 5. CiteSpace visualization of the multidimensional fields of BIM and a cluster analysis for SLR research [102–104].

Risk assessments, featured in the cluster #3, address the evaluation and mitigation of potential risks in BIM applications. When moving to cluster #4, the focus shifts to educational training, shedding light on the pedagogical aspects of BIM adoption and knowledge dissemination. The team-based learning in cluster #5 emphasizes collaborative practices within BIM, promoting effective teamwork in the industry. Lastly, cluster #6 delves into geometric quality inspections, unraveling the intricacies of BIM's role in ensuring precision and quality in geometric representations.

Based on the visualization in Figure 5, three of the most significant authors in the BIM category were identified. Succar (2009) [102] was notable in cluster #4 for educational training, Volk (2014) [103] in cluster #1 for FEMA p-58, and Azhar (2011) [104] in cluster #1 for building technology. These three authors were consistently the most mentioned in each visualization category.

3.3. Internet of Things (IoT)

The incorporation of IoT technology, including wireless sensors and computer networks, significantly advanced the concept of the smart environment during the recent technological revolution [105]. IoT technology has undergone extensive development and increased usage across various sectors such as social networks, infrastructure, security, business, and healthcare [106]. Integrating IoT-based intelligent monitoring systems, such as environmental and energy monitoring systems, with human involvement has emerged as a promising approach within the smart city framework, aiming to improve human health and overall well-being [107].

Amaxilatis et al. (2017) [108] conducted research focusing on measurements by IoT devices in two primary areas—power consumption and environmental comfort within school buildings. Power consumption meters, strategically placed on the general electricity distribution board of each building, assessed both the apparent and average power usage levels across the three-phase power supply. During another study, Amaxilatis et al. (2017) [108], using environmental comfort meters, evaluated factors crucial to occupants' well-being, including thermal satisfaction, visual comfort in terms of available light perception, and overall noise exposure. Room occupancy was tracked using passive infrared sensors (PIR). Beyond the building, weather and atmosphere stations provided comprehensive data on outdoor atmospheric conditions, including precipitation levels, wind dynamics, atmospheric pressure levels, and concentrations of specific pollutants. The insights from these atmospheric meters offered a clear view of pollution levels around the school buildings.

Hossain et al. (2020) [109] and Martínez et al. (2021) [110] both explored the monitoring of environmental parameters in educational and office settings using IoT technologies and sensor networks. They measured key factors such as CO_2, relative humidity, and temperature levels. Hossain et al. (2020) specifically targeted parameters such as the dry bulb temperature, illuminance, and sound pressure levels, while Martínez et al. (2021) focused on a broader set of measurements, including the light intensity, presence detection, and energy consumption. Both studies emphasized the importance of sensor placement for accurate data collection and recognized the value of historical and real-time data visualizations for managing building performance. Differences arose in the parameters measured, with Hossain et al. (2020) including sound pressure levels, which were not addressed by Martínez et al. (2021), who delved deeper into user-centric services and security protocols at the user level. Kamel et al. (2022) [111] adapted IoT technology for smart fire systems. Paganelli et al. (2019) [112] described IoT monitoring endpoints within classrooms, utilizing a mix of commercial hardware, sensor vendors, and open-source solutions, regularly collecting measurements such as power consumption values, environmental data, weather conditions, and air pollution levels.

In their comprehensive research, Mylonas et al. (2018) [113] implemented a network of 880 sensing points across various categories. These were strategically organized into four distinct groups, including sensors for monitoring classroom environments, assessing outdoor atmospheric conditions, gathering data from the weather stations on rooftops, and tracking power consumption through meters connected to the main building's electricity panels. This extensive deployment aimed to capture diverse data crucial for their study's comprehensive analysis and insights.

This section provides detailed insight into the major clusters within the network, incorporating both citing articles and references (Figure 6). The significance of network nodes is thoroughly examined using diverse metrics to measure their impact. Citation-based metrics, such as counts and bursts, highlight the scholarly influence of nodes, while network-based metrics such as the degree centrality and betweenness centrality offer insights into their structural importance. Additionally, the sigma metric combines the burst and betweenness centrality to provide a holistic measure of node importance.

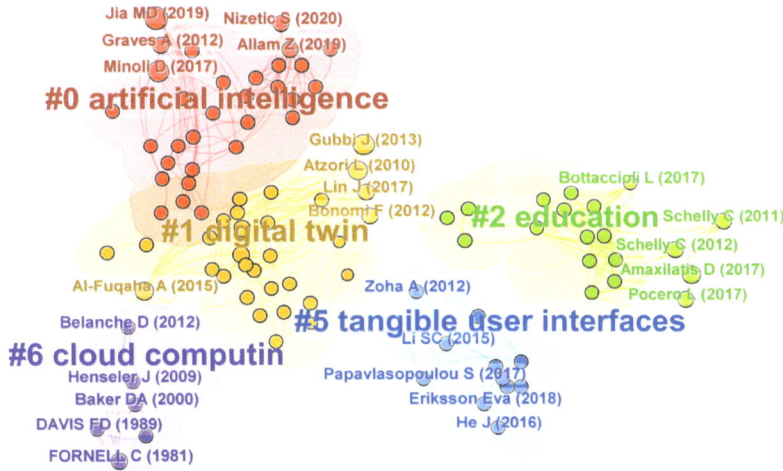

Figure 6. CiteSpace visualization of the multidimensional fields of IoT and cluster analyses for SLR research [114–116].

Figure 6 shows the main clusters of the IoT visualization, revealing the complex connections within the network. There were a total of 10 clusters in the network, each making a unique contribution to the extensive field of IoT-related research. These clusters covered various research areas, including artificial intelligence (#0), digital twins (#1), education (#2), big data analyses (#3), cloud computing (#4), tangible user interfaces (#5), digital libraries (#6), and educational materials (#7).

In the IoT field, several prominent authors appeared in the various visualization categories. For example, Jia (2019) [114] was a notable author in cluster #0, leading in citation counts (100in total). Additionally, Jia appeared in the degree and centrality categories. Such authors as Atzori (2010) [115] and Gubbi (2013) [116] from cluster #1 were also mentioned multiple times in various categories, including the citation count, degree, and centrality categories.

3.4. Digital Twins (DTs)

Digital twins represent a cutting-edge topic that is currently trending in various industries. However, within the scope of this SLR, it was observed that there is a lack of publications on school-related digital twins. Instead, the majority of publications on digital twins predominantly focus on aspects such as architecture, development, modeling, software platforms, and frameworks. This gap suggests a need for deeper and more comprehensive research and field studies to investigate how digital twins in school buildings affect student academic performance. Further exploration in this area could provide valuable insights into the application of digital twins in the educational field and the implications for enhancing the learning environment.

4. Discussion

4.1. Indoor Environment Factors

In this study, various authors examining factors related to students' academic performance considered different parameters. While many articles focused on external factors such as chemicals from streets and power plants, this study specifically looked at the indoor environment of schools and the factors influencing students within the school premises.

4.2. New-Generation vs. Fundamental Articles

When talking about the new-generation and fundamental articles, the first type focuses on modern technologies such as IoT, LCA, and BIM, while the second one concentrates on traditional aspects such as heating, ventilation, and IAQ. However, both types of articles remain relevant today. The research review indicated a significant growth in the number of articles on school ventilation, with notable surges over 2007–2011. The first increase occurred between 2007 and 2011; that is, around the time the era of Industry 4.0 began. Next, there was a decrease, possibly because many researchers moved on to the Industry 4.0 field and started writing about that instead. The second rise occurred in 2019, during the global impact of the COVID-19 pandemic, and no subsequent decrease was observed thereafter.

The comparison between fundamental and new-generation articles reveals a clear evolution in the research focus and methodology. Fundamental articles, dating from the late 19th century to the mid-20th century, primarily address basic and traditional aspects of IAQ, ventilation, heating, and microclimate. These studies provide essential environmental parameters such as CO2 levels, ventilation rates, and temperature values, offering foundational knowledge for maintaining healthy and comfortable indoor environments in school buildings. However, they are limited by the technology of their time, leading to challenges in implementation and a narrow focus that does not consider the broader impacts of technological integration and sustainability practices. Additionally, the earlier methods for collecting and analyzing data were less complex, potentially affecting the accuracy and comprehensiveness of the findings.

In contrast, new-generation articles, emerging from the late 20th century to the early 21st century, reflect the influence of Industry 4.0 and focus on modern trends such as

IoT, BIM, DTs, and LCAs. These studies emphasize sustainable infrastructure practices, eco-friendly designs, digitalization, and advanced modeling techniques. Despite their advantages, these articles face barriers in their technological adoption and resource allocation, with the complexity and cost associated with implementing advanced systems presenting significant difficulties. Nevertheless, integrating the foundational knowledge from fundamental articles with the innovative solutions from new-generation articles presents a comprehensive approach to improving school building environments. Future research should explore optimizing the integration of these technologies to overcome existing barriers, thereby enhancing the IAQ, energy efficiency, and overall sustainability in educational buildings.

4.3. Interdisciplinary Research Findings

This paper extensively combined diverse findings from clusters such as FEMA P-58, LCAs, environmental impacts, construction management, risk assessments, BIM education, and smart building. When analyzing citation metrics, bursts, centrality measures, and sigma metrics across these clusters, significant patterns and contributions by authors such as Su, Succar, Azhar, and Jia were identified, revealing their impact across various domains. This combination highlights the interdisciplinary nature of the research fields, showcasing interrelationships among environmental sciences, materials science, construction technology, and energy domains.

4.4. Focus on IAQ and Overlooked Aspects

Even though IAQ is a basic and vital subject, the contemporary new-generation articles are focused on digitalization and sustainability, often overlooking the impacts on students and teachers. It is interesting to note that there is a lack of scientific research on digital twins in schools in databases such as WoS and Scopus. The rise of new-generation articles began in the late 20th century and has continued into the early 21st century, while the fundamental articles can be traced back from the late 19th century to the mid-20th century.

Most of the articles that are related to LCA discuss important topics such as IAQ; heating, ventilation, and air conditioning (HVAC); and the global warming potential (GWP), and although the articles are different, all of them cover these listed topics [79,83,117–119]. However, it is interesting that not all of these papers thoroughly cover related subjects such as energy usage, sustainability, or the use of advanced modeling approaches such as 3D modeling and BIM, even though these are important in LCAs. This observation shows that the authors do not always grant equal importance to all aspects of am LCA. Some focus more on IAQ, HVAC, and GWP aspects, while not giving as much weight to things such as energy use, sustainability concerns, or the use of advanced modeling techniques such as 3D modeling or BIM.

4.5. Database Comparison and Article Selection

The research revealed that the WoS database contained a greater number of publications compared to the Scopus database when using the same search string. In particular, Scopus displayed 895 publications, whereas WoS exhibited 8993 publications, which amounted to a total of 9888. After applying various filters on the databases' search engines and identifying 59 duplicates using the Mendeley reference manager, a total of 1556 publications were included for the screening phase. Subsequently, during the screening phase, 1256 publications were excluded, leaving only 300 publications for detailed screening. From this detailed screening, half of the articles were selected for the full-text reading phase, also known as the eligibility phase. Within the full-text reading phase, 32 publications were excluded for various reasons, such as their focus on school surroundings rather than the schools themselves. Ultimately, only 118 publications were chosen for the systematic literature review (SLR).

When looking at scientific articles on databases such as Scopus and WoS, clear criteria must be established to distinguish next-generation articles from fundamental articles. This paper defines new-generation articles as those published from the late 20th century to the early 21st century. This distinction is based on a change in focus. Fundamental articles, from the late 19th century to the mid-20th century, mainly emphasized aspects such as CO_2, sustainability, IAQ, ventilation, and microclimates. On the other hand, new-generation articles focus on exploring modern trends and practical uses of technologies such as IoT, DTs, LCAs, and BIM. In order to identify these next-generation articles, publications that delve into the real-world impacts, progress, and integrative aspects of new technologies in various scientific fields are often sought. All of these factors differentiate these articles from the seminal discoveries and theoretical foundations found in the fundamental articles.

5. Conclusions

In terms of sustainability in schools, adopting eco-friendly methods is crucial. Programs such as LEED [44], CHPS [45], BREEAM [46], Green Star—Education V1 [47], and WELL [48] provide guidelines to ensure schools are more sustainable by emphasizing energy savings, efficient water usage, and maintaining clean IAQ.

The digitalization of schools has the potential to enhance students' cognitive performance. However, more research is needed to establish a connection between digitalization, IAQ, and the impact on academic performance.

The incorporation of new-generation approaches addresses various challenges in the education sector, such as insufficient energy use, high maintenance costs, and poor working conditions. These improvements are crucial, as they directly impact classroom air quality and consequently student learning outcomes. This study underscores the importance of creating healthier and more efficient learning environments through advanced technological integration.

The comparison between fundamental and new-generation articles shows a clear shift in research focus. The fundamental articles were focused on IAQ parameters, while the new-generation articles emphasize using broader technologies, such as LCAs, BIM, and IoT, to address environmental issues and improve building performance. This indicates a more comprehensive approach to solving the diverse challenges in educational building environments.

This study offers important insights into how the civil engineering research has evolved, highlighting the use of modern technologies to tackle both old and new challenges in school buildings. The findings suggest that a balanced approach, which mixes traditional methods with new technologies, can boost sustainability and improve student performance.

For future research, it is recommended to explore the influence of new-generation topics, including BIM, IoT, and LCAs, and their implementation in schools on students' academic performance. Additionally, there are no articles in the WoS and Scopus databases about digital twins in schools and how they affect students' academic performance, so looking deeper into the use of digital twins in schools is recommended.

Overall, it is implied that the current topics will not be forgotten. Instead, they will be integrated and further developed. As a result, future research efforts will need to cover a wider range of information to tackle the evolving dynamics within the field effectively.

Author Contributions: Conceptualization, P.V. and L.S.; methodology, P.V. and L.S.; software, P.V.; validation, P.V. and L.S.; formal analysis, P.V.; investigation, P.V.; resources, P.V. and L.S.; data curation, P.V. and L.S.; writing—original draft preparation, P.V.; writing—review and editing, L.S.; visualization, P.V.; supervision, L.S.; project administration, L.S.; funding acquisition, L.S. All authors have read and agreed to the published version of the manuscript.

Funding: This study is a part of the dissemination activities of the research project Boosting Research for a Smart and Carbon Neutral Built Environment with Digital Twins (SmartWins) (grant ID number 101078997), funded under the Horizon Europe call HORIZON-WIDERA-2021-ACCESS-03-01.

Data Availability Statement: The data presented in this study are available in this document.

Conflicts of Interest: The authors declare no conflicts of interest.

References

1. Klepeis, N.E.; Nelson, W.C.; Ott, W.R.; Robinson, J.P.; Tsang, A.M.; Switzer, P.; Behar, J.V.; Hern, S.C.; Engelmann, W.H. The National Human Activity Pattern Survey (NHAPS): A resource for assessing exposure to environmental pollutants. *J. Expo. Sci. Environ. Epidemiol.* **2001**, *11*, 231–252. [CrossRef]
2. Wang, D.; Song, C.; Wang, Y.; Xu, Y.; Liu, Y.; Liu, J. Experimental investigation of the potential influence of indoor air velocity on students' learning performance in summer conditions. *Energy Build.* **2020**, *219*, 110015. [CrossRef]
3. Currie, J.; Neidell, M.; Schmieder, J.F. Air pollution and infant health: Lessons from New Jersey. *J. Health Econ.* **2009**, *28*, 688–703. [CrossRef]
4. Currie, J.; Reed, W. Traffic Congestion and Infant Health: Evidence from E-ZPass. *Am. Econ. J. Appl. Econ.* **2011**, *3*, 65–90. [CrossRef]
5. Knittel, C.R.; Miller, D.L.; Sanders, N.J. Caution, drivers! children present: Traffic, pollution, and infant health. *Rev. Econ. Stat.* **2016**, *98*, 350–366. [CrossRef]
6. Schlenker, W.; Walker, W.R. Airports, Air Pollution, and Contemporaneous Health. *Rev. Econ. Stud.* **2016**, *83*, 768–809. [CrossRef]
7. Deryugina, T.; Heutel, G.; Miller, N.H.; Molitor, D.; Reif, J. The Mortality and Medical Costs of Air Pollution: Evidence from Changes in Wind Direction. *Am. Econ. Rev.* **2019**, *109*, 4178–4219. [CrossRef]
8. Heft-Neal, S.; Burney, J.; Bendavid, E.; Voss, K.; Burke, M. Air Pollution and Infant Mortality: Evidence from Saharan Dust. Working Paper 26107, National Bureau of Economic Research. 2019. Available online: https://www.nber.org/papers/w26107 (accessed on 20 February 2024).
9. Simeonova, E.; Currie, J.; Nilsson, P.; Walker, R. Congestion Pricing, Air Pollution, and Childrens Health. *J. Hum. Resour.* **2019**, *56*, 971–996. [CrossRef]
10. Ebenstein, A.; Lavy, V.; Roth, S. The Long-Run Economic Consequences of High-Stakes Examinations: Evidence from Transitory Variation in Pollution. *Am. Econ. J. Appl. Econ.* **2016**, *8*, 36–65. [CrossRef]
11. Persico, C.L.; Venator, J. The effects of local industrial pollution on students and schools. *J. Hum. Resour.* **2021**, *56*, 406–445. [CrossRef]
12. Heissel, J.A.; Persico, C.; Simon, D. Does Pollution Drive Achievement? The Effect of Traffic Pollution on Academic Performance. *J. Hum. Resour.* **2022**, *57*, 747–776. [CrossRef]
13. Gaihre, S.; Semple, S.; Miller, J.; Fielding, S.; Turner, S. Classroom Carbon Dioxide Concentration, School Attendance, and Educational Attainment. *J. Sch. Health* **2014**, *84*, 569–574. [CrossRef]
14. Kielb, C.; Lin, S.; Muscatiello, N.; Hord, W.; Rogers-Harrington, J.; Healy, J. Building-related health symptoms and classroom indoor air quality: A survey of school teachers in New York State. *Indoor Air* **2015**, *25*, 371–380. [CrossRef] [PubMed]
15. Mendell, M.J.; Heath, G.A. Do indoor pollutants and thermal conditions in schools influence student performance? A critical review of the literature. *Indoor Air* **2005**, *15*, 27–52. [CrossRef] [PubMed]
16. Shendell, D.; Prill, R.J.; Fisk, W.; Apte, M.; Blake, D.; Faulkner, D. Association between class room CO_2 concentrations and student attendance in Washington and Idaho. *Indoor Air* **2004**, *14*, 333–341. [CrossRef]
17. Wargocki, P.; Wyon, D.P. Ten questions concerning thermal and indoor air quality effects on the performance of office work and schoolwork. *Build. Environ.* **2017**, *112*, 359–366. [CrossRef]
18. Fisk, W. The ventilation problem in schools: Literature review. *Indoor Air* **2017**, *27*, 1039–1051. [CrossRef]
19. Madureira, J.; Paciência, I.; Pereira, C.; Teixeira, J.P.; de Oliveira Fernandes, E. Indoor air quality in Portuguese schools: Levels and sources of pollutants. *Indoor Air* **2015**, *26*, 526–537. [CrossRef] [PubMed]
20. Deng, S.; Lau, J. Seasonal variations of indoor air quality and thermal conditions and their correlations in 220 classrooms in the Midwestern United States. *Build. Environ.* **2019**, *157*, 79–88. [CrossRef]
21. Twardella, D.; Matzen, W.; Lahrz, T.; Burghardt, R.; Spegel, H.; Hendrowarsito, L.; Frenzel, A.; Fromme, H. Effect of classroom air quality on students' concentration: Results of a cluster-randomized cross-over experimental study. *Indoor Air* **2012**, *22*, 378–387. [CrossRef]
22. Durán-Narucki, V. School building condition, school attendance, and academic achievement in New York City public schools: A mediation model. *J. Environ. Psychol.* **2008**, *28*, 278–286. [CrossRef]
23. Berman, J.D.; McCormack, M.C.; Koehler, K.A.; Connolly, F.; Clemons-Erby, D.; Davis, M.F.; Gummerson, C.; Leaf, P.J.; Jones, T.D.; Curriero, F.C. School environmental conditions and links to academic performance and absenteeism in urban, mid-Atlantic public schools. *Int. J. Hyg. Environ. Health* **2018**, *221*, 800–808. [CrossRef] [PubMed]
24. Evans, G.W.; Yoo, M.J.; Sipple, J. The ecological context of student achievement: School building quality effects are exacerbated by high levels of student mobility. *J. Environ. Psychol.* **2010**, *30*, 239–244. [CrossRef]
25. Crampton, F.E. Spending on school infrastructure: Does money matter? *J. Educ. Adm.* **2009**, *47*, 305–322. [CrossRef]
26. Haverinen-Shaughnessy, U.; Shaughnessy, R.J.; Cole, E.C.; Toyinbo, O.; Moschandreas, D.J. An assessment of indoor environmental quality in schools and its association with health and performance. *Build. Environ.* **2015**, *93*, 35–40. [CrossRef]
27. Shaughnessy, R.; Haverinen-Shaughnessy, U.; Nevalainen, A.; Moschandreas, D. A preliminary study on the association between ventilation rates in classrooms and student performance. *Indoor Air* **2007**, *16*, 465–468. [CrossRef] [PubMed]
28. Godwin, C.; Batterman, S. Indoor air quality in Michigan schools. *Indoor Air* **2007**, *17*, 109–121. [CrossRef] [PubMed]

29. Mendell, M.; Ekaterina, E.; Davies, M.; Spears, M.; Lobscheid, A.; Fisk, W.; Apte, M. Association of Classroom Ventilation with Reduced Illness Absence: A Prospective Study in California Elementary Schools. *Indoor Air* **2013**, *23*, 515–528. [CrossRef] [PubMed]
30. Daisey, J.M.; Angell, W.J.; Apte, M.G. Indoor air quality, ventilation and health symptoms in schools: An analysis of existing information. *Indoor Air* **2003**, *13*, 53–64. [CrossRef]
31. Haverinen-Shaughnessy, U.; Shaughnessy, R. Effects of Classroom Ventilation Rate and Temperature on Students' Test Scores. *PLoS ONE* **2015**, *10*, e0136165. [CrossRef]
32. O'Neill, D.; Oates, A. The Impact of School Facilities on Student Achievement, Behavior, Attendance, and Teacher Turnover Rate in Central Texas Middle Schools. *Educ. Facil. Plan.* **2001**, *36*, 14–22.
33. Simons, E.; Hwang, S.-A.; Fitzgerald, E.F.; Kielb, C.; Lin, S. The Impact of School Building Conditions on Student Absenteeism in Upstate New York. *Am. J. Public Health* **2010**, *100*, 1679–1686. [CrossRef] [PubMed]
34. Toyinbo, O.; Shaughnessy, R.; Turunen, M.; Putus, T.; Metsämuuronen, J.; Kurnitski, J.; Haverinen-Shaughnessy, U. Building characteristics, indoor environmental quality, and mathematics achievement in Finnish elementary schools. *Build. Environ.* **2016**, *104*, 114–121. [CrossRef]
35. Wargocki, P.; Wyon, D.P. Providing better thermal and air quality conditions in school classrooms would be cost-effective. *Build. Environ.* **2013**, *59*, 581–589. [CrossRef]
36. Crosby, K.; Metzger, A.B. *Powering Down: A Toolkit for Behavior-Based Energy Conservation in K-12 Schools*; Technical Report; U.S. Green Building Council (USGBC): Washington, DC, USA, 2012.
37. Economidou, M.; Atanasiu, B.; Despret, C.; Maio, J.; Nolte, I.; Rapf, O.; Laustsen, J.; Ruyssevelt, P.; Staniaszek, D.; Strong, D.; et al. *Europe's Buildings under the Microscope: A Country-by-Country Review of the Energy Performance of Buildings*; Technical Report; Buildings Performance Institute Europe (BPIE): Brussels, Belgium, 2011; ISBN 9789491143014.
38. Shan, X.; Melina, A.N.; Yang, E.-H. Impact of indoor environmental quality on students' wellbeing and performance in educational building through life cycle costing perspective. *J. Clean. Prod.* **2018**, *204*, 298–309. [CrossRef]
39. Reed, R.H. The heating and ventilation of the mansfield schools and churches: A Lecture delivered before the Mansfield Lyceum, February 13, 1889. *J. Am. Med. Assoc.* **1889**, *XII*, 469–478. [CrossRef]
40. Reed, R.H. Original articles. original investigations on the heating and ventilation of school buildings: Read in the Section of State Medicine, at the Forty-second Annual Meeting of the American Medical Association, at Washington, D.C.; May, 1891. *J. Am. Med. Assoc.* **1891**, *XVII*, 389–396. [CrossRef]
41. Wheatley, J. The ventilation of schools. *Public Health* **1894**, *7*, 373–374. [CrossRef]
42. Rovira, C.; Codina, L.; Guerrero-Solé, F.; Lopezosa, C. Ranking by Relevance and Citation Counts, a Comparative Study: Google Scholar, Microsoft Academic, WoS and Scopus. *Futur. Internet* **2019**, *11*, 202. [CrossRef]
43. U.S. Environmental Protection Agency. Sustainability and the ROE. 2023. Available online: https://www.epa.gov/report-environment/sustainability-and-roe (accessed on 20 February 2024).
44. Center for Green Schools. Leed certification for schools. 2023. Available online: https://centerforgreenschools.org/about/leed-certification-schools (accessed on 20 February 2024).
45. The Collaborative for High Performance Schools (CHPS). 2021. Available online: https://chps.net/ (accessed on 20 February 2024).
46. Bre Group. BREEAM—Education. 2022. Available online: https://bregroup.com/asset_type/breeam-education/ (accessed on 20 February 2024).
47. Green building council Australia. Green Star—Education v1. 2008. Available online: https://www.gbca.org.au/uploads/226/17 62/Factsheet_Educationv1.pdf (accessed on 20 February 2024).
48. Kosasih, A. WELL v2 and Educational Spaces: Promoting Healthy Schools. 2018. Available online: https://resources.wellcertified.com/articles/well-v2-and-educational-spaces-promoting-healthy-schools/ (accessed on 20 February 2024).
49. Brink, H.W.; Krijnen, W.P.; Loomans, M.G.L.C.; Mobach, M.P.; Kort, H.S.M. Positive effects of indoor environmental conditions on students and their performance in higher education classrooms: A between-groups experiment. *Sci. Total Environ.* **2023**, *869*, 161813. [CrossRef]
50. Choi, H.-H.; van Merriënboer, J.J.G.; Paas, F. Effects of the Physical Environment on Cognitive Load and Learning: Towards a New Model of Cognitive Load. *Educ. Psychol. Rev.* **2014**, *26*, 225–244. [CrossRef]
51. Brink, H.W.; Loomans, M.G.L.C.; Mobach, M.P.; Kort, H.S.M. Classrooms' indoor environmental conditions affecting the academic achievement of students and teachers in higher education: A systematic literature review. *Indoor Air* **2021**, *31*, 405–425. [CrossRef]
52. Kim, J.; de Dear, R. Nonlinear relationships between individual IEQ factors and overall workspace satisfaction. *Build. Environ.* **2012**, *49*, 33–40. [CrossRef]
53. Xiong, L.; Huang, X.; Li, J.; Mao, P.; Wang, X.; Wang, R.; Tang, M. Impact of Indoor Physical Environment on Learning Efficiency in Different Types of Tasks: A $3 \times 4 \times 3$ Full Factorial Design Analysis. *Int. J. Environ. Res. Public Health* **2018**, *15*, 1256. [CrossRef] [PubMed]
54. Calderón-Garcidueñas, L.; Mora-Tiscareño, A.; Ontiveros, E.; Gómez-Garza, G.; Barragán-Mejía, G.; Broadway, J.; Chapman, S.; Valencia-Salazar, G.; Jewells, V.; Maronpot, R.R.; et al. Air polldogs. *Brain Cogn.* **2008**, *68*, 117–127. [CrossRef] [PubMed]
55. Gardin, T.N.; Requia, W.J. Air quality and individual-level academic performance in Brazil: A nationwide study of more than 15 million students between 2000 and 2020. *Environ. Res.* **2023**, *226*, 115689. [CrossRef] [PubMed]
56. Duque, V.; Gilraine, M. Coal use, air pollution, and student performance. *J. Public Econ.* **2022**, *213*, 104712. [CrossRef]

57. Kabirikopaei, A.; Lau, J.; Nord, J.; Bovaird, J. Identifying the K-12 classrooms' indoor air quality factors that affect student academic performance. *Sci. Total Environ.* **2021**, *786*, 147498. [CrossRef]
58. Requia, W.J.; Saenger, C.C.; Cicerelli, R.E.; Monteiro de Abreu, L.; Cruvinel, V.R.N. Air quality around schools and school-level academic performance in Brazil. *Atmos. Environ.* **2022**, *279*, 119125. [CrossRef]
59. Guo, H.; Morawska, L.; He, C.; Zhang, Y.L.; Ayoko, G.; Cao, M. Characterization of particle number concentrations and PM2.5 in a school: Influence of outdoor air pollution on indoor air. *Environ. Sci. Pollut. Res.* **2010**, *17*, 1268–1278. [CrossRef]
60. Richmond-Bryant, J.; Saganich, C.; Bukiewicz, L.; Kalin, R. Associations of PM2.5 and black carbon concentrations with traffic, idling, background pollution, and meteorology during school dismissals. *Sci. Total Environ.* **2009**, *407*, 3357–3364. [CrossRef] [PubMed]
61. Rivas, I.; Viana, M.; Moreno, T.; Pandolfi, M.; Amato, F.; Reche, C.; Bouso, L.; Àlvarez-Pedrerol, M.; Alastuey, A.; Sunyer, J.; et al. Child exposure to indoor and outdoor air pollutants in schools in Barcelona, Spain. *Environ. Int.* **2014**, *69*, 200–212. [CrossRef]
62. Martínez-Lazcano, J.; González Guevara, E.; Rubio, C.; Franco-Pérez, J.; Custodio, V.; Hernández-Cerón, M.; CARLOS, L.; Paz, C. The effects of ozone exposure and associated injury mechanisms on the central nervous system. *Rev. Neurosci.* **2013**, *24*, 337–352. [CrossRef] [PubMed]
63. Forns, J.; Dadvand, P.; Esnaola, M.; Alvarez-Pedrerol, M.; López-Vicente, M.; Garcia-Esteban, R.; Cirach, M.; Basagaña, X.; Guxens, M.; Sunyer, J. Longitudinal association between air pollution exposure at school and cognitive development in school children over a period of 3.5 years. *Environ. Res.* **2017**, *159*, 416–421. [CrossRef]
64. Benka-Coker, W.; Young, B.; Oliver, S.; Schaeffer, J.W.; Manning, D.; Suter, J.; Cross, J.; Magzamen, S. Sociodemographic variations in the association between indoor environmental quality in school buildings and student performance. *Build. Environ.* **2021**, *206*, 108390. [CrossRef]
65. Choi, N.; Yamanaka, T.; Takemura, A.; Kobayashi, T.; Eto, A.; Hirano, M. Impact of indoor aroma on students' mood and learning performance. *Build. Environ.* **2022**, *223*, 109490. [CrossRef]
66. Ryan, I.; Deng, X.; Thurston, G.; Khwaja, H.; Romeiko, X.; Zhang, W.; Marks, T.; Yu, F.; Lin, S. Measuring students' exposure to temperature and relative humidity in various indoor environments and across seasons using personal air monitors. *Hyg. Environ. Health Adv.* **2022**, *4*, 100029. [CrossRef]
67. Evidence for Policy and Practice Information and Co-ordinating Centre (EPPI-Centre). *EPPI-Centre Methods for Conducting Systematic Reviews*; EPPI-Centre, Social Science Research Unit, Institute of Education, University of London: London, UK, 2007.
68. Harden, A.; Thomas, J. Methodological Issues in Combining Diverse Study Types in Systematic Reviews. *Int. J. Soc. Res. Methodol.* **2005**, *8*, 257–271. [CrossRef]
69. Lamé, G. Systematic Literature Reviews: An Introduction. In Proceedings of the Design Society: International Conference on Engineering Design, Delft, The Netherlands, 5–8 August 2019; Volume 1. [CrossRef]
70. MacLure, M. 'Clarity bordering on stupidity': Where's the quality in systematic review? *J. Educ. Policy* **2005**, *20*, 393–416. [CrossRef]
71. Higgins, J.P.T.; Green, S. (Eds.) *Cochrane Handbook for Systematic Reviews of Interventions*; The Cochrane Collaboration: London, UK, 2009.
72. Chadegani, A.A.; Salehi, H.; Yunus, M.M.; Farhadi, H.; Fooladi, M.; Farhadi, M.; Ebrahim, N.A. A Comparison between Two Main Academic Literature Collections: Web of Science and Scopus Databases. *Asian Soc. Sci.* **2013**, *9*, 18–26. [CrossRef]
73. Chen, C.; Hu, Z.; Liu, S.; Tseng, H. Emerging trends in regenerative medicine: A scientometric analysis in CiteSpace. *Expert Opin. Biol. Ther.* **2012**, *12*, 593–608. [CrossRef] [PubMed]
74. Chen, C. Searching for intellectual turning points: Progressive knowledge domain visualization. *Proc. Natl. Acad. Sci. USA* **2004**, *101* (Suppl. S1), 5303–5310. [CrossRef] [PubMed]
75. Chen, C. CiteSpace II: Detecting and visualizing emerging trends and transient patterns in scientific literature. *J. Am. Soc. Inf. Sci. Technol.* **2006**, *57*, 359–377. [CrossRef]
76. Chen, C.; Song, M. Visualizing a Field of Research: A Methodology of Systematic Scientometric Reviews. *PLoS ONE* **2019**, *14*, e0223994. [CrossRef]
77. Chen, C. A Glimpse of the First Eight Months of the COVID-19 Literature on Microsoft Academic Graph: Themes, Citation Contexts, and Uncertainties. *Front. Res. Metr. Anal.* **2020**, *5*, 607286. [CrossRef]
78. Alshamrani, O.S.; Galal, K.; Alkass, S. Integrated LCA–LEED sustainability assessment model for structure and envelope systems of school buildings. *Build. Environ.* **2014**, *80*, 61–70. [CrossRef]
79. Ji, C.; Hong, T.; Jeong, J.; Kim, J.; Lee, M.; Jeong, K. Establishing environmental benchmarks to determine the environmental performance of elementary school buildings using LCA. *Energy Build.* **2016**, *127*, 818–829. [CrossRef]
80. Brás, A.; Gomes, V. LCA implementation in the selection of thermal enhanced mortars for energetic rehabilitation of school buildings. *Energy Build.* **2015**, *92*, 1–9. [CrossRef]
81. Pachta, V.; Giourou, V. Comparative Life Cycle Assessment of a Historic and a Modern School Building, Located in the City of Naoussa, Greece. *Sustainability* **2022**, *14*, 4216. [CrossRef]
82. Gamarra, A.R.; Herrera, I.; Lechón, Y. Assessing sustainability performance in the educational sector. A High Sch. Case Study. *Sci. Total Environ.* **2019**, *692*, 465–478. [CrossRef]
83. Gamarra, A.R.; Istrate, I.R.; Herrera, I.; Lago, C.; Lizana, J.; Lechon, Y. Energy and water consumption and carbon footprint of school buildings in hot climate conditions. *Results Life Cycle Assess. J. Clean. Prod.* **2018**, *195*, 1326–1337. [CrossRef]

84. Munoz, P.; Morales, P.; Letelier, V.; Munoz, L.; Morad, D. Implications of Life Cycle Energy Assessment of a new school building, regarding the nearly Zero Energy Buildings targets in EU: A case OF study. *Sustain. Cities Soc.* **2017**, *32*, 142–152. [CrossRef]
85. Su, S.; Zhu, C.; Li, X. A dynamic weighting system considering temporal variations using the DTT approach in LCA of buildings. *J. Clean. Prod.* **2019**, *220*, 398–407. [CrossRef]
86. Li, X.; Su, S.; Zhang, Z.; Kong, X. An integrated environmental and health performance quantification model for pre-occupancy phase of buildings in China. *Environ. Impact Assess. Rev.* **2017**, *63*, 1–11. [CrossRef]
87. Su, S.; Wang, Q.; Han, L.; Hong, J.; Liu, Z. BIM-DLCA: An integrated dynamic environmental impact assessment model for buildings. *Build. Environ.* **2020**, *183*, 107218. [CrossRef]
88. Su, S.; Zhang, H.; Zuo, J. Assessment models and dynamic variables for dynamic life cycle assessment of buildings: A review. *Environ. Sci. Pollut. Res.* **2021**, *28*, 26199–26214. [CrossRef] [PubMed]
89. Lee, S.; Yu, J.; Jeong, D. BIM acceptance model in construction organizations. *J. Manag. Eng.* **2015**, *31*, 04014048. [CrossRef]
90. Gheisari, M.; Irizarry, J. Investigating human and technological requirements for successful implementation of a BIM-based mobile augmented reality environment in facility management practices. *Facilities* **2016**, *34*, 69–84. [CrossRef]
91. Wang, J.; Wang, X.; Wang, J.; Yung, P.; Jun, G. Engagement of facilities management in design stage through BIM: Framework and a case study. *Adv. Civ. Eng.* **2013**, *2013*, 189105. [CrossRef]
92. Liu, Y.; van Nederveen, S.; Wu, C.; Hertogh, M. Sustainable infrastructure design framework through integration of rating systems and building information modeling. *Adv. Civ. Eng.* **2018**, *2018*, 8183536. [CrossRef]
93. Zhang, J.; Issa, R. Collecting fire evacuation performance data using BIM-based immersive serious games for performance-based fire safety design. In Proceedings of the 2015 International Workshop on Computing in Civil Engineering, Austin, TX, USA, 21–23 June 2015; pp. 612–619.
94. Vaughan, E. *"Elementary School", Whole Building Design Guide*; National Institute of Building Sciences: Washington, DC, USA, 2017. Available online: https://www.wbdg.org/buildingtypes/education-facilities/elementary-school (accessed on 20 February 2024).
95. Eastman, C.; Teicholz, P.; Sacks, R.; Liston, K. *BIM Handbook: A Guide to Building Information Modeling for Owners, Managers, Designers, Engineers, and Contractors*, 2nd ed.; John Wiley and Sons: Hoboken, NJ, USA, 2011.
96. Bynum, P.; Issa, R.R.A.; Olbina, S. Building information modeling in support of sustainable design and construction. *J. Constr. Eng. Manag.* **2013**, *139*, 24–34. [CrossRef]
97. Lee, K.; Choo, S. A hierarchy of architectural design elements for energy saving of tower buildings in Korea using green BIM simulation. *Adv. Civ. Eng.* **2018**, *2018*, 7139196. [CrossRef]
98. Zhuang, D.; Zhang, X.; Lu, Y.; Wang, C.; Jin, X.; Zhou, X.; Shi, X. A performance data integrated BIM framework for building life-cycle energy efficiency and environmental optimization design. *Autom. Constr.* **2021**, *127*, 103712. [CrossRef]
99. Neasden Primary School. *Application of BIM in Energy Management of Individual Departments Occupying University Facilities*; Neasden Primary School: Hull, UK, 2019. Available online: https://neasdenprimary.org.uk/school-new-build/ (accessed on 20 February 2024).
100. Konings, K.D.; Bovill, C.; Woolner, P. Towards an interdisciplinary model of practice for participatory building design in education. *Eur. J. Educ.* **2017**, *52*, 306–317. [CrossRef]
101. Van Merrienboer, J.J.G.; McKenney, S.; Cullinan, D.; Heuer, J. Aligning pedagogy with physical learning spaces. *Eur. J. Educ.* **2017**, *52*, 253–267. [CrossRef]
102. Succar, B. Building information modelling framework: A research and delivery foundation for industry stakeholders. *Autom. Constr.* **2009**, *18*, 357–375. [CrossRef]
103. Volk, R.; Stengel, J.; Schultmann, F. Building Information Modeling (BIM) for existing buildings—Literature review and future needs. *Autom. Constr.* **2014**, *38*, 109–127. [CrossRef]
104. Azhar, S. Building Information Modeling (BIM): Trends, Benefits, Risks, and Challenges for the AEC Industry. *Leadersh. Manag. Eng.* **2011**, *11*, 241–252. [CrossRef]
105. Gaur, A.; Scotney, B.; Parr, G.; McClean, S. Smart City Architecture and its Applications Based on IoT. *Procedia Comput. Sci.* **2015**, *52*, 1089–1094. [CrossRef]
106. Sethi, P.; Sarangi, S.R. Internet of Things: Architectures, Protocols, and Applications. *J. Electr. Comput. Eng.* **2017**, *2017*, 1–25. [CrossRef]
107. Li, S.; Xu, L.D.; Zhao, S. The internet of things: A survey. *Inf. Syst. Front.* **2015**, *17*, 243–259. [CrossRef]
108. Amaxilatis, D.; Akrivopoulos, O.; Mylonas, G.; Chatzigiannakis, I. An IoT-Based Solution for Monitoring a Fleet of Educational Buildings Focusing on Energy Efficiency. *Sensors* **2017**, *17*, 2296. [CrossRef] [PubMed]
109. Hossain, M.; Weng, Z.; Schiano-Phan, R.; Scott, D.; Lau, B. Application of IoT and BEMS to Visualise the Environmental Performance of an Educational Building. *Energies* **2020**, *13*, 4009. [CrossRef]
110. Martínez, I.; Zalba, B.; Trillo-Lado, R.; Blanco, T.; Cambra, D.; Casas, R. Internet of Things (IoT) as Sustainable Development Goals (SDG) Enabling Technology towards Smart Readiness Indicators (SRI) for University Buildings. *Sustainability* **2021**, *13*, 7647. [CrossRef]
111. Kamel, S.; Jamal, A.; Omri, K.; Khyyat, M. An IoT-based Fire Safety Management System for Educational Buildings: A Case Study. *Int. J. Adv. Comput. Sci. Appl.* **2022**, *13*, 765–771. [CrossRef]

112. Paganelli, F.; Mylonas, G.; Cuffaro, G.; Nesi, I. Experiences from Using Gamification and IoT-Based Educational Tools in High Schools Towards Energy Savings. In *Ambient Intelligence. AmI 2019*; Lecture Notes in Computer Science; Chatzigiannakis, I., De Ruyter, B., Mavrommati, I., Eds.; Springer: Cham, Switzerland, 2019; Volume 11912. [CrossRef]
113. Mylonas, G.; Amaxilatis, D.; Chatzigiannakis, I.; Paganelli, A.A.F. Enabling Sustainability and Energy Awareness in Schools Based on IoT and Real-World Data. *IEEE Pervasive Comput.* **2018**, *17*, 53–63. [CrossRef]
114. Jia, M.; Komeily, A.; Wang, Y.; Srinivasan, R.S. Adopting Internet of Things for the development of smart buildings: A review of enabling technologies and applications. *Autom. Constr.* **2019**, *101*, 111–126. [CrossRef]
115. Atzori, L.; Iera, A.; Morabito, G. The Internet of Things: A survey. *Comput. Netw.* **2010**, *54*, 2787–2805. [CrossRef]
116. Gubbi, J.; Buyya, R.; Marusic, S.; Palaniswami, M. Internet of Things (IoT): A vision, architectural elements, and future directions. *Future Gener. Comput. Syst.* **2013**, *29*, 1645–1660. [CrossRef]
117. Hoda, I.; Elsayed, M.S.; Wael, S.M.; Abdou, H.M. Functional analysis as a method on sustainable building design: A case study in educational buildings implementing the triple bottom line. *Alex. Eng. J.* **2023**, *62*, 63–73. [CrossRef]
118. Peng, C. Calculation of a building's life cycle carbon emissions based on Ecotect and building information modeling. *J. Clean. Prod.* **2016**, *112*, 453–465. [CrossRef]
119. Schwab, K. *The Fourth Industrial Revolution*; Penguin Books Limited: London, UK, 2017. Available online: https://books.google.lt/books?id=ST_FDAAAQBAJ (accessed on 20 June 2024).

Disclaimer/Publisher's Note: The statements, opinions and data contained in all publications are solely those of the individual author(s) and contributor(s) and not of MDPI and/or the editor(s). MDPI and/or the editor(s) disclaim responsibility for any injury to people or property resulting from any ideas, methods, instructions or products referred to in the content.

Review

Trends and Interdisciplinarity Integration in the Development of the Research in the Fields of Sustainable, Healthy and Digital Buildings and Cities

Lina Seduikyte [1,*], Indrė Gražulevičiūtė-Vileniškė [1], Ingrida Povilaitienė [1], Paris A. Fokaides [2] and Domantas Lingė [1]

1. Faculty of Civil Engineering and Architecture, Kaunas University of Technology, Studentu Str. 48, LT-51367 Kaunas, Lithuania; indre.grazuleviciute@ktu.lt (I.G.-V.); ingrida.povilaitiene@ktu.lt (I.P.); domantaslinge@gmail.com (D.L.)
2. School of Engineering, Frederick University, 7, Frederickou Str., Nicosia 1036, Cyprus; eng.fp@frederick.ac.cy
* Correspondence: lina.seduikyte@ktu.lt

Abstract: This article provides a thorough bibliometric analysis of significant research trends in sustainability from 1988 until now, focusing on sustainable, healthy and digital buildings and cities. It exemplifies how research emphasis has shifted from explicit ecological investigations to nature-based solutions and city greening programs, with a rising interest in the many responsibilities of urban stakeholders in attaining sustainability. Despite weak integration at the literature and author cooperation levels, the "healthy buildings and cities" topic indicates promise for multidisciplinary integration. The "digital buildings and cities" topic, on the other hand, presents a more particular concern with strong cross-cluster collaboration and significant integration possibilities. Global relevance has been demonstrated through research on "sustainable buildings and cities," mainly in journal papers. This topic's study clusters show remarkable synergy across management, transportation, ecology, remote sensing and environmental engineering domains. In comparison to "healthy buildings and cities" and "digital buildings and cities" topics, the study of "sustainable buildings and cities" demonstrates a deeper level of interdisciplinary integration, highlighting the significant potential for further exploration within sustainability science research. This study emphasizes the ongoing worldwide relevance of sustainability science research and identifies significant opportunities for multidisciplinary integration across the investigated subjects.

Keywords: sustainable buildings and cities; healthy buildings and cities; digital buildings and cities

Citation: Seduikyte, L.; Gražulevičiūtė-Vileniškė, I.; Povilaitienė, I.; Fokaides, P.A.; Lingė, D. Trends and Interdisciplinarity Integration in the Development of the Research in the Fields of Sustainable, Healthy and Digital Buildings and Cities. *Buildings* **2023**, *13*, 1764. https://doi.org/10.3390/buildings13071764

Academic Editor: Zhenjun Ma

Received: 16 June 2023
Revised: 3 July 2023
Accepted: 9 July 2023
Published: 11 July 2023

Copyright: © 2023 by the authors. Licensee MDPI, Basel, Switzerland. This article is an open access article distributed under the terms and conditions of the Creative Commons Attribution (CC BY) license (https://creativecommons.org/licenses/by/4.0/).

1. Introduction

The assessment of the sustainability of buildings and cities is at the forefront of the environmental analysis of the built environment. At this point, at the European and World level, a significant number of research projects, standards and methods have been developed, to assess the sustainability aspects. At the same time, all techniques to assess buildings are digitized, as we are moving fast to the Industry 4.0 era, where all buildings-related information should be easily communicated among designers, users and stakeholders.

Buildings are essential components of cities, where people spend a substantial part of their lives living, working, studying or relaxing. However, the same buildings and the whole construction industry are also responsible for considerable energy consumption, greenhouse gas emissions and waste generation. Therefore, how the buildings and entire cities are designed, constructed and configured to operate is crucial for reducing the environmental impact of urbanization, improving the health and well-being of the population and achieving the Sustainable Development Goals [1] adopted by 193 countries.

The concept of sustainable cities has evolved over time and has been defined in different ways by different scholars and organizations. According to one of the initial explanations of sustainable development, it is "a development that meets the needs of the present without compromising the ability of future generations to meet their own needs" [2]. In other words, sustainable development seeks to find "a balance between economic development, environmental protection and social improvement" [3,4], and the main aim of sustainable urban development is to create "beautiful, distinctive, secure, healthy and high-quality places for people to live and work in that foster a strong sense of community, pride, social equity, integration and identity" [5]. Recently, there has been a growing recognition that achieving this certain objective is not solely reliant on tangible measures but "based on the principles of democracy, gender equality, solidarity, the rule of law and respect for fundamental rights, including freedom and equal opportunities for all" [6].

2. Theoretical Background, Literature Overview
Approaches to Sustainable Buildings and Cities

Several approaches towards sustainable buildings and cities were identified throughout the analyzed literature, depending on the context, priorities and resources available. The most common approaches are presented below.

Compact city: it is one of the leading paradigms of both sustainable development [7] and the New Urbanism movement [8]. The compactness of the city can be defined in the following three aspects [9]. Firstly, the urban form should be defined by high-density settlements, fewer dependences on automobiles and clear boundaries from the surrounding areas. Then spatial features should encourage mixed land use, diversity of life as well as clear and unique identities. Finally, social functions should aim for social equality, self-sufficiency in daily life and independence of government. The main critique of this approach is that compact cities, in order to reduce sprawl and minimize environmental impact, promote densification, whereas low-density urban forms are often considered to be more livable. However, that critique has been denied [10] as residents from compact cities are more satisfied with their neighborhood because, despite high density, this model also provides a better public transport network, accessibility and a variety of land uses.

Eco-city: the concept of eco-city is also broad; there are many overlaps with other approaches. Still, the following ten critical eco-city dimensions can be distinguished [11]: compact and mixed-use urban form, an abundance of the natural environment, walking and cycling infrastructure, extensive environmental technologies for water, energy and waste management, the central city with subcenters, high-quality public realm, human scale physical environment, innovation and driven economy, visionary—"debate and decide" and sustainability-based decision making. Essentially, eco-cities focus on environmental sustainability, promoting green infrastructure, renewable energy and zero-waste strategies while addressing social and economic issues [12–15].

Resilient city: resilience, in terms of cities, refers to the ability to absorb, adapt and respond to changes in an urban system. For this reason, cities should be conceptualized as complex adaptive systems and divided into components and analytical elements. Hence, this systematic approach allows a better understanding of how urban system design, planning and management work towards resiliency enhancements [16]. Overall, resilient cities focus on the adaptation to the challenges posed by climate change, such as extreme weather, natural disasters or sea-level rise, ensuring the continuity of essential services and minimizing the impact on the population [17–20].

Digital city: offer innovative services based on broadband communication and service-oriented computing [21]. Digital cities were built and made operational throughout the developed countries between the 1990s and the 2000s. Digital cities are distinguished by activities based on online services [22]. The fast spread of developing digital technologies, digital service creation and delivery necessitate new and more organized approaches to service design, development and management. There are numerous perspectives and strategic elements for developing digital services from diverse stakeholders, with different solutions stating how to design the digital service inside IT infrastructures or how to reuse design techniques learnt from prior Digital City initiatives [23–25].

According to a literature review, Digital City and Smart City are the most commonly used terms to describe a city's smartness. Smart cities appear to be the inevitable successors of digital cities.

Smart city: there are many definitions of smart cities. The reason for that is the application of the term for two different kinds of domains: "hard" and "soft" [26]. The "hard" one includes buildings, energy grids, natural resources, water and waste management and mobility, while the "soft one" covers culture, education, policy innovations, social inclusion and governments. Depending on the domain, the role of information and communication technologies (ICTs) is also different—decisive for the "hard" and not so much in the "soft" domain. Anyway, the common conception of smart cities is the use of digital technologies to optimize urban systems, such as transport, energy, water and waste management, improving efficiency, reducing costs and enhancing citizens' quality of life [27–29].

Healthy city: the World Health Organization (WHO) evolved the concept of "Healthy Cities" to improve city-based public health and environmental hygiene with a special focus on marginalized urban areas [30]. "Healthy Cities Project" was launched in Europe in the 1980s and has spread globally. The main principle of the initiative was that "health can be improved by modifying living conditions, namely, the physical environment and the social and economic conditions of everyday life" [31]. Eventually, the tools to measure the index of healthy cities were developed [32], and the index's indicators fall into four main sectors: health, health services, socioeconomic indicators and environmental indicators. The latter is strongly dependent on urban development strategies. To succeed in the creation of healthy and sustainable cities, urban development has to promote access to green spaces, sports and leisure facilities, mitigation measures of air, water and noise pollution, walkability and cycling and public transportation modes. The observed spatial inequality can also reveal the existence of social inequality [33].

Differences in various approaches to sustainable buildings and cities are presented in Table 1.

This review aims to analyze patterns and trends of the scientific research related to sustainable, healthy and digital buildings and cities. Quantitative literature analysis and graphical visualization and analysis—knowledge mapping—were applied to understand better the current research situation and research frontiers [34] in the mentioned areas. According to Chen [35], the frontier of research reveals the emergence of theoretical trends and new topics. According to Price [34], the research frontier is the dynamic nature and ideological status of the research field; generally, the research frontier consists of approximately 40 or 50 recently published scientific papers. Knowledge mapping is part of the broader field of science metrology and is defined as a cross-disciplinary field of applied mathematics, information science and computer science; the purpose of it is to extract and visually reorganize the knowledge from a large number of previously published scientific research documents and to carry out knowledge discovery [36,37]. This study provides insights into research trends, identifies gaps and opportunities for multidisciplinary integration and highlights the global importance of sustainability science. By quantitatively analyzing the scientific literature, this analysis contributes to a deeper understanding of such areas as healthy, digital and sustainable buildings and cities and informs future research directions.

Table 1. Differences in various approaches to sustainable buildings and cities. Note: Approaches are not totally exclusive, there are overlaps in their goals, features and measures.

Approach	Goal	Conceptual Features						Measures
		Urban Form	Land Use	Social Realm	Governance	Technologies		
Compact city	to reduce sprawl, minimize car dependency and promote walkability and public transportation	Compact, high-density	Mixed use	Social equality, self-sufficiency	Integrating planning	Intelligent transport		population density, mixed land use ratios, walkability indices, public transportation usage, etc.
Eco-city	to achieve environmental sustainability, conserve resources and promote ecological balance.	Compact, sustainable	Mixed use, green infrastructure	Community-based	Sustainable policies	Environmental technologies		green space coverage, energy consumption per capita, waste management practices, sustainable building certifications, etc.
Resilient city	to enhance adaptability and resilience to climate change, natural disasters and social challenges	Adaptive	Green infrastructure	Social cohesion	Collaborative-participatory	Resilient infrastructure		climate adaptation plans, disaster preparedness indicators, social cohesion indices, infrastructure robustness, etc.
Digital city	to improve connectivity, access to digital services and enhance efficiency in urban operations	Digitally connected	Varied	Online communities	E-governance	Digital platforms		digital service accessibility, e-governance adoption and digital literacy rates
Smart City	to enhance quality of life, optimize resource management and foster innovation and economic development	Technologically advanced	Efficient	Data-driven	Smart governance	IoT, AI, sensors		IoT infrastructure deployment, data analytics usage, smart grid implementation and citizen engagement in smart initiatives
Healthy City	to improve public health, promote well-being and create a supportive and inclusive environment.	Health-oriented	Green infrastructure	Social wellbeing	Collaborative-participatory	Health monitoring		public health indicators, access to healthcare services, air and water quality and community engagement in health initiatives

3. Methods

Web of Science (WoS) database was selected as a source of information and bibliometric data for quantitative literature analysis and knowledge mapping. According to Su et al. [37], WoS provides more complete references, indexes and researcher relationships than other databases. Chen [38] also mentions this database's high standard and wide span: reference searches can be used to track previous research and monitor recent developments in content over the 100 years that are fully indexed.

CiteSpace 6.2.R2 Advanced was applied in this research. It is a citation visual analysis software developed from the background of scientometrics and knowledge visualization, which is specifically used to identify potential knowledge contained in the scientific literature [35,36,38,39]. According to its creator, this software can help researchers understand the basic knowledge of the discipline, find the classical literature in the field, discover research frontiers, and clarify the context of research evolution [39]. This software is widely used for bibliometric analysis and visualization. According to Su et al. [37], more than 15 000 papers have using the CiteSpace tool been published. The data format processed by CiteSpace software is based on the WoS data download format [38] and allows generating the merged networks, that characterize the development of the analyzed field over time, showing the most important footprints of the related research activities and performing cluster analysis, author cooperative analysis and co-citation (*the frequency with which two documents are cited together by other documents*) analysis [36,37]. The visual display of the bibliometric analysis of the CiteSpace tool can be characterized by network character with nodes (points in a network diagram at which links intersect or branch), links (the relationship between two nodes) and clusters (a group of similar findings that occur together).

This study collected the bibliometric data from the WoS core collection database, for the period 1988 until now, of the papers on the following topics:

- Healthy buildings and cities: word combination "healthy buildings and cities" was entered into WoS database search engine;
- Digital buildings and cities: word combination "digital buildings and cities" was entered into WoS database search engine;
- Sustainable buildings and cities: word combination "sustainable buildings and cities" was entered into WoS database search engine.

The data were collected on the 28 February 2023. Publications issued in the year 2023 were included in the analysis sample, as the aim of the research was to identify research frontiers. Considering this, excluding the newest publications would not allow comprehensive research outcomes. In total, the data of 13,804 papers were collected (Tables 2 and 3): 1064 papers on healthy buildings and cities, 2734 papers on digital buildings and cities and 10,006 papers on sustainable buildings and cities were identified.

Table 2. The type of records in the analyzed sample.

Search "Healthy Buildings and Cities"	Search "Digital Buildings and Cities"	Search "Sustainable Buildings and Cities"
Article—628	Article—1411	Article—6185
Proceeding paper—392	Proceeding paper—1289	Proceeding paper—3408
Review article—34	Review article—61	Review article—465
Book chapter—30	Book chapter—50	Book chapter—236
Editorial material—18	Early access—40	Early access—114
Early access—12	Editorial material—12	Editorial material—52
Data paper—1	Data paper—4	Book review—7
Meeting abstract—1	Book review—1	Data paper—6
Reprint—1	News item—1	Correction—2
		Meeting abstract—2
		Book—1
		Letter—1
		Retracted publication—1

Table 3. The yearly breakdown of the publications in the analyzed sample.

Publication Year	Search "Healthy Buildings and Cities"	Search "Digital Buildings and Cities"	Search "Sustainable Buildings and Cities"
1988	3	-	-
1989	-	-	-
1990	-	-	-
1991	1	-	-
1992	1	2	3
1993	-	1	-
1994	-	-	2
1995	1	1	6
1996	-	3	4
1997	1	5	10
1998	1	7	22
1999	2	6	10
2000	8	12	13
2001	1	14	15
2002	5	9	33
2003	6	18	29
2004	4	13	34
2005	13	29	222
2006	12	29	55
2007	13	28	84
2008	15	27	81
2009	17	45	144
2010	23	57	209
2011	19	56	287
2012	23	60	355
2013	26	97	416
2014	31	92	295
2015	39	117	440
2016	41	117	577
2017	51	175	791
2018	265	216	990
2019	89	394	1144
2020	86	317	1114
2021	130	391	1254
2022	125	366	1237
2023	13	30	130

This study performed the bibliometric analysis of this volume of the identified scientific literature and displayed the results visually.

Additionally, a separate topic of "sustainable cities" was analyzed, as the term "sustainable city" includes all previously mentioned fields. Result: 2060 articles selected for the systematic literature review.

4. Results

4.1. Healthy Buildings and Cities

The search on "healthy buildings and cities" provided the smallest number of results compared to other searches. During the time span 1988–February 2023, 1064 papers were published. The first three papers appeared in 1988, and the growth of published research in this field started in 2005. In 2021 and 2022, 130 and 125 publications were recorded, demonstrating the growing interest in the field. In 2018, 265 publications were recorded due to several important conferences focused on this topic held that year, including the 34th International Conference on Passive and Low Energy Architecture (PLEA)—Smart and Healthy Within the Two-Degree Limit. It is important to note that the predominant publication type in this search was conference papers and proceedings. The dominant fields of research according to WoS categories are Green Sustainable Science and Technology (288 publications), Architecture (228 publications), Public Environmental Occupational Health (212) and Environmental Sciences (179 publications). Civil Engineering has 103 recorded studies and Urban Studies—91 in the analyzed period. Three most cited publications in this research area were published in 2013, 2003 and 2017. The most cited contribution—multidisciplinary peer-reviewed research on contributions of nature or ecosystems to human well-being by Russell et al. [40]—has 328 citations in total; theoretical analysis of sense of place as a public health construct by Frumkin [41] has 305 citations in total, and the contribution by Nieuwenhuijsen et al. [42] from the field of epidemiology with 263 total number of citations deals with the involvement of environmental epidemiologists in better understanding of health effects of green spaces in urban environments. This demonstrates that the topic of "healthy buildings and cities" is under a wide umbrella of sustainability research and is strongly related to architecture, civil engineering, environmental and urban studies, and public health.

CiteSpace visualization for the search "healthy buildings and cities" are presented in Figures 1–3.

Figure 1. CiteSpace visualization of keywords and cluster analysis for the search "healthy buildings and cities" in the period 1988–2023. The nodes represent keywords, lines that connect nodes are keyword co-occurrence links. The top 10 dominant keywords were: city (58), impact (52), physical activity (51), environment (46), built environment (43), health (42), public health (40), exposure (34), quality (29) and performance (26). The cluster analysis of keyword distribution demonstrates close integration of disciplines and according to analyzed topics except of toxicology.

Figure 2. CiteSpace visualization of the reference co-citation network and cluster analysis for the search "healthy buildings and cities" in the period 1988–2023. The nodes are cited references, lines that connect nodes are co-citation links. Analysis reveals only two significant clusters of co-citation networks, labelled according to keywords: office building and air disinfection system. It is possible to conclude that there's very little integration in co-citation compared to keyword analysis.

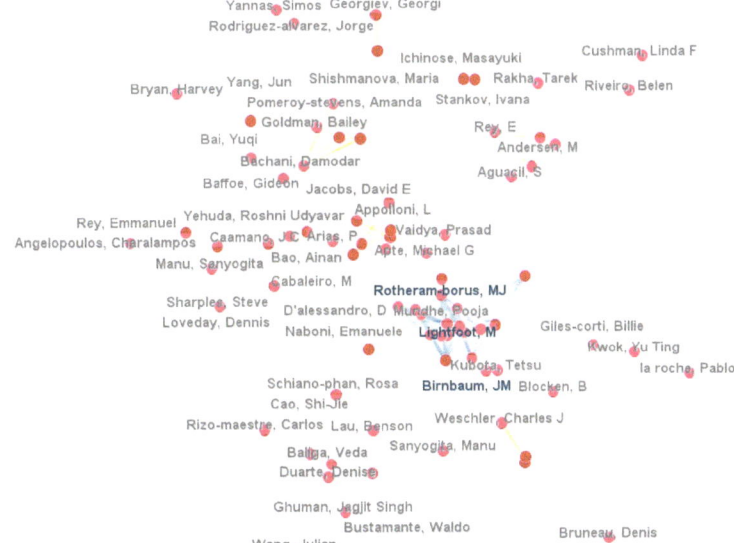

Figure 3. CiteSpace visualization of co-author network and cluster analysis for the search "healthy buildings and cities" in the period 1988–2023. The nodes represent the authors of publications in the author collaboration network. Analysis reveals only one significant cluster of co-authoring and several smaller ones. It is possible to conclude that there's much less integration in co-authoring compared to keyword analysis.

It is possible to summarize that knowledge mapping of the topic "healthy buildings and cities" as reflected in the WoS database in the period 1988–2023 reveals its untapped potential for interdisciplinary integration, as it shares multiple topics. However, the existing research appears little integrated from the points of view of the literature sources and authors collaboration.

4.2. Digital Buildings and Cities

During the time span 1992–2023, 2734 papers were published. The first two papers appeared in 1992, and the growth of published research in this field started in 2005, exceeding several dozens of publications per year. In 2020, 2021 and 2022, respectively, 317, 391 and 366 publications were recorded, demonstrating the growing interest in the field. In 2019 394 publications were recorded due to several important conferences focused on this topic held that year, including the International Conference on Climate Resilient Cities—Energy Efficiency and Renewables in the Digital Era (CISBAT). It is important to note that the predominant publication type in this search was conference papers and proceedings, similar to the previous search. The dominant fields of research, according to WoS categories, are Remote Sensing (474 publications), Construction Building Technology (470 publications), Green Sustainable Science Technology (465) and Energy Fuels (411 publications). Civil Engineering has 206 recorded studies in the analyzed period, Architecture—148 and Urban Studies—128. The most cited publication in this research area was published in 2014 by Zanella et al. [43] and is review of technologies of internet of things and their application to smart cities. This contribution has 3061 citations in total. Other two most cited contributions were published in 2014 and 2019, respectively. Publication by Neirotti et al. [44] has total number of 1081 citations and is analysis and classification of smart city initiatives aimed and policy makers and city managers. Publication by Sachs et al. [45] is cited 561 times in total; it is theoretical study of sustainability science distinguishing necessary transformations, including the digital revolution for sustainable development, for the achievement of sustainable development goals. This demonstrates that the topic of "digital buildings and cities" is more specialized compared to previously analyzed.

CiteSpace visualization for the search "digital buildings and cities" are presented in Figures 4–6.

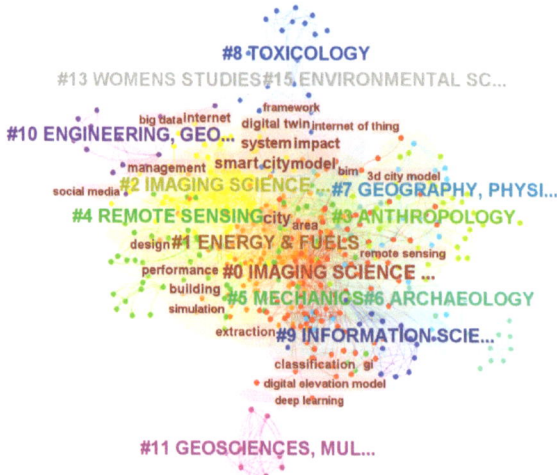

Figure 4. CiteSpace visualization of keywords and cluster analysis for the search "digital buildings and cities" in the period 1992–2023. The nodes represent keywords, lines that connect nodes are keyword co-occurrence links. The top 10 dominant keywords were: city (138), smart city (119), model (110), system (77), impact (60), area (49), digital twin (48), performance (47), management (44), building (44). The cluster analysis of keyword distribution demonstrates close integration of disciplines and according to analyzed topics.

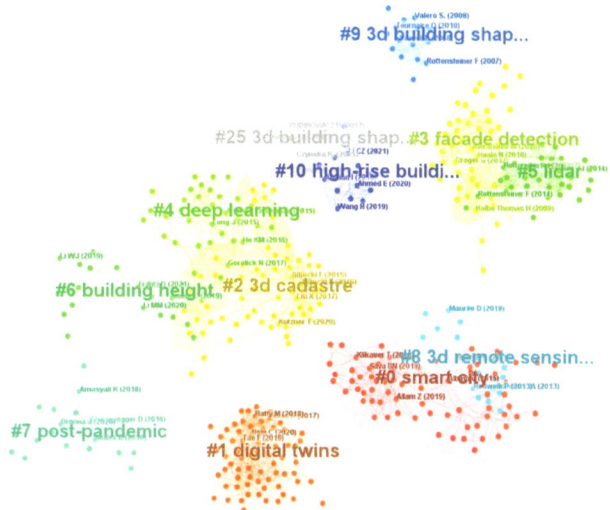

Figure 5. CiteSpace visualization of the reference co-citation network and cluster analysis for the search "digital buildings and cities" in the period 2000–2023. The nodes are cited references, lines that connect nodes are co-citation links. Analysis reveals 11 significant clusters of co-citation network, labelled according to keywords in the graph. It is visible that part of the clusters are separated and part are very closely merged. This graph reveals close integration between fields of remote sensing and chemical engineering, civil engineering and computer science and physical geography and imaging science according to cited references.

Figure 6. CiteSpace visualization of co-author network and cluster analysis for the search "healthy buildings and cities" in the period 1988–2023. The nodes represent the authors of publications in the author collaboration network. Analysis reveals several clusters of co-authors and wide cross-cluster author collaboration.

It is possible to summarize that knowledge mapping of the topic "digital buildings and cities" as reflected in the WoS database, reveals multiple shared topics and a wide array of research clusters with significant cases of cross-cluster collaboration and cluster integration. Compared to previously analyzed topic, research in the "digital buildings and cities" field appears more integrated, although with remaining untapped integration potential.

4.3. Sustainable Buildings and Cities

The search on "sustainable buildings and cities" provided the most significant number of results compared to other searches. During the time span 1988– 2023, 10 006 papers were published. The first three papers appeared in 1992, and the growth of published research in this field started in 2005. In 2019, 2020, 2021 and 2022, respectively 1144, 1114, 1254 and 1237 publications were recorded, demonstrating the growing interest in the field. It is important to note that the predominant publication type in this search is a journal article, with two leading journals—Sustainable Cities and Society and Sustainability. The dominant fields of research according to WoS categories are Green Sustainable Science and Technology (3650 publications), Construction Building Technology (3130 publications), Energy Fuels (2779) and Environmental Sciences (2193 publications). Civil Engineering has 1637 recorded studies in the analyzed period, Urban Studies—1002, Architecture—558. The most cited publications in this area of research were published in 2014, 2007 and 2015, respectively. Contribution by Cabeza et al. [46] has 730 citations in total; it is the literature review on life cycle assessment, life cycle energy analysis and life cycle cost analysis studies carried out for environmental evaluation of buildings and building related industry and sector. Publiaction by Kennedy et al. [47] is cited 730 times in total as well and is comparative study of metabolism in cities—water, materials, energy and nutrient flows. The study by Haaland and van den Bosch [48] is cited 608 times in total and provides a literature review on the effects of urban densification and compact city development on urban green space and its planning. The research areas of "digital buildings and cities" and "sustainable buildings and cities" share common most cited publication—the previously mentioned study on implementation of sustainable development goals by Sachs et al. [45]. This demonstrates that "sustainable buildings and cities" is under a broad umbrella of sustainability research with currently predominant technological disciplines.

CiteSpace visualization for the search "sustainable buildings and cities" are presented in Figures 7–9.

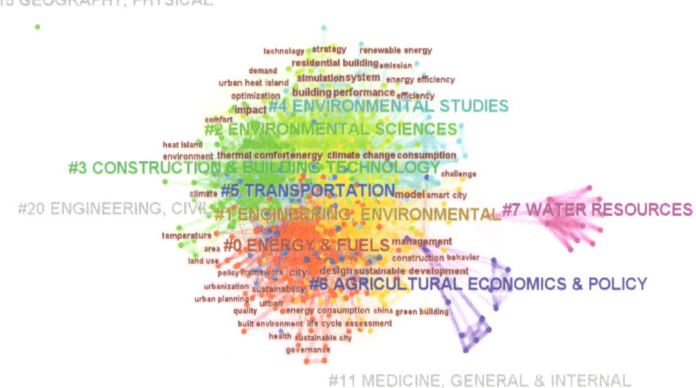

Figure 7. CiteSpace visualization of keywords and cluster analysis for the search "sustainable buildings and cities" in the period 1992–2023. The nodes represent keywords, lines that connect nodes are keyword co-occurrence links. The top 10 dominant keywords were: city (878), performance (636), impact (621), system (565), model (492), building (459), design (430), climate change (357), energy (350) and management (339). The cluster analysis of keyword distribution demonstrates close integration of disciplines and according to analyzed topics.

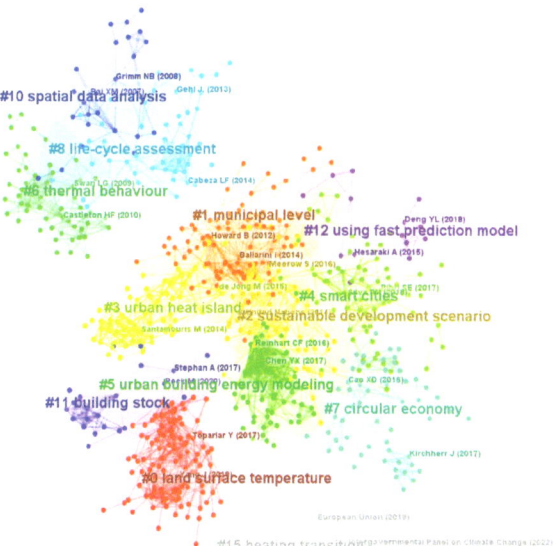

Figure 8. CiteSpace visualization of the reference co-citation network and cluster analysis for the search "sustainable buildings and cities" in the period 2006–2023. The nodes are cited references, lines that connect nodes are co-citation links. Analysis reveals 12 significant clusters of co-citation network, labelled according to keywords in the graph. It is visible that there are no separate clusters; they are merged at least with one or two clusters. Two groups of clusters can be identified, revealing close integration between fields of (1) management, transportation, ecology and (2) remote sensing and environmental engineering, according to cited references.

Figure 9. CiteSpace visualization of co-author network and cluster analysis for the search "sustainable buildings and cities" in the period 2006–2023. The nodes represent the authors of publications in the author collaboration network. Analysis reveals interlinked clusters of co-authors and cross-cluster author collaboration.

It is possible to summarize that knowledge mapping of the topic "sustainable buildings and cities" as reflected in the WoS database, reveals multiple shared topics and a wide array of interrelated research clusters with significant cases of cross-cluster collaboration and cluster integration. Compared to the topics "healthy buildings and cities" and "digital buildings and cities", research in "sustainable buildings and cities" appears more integrated, demonstrating that sustainability science research holds integration potential.

4.4. Sustainable Cities

As sustainable cities include healthy and digital buildings and cities, separate analyses were conducted for this topic. The data for that were collected on the 11 April 2023, and they include articles published until that time. Articles were selected based on the following inclusion and exclusion criteria: topic "*sustainable city*" or "*sustainable cities*", document type: "*article*", language: "*English*", categories: "*Construction Building Technology*", "*Engineering Civil*", "*Green Sustainable Science Technology*", "*Environmental Sciences*"; "*Environmental Studies*", "*Architecture*", "*Urban Studies*", "*Regional Urban Planning*" and "*Geography*". The number of published articles to some extent analyzing sustainable cities is growing each year. There are 64 articles published in 2023, 393 articles in 2022, 388 in 2021, 287 in 2020, 207 in 2019, 168 in 2018, etc.

The top 10 most-cited articles of the analyzed period are articles from 2004–2019 (see Table 4). They explore topics such as the importance of green urban spaces and ecosystems [49–52], the differences between concepts of sustainable, smart or digital cities [27,53,54], as well as transformative governance issues [45,55,56].

Table 4. Top 10 most-cited articles from the ones selected for the research.

No.	Citations	Author (Year)	Topics
1	1329 *	Chiesura (2004) [48]	role of urban parks in creating sustainable cities
2	730 *	Kennedy et al. (2007) [46]	changing metabolism of cities
3	647 *	Bulkeley and Betsill (2005) [49]	multilevel governance in sustainable cities
4	563 *	Sachs et al. (2019) [44]	transformations to achieve the SDGs
5	549 *	Ahvenniemi et al. (2017) [50]	differences between sustainable and smart cities.
6	507 *	De Jong et al. (2015) [51]	various concepts promoting sustainable urbanization
7	436 *	Hasse et al. (2014) [52]	ecosystem services in urban landscapes and their governance implications
8	436 *	Cocchia (2014) [53]	systematic literature review on smart and digital cities
9	375 *	Nevens et al. (2013) [54]	urban transition labs and their role in co-creating transformative actions for sustainable cities
10	366 *	Venter (2020) [55]	increased use of urban green spaces during the COVID-19 outbreak in Oslo

*—number of citations on 30 June 2023.

To not solely rely on the citation frequency, the co-citation analysis was carried out as it considers the collective influence of articles by examining the co-citation relationships between them, uncovers hidden connections, discovers emerging trends, overcomes biases and assesses research impact. In co-citation analysis conducted using CiteSpace, different elements are represented by colors, nodes and links to visualize and analyze the co-citation network. *Nodes* in the co-citation analysis represent individual articles or documents.

Links, also known as edges or lines, connect the nodes in the co-citation network. They represent the co-citation relationships between articles. The thickness or intensity of the links indicates the strength or frequency of co-citation between two articles. CiteSpace enables the identification of clusters of related articles based on their co-citation patterns. A cluster represents a group of articles that are conceptually or thematically similar and frequently cited together. Clusters are visually depicted as densely connected groups of nodes, often with a different color or shading to differentiate them from the overall co-citation network.

After the data processing procedure, 1181 nodes out of 2060 sources were identified as connected via 3992 co-citation links, i.e., average 3.38 links per node. Then the data were visualized using automatic cluster identification and computation of nodes centrality (see Figure 10). This visualization reveals the significance of different articles in the network, their links and the relationship between different clusters: some overlap, and others are almost isolated.

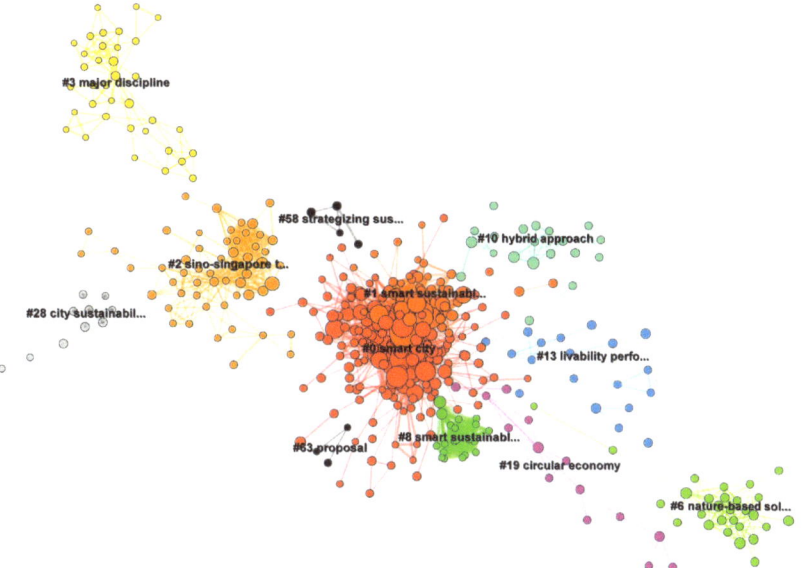

Figure 10. CiteSpace visualization of co-author network and cluster analysis for the search "sustainable cities" and its clusters in the period 1990–10 April 2023 Colours represent different clusters, nodes—different articles, links—co-citation relationships among articles.

The analyzed network consists of 12 major clusters. The ten largest clusters are presented in Table 5. Labels of the clusters are based on the top key terms, but those can be assigned using different algorithms—Latent Semantic Indexing (LSI) or Log-Likelihood Ratio (LLR). LSI is a more general summary of the concepts surrounding each cluster, while LLR provides a more concrete description of the topics within the cluster. Table 5 also reveals the main issues tackled in each of the most cited articles of the clusters.

Looking at the processed data (Figures 10 and 11 and Table 5), it is clear that a certain topic unites the articles in clusters, and there is not much of the overlap between the different clusters except for the first two—"smart cities" and "smart sustainable cities".

Cluster 1 is the largest and consists of 176 articles analyzing the relationship between sustainable and smart city concepts [27,28,50]. The shift from a more ecological approach (eco-city) to a more technocratic one (smart city) is recognized here. Yet, the limitations of both of those approaches are already highlighted as well.

Table 5. Research on Sustainable cities categorized by the clusters and the main articles distinguished (compiled by authors, using WoS data and the CiteSpace software).

No. ID	Cluster Label (LSI/LLR)	Size	Average Year	The Most Cited Articles in the Cluster	The Main Issues Analysed in Articles
1 #0	smart city/smart city	176	2017	Lim et al. (2018) [27]	smart cities, big data, value creation for different stakeholders
				Ahvenniemi et al. (2017) [50]	sustainable city vs. smart city, shift from sustainability assessment to smart city goals, concept of smart sustainable cities
				Doan et al. (2017) [56]	green building rating systems, sustainability of construction
				Yigitcanlar et al. (2019) [28]	smart vs sustainable, smart without sustainable, techno-centricity in smart development, systematic literature review (SLR)
2 #1	smart sustainable cities/smart city	58	2015	Bibri (2019) [57]	smart sustainable cities, big data, limitations of compact and eco-cities
				Bibri and Krogstie (2017) [58]	smart sustainable cities, urban sustainability, sustainable city models, smart city approach, big data
				Bibri (2018) [59]	smart sustainable cities, internet of things, big data analitics, environmental sustainability
				Höjer and Wangel (2015) [60]	smart sustainable city concept, five influential developments, challenges
3 #2	sino-singapore/eco-city china	55	2012	Caprotti et al. (2015) [13]	eco-city, urban environmental impacts, benefits for residents vs broader socio-environmental landscape
				Caprotti (2014) [12]	eco-city, future challenges, social resilience and emerging communities, new urban poor
				Cugurullo (2013) [14]	eco-city, sustainability ideology, case study, UAE, Masdar City
				Shwayri (2013) [15]	eco-city, global crisis, green infrastructure, South Korea, Songdo
4 #3	major discipline/ecological-infrastructural systems framework	36	2009	Ramaswami et al. (2012) [61]	sustainable city systems, social–ecological–infrastructural systems, social actors, multidisciplinarity
				Grimm et al. (2008) [62]	global change, ecology of cities, land use and cover, environmental changes, urban socioecosystems
				Pickett et al. (2011) [63]	socioecology, humane metropolis, urban system, human ecosystem, eco system services
				Ernstson et al. (2010) [18]	urban resilience, human dominated ecosystems, case studies, urban governance, system of cities
5 #6	nature-based solution/nature-based solution	28	2017	Brokking et al. (2021) [64]	green infrastructure, municipal practices, governance, urban development, case studies, Stockholm
				Raymond et al. (2017) [65]	shift from eco-based to nature-based solutions, 10 societal challenges, co-benefits and costs of sustainability
				Haase et al. (2017) [66]	city greening, social inclusivity, well-being, social effects of greening
				Andersson (2019) [17]	green and blue infrastructure, environmental justice, resilience

Table 5. Cont.

No. ID	Cluster Label (LSI/LLR)	Size	Average Year	The Most Cited Articles in the Cluster	The Main Issues Analysed in Articles
6 #8	smart sustainable cities/bridging stakeholder value creation	20	2019	Beck and Ferasso (2023) [67]	urban stakeholders, stakeholder value creation (SVC), urban sustainability, significance of environmental dimension
				Macke et al. (2019) [68]	smart sustainable city, sense of community, Brazil, residents satisfaction, social capital, shared values
				Camboim et al. (2019) [69]	smart city dimensions: governance; environ-urban; techno-economic; socio-institutional, urban innovation ecosystems
				Beck and Storopoli (2021) [70]	stakeholder theory, urban governance, literature review, urban strategy (stakeholders expectations) and urban marketing (urban image attractiveness)
7 #10	hybrid approach/uk context	20	2018	Stevenson et al. (2021) [71]	climate action (SDG 13) interaction with other SDGs—synergies and trade-offs, expert survey, UK
				Nilsson et al. (2016) [72]	interactions among SDGs, goals scoring
				Pradhan et al. (2017) [73]	SDGs interaction, SDG indicator data, sinergies (SDG 1—no poverty) and trade-offs (SDG 12—responsible consumption and production)
				Klopp and Petretta (2017) [74]	urban sustainable development goal (USDG), indicators, politics of measurement
8 #13	livability performance/learning approach	19	2018	Kutty et al. (2022) [19]	city resilience, urban livability, machine learning, European smart cities
				Sharifi and Khavarian-Garmsir (2020) [75]	smart cities, pandemics, environmental factors, air quality, urban design
				Brelsford et al. (2017) [76]	heterogeneity, scale of sustainable development, sustainable development indices, Brazil, South Africa
				Ugolini et al. (2020) [77]	pandemics, urban green spaces, people perception, European countries
9 #19	circular economy/circular economy	14	2018	Rama et al. (2021) [78]	key sustainability indicators, unemployment rates, waste collection, Spanish cities
				Feleki et al. (2018) [79]	systems of indicators, "traditional" dimensions of sustainability, European urban areas
				Azunre et al.(2019) [80]	urban agriculture, indicators of sustainability, economic, social and environmental benefits of urban agriculture
				Meerow et al. (2016) [20]	definition of urban resilience, climate change, review
10 #28	city sustainability index/city sustainability index	10	2012	Mori and Yamashita (2015) [81]	city sustainability index, concept of constraint and maximization indicators, limitations and benefits
				Haghshenas and Vaziri (2012) [82]	9 sustainable transportation indicators, millennium cities database for sustainable mobility
				McCormic et al. (2013) [83]	urban initiatives on sustainability, 35 cases, Europe and some other locations, sustainable urban transformation, governance
				Shen et al. (2011) [84]	sustainability indicators, comparison of different practices

Figure 11. CiteSpace Timeline visualization of the co-citation network of scientific research on sustainable cities and its clusters. Colors represent different clusters, nodes—different articles, links—co-citation relationships among articles.

Cluster 2 is closely related (almost overlapping) to the previously described cluster "smart city". The main difference is that "sustainable cities" are not compared with the "smart ones" anymore. Instead, the term "smart sustainable city "(SSC) is established and conventionally used. Thus, there is also a change in the research strategies: it shifted from the concept definition towards studying the possibilities of using big data and information technologies to improve SSG development [57–60].

Cluster 3 is not linked to the first two but reveals a group of articles focused on sustainable development of eco-cities in Asia (i.e., Tianjin in China [12,13], Masdar City in the United Arab Emirates [14] or Songdo in South Korea [15].

Cluster 4 unifies articles about social–ecological–infrastructural systems [61], human-dominated [18] or human ecosystems [62,63]. The main idea of those articles is that cities or their parts do not function in isolation, but they are connected through the flows of energy, matter and information.

Cluster 5 consists of articles related to nature-based solutions (NBS) for urban development. They discuss the role of appropriate governance [67] and the benefits of greening the cities, such as social effects, increased inclusivity and environmental justice [17,65,66].

Cluster 6 brings together articles that examine the role and significance of urban stakeholders (i.e., municipalities, housing corporations, developers, city inhabitants and urban designers) [67,69]. Those articles also highlight the importance of social capital, a sense of community and attractiveness [68,70].

Cluster 7 includes articles analyzing the SDGs [1] and their interrelations, as the interactions among them did not always cause positive results [71,72]. For example, a systematic study [73] revealed that SDG 1 (no poverty) has a synergy with most of the other goals, while SDG 12 (Responsible consumption and production) is most commonly associated with trade-offs.

Cluster 8 includes articles aiming to understand how it is possible to create a more livable environment, considering such issues as pandemics [75,78] in both highly developed cities [19] as well as in developing ones [76].

Cluster 9 covers articles to some extent related to a circular economy, either it would be analysis of different indicators such as unemployment or waste management [78,79] or the role of urban agriculture [80], or the theoretical clarification of urban resilience [20].

Cluster 10 consists of articles tackling the issues of sustainability index construction and measurement—formulation of methodological frameworks [81,82] and overviewing and evaluating the existing practices [83,84].

From a temporal standpoint (see Figure 11), the oldest topic that emerged in the analyzed data set of the sustainability literature is the one that examines the social–ecological–infrastructural systems. It is followed by a group of research papers presenting the eco-city cases and their analysis. Subsequently, there was a shift in research emphasis towards nature-based solutions and the additional values generated by city greening initiatives, and the most recent cluster of articles deals with more intangible matters—the roles and contributions of various urban stakeholders in the process of reaching sustainability. It is also important to note that after 2015, when the UN General Assembly adopted SDGs [1], a group of articles examined the compatibility and mutual impact of different SDGs. Moreover, even though the idea of a circular economy has been known as far as the 1980s [85], in the sustainability literature, its importance also grew more significantly after the adoption of SDGs, as this particular economic system is based on the reuse and regeneration of materials or products, especially as a means of continuing production in a sustainable or environmentally friendly way. The quality of life and the requirements for the design of livable urban spaces is another cluster of articles relevant at that time. Still, the most enduring trend in sustainability research is the theme of "smart cities," which emerged circa 2012, peaked in 2016, and continues to be a prominent topic of investigation in the contemporary literature. Almost parallel to this trend, the development of interest in the "smart sustainable cities" topic is also observed, although in the most recent years, there has been a decrease in scientific articles on the latter topic.

5. Conclusions and Discussion

This study is a general quantitative literature overview with qualitative insights; further research is required in order to understand deeper each distinguished subtopic. According to some researchers, CiteSpace bibliometric analysis has some limitations, for example, some maps and clusters are complex and require specialized domain knowledge for interpretation or the learning curve required to set accurate visualization parameters [86]; however it proved to be a valuable tool in this research.

The earliest theme that emerged from the examined data set of the sustainability literature is the one that investigates social–ecological–infrastructural systems. The research emphasis has shifted toward nature-based solutions and the additional values generated by city greening initiatives, and the most recent cluster of articles addresses more intangible issues—the roles and contributions of various urban stakeholders in achieving sustainability.

Following the adoption of the SDGs by the UN General Assembly in 2015, a set of papers appeared explicitly investigating the concerns of compatibility and mutual impact of multiple SDGs.

The concept of "smart cities," which peaked in 2016 and remained a popular examination area in modern literature, is the most durable trend in sustainability research. Almost simultaneous to this trend is the growth of interest in the concept of "smart sustainable cities;" however, in recent years, there has been a reduction in research on the latter.

Research showed that studies on "healthy buildings and cities" yielded the fewest findings. 1064 articles were published in the WoS database between 1988 and 2023. The first articles were published in 1988, and the increase in published research on this topic began in 2005. In this search, conference papers and proceedings were the most common kind of publishing. "Healthy buildings and cities" fall under the broad banner of sustainability study and are closely connected to architecture, civil engineering, environmental and urban studies and public health. The knowledge mapping of the topic "healthy buildings and cities" demonstrates its latent potential for interdisciplinary integration, as it shares various issues, despite the fact that the present study appears to be poorly integrated from the standpoint of literature sources and author collaboration.

Between 1992 and 2023, 2734 papers on "digital buildings and cities" were published. The first articles were published in 1992, and the increase in published research on this topic began in 2005. Similar to the "healthy buildings and cities" search, the most common publication type in this search was conference papers and proceedings. Compared to prior studies, the issue of "digital buildings and cities" is more specific. This topic offers a wide range of research clusters with notable examples of cross-cluster collaboration and cluster integration. Compared to prior topics studied, research in "digital buildings and cities" looks more integrated, yet with unrealized integration potential.

Research showed that the research topic of "sustainable buildings and cities" was and still is very important worldwide, as from 1988 to 2023, 10 006 papers were identified in WoS database. The first articles were published in 1992, and the increase in published research in this topic began in 2005. The most common type of publication in this search is a journal article. The issue of "sustainable buildings and cities" falls within the broad umbrella of sustainability research, which includes the currently leading technical fields. According to our study, two sets of clusters were discovered, demonstrating tight integration across the domains of:

- Management, transportation, ecology;
- Remote sensing and environmental engineering, according to cited references.

In comparison to other analyzed topics ("healthy buildings and cities" and "digital buildings and cities,") research in the field of "sustainable buildings and cities" looks more integrated, demonstrating that integration potential exists in sustainability science research.

This study revealed the ongoing worldwide relevance of sustainability science research and identified significant opportunities for multidisciplinary integration across the investigated subjects. As such, it sets the path for further study into these tendencies and prospective partnership opportunities in pursuing sustainability in the cities.

Even though the potential future development of sustainable cities remains complicated and not fully answered question; but, based on the results of the literature review, to successfully develop future cities, a multidisciplinary and integrated approach is essential. It involves embracement of compact city planning and development practices, integration of green and blue infrastructure, and promotion of resilience. Smart city concepts, supported by digital technologies and data-driven approaches, can also enhance urban livability and efficiency. Additionally, the creation of healthy cities requires considering factors such as urban sports facilities, access to nature and the well-being of residents. It is crucial to address spatial and social inequalities. Overall, the development of future cities should aim to achieve the balance between environmental, social and economic aspects to create livable, resilient and sustainable urban environments.

Author Contributions: Conceptualization, L.S.; data curation, D.L.; formal analysis, I.G.-V. and I.P.; investigation, I.G.-V. and I.P.; supervision, L.S.; validation, D.L.; writing—original draft, I.G.-V. and I.P.; writing—review and editing, L.S. and P.A.F. All authors have read and agreed to the published version of the manuscript.

Funding: This research received no external funding.

Data Availability Statement: Not applicable.

Conflicts of Interest: The authors declare no conflict of interest.

References

1. United Nations. Transforming Our World: The 2030 Agenda for Sustainable Development. 2015. Available online: https://wedocs.unep.org/20.500.11822/9814 (accessed on 10 April 2023).
2. World Commission on Environment and Development (Ed.) *Our Common Future*; Oxford University Press: Oxford, UK, 1987.
3. Council of Europe. *The European Urban Charter*; Council of Europe: Strasbourg, France, 1992; p. 60.
4. Grazuleviciute-Vilniske, I.; Seduikyte, L.; Teixeira-Gomes, A.; Mendes, A.; Borodinecs, A.; Buzinskaite, D. Aging, living environment, and sustainability: What should be taken into account? *Sustainability* **2020**, *12*, 1853. [CrossRef]
5. Commission of the European Communities. Towards a Thematic Strategy on the Urban Environment. *COM* **2004**, *60*, 57.
6. Council of European Union. Renewed EU Sustainable Development Strategy (No. 10917/06; p. 29). 2006. Available online: https://register.consilium.europa.eu/doc/srv?l=EN&f=ST%2010917%202006%20INIT (accessed on 10 April 2023).
7. Bibri, S.E.; Krogstie, J.; Kärrholm, M. Compact city planning and development: Emerging practices and strategies for achieving the goals of sustainability. *Dev. Built Environ.* **2020**, *4*, 100021. [CrossRef]
8. Duany, A.; Sorlien, S.; Wright, W. *The SmartCode Version 9.2*; The TownPaper Publisher: Orlando, FL, USA, 2003.

9. Dantzig, G.B.; Saaty, T.L. *Compact City: A Plan for a Liveable Urban Environment*; W. H. Freeman: New York, NY, USA, 1973.
10. Mouratidis, K. Is compact city livable? The impact of compact versus sprawled neighbourhoods on neighbourhood satisfaction. *Urban Stud.* **2018**, *55*, 2408–2430. [CrossRef]
11. Kenworthy, J.R. The eco-city: Ten key transport and planning dimensions for sustainable city development. *Environ. Urban.* **2006**, *18*, 67–85. [CrossRef]
12. Caprotti, F. Critical research on eco-cities? A walk through the Sino-Singapore Tianjin Eco-City, China. *Cities* **2014**, *36*, 10–17. [CrossRef]
13. Caprotti, F.; Springer, C.; Harmer, N. 'Eco' For Whom? Envisioning Eco-urbanism in the Sino-Singapore Tianjin Eco-city, China: 'ECO' FOR WHOM? *Int. J. Urban Reg. Res.* **2015**, *39*, 495–517. [CrossRef]
14. Cugurullo, F. How to Build a Sandcastle: An Analysis of the Genesis and Development of Masdar City. *J. Urban Technol.* **2013**, *20*, 23–37. [CrossRef]
15. Shwayri, S.T. A Model Korean Ubiquitous Eco-City? The Politics of Making Songdo. *J. Urban Technol.* **2013**, *20*, 39–55. [CrossRef]
16. Desouza, K.C.; Flanery, T.H. Designing, planning, and managing resilient cities: A conceptual framework. *Cities* **2013**, *35*, 89–99. [CrossRef]
17. Andersson, E.; Langemeyer, J.; Borgström, S.; McPhearson, T.; Haase, D.; Kronenberg, J.; Barton, D.N.; Davis, M.; Naumann, S.; Röschel, L.; et al. Enabling Green and Blue Infrastructure to Improve Contributions to Human Well-Being and Equity in Urban Systems. *Bioscience* **2019**, *69*, 566–574. [CrossRef] [PubMed]
18. Ernstson, H.; Van Der Leeuw, S.E.; Redman, C.L.; Meffert, D.J.; Davis, G.E.; Alfsen, C.; Elmqvist, T. Urban Transitions: On Urban Resilience and Human-Dominated Ecosystems. *AMBIO* **2010**, *39*, 531–545. [CrossRef] [PubMed]
19. Kutty, A.A.; Wakjira, T.G.; Kucukvar, M.; Abdella, G.M.; Onat, N.C. Urban resilience and livability performance of European smart cities: A novel machine learning approach. *J. Clean. Prod.* **2022**, *378*. [CrossRef]
20. Meerow, S.; Newell, J.P.; Stults, M. Defining urban resilience: A review. *Landsc. Urban Plan.* **2016**, *147*, 38–49. [CrossRef]
21. Nam, T.; Pardo, T.A. Conceptualizing Smart City with Dimensions of Technology, People, and Institutions. In Proceedings of the 12th Annual International Digital Government Research Conference: Digital Government Innovation in Challenging Times, College Park, MD, USA, 12–15 June 2011; pp. 282–291.
22. Ishida, T. Digital City, Smart City and Beyond. In Proceedings of the 2017 International World Wide Web Conference Committee (IW3C2), published under Creative Commons CC BY 4.0 License, Perth, Australia, 3–7 April 2017.
23. Keenahan, J.; MacReamoinn, R.; Paduano, C. Sustainable Design using Computational Fluid Dynamics in the Built Environment—A Case Study. *J. Sustain. Archit. Civ. Eng.* **2017**, *19*, 92–103. [CrossRef]
24. Štěpánek, P.; Ge, M.; Walletzký, L. IT-Enabled Digital Service Design Principles—Lessons Learned from Digital Cities. In *Information Systems. EMCIS 2017. Lecture Notes in Business Information Processing*; Themistocleous, M., Morabito, V., Eds.; Springer: Cham, Switzerland, 2017; Volume 299. [CrossRef]
25. Pupeikis, D.; Morkūnaitė, L.; Daukšys, M.; Navickas, A.A.; Abromas, S. Possibilities of Using Building Information Model Data in Reinforcement Processing Plant. *J. Sustain. Archit. Civ. Eng.* **2021**, *28*, 80–93. [CrossRef]
26. Albino, V.; Berardi, U.; Dangelico, R.M. Smart Cities: Definitions, Dimensions, Performance, and Initiatives. *J. Urban Technol.* **2015**, *22*, 3–21. [CrossRef]
27. Lim, C.; Kim, K.-J.; Maglio, P.P. Smart cities with big data: Reference models, challenges, and considerations. *Cities* **2018**, *82*, 86–99. [CrossRef]
28. Yigitcanlar, T.; Kamruzzaman, M.; Foth, M.; Sabatini-Marques, J.; da Costa, E.; Ioppolo, G. Can cities become smart without being sustainable? A systematic review of the literature. *Sustain. Cities Soc.* **2019**, *45*, 348–365. [CrossRef]
29. Goli, S.; Arokiasamy, P.; Chattopadhayay, A. Living and health conditions of selected cities in India: Setting priorities for the National Urban Health Mission. *Cities* **2011**, *28*, 461–469. [CrossRef]
30. Werna, E.; Harpham, T.; Blue, I.; Goldstein, G. From healthy city projects to healthy cities. *Environ. Urban.* **1999**, *11*, 27–40. [CrossRef]
31. Webster, P.; Sanderson, D. Healthy Cities Indicators—A Suitable Instrument to Measure Health? *J. Urban Health* **2013**, *90*, 52–61. [CrossRef] [PubMed]
32. Shen, J.; Cheng, J.; Huang, W.; Zeng, F. An Exploration of Spatial and Social Inequalities of Urban Sports Facilities in Nanning City, China. *Sustainability* **2020**, *12*, 4353. [CrossRef]
33. Price, D.J.D.S. Networks of scientific papers. *Science* **1965**, *145*, 510–515. [CrossRef] [PubMed]
34. Chen, C. CiteSpace II: Detecting and visualizing emerging trends and transient patterns in scientific literature. *J. Am. Soc. Inf. Sci. Technol.* **2006**, *57*, 359–377. [CrossRef]
35. Chen, C. *Mapping Scientific Frontiers*; Springer-Verlag: London, UK, 2003.
36. Su, X.; Li, X.; Kang, Y. A bibliometric analysis of research on intangible cultural heritage using CiteSpace. *Sage Open* **2019**, *9*, 2158244019840119. [CrossRef]
37. Chen, C. *How to Use CiteSpace*; Leanpub: Victoria, BC, Canada, 2015.
38. Chen, C. *CiteSpace: A Practical Guide for Mapping Scientific Literature*; Nova Science Publishers: New York, NY, USA, 2016.
39. Russell, R.; Guerry, A.D.; Balvanera, P.; Gould, R.K.; Basurto, X.; Chan, K.M.; Klain, S.; Levine, J.; Tam, J. Humans and nature: How knowing and experiencing nature affect well-being. *Annu. Rev. Environ. Resour.* **2013**, *38*, 473–502. [CrossRef]
40. Frumkin, H. Healthy places: Exploring the evidence. *Am. J. Public Health* **2003**, *93*, 1451–1456. [CrossRef]

41. Nieuwenhuijsen, M.J.; Khreis, H.; Triguero-Mas, M.; Gascon, M.; Dadvand, P. Fifty Shades of Green: Pathway to Healthy Urban Living. *Epidemiology* **2017**, *28*, 63–71. [CrossRef]
42. Zanella, A.; Bui, N.; Castellani, A.; Vangelista, L.; Zorzi, M. Internet of things for smart cities. *IEEE Internet Things J.* **2014**, *1*, 22–32. [CrossRef]
43. Neirotti, P.; De Marco, A.; Cagliano, A.C.; Mangano, G.; Scorrano, F. Current trends in Smart City initiatives: Sotylizedsed facts. *Cities* **2014**, *38*, 25–36. [CrossRef]
44. Sachs, J.D.; Schmidt-Traub, G.; Mazzucato, M.; Messner, D.; Nakicenovic, N.; Rockström, J. Six transformations to achieve the sustainable development goals. *Nat. Sustain.* **2019**, *2*, 805–814. [CrossRef]
45. Cabeza, L.F.; Rincón, L.; Vilariño, V.; Pérez, G.; Castell, A. Life cycle assessment (LCA) and life cycle energy analysis (LCEA) of buildings and the building sector: A review. *Renew. Sustain. Energy Rev.* **2014**, *29*, 394–416. [CrossRef]
46. Kennedy, C.; Cuddihy, J.; Engel-Yan, J. The changing metabolism of cities. *J. Ind. Ecol.* **2007**, *11*, 43–59. [CrossRef]
47. Haaland, C.; van Den Bosch, C.K. Challenges and strategies for urban green-space planning in cities undergoing densification: A review. *Urban For. Urban Green.* **2015**, *14*, 760–771. [CrossRef]
48. Chiesura, A. The role of urban parks for the sustainable city. *Landsc. Urban Plan.* **2004**, *68*, 129–138. [CrossRef]
49. Bulkeley, H.; Betsill, M. Rethinking Sustainable Cities: Multilevel Governance and the 'Urban' Politics of Climate Change. *Environ. Politics* **2005**, *14*, 42–63. [CrossRef]
50. Ahvenniemi, H.; Huovila, A.; Pinto-Seppä, I.; Airaksinen, M. What are the differences between sustainable and smart cities? *Cities* **2017**, *60*, 234–245. [CrossRef]
51. De Jong, M.; Joss, S.; Schraven, D.; Zhan, C.; Weijnen, M. Sustainable–smart–resilient–low carbon–eco–knowledge cities; making sense of a multitude of concepts promoting sustainable urbanization. *J. Clean. Prod.* **2015**, *109*, 25–38. [CrossRef]
52. Haase, D.; Frantzeskaki, N.; Elmqvist, T. Ecosystem Services in Urban Landscapes: Practical Applications and Governance Implications. *AMBIO* **2014**, *43*, 407–412. [CrossRef] [PubMed]
53. Cocchia, A. Smart and Digital City: A Systematic Literature Review. In *Smart City*; Dameri, R.P., Rosenthal-Sabroux, C., Eds.; Springer International Publishing: Berlin/Heidelberg, Germany, 2014; pp. 13–43. [CrossRef]
54. Nevens, F.; Frantzeskaki, N.; Gorissen, L.; Loorbach, D. Urban Transition Labs: Co-creating transformative action for sustainable cities. *J. Clean. Prod.* **2013**, *50*, 111–122. [CrossRef]
55. Venter, Z.S.; Barton, D.N.; Gundersen, V.; Figari, H.; Nowell, M. Urban nature in a time of crisis: Recreational use of green space increases during the COVID-19 outbreak in Oslo, Norway. *Environ. Res. Lett.* **2020**, *15*, 104075. [CrossRef]
56. Doan, D.T.; Ghaffarianhoseini, A.; Naismith, N.; Zhang, T.; Ghaffarianhoseini, A.; Tookey, J. A critical comparison of green building rating systems. *Build. Environ.* **2017**, *123*, 243–260. [CrossRef]
57. Bibri, S.E. Advancing Sustainable Urbanism Processes: The Key Practical and Analytical Applications of Big Data for Urban Systems and Domains. In *Big Data Science and Analytics for Smart Sustainable Urbanism*; Bibri, S.E., Ed.; Springer International Publishing: Berlin/Heidelberg, Germany, 2019; pp. 221–252. [CrossRef]
58. Bibri, S.E.; Krogstie, J. Smart sustainable cities of the future: An extensive interdisciplinary literature review. *Sustain. Cities Soc.* **2017**, *31*, 183–212. [CrossRef]
59. Bibri, S.E. The IoT for smart sustainable cities of the future: An analytical framework for sensor-based big data applications for environmental sustainability. *Sustain. Cities Soc.* **2018**, *38*, 230–253. [CrossRef]
60. Höjer, M.; Wangel, J. Smart Sustainable Cities: Definition and Challenges. In *ICT Innovations for Sustainability*; Hilty, L.M., Aebischer, B., Eds.; Springer International Publishing: Berlin/Heidelberg, Germany, 2015; Volume 310, pp. 333–349. [CrossRef]
61. Ramaswami, A.; Weible, C.; Main, D.; Heikkila, T.; Siddiki, S.; Duvall, A.; Pattison, A.; Bernard, M. A Social-Ecological-Infrastructural Systems Framework for Interdisciplinary Study of Sustainable City Systems: An Integrative Curriculum Across Seven Major Disciplines. *J. Ind. Ecol.* **2012**, *16*, 801–813. [CrossRef]
62. Grimm, N.B.; Faeth, S.H.; Golubiewski, N.E.; Redman, C.L.; Wu, J.; Bai, X.; Briggs, J.M. Global Change and the Ecology of Cities. *Science* **2008**, *319*, 756–760. [CrossRef]
63. Pickett, S.T.A.; Buckley, G.L.; Kaushal, S.S.; Williams, Y. Social-ecological science in the humane metropolis. *Urban Ecosyst.* **2011**, *14*, 319–339. [CrossRef]
64. Brokking, P.; Mörtberg, U.; Balfors, B. Municipal Practices for Integrated Planning of Nature-Based Solutions in Urban Development in the Stockholm Region. *Sustainability* **2021**, *13*, 10389. [CrossRef]
65. Raymond, C.M.; Frantzeskaki, N.; Kabisch, N.; Berry, P.; Breil, M.; Nita, M.R.; Geneletti, D.; Calfapietra, C. A framework for assessing and implementing the co-benefits of nature-based solutions in urban areas. *Environ. Sci. Policy* **2017**, *77*, 15–24. [CrossRef]
66. Haase, D.; Kabisch, S.; Haase, A.; Andersson, E.; Banzhaf, E.; Baró, F.; Brenck, M.; Fischer, L.K.; Frantzeskaki, N.; Kabisch, N.; et al. Greening cities—To be socially inclusive? About the alleged paradox of society and ecology in cities. *Habitat Int.* **2017**, *64*, 41–48. [CrossRef]
67. Beck, D.; Ferasso, M. Bridging 'Stakeholder Value Creation' and 'Urban Sustainability': The need for better integrating the Environmental Dimension. *Sustain. Cities Soc.* **2023**, *89*, 104316. [CrossRef]
68. Macke, J.; Rubim Sarate, J.A.; de Atayde Moschen, S. Smart sustainable cities evaluation and sense of community. *J. Clean. Prod.* **2019**, *239*, 118103. [CrossRef]

69. Camboim, G.F.; Zawislak, P.A.; Pufal, N.A. Driving elements to make cities smarter: Evidences from European projects. *Technol. Forecast. Soc. Change* **2019**, *142*, 154–167. [CrossRef]
70. Beck, D.; Storopoli, J. Cities through the lens of Stakeholder Theory: A literature review. *Cities* **2021**, *118*, 103377. [CrossRef]
71. Stevenson, S.; Collins, A.; Jennings, N.; Köberle, A.C.; Laumann, F.; Laverty, A.A.; Vineis, P.; Woods, J.; Gambhir, A. A hybrid approach to identifying and assessing interactions between climate action (SDG13) policies and a range of SDGs in a UK context. *Discover Sustainability* **2021**, *2*, 43. [CrossRef]
72. Nilsson, M.; Griggs, D.; Visbeck, M. Policy: Map the interactions between Sustainable Development Goals. *Nature* **2016**, *534*, 320–322. [CrossRef]
73. Pradhan, P.; Costa, L.; Rybski, D.; Lucht, W.; Kropp, J.P. A Systematic Study of Sustainable Development Goal (SDG) Interactions: A Systematic Study of SDG Interactions. *Earth's Future* **2017**, *5*, 1169–1179. [CrossRef]
74. Klopp, J.M.; Petretta, D.L. The urban sustainable development goal: Indicators, complexity and the politics of measuring cities. *Cities* **2017**, *63*, 92–97. [CrossRef]
75. Sharifi, A.; Khavarian-Garmsir, A.R. The COVID-19 pandemic: Impacts on cities and major lessons for urban planning, design, and management. *Sci. Total Environ.* **2020**, *749*, 142391. [CrossRef]
76. Brelsford, C.; Lobo, J.; Hand, J.; Bettencourt, L.M.A. Heterogeneity and scale of sustainable development in cities. *Proc. Natl. Acad. Sci. USA* **2017**, *114*, 8963–8968. [CrossRef]
77. Ugolini, F.; Massetti, L.; Calaza-Martínez, P.; Cariñanos, P.; Dobbs, C.; Ostoić, S.K.; Marin, A.M.; Pearlmutter, D.; Saaroni, H.; Šaulienė, I.; et al. Effects of the COVID-19 pandemic on the use and perceptions of urban green space: An international exploratory study. *Urban For. Urban Green.* **2020**, *56*, 126888. [CrossRef] [PubMed]
78. Rama, M.; Andrade, E.; Moreira, M.T.; Feijoo, G.; González-García, S. Defining a procedure to identify key sustainability indicators in Spanish urban systems: Development and application. *Sustainable Cities Soc.* **2021**, *70*, 102919. [CrossRef]
79. Feleki, E.; Vlachokostas, C.; Moussiopoulos, N. Characterization of sustainability in urban areas: An analysis of assessment tools with emphasis on European cities. *Sustain. Cities Soc.* **2018**, *43*, 563–577. [CrossRef]
80. Azunre, G.A.; Amponsah, O.; Peprah, C.; Takyi, S.A.; Braimah, I. A review of the role of urban agriculture in the sustainable city discourse. *Cities* **2019**, *93*, 104–119. [CrossRef]
81. Mori, K.; Yamashita, T. Methodological framework of sustainability assessment in City Sustainability Index (CSI): A concept of constraint and maximization indicators. *Habitat Int.* **2015**, *45*, 10–14. [CrossRef]
82. Haghshenas, H.; Vaziri, M. Urban sustainable transportation indicators for global comparison. *Ecol. Indic.* **2012**, *15*, 115–121. [CrossRef]
83. McCormick, K.; Anderberg, S.; Coenen, L.; Neij, L. Advancing sustainable urban transformation. *J. Clean. Prod.* **2013**, *50*, 1–11. [CrossRef]
84. Shen, L.-Y.; Jorge Ochoa, J.; Shah, M.N.; Zhang, X. The application of urban sustainability indicators—A comparison between various practices. *Habitat Int.* **2011**, *35*, 17–29. [CrossRef]
85. Stahel, W.R.; Reday-Mulvey, G. *Jobs for Tomorrow: The Potential for Substituting Manpower for Energy*; Vantage Press: Burlington, VT, USA, 1981.
86. Synnestvedt, M.B.; Chen, C.; Holmes, J.H. CiteSpace II: Visualization and knowledge discovery in bibliographic databases. In *AMIA Annual Symposium Proceedings*; American Medical Informatics Association: Washington, WA, USA, 2005; Volume 2005, p. 724.

Disclaimer/Publisher's Note: The statements, opinions and data contained in all publications are solely those of the individual author(s) and contributor(s) and not of MDPI and/or the editor(s). MDPI and/or the editor(s) disclaim responsibility for any injury to people or property resulting from any ideas, methods, instructions or products referred to in the content.

MDPI AG
Grosspeteranlage 5
4052 Basel
Switzerland
Tel.: +41 61 683 77 34

Buildings Editorial Office
E-mail: buildings@mdpi.com
www.mdpi.com/journal/buildings

Disclaimer/Publisher's Note: The title and front matter of this reprint are at the discretion of the Guest Editors. The publisher is not responsible for their content or any associated concerns. The statements, opinions and data contained in all individual articles are solely those of the individual Editors and contributors and not of MDPI. MDPI disclaims responsibility for any injury to people or property resulting from any ideas, methods, instructions or products referred to in the content.